Computer Communications and Networks

T0137993

For other titles published in this series, go to
www.springer.com/series/4198

The **Computer Communications and Networks** series is a range of textbooks, monographs and handbooks. It sets out to provide students, researchers and non-specialists alike with a sure grounding in current knowledge, together with comprehensible access to the latest developments in computer communications and networking.

Emphasis is placed on clear and explanatory styles that support a tutorial approach, so that even the most complex of topics is presented in a lucid and intelligible manner.

Sid Katzen

The Essential PIC18® Microcontroller

 Springer

Sid Katzen
School of Engineering
University of Ulster at Jordanstown
Newtownabbey BT37 0QB
Northern Ireland, UK
sj.katzen@ulster.ac.uk

Series Editor
Professor A.J. Sammes, BSc, MPhil, PhD, FBCS, CEng
Centre for Forensic Computing
Cranfield University
DCMT, Shrivenham
Swindon SN6 8LA
UK

ISBN 978-1-4471-2574-7   ISBN 978-1-84996-229-2 (eBook)
DOI 10.1007/978-1-84996-229-2
Springer London Dordrecht Heidelberg New York

British Library Cataloguing in Publication Data
A catalogue record for this book is available from the British Library

Cover design: VTEX, Vilnius

Printed on acid-free paper

Springer is part of Springer Science+Business Media (www.springer.com)

In memory of my father Louis Cahill

Preface

In December 2009[1] Federico Faggin (see p. 55) wrote "TWO inventions have shaped our modern world more that any other: the engine and the computer. Where the engine captured and extended the human capacity to do physical work, the computer did the same for the capacity of the human brain to think, organize and control...". Federico Faggin, who was part of the team that developed the Intel 4004, the world's first commercially successful microprocessor released in 1971, was commissioned to write this article by *New Scientist* to celebrate the microprocessor's victory in a poll to find the discovery that has had the greatest impact on the world in the past 50 years.[2] That having been said, this was from a field of only ten nominated discoveries, but nevertheless it won with 48% of the vote, followed by the World Wide Web at 31%; which in reality is only made possible by the microprocessor.

Microprocessors and their microcontroller derivatives are a widespread, if rather invisible, part of the infrastructure of our twenty-first-century electronic information-based society. In 1998, it was estimated[3] that hidden in every home there were about 100 microcontrollers and microprocessors: in the singing birthday card, washing machine, microwave oven, television controller, telephone, personal computer and so on. About 20 more lurked in the average family car, for example, monitoring in-tire radio pressure sensors and displaying critical data through a control area network (CAN). As anyone will testify if they have driven a modern car, these figures can be dramatically revised upwards. Your pocket alone will carry several disguised as smart credit, debit and employment cards, bus passes and so on. Even your pets may be chipped in case they stray too far. Indeed, there is more computing power in a singing birthday card than there was in the world in 1948, when the first von-Neumann computer was commissioned at the University of Manchester.

---

[1] *New Scientist*, Computers top the poll of modern discoveries, vol. 70, no. 2737, 2 December, 2009.

[2] http://newscientist.com/special/big-impact.

[3] *New Scientist*, vol. 59, no. 2141, 4 July 1998, p. 139.

Over 4 billion such devices are sold each year to implement the intelligence of these 'intelligent' electronic devices, ranging from smart eggtimers through to aircraft management systems. The evolution of the microprocessor from the first Intel device introduced in 1971 has revolutionised the structure of society, effectively creating the second Industrial Revolution at the beginning of the twenty-first century. Although the microprocessor is better known for its role in powering the ubiquitous PC, where raw computing power is the goal, sales of microprocessors such as the Intel Pentium represent only around 2% of the total volume. The vast majority of sales are of low-cost microcontrollers embedded into a dedicated-function digital electronic device, such as the smart card. Here the emphasis is on the integration of the core processor with memory and input/output resources in the one chip. This integrated computing system is known as a *microcontroller*.

In seeking to write a book in this area, the overall objective was to get the reader up-to-speed in designing small embedded microcontroller-based systems, rather than using microcontrollers as a vehicle to illustrate computer architecture in the traditional manner. This will hopefully give the reader confidence that, even at such an introductory level, he/she can design, construct, and program a complete working embedded system.

Given the practical nature of this material, real-world hardware and software products are used throughout to illustrate the material. The microcontroller market is still dominated by devices that operate on 8-bit data (although 4–16 and 32-bit examples are readily available) like early microprocessors and unlike the 64-bit Intel Pentium family 'heavy brigade'. In contrast, the essence of the microcontroller lies in its high system-integration/low-cost profile. Power can be increased by distributing processors throughout the system. Thus, for example, a robot arm may have a microcontroller for each joint implementing simple local processes and communicating with a more powerful processor making overall executive decisions.

In choosing a target architecture, acceptance in the industrial market, easy availability, and low-cost development software have made the Microchip family one of the most popular choices as the pedagogic vehicle in learning microprocessor/microcontroller technology at all levels of electronic engineering from grade school to university. In particular, the reduced instruction set, together with the relatively simple innovative architecture, reduces the learning curve. In addition to their industrial and educational roles, the PIC® MCU families are also the mainstay of hobbyist projects, as a leaf through any electronics magazine will show.

Microchip, Inc., is a relatively recent entrant to the microcontroller market with its family of Harvard architecture PIC devices introduced in 1989. By 2006, Microchip was the largest producer of 8-bit units—after a 20-year tousle with the market leaders Motorola; which is now spun-off its activities in this field to Freescale Semiconductor.

In 2001 (2nd edn. 2005) Springer published by first book on PIC microcontrollers; *The Quintessential PIC Microcontroller*. This is based on what was the mainstream mid-range 8-bit family, the PIC16 series. Although this range still continues to expand, with a new enhanced core introduced in 2009, the enhanced-range PIC18 has now become the mainstream as well as the most advanced 8-bit family.

With this in mind, a complete rewrite was in order. A new edition would not fit the bill, both because the PIC16 family is still very much alive and also although the PIC18 family is upwards compatible, there are considerable changes. However, the aims and structure very much follow the *Quintessential* format.

This book is split into three parts. Part I covers sufficient digital, logic and computer architecture to act as a foundation for the microcontroller engineering topics presented in the rest of the text. Inclusion of this material makes the text suitable for stand-alone usage, as it does not require a prerequisite digital systems module.

Part II looks mainly at the software aspects of the enhanced-range PIC microcontroller family, its instruction set, how to program it at assembly and high-level **C** coding levels, and how the microcontroller handles subroutines and interrupts. Although the PIC18 family is the exemplar, both architecture and software are comparable to earlier families and indeed the 16-bit PIC24 device range.

Part III moves on to the hardware aspects of interfacing and interrupt handling, with the integration of the hardware and software being a constant theme throughout. Parallel and serial input/output, timing, analog, and EEPROM data-handling techniques are covered. A practical build and program case study integrates the previous material into a working system, as well as illustrating simple testing strategies.

With the exception of the first two and last chapter, all chapters have both fully worked examples and self-assessment questions. As an extension to this, an associated Web site has the following facilities:

- Solutions to self-assessment questions.
- Further self-assessment questions.
- Additional material.
- Source code for all examples and questions in the text.
- Pointers to development software and data sheets for devices used in the book.
- Errata.
- Feedback from readers.

You can visit http://www.springer.com, search for the text and click on *Author's Manual* to find the current web location of this site.

Hopefully, any gremlins have been exorcised, but if you find any or have any other suggestions, I will be happy to acknowledge such communications via the Web site.

University of Ulster at Jordanstown   Sid Katzen

Contents

Part I
The Fundamentals

This book is about microcontrollers (MCUs). These are digital engines modeled after the architecture of a stored-program computer and integrated onto a single very large-scale integrated circuit together with support circuitry, memories and peripheral interface devices. Although the microcontroller is often confused with its better-known cousin, the microprocessor, in its role as the driving force of the ubiquitous personal computer, the vast majority of both microprocessors and microcontrollers are embedded into a variety of other digital components. The first microprocessors in the early 1970s were marketed as an alternative way of implementing digital circuitry. Here the task would be determined by a series of instructions encoded as binary code groups in read-only memory. This is more flexible than the alternative approach of wiring hardware integrated circuits in the appropriate manner. The microcontroller is simply the embodiment of this original role of the integrated computer.

We will look at embedded microcontrollers in a general digital processing context in Parts II and III. Here our objective is to lay the foundation for this material. We will be covering:

- Digital code patterns.
- Binary arithmetic.
- Digital circuitry.
- Computer architecture and programming.

This will by no means be a comprehensive review of the subject, but there are many other excellent texts in this area[1] which will launch you into greater depths.

---

[1] Such as S.J. Cahill's *Digital and Microprocessor Engineering*, 2nd edn., Prentice Hall, Englewood Cliffs, NJ, 1993.

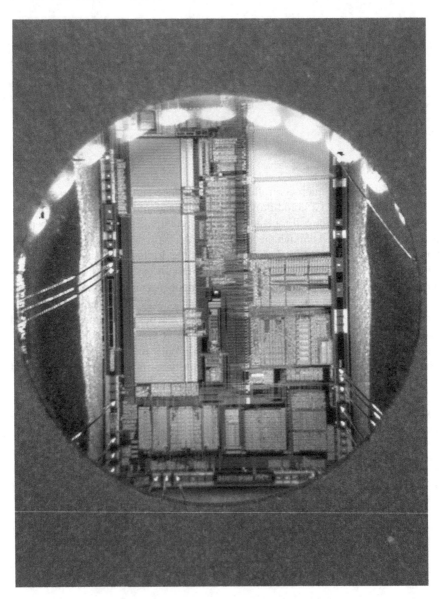

Peeking into the silicon of the PIC18C252

Chapter 1
Digital Representation

To a computer or microcontroller, the world is seen in terms of patterns of digits. The **decimal** (or denary) system represents quantities in terms of the ten digits $0, \ldots, 9$. Together with the judicious use of the symbols $+$, $-$ and $\cdot$ any quantity in the range $\pm\infty$ can be depicted. Indeed non-numeric concepts can be encoded using numeric digits. For example the American Standard Code for Information Interchange (ASCII) defines the alphabetic (alpha) characters A as 65, B $= 66, \ldots,$ Z $= 90$ and a $= 97$, b $= 98, \ldots,$ z $= 122$, etc. Thus the string "Microcontroller" could be encoded as "77, 105, 99, 114, 111, 99, 111, 110, 116, 114, 111, 108, 108, 101, 114". Provided you know the context—that is, what is a pure quantity and what is text—just about any symbol can be coded as numeric digits.[1]

Electronic circuits are not very good at storing and processing a multitude of different values. It is true that the first American digital computer, the Electronic Numerical Integrator And Calculator (ENIAC) in 1946 did its arithmetic in decimal form,[2] but all computers since then have handled data in **binary** (base 2) form. The decimal (base 10) system is really only convenient for humans, in that we have ten fingers.[3] Thus, in this chapter we will solely look at the properties of binary digits, their groupings and processing. After reading it you will:

- Understand why a binary data representation is the preferred base for digital circuitry.
- Know how a quantity can be depicted in natural binary, hexadecimal and binary coded decimal.
- Be able to apply the rules of addition and subtraction for both signed and unsigned natural binary quantities.
- Know how to multiply by shifting left.
- Know how to divide by shifting right and propagating the sign bit.
- Understand the Boolean operations of NOT, AND, OR and XOR.

---

[1] Of course, there are lots of digital encoding standards; for instance, the 6-dot Braille code for the visually impaired.

[2] As did Babbage's mechanical computer of a century earlier.

[3] And ten toes, but base-20 systems are rare though not unknown.

S. Katzen, *The Essential PIC18® Microcontroller,*
Computer Communications and Networks,
DOI 10.1007/978-1-84996-229-2_1, © Springer-Verlag London Limited 2010

The information technology revolution is based on the manipulation, computation and transmission of digitized information. This information is virtually universally represented as aggregates of *bi*nary digi*ts* (**bits**).[4] Most of this processing is effected using microprocessors[5] and microcontrollers, and it is sobering to reflect that there is more computing power in a singing birthday card than existed on the entire planet in 1950!

Binary is the universal choice for data representation, as an electronic switch is just about the easiest device that can be implemented using a transistor. Such 2-state switches are very small; they change state very quickly and consume little power. Furthermore, as there are only two states to distinguish between, a binary depiction is likely to be resistant to the effects of noise. The upshot of this is that both the packing density on a silicon chip and switching rate can be very high. Although a switch on its own does not represent much computing power, 5 million switches changing at 100 million times a second manage to present at least a façade of intelligence!

The two states of a bit are conventionally designated **logic 0** and **logic 1**, or just 0 and 1. A bit may be represented by two states of any number of physical quantities; for instance, electric current or voltage, light, or pneumatic pressure. Most microcontrollers use 0 V (or ground) for state 0 and 3–5 V for state 1, but this is not universal. For instance, the RS232 serial port on your computer uses nominally $+12$ V for state 0 and -12 V for state 1.

A single bit on its own can only represent two states. By dealing with groups of bits, rather more complex entities can be coded. For example, the standard alphanumeric characters can be coded using 7-bit groups of digits, as listed in Table 1.1. Thus the ASCII code for "Microcontroller" becomes:

```
1001101 1101001 1100011 1110010 1101111 1100011 1101111 1101110
1110100 1110010 1101111 1101100 1101100 1100101 1110010
```

Unicode is an extension of ASCII and with its 16-bit code groups is able to represent characters from many languages and mathematical symbols.

The ASCII code is **unweighted**, as the individual bits do not signify a particular quantity; only the overall pattern has any significance. Other examples are the die code on gaming dice and the 7-segment code of Fig. 6.8 on p. 173. Here we will deal with **natural binary** weighted codes, where the position of a bit within the number field determines its value or weight. In an integer binary number the rightmost digit

---

[4]The binary base is not a new fangled idea invented for digital computers; some cultures have used base 2 numeration in the past. The Harappān civilization existed more than 4000 years ago in the Indus River basin. Found in the ruins of the Harappān city of Mohenjo-Daro, in the beadmakers' quarter, was a set of stone pebble weights. These were in ratios that doubled in the pattern, $1, 1, 2, 4, 8, 16, \ldots$, with the base weight of around 25 g ($\approx$1 oz). Thus bead weights were expressed by digits which represented powers of 2; that is, in binary.

[5]*Microprocessors* and *microcontrollers* are closely related (see Fig. 3.8 on p. 59) and so we will often use the terms interchangeably.

Table 1.1 7-bit ASCII characters

MS nybble → LS nybble	h'0' b'000'	h'1' b'001'	h'2' b'010'	h'3' b'011'	h'4' b'100'	h'5' b'101'	h'6' b'110'	h'7' b'111'
h'0' b'0000'	NUL	DLE	SP	0	@	P	`	p
h'1' b'0001'	SOH	XON	!	1	A	Q	a	q
h'2' b'0010'	STX	DC2	"	2	B	R	b	r
h'3' b'0011'	ETX	XOFF	#	3	C	S	c	s
h'4' b'0100'	EOT	DC4	$	4	D	T	d	t
h'5' b'0101'	ENQ	NAK	%	5	E	U	e	u
h'6' b'0110'	ACK	SYN	&	6	F	V	f	v
h'7' b'0111'	BEL	ETB	'	7	G	W	g	w
h'8' b'1000'	BS	CAN	(	8	H	X	h	x
h'9' b'1001'	HT	EM	)	9	I	Y	i	y
h'A' b'1010'	LF	SUB	*	:	J	Z	j	z
h'B' b'1011'	VT	ESC	+	;	K	[	k	{
h'C' b'1100'	FF	FS	,	<	L	\	l		
h'D' b'1101'	CR	GS	-	=	M	]	m	}
h'E' b'1110'	SO	RS	.	>	N	^	n	~
h'F' b1111'	SI	US	/	?	O	_	o	DEL

is worth $2^0 = 1$, the next left column $2^1 = 2$, and so on to the nth column which is worth 2^{n-1}. For instance, the decimal number 1998 is represented as:

$$10^3 \quad 10^2 \quad 10^1 \quad 10^0$$
$$1 \quad\quad 9 \quad\quad 9 \quad\quad 8$$

i.e., $1 \times 10^3 + 9 \times 10^2 + 9 \times 10^1 + 8 \times 10^0$, or just 1998. In **natural binary** the same quantity is:

$$2^{10} \; 2^9 \; 2^8 \; 2^7 \; 2^6 \; 2^5 \; 2^4 \; 2^3 \; 2^2 \; 2^1 \; 2^0$$
$$1 \; \; 1 \; \; 1 \; \; 1 \; \; 1 \; \; 0 \; \; 0 \; \; 1 \; \; 1 \; \; 1 \; \; 0$$

i.e., $1 \times 2^{10} + 1 \times 2^9 + 1 \times 2^8 + 1 \times 2^7 + 1 \times 2^6 + 0 \times 2^5 + 0 \times 2^4 + 1 \times 2^3 + 1 \times 2^2 + 1 \times 2^1 + 0 \times 2^0$, or b'11111001110'.[6] Fractional numbers may equally well be represented by columns to the right of the binary point using negative powers

---

[6]The b'···' notation is not universal; for example, $(1111011110)_2$ is an alternative. If the base is unambiguous then the base indicator may be omitted.

Table 1.2 Some common bit groupings

Bit (1 bit) 0 – 1 (0 – 1)		
Nybble (4 bits) 0 – 15 (0000 – 1111)		
Byte (8 bits) 0 – 255 (0000 0000 – 1111 1111)		
Word (16 bits) 0 – 65,535 (0000 0000 0000 0000 – 1111 1111 1111 1111)		
Long-word (32 bits) 0 – 4,294,967,295 (0000 0000 0000 0000 0000 0000 0000 0000 – 1111 1111 1111 1111 1111 1111 1111 1111)		

of 2. Thus b'1101.11' is equivalent to 13.75. As can be seen from this example, binary numbers are rather longer than their decimal equivalent—on average, a little over three times longer. Nevertheless, 2-way switches are considerably simpler than 10-way devices, so the binary representation is preferable.

An n-digit binary number can represent up to 2^n patterns. Most computers store and process groups of bits. For instance, the first microprocessor, the Intel 4004, handled its data four bits (a **nybble**) at a time. Many current processors cope with blocks of 8 bits (a **byte**), 16 bits (a **word**), 32 bits (a **long-word**) or 64-bits (a **quad-word**). Some of these groupings are shown in Table 1.2. The names illustrated are somewhat de facto, and variations are sometimes encountered.

As in the decimal number system, large binary numbers are often expressed using the prefixes k (kilo), M (mega) and G (giga). A binary kilo (officially called a kibibyte) is $2^{10} = 1024$; for instance, 64 kbyte of memory. In an analogous way, a binary mega (mebibyte) is $2^{20} = 1,048,576$ and a binary giga (gibibyte) is $2^{30} = 1,073,741,824$; thus a 4 Gbyte (or 4 GB) memory stick has a nominal storage capacity of $4 \times 2^{30} = 6,442,450,944$ bytes. The former representation is certainly preferable.

Long binary numbers are not very human friendly. In Table 1.2, binary numbers were zoned into fields of four digits to improve readability. Thus the address of a data unit stored in memory might be b'1000 1100 0001 0100 0000 1010'. If each group of four can be given its own symbol, 0, ..., 9 and A, ..., F, as shown in Table 1.3, then the address becomes h'8C140A';[7] a rather more manageable characterization. This code is called **hexadecimal**, as there are 16 symbols. Hexadecimal (base-16) numbers are a viable number base in their own right, rather than just being a convenient binary representation. Each column is worth $16^0, 16^1, 16^2, \ldots, 16^n$ in the normal way.[8]

Binary coded decimal (BCD) is a hybrid binary/decimal code widely used at the input/output ports of a digital system (see Example 11.6 on p. 370). Here each decimal digit is individually replaced by its 4-bit binary equivalent. Thus 1998 is coded

---

[7]Other representations for the hexadecimal base are 8C140A*h* and 0x8C140A.

[8]Many scientific calculators, including that in the Accessories group under Microsoft's Windows, can do hexadecimal (and binary) arithmetic.

Table 1.3 Different ways of representing the quantities decimal $0, \ldots, 20$

Decimal	Natural binary	Hexadecimal	Binary coded decimal
00	00000	00	0000 0000
01	00001	01	0000 0001
02	00010	02	0000 0010
03	00011	03	0000 0011
04	00100	04	0000 0100
05	00101	05	0000 0101
06	00110	06	0000 0110
07	00111	07	0000 0111
08	01000	08	0000 1000
09	01001	09	0000 1001
10	01010	0A	0001 0000
11	01011	0B	0001 0001
12	01100	0C	0001 0010
13	01101	0D	0001 0011
14	01110	0E	0001 0100
15	01111	0F	0001 0101
16	10000	10	0001 0110
17	10001	11	0001 0111
18	10010	12	0001 1000
19	10011	13	0001 1001
20	10100	14	0010 0000

as $(0001\ 1001\ 1001\ 1000)_{BCD}$. This is very different from the equivalent natural binary code, even if it is represented by 0s and 1s. As might be expected, arithmetic in
such a hybrid system is difficult, and BCD is normally converted to natural binary
at the system input, and processing is done in natural binary before being converted
back (see Program 5.11 on p. 149).

The rules of arithmetic are the same in natural binary[9] as they are in the more
familiar base 10 system, or indeed in any base-n radix scheme. The simplest of these
is **addition**, which is a shorthand way of totaling quantities, as compared to the
more primitive counting or incrementation process. Thus $2 + 4 = 6$ is rather more
efficient than $2 + 1 = 3; 3 + 1 = 4; 4 + 1 = 5; 5 + 1 = 6$. However, it does involve
memorizing the rules of addition.[10] In decimal this involves 45 rules, assuming that
order is irrelevant; from $0 + 0 = 0$ to $9 + 9 = 18$. Binary addition is much simpler
as it is covered by only three rules:

---

[9]Sometimes called 8-4-2-1 code after the weightings of the first four lowest columns.

[10]Which you had to do way back in the mists of time in primary/elementary school!

$$0 + 0 = 0$$
$$\left.\begin{array}{c} 0+1 \\ 1+0 \end{array}\right\} = 1$$
$$1 + 1 = 10 \quad (0 \text{ carry } 1)$$

Based on these rules, the least significant bit (LSB) is totalled first, passing a **carry** if necessary to the next left column. The process ends with the most significant bit (MSB) column, its carry being the new MSD of the sum. For example:

1			1	
0 1			2 6 3 1	
0 0 1			8 4 2 6 8 4 2 1	
96	Augend		1100000	Augend
+ 37	Addend		+ 0100101	Addend
1 1	Carries		1 1	Carries
133	Sum		10000101	Sum

(a) Decimal   *(b) Binary*

Just as addition implements an up count, **subtraction** corresponds to a down count, where units are removed from the total. Thus $8 - 5 = 3$ is the equivalent of $8 - 1 = 7; 7 - 1 = 6; 6 - 1 = 5; 5 - 1 = 4; 4 - 1 = 3$.

The technique of decimal subtraction you are familiar with applies the subtraction rules commencing from LSB and working through to the MSB. In any given column where a larger quantity is to be taken away from a smaller quantity, a unit digit is **borrowed** from the next higher column and given back after the subtraction is completed. Based on this borrow principle, the subtraction rules are given by:

$$0 - 0 = 0$$
$${}^1 0 - 1 = 1 \quad \text{Borrowing 1 from the higher column}$$
$$1 - 0 = 1$$
$$1 - 1 = 0$$

For example:

1			6 3 1	
0 1			4 2 6 8 4 2 1	
96	Minuend		1100000	Minuend
− 37	Subtrahend		− 0100101	Subtrahend
1	Borrows		1 1 1 1 1 1	Borrows
59	Difference		0111011	Difference

(a) Decimal   *(b) Binary*

Although this familiar method works well, there are several problems implementing it in digital circuitry.

- How can we deal with situations where the subtrahend is larger than the minuend?
- How can we distinguish between positive and negative quantities?
- Can a digital system's adder circuits be coerced into subtracting?

To illustrate these points, consider the following example:

37	Minuend	0100101	Minuend
- 96	Subtrahend	-1100000	Subtrahend
—1—		—1—	
41	Difference (-59)	1000101	Difference (-0111011)

(a) Decimal   *(b) Binary*

Normally when we know that the minuend is greater than the subtrahend, the two operands are interchanged and a minus sign is appended to the outcome; that is—(subtrahend—minuend). If we do not swap, as in (a) above, then the outcome appears to be incorrect. In fact, 41 is correct, in that this is the difference between 59 (the correct outcome) and 100; 41 is described as the **10's complement** of 59. Furthermore, the fact that a borrow digit was generated from the MSD indicates that the difference is negative, and therefore will be in this 10's complement form. Converting from 10's complement decimal numbers to the 'normal' magnitude form is simply a matter of inverting each digit and then adding one to the outcome. A decimal digit is inverted by computing its difference from 9. Thus the 10's complement of 3941 is −6059:

$$\overline{3941} \mapsto 6058; \qquad +1 = -6059$$

However, there is no reason why negative numbers should not remain in this complement form just because we are not familiar with this type of notation.

The complement method of negative quantity representation of course applies to binary numbers. Here the ease of inversion ($0 \rightarrow 1; 1 \rightarrow 0$) makes this technique particularly attractive. Thus in our example above:

$$\overline{1000111} \mapsto 0111000; \qquad +1 = -0111001$$

Again, negative numbers should remain in a **2's complement** form.[11] This complement process is reversible. Thus:

$$\text{complement} \quad \Longleftrightarrow \quad \text{normal}$$

Signed decimal numeration has the luxury of using the symbols + and − to denote positive and negative quantities. A 2-state system is stuck with 1s and 0s. However, looking at the last example gives us a clue about how to proceed. A negative outcome gives a borrow back out from the highest column. Thus we can use this MSD as a **sign bit**, with 0 for + and 1 for −. This gives b'1,1000101' for −59 and b'0,0111011' for +59. Although for clarity the sign bit has been highlighted above using a comma delimiter, the advantage of this system is that it can be treated in all arithmetic processes in the same way as any other ordinary bit. Doing this, the outcome will give the correct sign:

---

[11]If you enter a negative decimal number in the Microsoft Windows calculator and change base to Binary, the number will be displayed in 2's complement form.

```
0,1100000  (+96)              0,0100101  (+37)
1,1011011  (-37)              1,0100000  (-96)
  1                             1
────────────────             ────────────────
0,0111011  (+59)              1,1000101  (-59)
```

(a) Minuend less than subtrahend   *(b) Minuend greater than subtrahend*

From this example we see that if negative numbers are in a signed 2's complement form, then we no longer have the requirement to implement hardware sub-
tractors, as adding a negative number is equivalent to subtracting a positive number.
Thus $A - B = A + (-B)$. Furthermore, once numbers are in this form, the outcome of
any subsequent processing will always remain 2's complement signed throughout.

There are two difficulties associated with signed 2's complement arithmetic. The
first of these is **overflow**. It is possible that adding two positive or two negative
numbers will cause overflow into the sign bit; for instance:

```
0,1000  (+8)                  1,1000  (-8)
0,1011  (+11)                 1,0101  (-11)
  1                             1
──────────                    ──────────
1,0011  (-13!!!)              0,1101  (+13!!!)
```

(a) Sum of two +ve numbers gives -ve   *(b) Sum of two -ve numbers gives +ve*

In (a) the outcome of $(+8) + (+11)$ is -13! The 2^4 numerical digit has overflowed
into the sign position (actually b'10011' = 19 is the correct outcome). Example (b)
shows a similar problem for the addition of two signed negative numbers. Overflow
can only happen if both operands have the *same* sign bits. Detection is then a matter
of determining this situation with an outcome that differs. See Fig. 1.5 for a logic
circuit to implement this overflow condition.

The final problem concerns arithmetic on signed operands with different sized
fields. For instance:

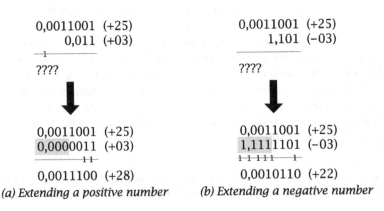

```
0,0011001  (+25)              0,0011001  (+25)
     0,011  (+03)                  1,101  (-03)
   1                             
──────────                    ──────────
????                          ????
```

```
0,0011001  (+25)              0,0011001  (+25)
0,0000011  (+03)              1,1111101  (-03)
      11                       11111   1
────────────────             ────────────────
0,0011100  (+28)              0,0010110  (+22)
```

(a) Extending a positive number   *(b) Extending a negative number*

Both the examples involve adding an 8-bit to a 16-bit operand. Where the former is positive, the data may be increased to 16 bits by padding with 0s—see also p. 120. The situation is slightly less intuitive where negative data requires extension. Here the prescription is to extend the data by padding out with 1s. In the general case the rule is simply to pad out data by propagating the sign bit left. This technique is known as **sign extension**.

Multiplication by the nth power of two is simply implemented by shifting the data left n places. Thus 00110(6) $<<$ 01100(12) $<<$ 11000(24) multiplies 5 by 2^2, where the $<<$ operator is used to denote shifting left. The process works for signed numbers as well:

```
0,00000011  ( 3)        1,11111101  (-3)          0,00000110  (3 x 2)
    <<                       <<                 +  0,00011000  (3 x 8)
0,00000110  ( 6)        1,11111010  (-6)          ────────────────────
    <<                       <<                    0,00011110  (3 x 10 = 30)
0,00001100  (12)        1,11110100  (-12)
    <<                       <<
0,00011000  (24)        1,11101000  (-24)
```

    (a) +3 x 8 = +24   *(b) – 3 x 8 = – 24*   *(c) +3 x 10 = 30*

Should the sign bit change polarity, then a magnitude bit has overflowed giving an overflow error.

Multiplication by nonpowers of 2 can be implemented by a combination of shifting and adding. Thus, as shown in (c) above, 3×10 is implemented as $(3 \times 8) + (3 \times 2) = (3 \times 10)$ or $(3 << 3) + (3 << 1)$.

In a similar fashion, division by powers of 2 is implemented by shifting right n places. Thus 1100(12) $>>$ 0110(6) $>>$ 0011(3) $>>$ 0001.1(1.5). This process also works for signed numbers:

```
0,1111.000  (+15)       1,0001.000  (-15)               0001.1
    >>                       >>                    1010⌐1111.0
0,0111.100  (+7.5)      1,1000.100  (-7.5)            -1010
    >>                       >>                       ─────
0,0011.110  (+3.75)     1,1100.010  (-3.75)            0101
    >>                       >>                        -101.0
0,0001.111  (+1.875)    1,11110.001  (-1.875)         ──────
                                                       000.0
```

    (a) +15/8 = 1.875   *(b) –15/8 = –1.875*   *(c) 15/10 = 1.5*

Notice that rather than always shifting in 0s, the sign bit should be propagated in from the left. Thus positive numbers shift in 0s and negative numbers shift in 1s. This is known as **arithmetic shift right** as opposed to **logic shift right** which always shifts in 0s.

Fig. 1.1 The NOT operation

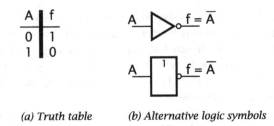

A	f
0	1
1	0

(a) Truth table   (b) Alternative logic symbols

Division by nonpowers of 2 is illustrated in (c) above. This shows the familiar long division process used in decimal division. This is an analogous process to the shift and add technique for multiplication, using a combination of shifting and subtracting.

Arithmetic is not the only way to manipulate binary patterns. George Boole[12] in the mid-nineteenth century developed an algebra dealing with symbolic processing of logic propositions. This **Boolean algebra** deals with variables which can be *true* or *false*. In the 1930s it was realized that this mathematical system could equally well be used to analyze switching networks and thus binary logic systems. Here we will confine ourselves to looking at the fundamental logic operations of this switching algebra.

The inversion or **NOT** operation is represented by overscoring. Thus $f = \overline{A}$ states that the variable f is the inverse of A; that is if $A = 0$ then $f = 1$ and if $A = 1$ then $f = 0$. In Fig. 1.1(a) this transfer characteristic is presented in the form of a **truth table**. By definition, inverting twice returns a variable to its original state; thus $\overline{\overline{f}} = f$.[13]

Logic function implementations are normally represented in an abstract manner rather than as a detailed circuit diagram. The **NOT gate** is symbolized as shown in Fig. 1.1(b). The circle *always* represents inversion in a logic diagram, and is often used in conjunction with other logic elements, such as in Fig. 1.2(c).

The **AND operator** gives an *all or nothing* function. The outcome will only be true when *every* one of the n inputs are true. In Fig. 1.2 two input variables are shown, and the output is symbolized as $f = B \cdot A$, where · is the Boolean AND operator. The number of inputs is not limited to two, and in general $f = A(0) \cdot A(1) \cdot A(2) \cdots A(n)$. The AND operator is sometimes called a logic product, as ANDing (cf. multiplying) any bit with logic 0 always yields a 0 output.

If we consider B as a control input and A as a stream of data, then consideration of the truth table shows that the output follows the data stream when $B = 1$ and

---

[12]The first professor of mathematics at Queen's College, Cork.

[13]A true story from Dr. Seamus Laverty. In days of yore when logic circuits were built out of discrete devices, such as diodes, resistors and transistors, problems arising from sneak current paths were rife. In one such laboratory experiment the output lamp was rather dim, and the lecturer in charge suggested that two NOTs in series in a suspect line would not disturb the logic but would block off the unwanted current leak. On returning sometime later, the students complained that the remedy had no effect. On investigation the lecturer discovered two knots in the offending wire—obviously not tied tightly enough!

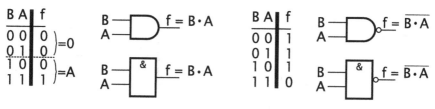

(a) Truth table   (b) Alternative logic symbols (c) NAND

Fig. 1.2 The AND function

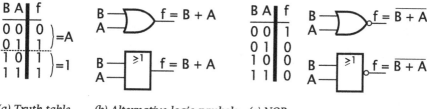

(a) Truth table   (b) Alternative logic symbols   (c) NOR

Fig. 1.3 The inclusive-OR operation

is always 0 when B = 0. Thus the circuit can be considered to be acting as a valve, gating the data through on command. The term **gate** is generally applied to any logic circuit implementing a fundamental Boolean operator.

Most practical AND gate implementations have an inverting output. The logic of such implementations is NOT AND, or **NAND** for short, and is symbolized as shown in Fig. 1.2(c).

The **Inclusive-OR (IOR) operator** gives an *anything* function. Here the outcome is true when *any* input or inputs are true (hence the ≥1 label in the logic symbol). In Fig. 1.3 two inputs are shown, but any number of variables may be ORed together. ORing is sometimes referred to as a logic sum, and the + used as the mathematical operator; thus $f = B + A$. In an analogous manner to the AND gate detecting all ones, the OR gate can be used to detect all zeros. This is illustrated in Fig. 2.20 on p. 34 where an 8-bit zero outcome brings the output of the NOR gate to 1. Inclusive-ORing any bit with a logic 1 always yields a 1 output.

Considering B as a control input and A as data (or vice versa), then from Fig. 1.3(a) we see that the data is gated through when B is 0 and inhibited (always 1) when B is 1. This is a little like the inverse of the AND function. In fact, the OR function can be expressed in terms of AND using the duality relationship $\overline{A + B} = \overline{B} \cdot \overline{A}$. This states that the NOR function can be implemented by inverting all inputs into an AND gate.

The three fundamental Boolean operators are AND, OR and NOT. There is one more operation commonly available as an electronic gate; the **eXclusive-OR operator** (**XOR**). The XOR function is true if *only one* input is true (hence the = 1 label in the logic symbol). Unlike the inclusive-OR, the situation where both inputs are true gives a false outcome.

B A	f
0 0	0
0 1	1
1 0	1
1 1	0

$\left.\begin{matrix} \\ \\ \end{matrix}\right) = A$ $\left.\begin{matrix} \\ \\ \end{matrix}\right) = \overline{A}$

$f = B \oplus A$

$f = B \oplus A$

B A	f
0 0	1
0 1	0
1 0	0
1 1	1

$f = \overline{B \oplus A}$

$f = \overline{B \oplus A}$

(a) Truth table   (b) Alternative logic symbols   (c) ENOR

Fig. 1.4 The XOR operation

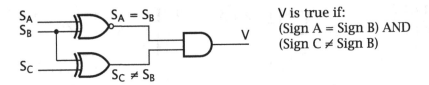

$S_A = S_B$

$S_C \neq S_B$

V

V is true if:
(Sign A = Sign B) AND
(Sign C ≠ Sign B)

Fig. 1.5 Detecting sign overflow

If we consider B is a control input and A as data (they are fully interchangeable) then:

- When B = 0 then f = A; that is, the output follows the data input.
- When B = 1 then f = $\overline{A}$; that is, the output is the inverse of the data input.

Thus an XOR gate can be used as a programmable inverter.

Another useful property considers the XOR function as a logic differentiator. The XOR truth table shows that the gate gives a true output if the two inputs *differ*. Alternatively, the XNOR truth table of Fig. 1.4(c) shows a true output when the two inputs are the same. Thus an XNOR gate can be considered to be a 1-bit equality detector. The equality of two n-bit words can be tested by ANDing an array of XNOR gates (see Fig. 2.7 on p. 23), each generating the function $\overline{B_k \oplus A_k}$; that is:

$$f_{B=A} = \sum_{k=0}^{n-1} \overline{B_k \oplus A_k}$$

As a simple example of the use of the XOR/XNOR gates, consider the problem of detecting sign overflow (see p. 10). This occurs if both the sign bits of word B and word A are the same ($\overline{S_B \oplus S_A}$) AND the sign bit of the outcome word C is not the same as either of these sign bits, say $S_B \oplus S_C$. The logic diagram for this detector is shown in Fig. 1.5 and implements the Boolean function:

$$(\overline{S_B \oplus S_A}) \cdot (S_B \oplus S_C)$$

Finally, the XOR function can be considered as detecting when the number of true inputs are odd. By cascading $n + 1$ XOR gates, the overall **parity** function is true if the n-bit word has an odd number of ones. Some measure of error protection can be obtained by adding an additional bit to each word, so that overall the num-

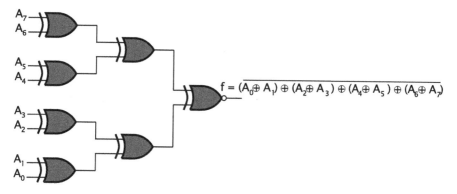

$$f = \overline{(A_0 \oplus A_1) \oplus (A_2 \oplus A_3) \oplus (A_4 \oplus A_5) \oplus (A_6 \oplus A_7)}$$

Fig. 1.6 Generating parity for a byte datum

ber of bits is odd. This oddness can be checked at the receiver and any deviation indicates corruption.

Figure 1.6 uses a tree of 2-I/P Exclusive-OR/NOR gates to effectively create an 8-I/P XNOR gate. The output of this circuit will be logic 1 when there is an even number of 1s. Adding this to the original 8-bit byte gives a 9-bit datum which will always have an odd number of ones. A similar 9-I/P XOR array at the receiver will then act in a similar way to indicate the oddness of the datum at that point. The principle can be extended with additional parity checks to not only detect an error, but to determine which bit is at fault and therefore correct it.

Chapter 2
Logic Circuitry

We have noted that digital processing is all about transmission, manipulation and storage of binary word patterns. Here we will extend the concepts introduced in the last chapter as a lead into the architecture of the computer and microcontroller. We will look at some relevant logic functions, their commercial implementations and some practical considerations.

After reading this chapter you will:

- Understand the properties and use of active pull-up, open-collector and 3-state output structures.
- Appreciate the logic structure and function of the natural decoder.
- See how a MSI implementation of an array of XNOR gates can compare two words for equality.
- Understand how a 1-bit adder can be constructed from gates, and can be extended to deal with the addition and subtraction of two n-bit words.
- Appreciate how the function of an ALU is so important to a programmable system.
- Be aware of the structure and utility of a read-only memory (ROM).
- Understand how two cross-coupled gates can implement a R S latch.
- Appreciate the difference between a D latch and a D flip flop.
- Understand how an array of D flip flops or latches can implement a register.
- See how a serial cascade of D flip flops can perform a shifting function.
- Understand how a D flip flop can act as a frequency divide by two, and how a cascade of these can implement a binary count.
- See how an ALU/PIPO register can implement a programmable accumulator processor unit.
- Appreciate the function of a RAM.

The first digital integrated circuits, available at the end of the 1960s, were mainly NAND, NOR and NOT gates. The most popular family of logic functions was the 74 series Transistor Transistor Logic (TTL) introduced by Texas Instruments and soon copied by all the major semiconductor manufacturers. In various forms TTL still represents the de facto standard.

S. Katzen, *The Essential PIC18® Microcontroller,*
Computer Communications and Networks,
DOI 10.1007/978-1-84996-229-2_2, © Springer-Verlag London Limited 2010

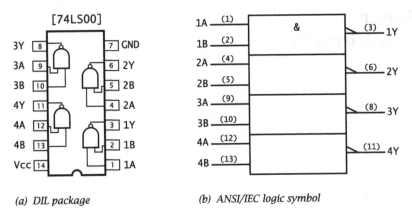

(a) DIL package   (b) ANSI/IEC logic symbol

Fig. 2.1 The 74LS00 quad 2-I/P NAND package

The 74LS00[1] comprises four 2-input NAND gates in a 14-pin package. The integrated circuit (IC) is powered with a 5 ± 0.25 V supply between V_{CC}[2] (usually about 5 V) and GND. The logic outputs are 2.4–5 V for the High state and 0–0.4 V for the Low state. Most IC logic families require a 5 V supply, but 3 V versions are available, and some CMOS implementations can operate with a range of supplies between 3 V and 15 V.

The 74LS00 IC is shown in Fig. 2.1(a) in its dual in-line (DIL) package. Strictly it should be described as a positive-logic quad 2-I/P NAND, as the electrical equivalent for the two logic levels 0 and 1 are Low (L is around ground potential) and High (H is around V_{CC}, usually about 5 V). If the relationship $0 \rightsquigarrow H$; $1 \rightsquigarrow L$ is used (negative logic) then the 74LS00 is actually a quad 2-I/P NOR gate. The ANSI/IEC[3] logic symbol of Fig. 2.1(b) denotes a Low electrical potential by using the polarity symbol. The ANSI/IEC NAND symbol shown is thus based on the *real* electrical operation of the circuit. In this case the logic coincides with a positive-logic NAND function. The & operator shown in the top block is assumed applicable to the three lower gates.

The output structure of a 74LS00 NAND gate is **active pull-up**. Here both the High and Low states are generated by connection via a low-resistance switch to V_{CC} or GND, respectively. In Fig. 2.2(a) these switches are shown for simplicity as metallic contacts, but they are of course transistor derived.

---

[1]The LS stands for "low-power schottky transistor". There are very many other versions, such as ALS (advanced LS), AS (advanced schottky) and HC (high-speed complementary metal-oxide transistor, CMOS). These family variants differ in speed and power consumption, but for a given number designation have the same logic function and pinout.

[2]For historical reasons the positive supply on logic ICs are usually designated as V_{CC}; the C referring to a bipolar's transistor collector supply. Similarly field-effect circuitry sometimes use the designation V_{DD} for drain voltage. The zero reference pin is normally designated as the ground point (GND), but sometimes the V_{EE} (for emitter) or V_{SS} (for source) label is employed.

[3]The American National Standards Institution/International Electrotechnical Commission.

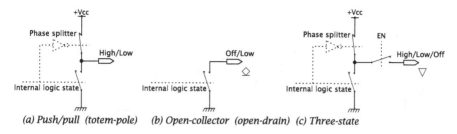

(a) Push/pull (totem-pole)   (b) Open-collector (open-drain) (c) Three-state

Fig. 2.2   Output structures

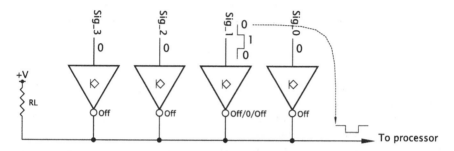

Fig. 2.3   Open-collector buffers driving a party line

Logic circuits, such as the 74LS00, change output state in around 10 nano-seconds.[4] To be able to do this, the capacitance of any interconnecting conductors and other logic circuits' inputs must be rapidly discharged. Mainly for this reason, active pull-up (sometimes called totem-pole) outputs are used by most logic circuits. There are certain circumstances where alternative output structures have some advantages. The **open-collector** (or open-drain) configuration of Fig. 2.2(b) provides a 'hard' Low state, but the High state is in fact an open circuit. The High-state voltage can be generated by connecting an external resistor to either V_{CC} or indeed to a different power rail. Nonorthodox devices, such as relays, lamps or light-emitting diodes, can replace this pull-up resistor. The output transistor is often rated with a higher than usual current and/or voltage rating for such purposes.

The application of most interest to us here is illustrated in Fig. 2.3. Here four open-collector gates share a *single* pull-up resistor. Note the use of the ⌂ symbol to denote an open-collector output. Assume that there are four peripheral devices, any of which may wish to attract the attention of the processor, e.g., computer or microcontroller. If this processor has only one Attention pin, then the four Signal lines must be **wire-ORed** together as shown. With all Signal lines inactive (logic 0) the outputs of all buffer NOT gates are off (state H), and the party line is pulled up to +V by RL. If *any* Signal line is activated (logic 1), as in Sig_1, then the output of

---

[4]A nanosecond is 10^{-9} s, so 100,000,000 transitions each second are possible.

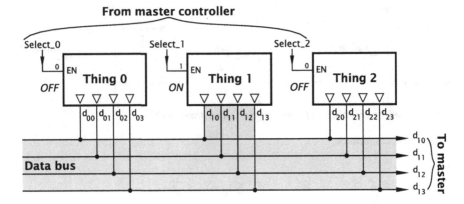

Fig. 2.4 Sharing a bus

the corresponding buffer gate goes hard Low. This pulls the party line Low and thus interrupts the processor.

The **three-state** structure of Fig. 2.2(c) has the properties of both preceding output structures. When enabled, the two logic states are represented in the usual way by high and low voltages. When disabled, the output is open circuit irrespective of the activities of the internal logic circuitry and any change in input state. A logic output with this three-state is indicated by the ∇ symbol.

As an example of the use of this structure, consider the situation depicted in Fig. 2.4. Here a master controller wishes to read one of several devices, all connected to this master over a set of party lines. As this data highway or **data bus** is a common resource, only the selected device can be allowed access to the bus at any one time. The access has to be withdrawn immediately after the data has been read, so that another device can use the resource. As shown in the diagram, each 'thing' connected to the bus outputs is designated by the ∇ symbol. When selected, *only* the active logic levels will drive the bus lines. The 74LS244 octal ($\times 8$) 3-state (sometimes called tristate or TRIS) buffer has high-current outputs (designated by the $\triangleright$ symbol) specifically designed to charge/discharge the capacitance associated with long bus lines.

Integrated circuits with a complexity of up to 12 gates are categorized as small-scale integration (SSI). Gate counts upwards to 100 on a single IC are medium-scale integration (MSI); up to 1000 are known as large-scale integration (LSI) and over this, very large scale integration (VLSI). Memory chips and microcontrollers are examples of this latter category.

The NAND gate networks shown in Fig. 2.5 are typical MSI-complexity ICs. Remembering that the output of a NAND gate is logic 0 only when *all* its inputs are logic 1 (see Fig. 1.2(c) on p. 13) then we see that for any combination of the *select* inputs B A ($2^1 2^0$) in Fig. 2.5(a) only *one* gate will go to logic 0. Thus output $\overline{Y_2}$ will be activated when B A = 10. The associated truth table shows the circuit *decodes* the binary address B A so that address n selects output $\overline{Y_n}$. The 74LS139 is described as a dual 2- to 4-line **natural decoder**. Dual because there are two

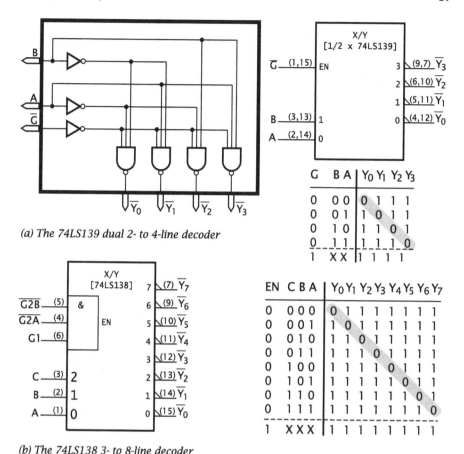

(a) The 74LS139 dual 2- to 4-line decoder

G	B A	Y_0 Y_1 Y_2 Y_3
0	0 0	0 1 1 1
0	0 1	1 0 1 1
0	1 0	1 1 0 1
0	1 1	1 1 1 0
1	X X	1 1 1 1

EN	C B A	Y_0 Y_1 Y_2 Y_3 Y_4 Y_5 Y_6 Y_7
0	0 0 0	0 1 1 1 1 1 1 1
0	0 0 1	1 0 1 1 1 1 1 1
0	0 1 0	1 1 0 1 1 1 1 1
0	0 1 1	1 1 1 0 1 1 1 1
0	1 0 0	1 1 1 1 0 1 1 1
0	1 0 1	1 1 1 1 1 0 1 1
0	1 1 0	1 1 1 1 1 1 0 1
0	1 1 1	1 1 1 1 1 1 1 0
1	X X X	1 1 1 1 1 1 1 1

(b) The 74LS138 3- to 8-line decoder

Fig. 2.5 The 74LS138 and 74LS139 MSI natural decoders

such circuits in the one chip. The symbol X/Y denotes converting code X (natural binary) to code Y (unary—one of *n*). The enabling input $\overline{G}$ is connected to all gates in parallel. Thus the decoder function only operates if $\overline{G}$ is Low (logic 0). If $\overline{G}$ is High, then irrespective of the state of B A (the X entries in the truth table denote a 'don't care' situation) all outputs remain deselected—logic 1. An example of the use of the 74LS139 is given in Fig. 2.25 on p. 38.

The 74LS138 of Fig. 2.5(b) is similar, but implements a 3- to 8-line decoder function. The state of the three address lines C B A ($2^2\,2^1\,2^0$) *n* selects only one of the eight outputs $\overline{Y_n}$. The 74LS138 has three Gate inputs which generate an internal enabling signal $\overline{G2B} \cdot \overline{G2A} \cdot G1$. Only if both $\overline{G2A}$ and $\overline{G2B}$ are Low and G1 is High will the device be enabled.

The **priority encoder** illustrated in Fig. 2.6 is a sort of reverse decoder. Bringing one of the eight input lines Low results in the active-Low three-bit binary equivalent appearing at the output. Thus if $\overline{5}$ is Low, then $\overline{a_2}\,\overline{a_1}\,\overline{a_0} = 010$ (active Low 101).

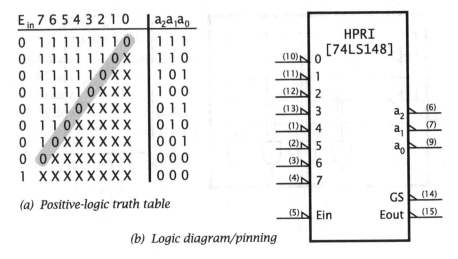

E_{in} 7 6 5 4 3 2 1 0	$a_2a_1a_0$
0 1 1 1 1 1 1 1 0	1 1 1
0 1 1 1 1 1 1 0 X	1 1 0
0 1 1 1 1 1 0 X X	1 0 1
0 1 1 1 1 0 X X X	1 0 0
0 1 1 1 0 X X X X	0 1 1
0 1 1 0 X X X X X	0 1 0
0 1 0 X X X X X X	0 0 1
0 0 X X X X X X X	0 0 0
1 X X X X X X X X	0 0 0

(a) Positive-logic truth table

(b) Logic diagram/pinning

Fig. 2.6 The 74LS148 highest-priority encoder

If more than one input line is active, then the output code reflects the highest. Thus if both $\overline{5}$ and $\overline{3}$ are Low, the output code is still 010. Hence the label HPRI for Highest PRIority. The device is enabled when Enable_IN ($\overline{E_{in}}$) is Low. Enable_OUT ($\overline{E_{out}}$) and Group_Strobe ($\overline{GS}$) are used to cascade 74LS148s to expand the number of lines.

A large class of ICs implement arithmetic operations. The gate array illustrated in Fig. 2.7 detects when the 8-bit byte P7, . . . , P0 is identical to the byte Q7, . . . , Q0. Eight XNOR gates each give a logic 1 when its two input bits Pn & Qn are identical, as described on p. 14. Only if *all* 8-bit pairs are the same, will the output NAND gate go Low. The 74LS688 **equality comparator** also has a direct input $\overline{G}$ into this NAND gate, acting as an overall enabling signal.

The ANSI/IEC logic symbol, shown in Fig. 2.7(b), uses the COMP label to denote the arithmetic comparator function. The output is prefixed with the numeral 1, indicating that its operation P=Q is dependent on any input qualifying the same numeral; that is G1. Thus the active-Low enabling input G1 gates the active-Low output, 1 P=Q.

One of the first functions beyond simple gates to be integrated into a single IC was that of addition. The truth table of Fig. 2.8(a) shows the sum (S) and carry-out (C_1) resulting from the addition of the two bits A and B and any carry-in (C_0).

For instance, row 6 states that adding two 1s with a carry-in of 0 gives a sum of 0 and a carry-out of 1 ($1 + 1 + 0 = {}^1 0$). To implement this row we need to detect the pattern 1 1 0; that is, $A \cdot B \cdot \overline{C_0}$; which is gate 6 in the logic diagram. Thus we have by ORing all applicable patterns together for each output:

$$S = (\overline{A} \cdot \overline{B} \cdot C_0) + (\overline{A} \cdot B \cdot \overline{C_0}) + (A \cdot \overline{B} \cdot \overline{C_0}) + (A \cdot B \cdot C_0),$$
$$C_1 = (\overline{A} \cdot B \cdot C_0) + (A \cdot \overline{B} \cdot C_0) + (A \cdot B \cdot \overline{C_0}) + (A \cdot B \cdot C_0).$$

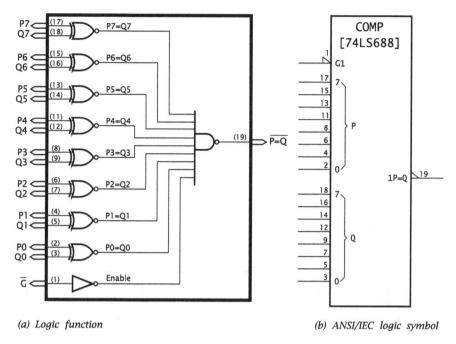

(a) Logic function   (b) ANSI/IEC logic symbol

Fig. 2.7   The 74LS688 octal equality detector

Using such a circuit for *each* column of a binary addition, with the carry-out from column $k-1$ feeding the carry-in of column k means that the addition of any two n-bit words can be implemented simultaneously.

As shown in Fig. 2.8(b), the 74LS283 adds two 4-bit nybbles in 25 ns. In practice the final carry-out C_4 is generated using additional circuitry to avoid the delays inherent on the carries rippling through each stage from the least to the most significant digit. n 74LS283s can be cascaded to implement addition for words of $4 \times n$ width. Four 74LS283s can perform a 16-bit addition in 45 ns, the extra time being accounted for by the carry propagation between the two units.

Adders can, of course, be coaxed into subtraction by inverting the minuend and adding one, that is 2's complementation—as described on p. 9. An adder/subtractor circuit could be constructed by feeding the minuend word through an array of XOR gates acting as programmable inverters (see p. 14). The mode line $\overline{\text{ADD}}/\text{SUB}$ in Fig. 2.9 that controls these inverters also feeds the Carry-In C_0, effectively adding one when in the Subtract mode.

Extending this line of argument leads to the **arithmetic logic unit (ALU)**. An ALU is a circuit which can undertake a selection of arithmetic and logic processes on input data as controlled by mode inputs. The 74LS382 in Fig. 2.10 processes two 4-bit operands in eight ways, as controlled by the three Mode Select bits $S_2 S_1 S_0$ and tabulated in Fig. 2.10(a). Besides addition and subtraction, the logic operations of AND, OR and XOR are supported. The 74LS382 also generates the 2's complement overflow function—see p. 10.

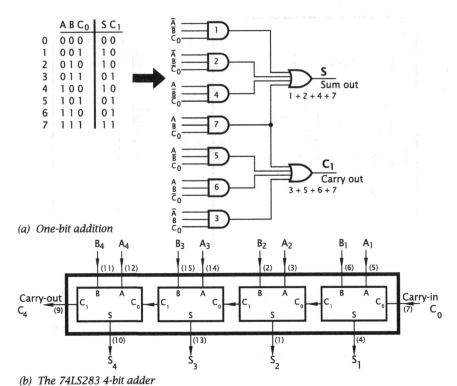

(a) One-bit addition

(b) The 74LS283 4-bit adder

Fig. 2.8 Addition

As we shall see, the ALU is at the heart of the computer and microcontroller architectures. By feeding the Mode Select inputs with a series of binary words, a program of operations can be performed by the ALU. Such **operation codes** are stored in an external memory, and are accessed sequentially by the computer's control circuits.

Sequences of program operation codes are normally stored in some kind of LSI read-only memory. Consider the architecture illustrated in Fig. 2.11. This is essentially a 3- to 8-line decoder driving an 8×2 array of diodes. The 3-bit address selects only row n for each input combination n. If a diode is connected to this row, then it conducts and brings the appropriate column Low. The inverting 3-state output buffer consequently gives a High for each connected diode and Low where the link is broken. The pattern of diode links then defines the output code for each input. For illustrative purposes, the structure has been programmed to implement the 1-bit full adder of Fig. 2.8(a), but *any* two functions of three variables can be generated.

The diode matrix look-up table shown here is known as a **read-only memory (ROM)**, as its 'memory' is in the diode pattern, which is programmed in when the device is manufactured. Early devices, which were typically decoder/32×8 matrices, usually came in user-programmable versions in which the interconnections were implemented with fusible links. By using a high voltage, a selection of

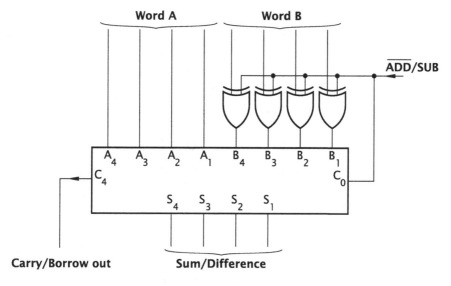

Fig. 2.9 Implementing a programmable adder/subtractor

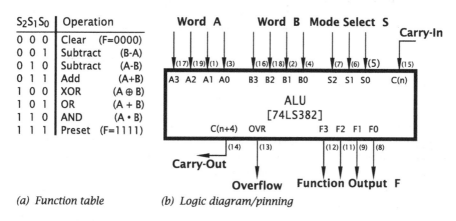

$S_2 S_1 S_0$	Operation
0 0 0	Clear (F=0000)
0 0 1	Subtract (B-A)
0 1 0	Subtract (A-B)
0 1 1	Add (A+B)
1 0 0	XOR (A ⊕ B)
1 0 1	OR (A + B)
1 1 0	AND (A · B)
1 1 1	Preset (F=1111)

(a) Function table   *(b) Logic diagram/pinning*

Fig. 2.10 The 74LS382 ALU

diodes could be taken out of contact. Such devices are called **programmable ROMs (PROMs)**.

Fuses are messy when implementing the larger sizes of VLSI PROMs necessary to store computer programs. For instance, the small 27C64 PROM shown in Fig. 2.12 has the equivalent of 65,536 fuse/diode pairs, and this is a relatively small device capable of storing 8192 bytes of memory. The 27C64 uses the electrical charge on the floating gate of a metal-oxide field-effect transistor (MOSFET) as the programmable link, with another MOSFET to replace the diode. Charge can be tunneled onto this isolated gate by, again, using a high voltage. Once on the gate, the electric field keeps the link MOSFET conducting. This charge takes many

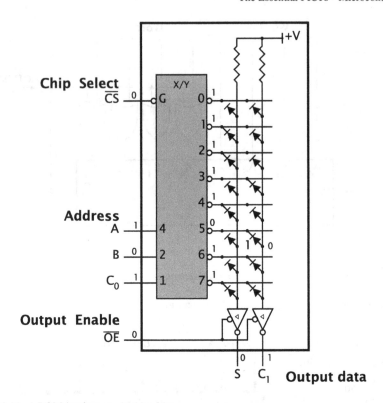

Fig. 2.11 A ROM-implemented 1-bit adder

decades to leak away, but this can be dramatically reduced to about 20 minutes
by exposure to intense ultraviolet radiation. For this reason the 27C64 is known
as an **erasable PROM (EPROM)**. When an EPROM is designed for reusability, a
quartz window is integrated into the package, as shown in Fig. 2.12 and on p. 2.
Programming is normally done externally with special equipment, known as PROM
programmers, or colloquially as PROM blasters. Versions without windows are referred to as one-time programmable (OTP) ROMs, as they cannot easily be erased
once programmed. They are, however, much cheaper to produce and are thus suitable for small- to medium-scale production runs. However, as a general rule flash
EEPROM has a more limited lifetime, as measured as the number of times a cell
can be written to.

Figure 2.13 shows a simplified representation of such a floating-gate MOSFET
link. The cross-point device is a metal-oxide enhancement n-channel field-effect
transistor TR1, rather than a diode. This MOSFET has its gate G1 connected to the
X line and its source S1 to the Y line. If its drain D1 is connected to the positive
supply and the X line is selected (positive), then the Y line too becomes positive
(positive-logic 1) as TR1 is conducting (switch is on). However, if TR1 is disconnected from V_{DD} then it does not conduct and the output on the Y line is logic 0.
Transistor TR2 is in series with V_{DD} and thus acts as the programmable element.

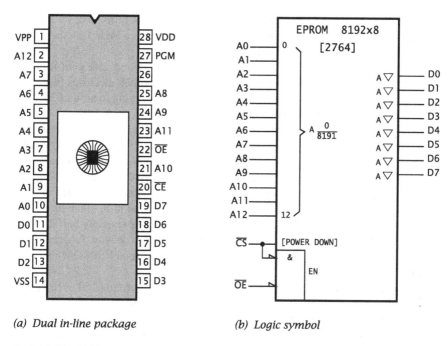

(a) Dual in-line package   (b) Logic symbol

Fig. 2.12 The 2764 erasable PROM (EPROM)

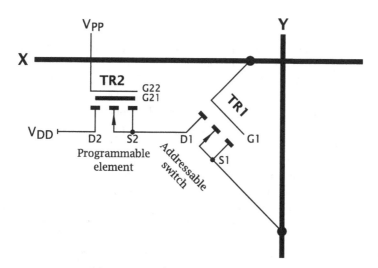

Fig. 2.13 Floating-gate MOSFET link

Transistor TR2 has an extra unconnected gate buried in the silicon dioxide insulation layer. Normally there is no charge on this gate and TR2 is off. If the programming voltage V_{PP} is pulsed high to typically 20–25 V, negative charges tunnel across the extremely thin insulation surrounding the buried gate. This turns TR2 on permanently and thus connects TR1 to its supply. This shows up as a logic 1 on the Y line when selected by the internal memory decoder.

This charge remains more or less permanently on the buried gate until it is exposed to ultraviolet light. The high-energy light photons knock electrons (negative charges) out of the buried (floating) gate[5] effectively discharging in around 20 minutes and wiping out all stored information.

There are PROM structures which can be erased electrically, often *in situ* in the circuit. These are known variously as electrically-erasable PROMs (EEPROMs) or flash memories. In the former case a large negative pulse at V_{PP} causes the captured electrons on the buried gate to tunnel back out. Generally the negative voltage is generated on the chip, which saves having to provide an additional external supply. The **flash** variant of EEPROM relies on hot electron injection rather than tunneling to charge the floating gate. The geometry of the cell is approximately half the size of a conventional EEPROM cell which increases the memory density. Programming voltages are also somewhat lower. An example of a commercial EEPROM memory is given in Fig. 12.28 on p. 443.

Most modern EPROMs/EEPROMs are fairly fast, taking around 150 ns to access and read. Programming is slow, at perhaps 10 ms per word, but this is an infrequent activity. Flash EEPROMs program around 100 times faster, in around 100 μs per cell. However, as a rule they have a more limited lifetime, as measured by the number of times they can be successfully written to. Typically this may be around 100,000 times[6] as compared to over a million.

All the circuits shown thus far are categorized as **combinational logic**. They have no memory in the sense that the output depends only on the present input, and not the sequence of events leading up to that input. Logic circuits, such as latches, counters, registers and read/write memories are described as **sequential logic**. Their output not only depends on the current input, but the sequence of prior inputs.

Consider a typical doorbell pushswitch. When you press such a switch the bell rings, and it stops as soon as you release it. This switch has no memory.

Compare this with a standard light switch. Set the switch and the light comes on. Moreover, it remains on when you remove the stimulus (usually your finger!). To turn the light off you must reset the switch. Again it remains off when the input is taken away. This type of switch is known as a **bistable**, as it has two stable states. Effectively it is a 1-bit memory cell, that can store either an on or off state indefinitely.

A read/write memory, such as the 6264 device of Fig. 2.26, implements each bistable cell using two cross-coupled transistors. Here we are not concerned with

---

[5]This is called the Einstein effect. Einstein was awarded his Nobel prize for this discovery and not for his theories of relativity, as these were considered too revolutionary!

[6]There are around 600,000 seconds in a week and so if a cell is written into once every six seconds the entire lifetime could be used up in a week!

R S	Q
0 0	Q (no change)
0 1	1 (set)
1 0	0 (reset)

(a) Defining RS latch truth table

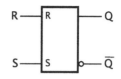

(b) Logic symbol with true/complement outputs

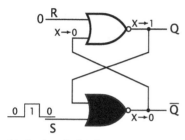

(c) Setting the latch

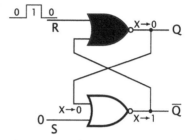

(d) Resetting the latch

Fig. 2.14 The R S latch

this microscopic view. Instead, consider the two cross-coupled NOR gates of Fig. 2.14. Remembering from Fig. 1.3(c) on p. 13 that any logic 1 into a NOR gate will always give a logic 0 output irrespective of the state of the other inputs, allows us to analyse the circuit:

- If the S input goes to 1, then output $\overline{Q}$ goes to 0. Both inputs to the top gate are now 0 and thus output Q goes to 1. If the S input now goes back to 0, then the lower gate remains 0 (as the Q feedback is 1) and the top gate output also remains unaltered. Thus the latch is *set* by pulsing the S input.
- If the R input goes to 1, then output Q goes to 0. Both inputs to the bottom gate are now 0 and thus output $\overline{Q}$ goes to 1. If the R input now goes back to 0, then the upper gate remains 0 (as the $\overline{Q}$ feedback is 1) and the bottom gate output also remains unaltered. Thus the latch is *reset* by pulsing the R input.

In the normal course of events—that is assuming that the R and S inputs are not both active at the same time[7]—then the two outputs are always complements of each other, as indicated by the logic symbol of Fig. 2.14(b).

There are many bistable implementations. For example, replacing the NOR gates by NAND gives a $\overline{R}\,\overline{S}$ latch, where the inputs are active on a logic 0. The circuit illustrated in Fig. 2.15 shows such a latch used to debounce a mechanical switch. Manual

---

[7]If they were, then both Q and $\overline{Q}$ would go to 0. On relaxing the inputs, the latch would end up in one of its stable states, depending on the relaxation sequence. The response of a latch to a simultaneous Set and Reset input signal is not part of the latch definition, shown in Fig. 2.14(a), but depends on its implementation. For instance, trying to turn a light switch on and off together could end in splitting it in two!

Fig. 2.15 Using a $\overline{R}\,\overline{S}$ latch
to debounce a switch

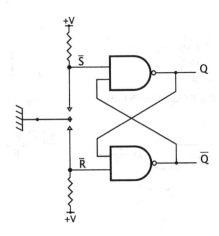

switches are frequently used as inputs to logic circuits. However, most metallic contacts will bounce off the destination contact many times over a period of several tens of milliseconds before settling. For instance, using a mechanical switch to interrupt a computer/microcontroller will give entirely unpredictable results.

In Fig. 2.15, when the switch is moved up and hits the contact the latch move into its Set state. When the contact is broken, the latch remains unchanged, provided that the switch does not bounce all the way back to the lower contact. The state will remain Set no matter how many bounces occur. By symmetry, the latch will reset when the switch is moved to the bottom contact, and remain in this Reset state on subsequent bounces.

The **D latch** is an extension to the R S latch, where the output follows the D (Data) input when the C (Control) input is active (logic 1 in our example) and freezes when C is inactive. The D latch can be considered to be a 1-bit memory cell where the datum is retained at its value at the end of the sample pulse.

In Fig. 2.16(b) the dependency of the Data input with its Control signal is shown by the symbols C1 and 1D. The 1 prefix to D shows that it depends on any signal with a 1 suffix, in this case the C input. That is, C1, clocks in the 1D data.

A flip flop is also a 1-bit memory cell, but the datum is only sampled on an *edge* of the control (known here as the **Clock**) input.

The **D flip flop** described in Fig. 2.16(c) is triggered on a ⟋ (as illustrated in the truth table as ↑), but ⎺⎳ clocked flip flops are common. The edge-triggered activity is denoted as > on a logic diagram, as shown in Fig. 2.16(d).

The 74LS74 shown in Fig. 2.17 has two D flip flops in the one SSI circuit. Each flip flop has an overriding Reset ($\overline{R}$) and Set ($\overline{S}$) input, which are asynchronous, that is, not controlled by the Clock input. MSI functions include arrays of four, six and eight flip flops all sampling simultaneously with a common Clock input.

The 74LS377 shown in Fig. 2.18 consists of eight D flip flops all clocked by the same single Clock input C, which is gated by input $\overline{G}$. Thus the 8-bit data 8D, ..., 1D is clocked in on the ⟋ of C if $\overline{G}$ is Low. In the ANSI/ISO logic diagram shown

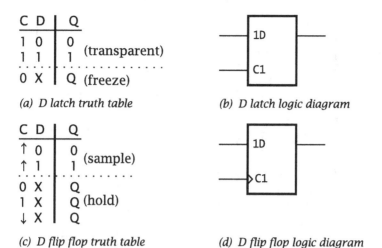

C	D	Q	
1	0	0	(transparent)
1	1	1	
0	X	Q	(freeze)

(a) D latch truth table

(b) D latch logic diagram

C	D	Q	
↑	0	0	(sample)
↑	1	1	
0	X	Q	
1	X	Q	(hold)
↓	X	Q	

(c) D flip flop truth table

(d) D flip flop logic diagram

Fig. 2.16 The D latch and flip flop

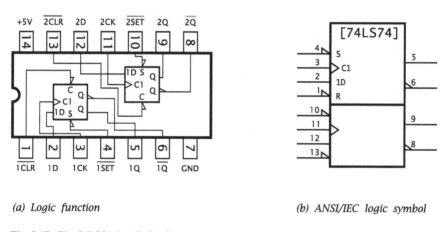

(a) Logic function

(b) ANSI/IEC logic symbol

Fig. 2.17 The 74LS74 dual D flip flop

in Fig. 2.18(b), this dependency is indicated as $G1 \rightarrow 1C2 \rightarrow 2D$, which states that $\overline{G}$ enables the Clock input, which in turn acts on the Data inputs.

Arrays of D flip flops are known as **registers**, that is, read/write memories that hold a single word. The 74LS377 is technically known as a parallel-in parallel-out (PIPO) register, as data is entered in parallel (that is, all in one go) and is available to read at one go. D latch arrays are also available, such as the 74LS373 octal PIPO register shown in Fig. 2.19, in which the eight D flip flops are replaced by D latches. In addition, the latch outputs have a 3-state capability. This is useful if data is to be captured and later put onto a common data bus to be read subsequently as desired by a computer.

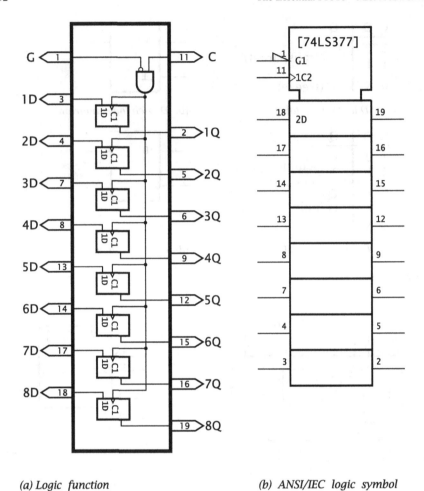

(a) Logic function   (b) ANSI/IEC logic symbol

Fig. 2.18 The 74LS377 octal D flip flop array

A pertinent example of the use of a PIPO register is shown in Fig. 2.20. Here an 8-bit ALU is coupled with an 8-bit PIPO register, accepting as its input the ALU output, and in turn feeding one input word back to the ALU. This register accumulates the outcome of a series of operations, and is sometimes called an **Accumulator** or **Working register**. To describe the operation of this circuit, consider the problem of adding two words A and B. The sequence of operations, assuming the ALU is implemented by cascading two 74LS382s, might be:

1. Program step.

   - Mode = 000 (Clear).
   - Pulsing Execute loads the ALU output (0000 0000) into the register.
   - Data out is zero (0000 0000).

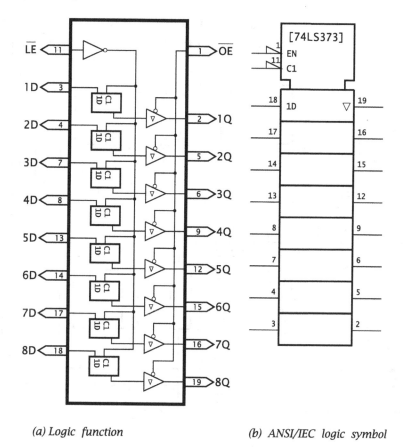

(a) Logic function   (b) ANSI/IEC logic symbol

Fig. 2.19 The 74LS373 octal D latch array

2. Program step.

- Fetch word A down to the ALU input.
- Mode = 011 (Add).
- Pulse ⎍ Execute to load the ALU output (word A + zero) into the register.
- Data out is word A.

3. Program step.

- Fetch word B down to the ALU input.
- Mode = 011 (Add).
- Pulse ⎍ Execute to load the ALU output (word B + word A) into the register.
- Data out is word B plus word A.

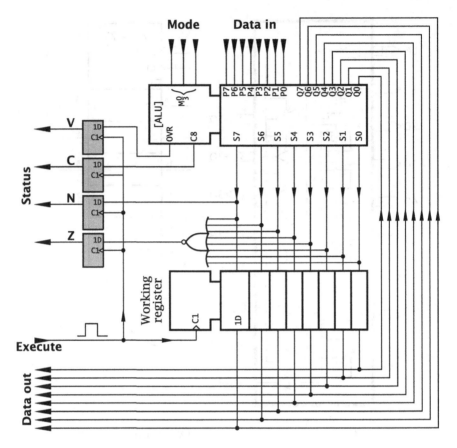

Fig. 2.20 An 8-bit ALU-accumulator processor

The sequence of operation codes, that is 000–100–100 constitutes the program. In practice each instruction would also contain the address (where relevant) in memory of the data to be processed; in this case the locations of word **A** and word **B**.

Each outcome of a process will have associated properties. For instance, it may be zero, be negative (most-significant bit is 1), have a carry-out or 2's complement overflow.

Such properties may be significant in the future progress of the program. In the diagram, four D flip flops, clocked by Execute, are used to grab this status information. In this situation the flip flops are usually known as **flags** (or sometimes semaphores). Thus we have **C**, **N**, **Z** and **V** flags, which form a Status register.

As we will see in the next chapter, the ALU/Working register processor is the heart of digital computing engines. In complex systems, such as a computer or microcontroller, the detail of a diagram like Fig. 2.20 is not necessary and will hide the underlying system process from the observer. Figure 2.21 shows the same process at a higher level of abstraction. For instance, the various multiple wire data connections or **buses** are shown as a single thick path; the actual details are unimportant. The

Fig. 2.21 A system look at our ALU-accumulator processor

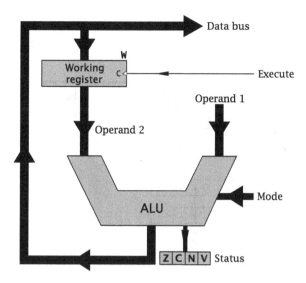

number of connections in a path is not shown, but if important, is usually indicated by a diagonal tick, thus ![8](.).

The ALU, with its distinctive shape, is at the center of our system. Its two data inputs, or operands, are processed according to the Mode input. Operand 1 comes from outside our system, whilst Operand 2 is connected from the Working register. In a computer, the Mode input codes normally come from the program memory, whilst Operand 1 is obtained from the data memory.

The ALU output can be either latched back into the Working register when sampled by the Execute signal, or it can be fed outside into a data memory via the bus. This enhancement is shown in Fig. 3.2 on p. 43.

There are various other forms of register. The 4-bit **shift register** of Fig. 2.22(a) is an example of a serial-in serial-out (SISO) structure. In this instance the data held in the nth D flip flop is presented to the input of the $(n + 1)$th stage. On receipt of a clock pulse (or shift pulse in this context), this data moves into this $(n + 1)$th flip flop, i.e., effectively moving from stage n to stage $n + 1$. As all flip flops are clocked simultaneously, the entire word moves once to the right on each shift pulse.

In the example of Fig. 2.22 a 4-bit external data nybble is fed into the leftmost stage bit-by-bit as synchronized by the clock. After four shift pulses the serial 4-bit word is held in the register. To get it out again, four further shifts move the word bit-by-bit out of the shift register; this is SISO. If the individual flip flops are accessible then the data can be accessed at one go, that is, serial-in parallel-out.

The logic diagram of Fig. 2.22(b) uses the $\rightarrow$ symbol prefixed by the clock input to indicate the shift action; i.e., C1 $\rightarrow$. SRG4 indicates a Shift ReGister 4-stage architecture. An example of an 8-stage shift register is given in Fig. 12.2 on p. 381.

Other architectures include parallel-in serial-out, which is useful for parallel to serial conversion. Counting registers (counters) increment or decrement on each

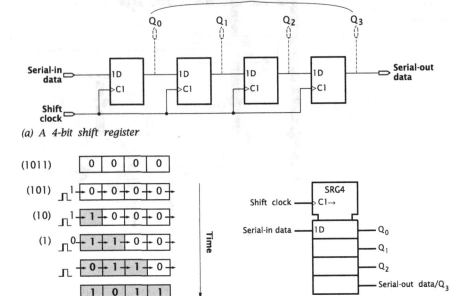

(a) A 4-bit shift register

(b) Shifting 1011 into the register   (c) The ANSII/IEC symbol for a SIPO register

Fig. 2.22 The SISO shift register

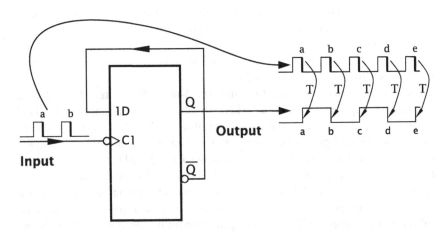

Fig. 2.23 The T flip flop

clock pulse, according to a binary sequence. Typically an n-bit counter can perform a count of 2^n states. Some can also be loaded in parallel and thus act as a store.

Consider the negative-edge triggered D flip flop shown in Fig. 2.23 where its $\overline{Q}$ output is connected back to the 1D input. On each $\sim\!\!\_$ at the Clock input C1,

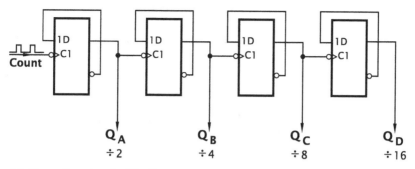

(a) Cascading toggle flip flops

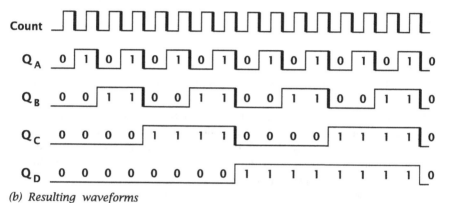

(b) Resulting waveforms

Fig. 2.24 A modulo-16 ripple counter

the data at the 1D input will be latched in to appear at the Q output. As it is the complement of this output that is fed back to the input, then the next time the flip flop is clocked the *opposite* logic state will be latched in. This constant alternation is called *toggling* and is depicted on the diagram by T. The output waveform resulting from a constant-frequency input pulse train is half this frequency. This waveform is a precision squarewave, provided that the input frequency remains constant. This **T flip flop** is sometimes known as a binary or a divide-by-2.

T flip flops can be cascaded, as shown in Fig. 2.24(a). Here four ⌐_ triggered flip flops are chained, with the output of binary n clocking binary $n + 1$. Thus if the input Count frequency was 8 kHz, then Q_A would be a 4 kHz square waveform and similarly Q_B would measure in at 2 kHz, Q_C at 1 kHz, Q_D at 500 Hz.

The waveform Q_A of Fig. 2.24(b) was derived in the same manner as in Fig. 2.23. Q_B is toggled on each ⌐_ of Q_A and likewise for the subsequent outputs. Marking a High as logic 1 and a Low as logic 0 gives the 2^4 (16) positive-logic binary patterns as time advances, with the count rolling over back to state 0 on a continual basis. Each pattern remains in the register until the next event clocks the chain; an event being defined in our example as a ⌐_ at Count. Examining the sequence shows

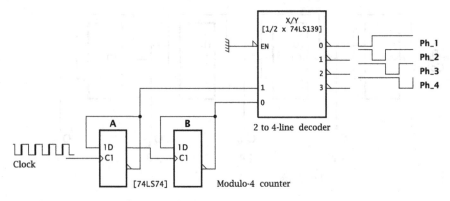

Fig. 2.25 Generating timing waveforms

it to be a natural 8-4-2-1 binary up count, incrementing from b'0000' to b'1111'. In fact, the circuit is a modulo-16 **binary counter** or **timer**. A modulo-n count is the sequence taking only the first n numbers into account.[8]

In theory there is no limit to the number of stages that can be cascaded. Thus using eight T flip flops would give a modulo-256 (2^8) counter. In practice there is a small propagation delay through each stage and this limits the ultimate frequency. For instance, the 74LS74 dual D flip flop of Fig. 2.17 has a maximum propagation from an event at its Clock input to Q output of 25 ns (the maximum toggling frequency for a single stage, such as in Fig. 2.23, is quoted as 25 MHz). An 8-stage counter thus has a maximum ripple-through time of 200 ns (8×25). If such a **ripple counter** were clocked at the resulting 5 MHz ($\frac{1}{200 \text{ ns}}$) then no sooner would one particular code pattern stabilize then the next one would begin to appear. This is only really a problem if the various states of the counter are to be decoded and used to control other logic. The decoding logic, such as shown in Fig. 2.25, may inadvertently respond to these short transient states and cause havoc. In such cases more sophisticated synchronous counter configurations are more applicable where the flip flops are clocked simultaneously and steered by the appropriate logic configuration to count in the desired sequence.

The circuit illustrated here implements an up count. If the complement $\overline{Q}$ lines are used as the outputs, but with the clocking arrangements remaining the same, then the count sequence will decrement, that is a down count. Likewise, if ⌐ triggered flip flops, such as the 74LS74 dual flip flop (see Fig. 2.25), are used as the storage element, then the count will be down. It is easily possible to use some simple logic to combine the two functions to produce a programmable up/down counter. It is also feasible to provide logic to load the flip flop array in parallel with any number

---

[8]Mathematically any number can be converted to its modulo-n equivalent by dividing by n. The remainder, or modulus, will be a number from 0 to $n - 1$.

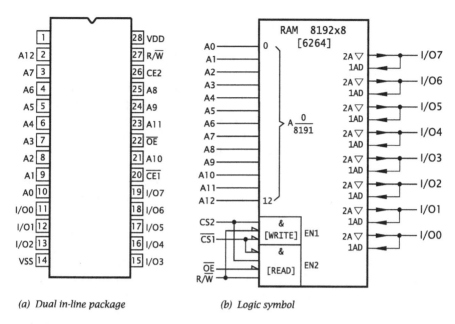

(a) Dual in-line package   (b) Logic symbol

Fig. 2.26 The 6264 8196 × 8 RAM

and then count up or down from that point. Such an arrangement can be thought of as a parallel-in counting register.

In addition to the more obvious uses of a counter register to add up the number of events, such as cans of peas coming along a conveyor belt, there are other uses. One of these is to time a sequence of operations. In Fig. 2.25 a modulo-4 counter is used to address one section of a 74LS139 2- to 4-line decoder; see Fig. 2.5(a). This detects each of the four states of the counter, and the outcome is four time-separated outputs that can be used to sequence, say, the operation of a computer's control section logic—such as that in Fig. 4.5 on p. 76. As a practical point, the complement $\overline{Q}$ flip flop outputs have been used to address the decoder to compensate for the $\underline{\hspace{0.3em}/}$ triggered action that would normally give a down count. Larger counters with the appropriate decoding circuitry can be used to generate fairly sophisticated sequences of control operations.

The term register is commonly applied to a read/write memory that can store a single binary word, typically 4–64 bits. Larger memories can be constructed by grouping n such registers and selecting one of n. Such a structure is sometimes known as a register file. For example, the 74LS670 is a 4 × 4 register file with a separate 4-bit data input and data output and separate 2-bit address. This means that any register can be read at any time, independently of any concurrent writing process.

Larger read/write memories are customarily known as **read/write random-access memories**, or **RAMs** for short. The term random-access indicates that any memory word may be selected with the same access time, irrespective of its position in the memory matrix.[9] This contrasts with a magnetic tape memory, where the reel must be wound to the sector in question—and if this is at the end of the tape . . . .

For our example, Fig. 2.26 shows the 6264 RAM. This has a matrix of 65,536 (2^{16}) bistables organized as an array of 8192 (2^{13}) words of 8 bits. Word n is accessed by placing the binary pattern of n on the 13-bit Address pins A12, . . . , A0.

When in the Read mode (Read/$\overline{\text{Write}}$ = 1), word n will appear at the eight data outputs (I/O7, . . . , I/O0) as determined by the state n of the address bits. The A symbol at the input/outputs (as was the case in Fig. 2.12) indicates this addressability. In order to enable the 3-state output buffers, the $\overline{\text{Output Enable}}$ input must be Low.

The addressed word is written into if R/$\overline{\text{W}}$ is Low. The data to be written into word n is applied by the outside controller to the eight I/O pins. This bidirectional traffic is a feature of computer buses.

In both cases, the RAM chip as a whole is enabled when $\overline{\text{CS1}}$ is Low and CS2 is High. Depending on the version of the 6264, this access from enabling takes around 100–150 ns. There is no upper limit to how long the data can be held, provided power is maintained. For this reason, the 6264 is described as static (SRAM). Rather than using a transistor pair bistable to implement each bit of storage, data can be stored as charge on the gate-source capacitance of a single field-effect transistor. Such charge leaks away in a few milliseconds, so needs to be refreshed on a regular basis. Dynamic RAMs (DRAMs) are cheaper to fabricate than SRAM equivalents and obtainable in larger capacities. They are usually found where very large memories are to be implemented, such as found in a personal computer. In such situations, the expense of refresh circuitry is more than amortized by the reduction in cost of the memory devices.

Both types of read/write memories are volatile, that is, they do not retain their contents if power is removed. Some SRAMs can support existing data at a very low holding current and lower than normal power supply voltage. Thus a backup battery can be used in such circumstances to keep the contents intact for many months. The advantage of this strategy over EEPROM technology is the unlimited number of writes to memory—see Footnote 2.

---

[9]Strictly speaking, ROMs should also be described as random access, but custom and practice has reserved the term for read/write memories.

Chapter 3
Stored Program Processing

In Chap. 2 we designed a simple computing engine based on an arithmetic logic unit (ALU) paired with a parallel-in parallel-out register. The ALU did the number crunching and the Working register held one of the operands and also stored any outcome. In our example, shown on p. 32, we added three numbers together, with the result accumulating in the Working register. If the ALU's mode code is set up before each step, then we can potentially make the computing engine carry out any task that can be described as a sequence of arithmetic and logic operations. This set of command codes (e.g., Add, Subtract, AND, ...) can be stored in digital memory, as can the various operands fed to the ALU and likewise any outcomes. These codes constitute both the **software** of the programmable machine and the various operands or **data**. By *fetching* these **instructions** down one at a time, we can *execute* the system's **program**. This structure, together with its associated data paths, decoders and logic circuitry is known as a digital **computer**.

As we will see, microcontroller architecture is modeled on that of a computer. With this in mind, this chapter looks at the architecture and operating rhythm of the computer structure. Although this computer is strictly hypothetical, it has been very much 'designed' with our book's target microcontroller in mind.

After reading this chapter you will have an understanding of:

- The von Neumann computer structure and recognise its weakness.
- The Harvard architecture with its parallel fetch and execute units, and separate memory spaces.
- The relationship between a digital computer, microprocessor and a microcontroller.
- The structure of a Program store and its interaction with the Program Counter and Pipeline.
- The binary anatomy of typical program instructions.
- The function and structure of a Data store.

Historically the electronic digital computer that we know today was an indirect outcome of the Second World War. Several experimental computers were designed,

S. Katzen, *The Essential PIC18® Microcontroller,*
Computer Communications and Networks,
DOI 10.1007/978-1-84996-229-2_3, © Springer-Verlag London Limited 2010

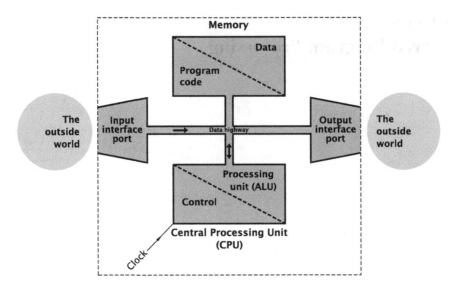

Fig. 3.1 An elementary von Neumann computer; Address bus not shown

and some actually functioned in that period.[1] These computing machines were either special-purpose structures, mainly designed to do a single task on various data, or else needed to be partly rewired to change their behavior.

Given the feasibility of building such computing structures, a major breakthrough by a team of engineers working with von Neumann[2] was to recognize that the program could be stored in memory along with any data. The advantage of this approach is flexibility. To alter the software the new program bit patterns are simply loaded into the appropriate area of memory. In essence, the von Neumann architecture, shown in Fig. 3.1, comprises a **Central Processing Unit** (CPU), a memory and a common connecting bus (or highway) carrying data back and forth. In practice the CPU must also communicate with the environment outside the computer. For this purpose, data to and from suitable interface ports are also funneled through this *single* highway or **data bus**.

---

[1] A prime example was the British Colossus which spent several years breaking Enigma codes. See the book's website for more historical and technical details of these early machines.

[2] Von Neumann was a Hungarian mathematician working for the American Manhattan nuclear weapons program during the Second World War. After the war he became a consultant for the Moore School of Electrical Engineering at the University of Pennsylvania's EDVAC computer project, for which he was to employ his new concept that the program was to be stored in memory along with its data. He published his ideas in 1946 and EDVAC became operational in 1951. Ironically, a somewhat lower key project at Manchester University, UK made use of this approach and the Mark 1 executed its first stored program in June 1948! This was closely followed by Cambridge University's EDSAC which ran its program in May 1949, almost two years ahead of EDVAC.

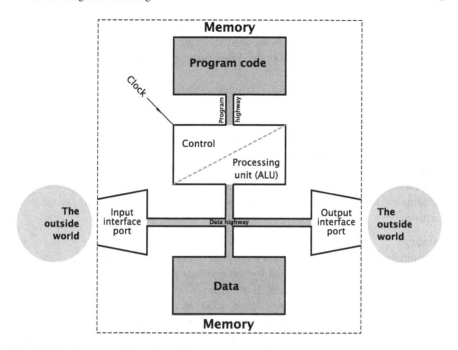

Fig. 3.2 An elementary Harvard architecture computer; Address bus not shown

The great advantage of the von Neumann architecture is simplicity, and because of this the majority of general-purpose computers are modeled after this concept. However, the use of a common bus means that only one thing can happen at time. Thus an execution transaction between the CPU and the **Data store** cannot occur at the same time that an instruction is being fetched from the **Program store**. This phenomena is sometimes known as the von Neumann bottleneck.

In the first decade after the war, Harvard University designed and implemented the Mark 1 through Mark 4 series of computers, which used a variation of this structure where the program memory was completely separate from the data memory. In the original Mark 1 and Mark 2 machines the program was physically implemented as patterns of holes on punched paper tape, which was read as required. This strategy was more efficient then the von Neumann (or, as it is sometimes known, Princeton) architecture, since code could be fetched from program memory concurrently with activity between the CPU and the data memory or input/output transactions. However, such machines were more complex and expensive, and with 1950s technology never became widely accepted after loosing out in a Department of Defence competition to build a computer to monitor the far-flung radar stations in North America. With the evolution of complex integrated circuits this **Harvard** architecture has made a reappearance, with the additional bus connections being subsumed onto the silicon.

Figure 3.2 shows the two physically distinct buses used to carry information to the CPU from these disjoint memories. Each memory has its own address bus and thus there is no interaction between a Program store cell's address and a Data store cell's address. The two memories are said to lie in *separate memory spaces*. The Data store is sometimes known as the **File store**, with a cell at location *n* being described as **File n**.

Let us look at these elements in a little more detail.

The Central Processing Unit

The CPU consists of the ALU/Working register together with the associated control logic. Under the management of the control circuitry, program instructions are fetched from memory, decoded and executed. Data resulting from, or used by, the program is also accessed from memory. This fetch-and-execute cycle constitutes the operating rhythm of the computer and continues indefinitely, as long as the system is active.

Memory

All computing structures use memory to hold both program code and data. Random access memories are characterised by the *contents* they hold in a set of cells and the location or *address* of each cell. In the case of von Neumann type architectures these are held in a single memory space, whereas in Harvard structures they are located in completely separate memory spaces. That is, the addresses of one type of memory do not relate in any way to the addresses of the other memory. In all cases the data stored in a memory is transported to the CPU via a data bus. The CPU sends the address code of the cell it wishes to communicate with via an address bus. In Harvard structures, there will be separate data and address buses for each memory space; see Fig. 3.4.

Most computers have large backup memories, usually magnetic or optical disk-based, in which case access time depends on the cell's physical position in the memory rather than being random access. Apart from this sequential access problem, such media are normally too slow to act as the main memory and are used for backup storage of large arrays of data (e.g., student exam records) or programs that must be loaded (or swapped) into main memory before execution.

Program Memory

The Program store holds the bit patterns which define the program or software. The word is a play on the term hardware; as such patterns do not correspond to any physical rearrangement of the circuitry. Memory holding software should ideally be as fast as the CPU, and normally use semiconductor technologies, such as that described in the last chapter.[3]

Data Memory

The Data store holds data being processed by the program. Again, this memory

---

[3]This wasn't always so; the earliest practical large high-speed program memories used miniature ferrite cores (donuts) that could be magnetized in any one of two directions. Core memories were in use from the 1950s to the early 1970s, and program memory is sometimes still referred to as core.

is normally as fast as the CPU. The processor may also locate special-purpose registers in this memory space; for instance, input/output ports.

The Interface Ports

To be of any use, a computer must be able to interact with its environment. Although conventionally one thinks of a keyboard and screen, any of a range of physical devices may be read and controlled. Thus the flow of fuel injected into a cylinder together with engine speed may be used to alter the instant of spark ignition in the combustion chamber of an internal combustion engine.

Data Highway

All the elements of the von Neumann computer are wired together with the one *common* data highway, or bus (see Fig. 2.4 on p. 20 for a definition of a bus). With the CPU acting as the master controller, all information flow is back and forward along these shared wires. A Harvard computer has a separate data bus for the Program store allowing the instruction codes to be fetched in parallel with activity on the Data store's data bus. Other buses carry addresses to the various memories and control/status information; see Fig. 3.4.

Our target microcontroller is a Harvard computing engine and so we will concentrate on this structure from now on. Based on the CPU of Fig. 2.20 on p. 34, we can add program and data memory with some control and decoding logic to give us

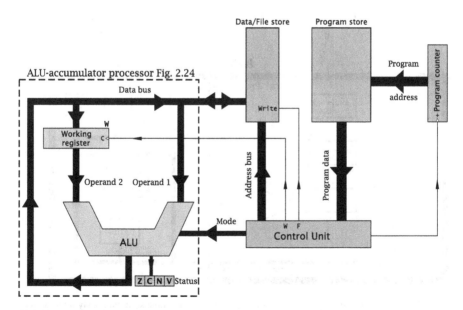

Fig. 3.3 A system look at a rudimentary Harvard computer

the elementary Harvard computer of Fig. 3.3. The delineated portion of the diagram is the original circuit of Fig. 2.21 on p. 35.

By extending the data bus out to the Data store, we can both source operand 1 from this memory and also optionally put the outcome back there. The address of this operand is part of the instruction fetched from the Program store and decoded by the control circuitry. This Control Unit also extracts the mode bits for the ALU, which depends on the current instruction. The outcome from the ALU can either be loaded into the Working register (the control unit pulses WREG) or back into the File in the Data store where the operand originated (the Control Unit pulses F). Again, this destination information is part of the instruction code.

Instructions are normally located as sequential code words in the Program store. A binary up counter (see Fig. 2.24 on p. 37) is used to address each instruction word in turn. If we assume that this **Program Counter** is zeroed when the computer is reset, then the first instruction is located at address h'000' in the Program store, the

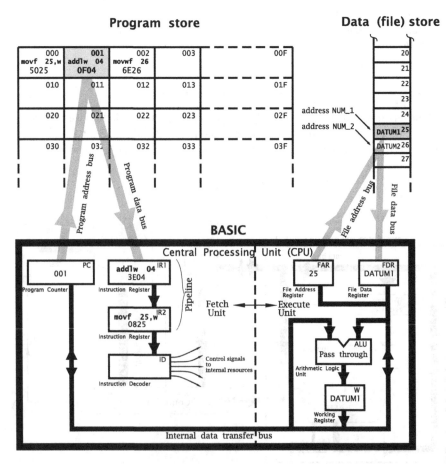

Fig. 3.4 A snapshot of the CPU executing the first instruction whilst simultaneously fetching the second instruction. All addresses/data are in hexadecimal

second at h'001' and so on; see Fig. 3.4. The Control Unit simply increments the counter after each instruction has been fetched. By parallel loading a new address into the Program Counter, overriding this incrementation, the program can be forced to jump to another routine.

The fetch instruction down/decode it/execute sequence, the so-called **fetch-and-execute cycle**, is fundamental to the understanding of the operation of the computer. To illustrate this operating rhythm we are going to look at a simple program that takes a variable datum which is stored in location h'25' in the Data store, then adds the constant four to it and finally assigns the resultant outcome to File h'26'. For reasons of clarity, programmers give names to addresses, and thus NUM_1 and NUM_2 are actually the addresses h'25' and h'26' in which the data are stored. In the high-level language **C** this may be written as:

```
NUM_2 = NUM_1 + 4;
```

A rather more detailed close-up of our computer, which I have named BASIC (for Basic All-purpose Stored Instruction Computer) is shown in Fig. 3.4. This shows the CPU and memories, together with the two data buses and corresponding address buses.

The CPU can broadly be partitioned into two sectors. The leftmost circuitry deals with *fetching* the instruction codes and sequentially presenting them to the Instruction decoder. The rightmost sector *executes* each instruction, as controlled by this Instruction decoder ID.

Looking first at the fetch process:

Program Counter

Instructions are normally stored sequentially in program memory, and the PC is the counter register that keeps track of the current instruction word. This up-counter is sometimes called (perhaps more sensibly) an Instruction Pointer.

As the PC is connected to the execution unit—via an internal data bus—the ALU can be used to manipulate this register and disrupt the orderly execution sequence. In this way various Goto and Skip to another part of the program operations can be implemented.

Pipeline

Two instruction registers hold instruction codes from the Program store. At the top, instruction word n is latched into Instruction Register 1 (IR1) and held for processing during the next cycle. This enables instruction $n-1$ at the bottom of the Pipeline (Instruction Register 2, IR2) to be executed at the same time as instruction n is being fetched into the top of the Pipeline. The Pipeline operation is illustrated in Fig. 3.7.

Instruction Decoder

The ID is the 'brains' of the CPU, deciphering the instruction word in IR2 and sending out the appropriate sequence of signals to the execution unit as necessary to locate any operand in the Data store and to configure the ALU to its appropriate mode. In the diagram, the instruction shown is movf h'25',w (MOVe contents of File h'25' to the Working register).

The execution section deals with accesses to the Data store and configuring the ALU. Execution circuitry is controlled from the Instruction decoder, which is in turn commanded by instruction word $n - 1$ at the bottom of the pipeline in IR2.

All number crunching in the execute unit is done eight bits at a time, and all the registers and Data store likewise hold data in byte-sized chunks. Because of this, the computer would usually be described as an 8-bit processor.

File Address Register
When the CPU wishes to access a cell (or File) in the Data store, it places the File address in the FAR. This directly addresses the memory via the File address bus. As shown in the diagram, File h'25' is being read from the Data store and the resulting datum is latched into the CPU's File Data register.

File Data Register
This is a bidirectional register which either:

- Holds the contents of an addressed File if the CPU is executing a **read cycle**. This is the case for instruction 1 (movf h'25',w) that MOVes (copies or reads from) a datum from File h'25' into the Working register.
- Holds the datum that a CPU wishes to send out (write to) to an addressed File. This **write cycle** is implemented for the movwf h'26' instruction that moves (writes) out the contents of the Working register to File h'26'.

Arithmetic Logic Unit
The ALU carries out an 8-bit arithmetic or logic operation as commanded by its mode code (see Fig. 2.10 on p. 25) which is extracted from the instruction code by the Instruction decoder.

Status Register
This holds the **C**, **Z**, **N** and **V** flags, as described in p. 34.

Working Register
WREG is the ALU's Working register, generally holding one of an instruction's operands, either source or destination. For instance, addwf h'20',w ADDs the contents of the Working register to the contents of File h'20' and places the sum back in WREG. Some computers call this a data or accumulator register.

In addition to the CPU, our BASIC computer has two stores to hold the program code and data.

Program Store
Each location (or cell) in the Program store holds one instruction which is coded as a 16-bit word. In Fig. 3.4 each one of these cells has an address, which originates from the Program Counter via the Program store's address bus. In the diagram the contents of the PC are h'001' (or b'0000000000001'), and this enables the contents of cell h'001' to be placed on the Program store's data bus and hence read into the top

of the Pipeline IR1. In the illustrated case this is h'0F04' (or b'0000111100000100'), which is the machine code for the instruction addlw 04.addlw This will eventually be interpreted by the Instruction decoder as a command to add the constant four to the Working register.

Data Store

Each cell (or File) in the Data store holds one byte (eight bits) of data. The File address is generated by the execute unit via the File Address Register (FAR) and the Data store's address bus. The contents of the addressed File is either read into the File Data Register (FDR) or written from it.

The File address and data busses are completely separate from the Program store counterparts and so processes can proceed on both stores at the same time. Also, Program and Data store addresses are not the same; e.g., the Program store address h'26' is completely different from the Data store address h'26'—that is File h'26'.

Now that we have our CPU with its Program and Data stores, we need to look in more detail at the program itself. There are three instructions in our illustrative software, and as we have already observed, the task is to copy the contents of a byte-sized variable located at the address we have called NUM_1 plus 4 into the location at the address we have called NUM_2.

movf

movf The instruction MOVe File copies the *contents of* the specified File, usually down to the Working register. Thus movf NUM_1,w loads the byte out in data memory at location File h'25' into the Working register. This will set the **Z** flag if the contents of the specified File are all zero (b'0000000') and the **N** if bit 7 is 1. This can be used as a simple test for zero or for negative.

addlw

addlw The ADD Literal to Working register instruction adds a byte-sized literal (constant) to the Working register. Thus addlw 04 adds four to the byte in the Working register and overwrites it with the outcome. The **C** flag is set if a carry-out is generated and the **Z** flag is set if the outcome is zero. Also the **N** is set if bit 7 is set and the **V** is set if there is 2's complement overflow.

movwf

movwf The MOV Working register to File instruction copies the contents of the Working register to the specified File in the Data store. Thus movwf NUM_2 stores the byte in the Working register in File h'26'. None of the flags are altered by this instruction.

In our description of the instructions we have used mnemonics, such as addlw. Of course the actual digital logic decoding these instructions only operate with binary patterns. Mnemonics are simply just a symbolic *aide-mémoire* for the programmer. Although it is unlikely he/she will ever program in **machine code**, the binary

Fig. 3.5 Machine-code
structure for Direct
instructions

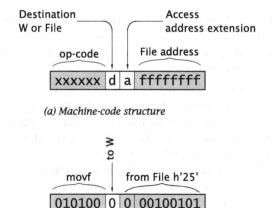

(a) Machine-code structure

(b) For instance, movf h'25',w

structure of instructions are logical and a working knowledge of this will be useful in understanding the foibles and limitations of the instruction set and the real
hardware we will discuss in the next two chapters.

Here we will look at two categories of instructions:

File Direct

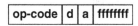

Instructions that specify the File address where their target operand is located use
this type of addressing; e.g., movf h'25',w designates File h'25' as the target.

From Fig. 3.5 we see that the 16-bit instruction code is split into four zones.

- The leftmost six bits (bits 15 through 10) are known as the **operation code**, or
  op-code for short. Every instruction has a unique op-code, and it is this pattern
  that the decoding circuits use to define what type of instruction it is.
- The middle bit (bit 9) labeled d defines the destination of the outcome. For
  instance, addwf h'30',w means "add the contents of the Working register to File h'30' and put the answer back in the Working register", whereas
  addwf h'30',f means "add the contents of the Working register to File h'30'
  and put the answer back in File h'30'". In the former case the destination is WREG
  and the d bit is 0, and in the latter case the destination is the File and the d bit is 1.
  We will look at this instruction in Chap. 5, p. 112. In our symbolic representation
  of instructions, ,w symbolizes a destination bit of 0 whereas, f means d = 1.
- The rightmost eight bits (bits 7 through 0) define the File address. Thus in our
  example File h'25' is b'00100101'. The fact that the address field is just eight bits
  wide means that only a *bank* of $2^8 = 256$ Files can be directly addressed.
- Bit 8 is labeled a. This is really an expansion of the address field. If a is 0 then the
  8-bit address to its right is the absolute address of the target datum. However, if a
  is 1 then the 8-bit address field is augmented with four additional bits located in a
  special function register called the Bank Switch Register (BSR). More details are
  given in Fig. 4.8 on p. 80.

Fig. 3.6 Machine-code
structure for Literal
instructions `addlw`

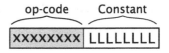

(a) Machine-code structure

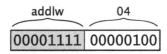

(b) For instance `addlw 04`

Literal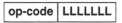

Instructions that deal with constants or **literals** are coded in a slightly different manner, as shown in Fig. 3.6. The upper 8-bit zone defines the instruction op-code. The lower 8-bit zone is the byte constant itself. The outcome is always in the Working register and so there is no need for a destination bit nor any information on a Data store address.

In our exemplar instruction `addlw 04`, the op-code is b'00001111' and the literal is b'00000100'. As the instruction has only an 8-bit data zone, the literal is limited to the range b'00000000'–b'11111111' (h'00'–h'FF', or decimal 0–255), which makes sense since the Working register, like all registers relating to the execution unit, is only eight bits wide.[4]

In addition to using symbolic instruction mnemonics, we have already seen that locations in the Data store can also be given names. Thus, in Fig. 3.4, NUM_1 is our name for "the contents of File h'25'" and NUM_2 names File h'26'. We thus can symbolise our program as:

        NUM_2 = NUM_1 + 4;

Now as far as the computer is concerned, starting at location h'000' our program is:

        0101000000100101
        0000111100000100
        0110111000100110

Unless you are a CPU this is not much fun![5]

---

[4]One of the more frequent mistakes is to forget the 8-bit size restriction and try and use instructions, such as `addlw d'500'` in our program. That makes as much sense as trying to fill a liter (quart) bottle with the contents of a 4-liter (gallon) bucket!

[5]I know; I programmed this way back in the primitive middle 1970s.

Using hexadecimal[6] is a little better.

```
5025
0F04
6E26
```

but is still instantly forgettable. Furthermore, the CPU still only understands binary, so you are likely to have to use a translator program running on, say a PC, to translate from hexadecimal to binary.

If you are going to use a computer as an aid to translate your program, known as **source code**, to binary machine code, known as **object code**, then it makes sense to go whole hog and express the program symbolically. Here the various instructions are represented by mnemonics and variables' addresses are given names. Doing this our program becomes:

```
movf   NUM_1,w     ; Copy the variable NUM_1 to W
addlw  4           ; Add the literal constant 4 to it
movwf  NUM_2       ; Copy NUM_1 + 4 into NUM_2
```

where the text after a semicolon is comment, which makes the program easier to understand by the tame human programmer.

Code written in this symbolic manner is known as **assembly-level** programming. Chapter 8 is completely devoted to the syntax of assembly-level language, and its translation to machine-executable binary.

In writing programs using assembly-level symbolic representation, it is important to remember that each instruction has a one-to-one correspondence to the underlying machine instructions and its binary code. In Chap. 9 we will see that high-level languages lose that 1:1 relationship.

The core of computer operation is the rhythm of the **fetch-and-execute cycle**. Here each instruction is successively fetched from the Program store, interpreted and then executed. Because any execution memory access will be on the Data store and as each store has its own busses, then the fetch and execution processes can progress *in parallel*. Thus while instruction n is being fetched, instruction $n - 1$ is being executed. In Fig. 3.4 the instruction codes for both the imminent and current instructions are held in the two Instruction registers IR1 and IR2 respectively. Instructions are fetched into one end of this Pipeline and 'popped out', into the Instruction decoder at the other end. Figure 3.7 shows the timeline of our 3-instruction exemplar program, quantized in instruction cycles. During each cycle, except for the first, both a fetch and an execution is proceeding simultaneously.

---

[6]Remember that we are only using hexadecimal notation as a human convenience. If you took an electron microscope and looked inside these cells you would only 'see' the binary patterns indicated.

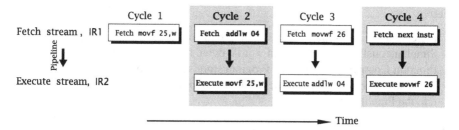

Fig. 3.7 Parallel fetch and execute streams

In order to illustrate the sequence in a little more detail, let us trace through our specimen program. We assume that our computer (that is, the Program Counter) has been reset to h'000' and has just finished the Cycle 1 fetch.

Fetch (Fig. 3.4) .... Cycle 2

• Increment the Program Counter to point to instruction 2.
• Simultaneously move the instruction word 1 down the Pipeline (from Instruction register 1 to Instruction register 2).
• Program Counter (h'001') to the Program store's address bus.
• The instruction word 2 then appears on the Program store's data bus and is loaded into Instruction register 1.

Execute (Fig. 3.4) .... Cycle 2

• The operand address h'25' (i.e., NUM_1) moves to the File Address register and out onto the File address bus.
• The resulting datum at NUM_1 is read onto the Data store's data bus and loaded into the File Data register.
• The ALU is configured to the Pass Through mode, which feeds the datum through to the Working register.

Fetch .... Cycle 3

• Increment the Program Counter to point to instruction 3.
• Simultaneously move the instruction word 2 down the Pipeline (from Instruction register 1 to Instruction register 2).
• Program Counter (h'002') to the Program store's address bus.
• The instruction word 3 then appears on the Program store's data bus and is loaded into the Pipeline at Instruction register 1.

Execute .... Cycle 3

• The ALU is configured to the Add mode and the literal (which is part of instruction word 2) is added to the datum in WREG.
• The ALU output, NUM_1 + 4, is placed in WREG.

Fetch .... Cycle 4

- Increment the Program Counter to point to instruction 4.
- Simultaneously move instruction word 3 down the Pipeline to IR2.
- Program Counter (h'003') to the Program store's address bus.
- The instruction word 4 then appears on the Program store's data bus and is loaded into the Pipeline at IR1.

Execute .... Cycle 4

- The operand address (i.e., NUM_2) h'26' to the File Address register and out onto the File address bus.
- The ALU is configured to the Pass Through mode, which feeds the contents of WREG through to the File Data register and onto the Data store's data bus.
- The datum in the File Data register is written into the Data store at the address on the Data store's address bus and becomes the new datum in NUM_2.

Notice how the Program Counter is automatically advanced during each fetch cycle. This sequential advance will continue indefinitely unless an instruction to modify the PC occurs, such as goto h'200'. This would place the address h'200' into the PC, overwriting the normal incrementing process, and effectively causing the CPU to jump to whatever instruction was located at h'200'. Thereafter, the linear progression would continue.

Although our program doesn't do very much, it only takes around 1 μs to implement each instruction. A million unimpressive operations each second can amount to a great deal! In essence, all computers, no matter how intelligent they may appear, are executing relatively simple instructions very rapidly. The skill of course lies with the programmer in deciding what sequence of instructions and data structures are needed to implement the appropriate task!

Up to now we have referred specifically to computerlike structures. To finish the chapter we have to link the subject of this text to this material—that is, the microcontroller.

What exactly is a **microcontroller unit (MCU)**? In a nutshell, a microcontroller is a microprocessor unit (MPU) which is integrated with memory and input/output peripheral interface functions on the (usually) one integrated circuit. In essence it is a MPU with on-board system support circuitry. Thus we begin by investigating the origins of the MPU. From a historical perspective the story begins in 1968 when Robert Noyce (one of the inventors of the integrated circuit), Gordon Moore[7] and Andrew Grove left the Fairchild Corporation and founded their own company, which

---

[7]Moore's law stated in 1965 (when ICs had around 50 transistors per chip) that the number of elements on a chip would double every 18 months. This was based on an extrapolation of growth from 1959 and this was subsequently revised to 2 years.

they called Intel.[8] Within three years, Intel had developed all the basic types of semiconductor memories used today—dynamic and static RAMs and EPROMs.

As a sideline, Intel also designed large-scale integrated circuits to customers' specifications. In 1970 they were approached by the Busicom corporation of Japan, and asked to manufacture a suitable chip set for a line of calculators. At that time calculators were a fast-evolving product and any LSI devices were likely to be superseded within a few years. This of course would reduce an LSI product's profitability and increase its cost. Engineer Ted Hoff—reputedly while on a topless beach in Tahiti—came up with a revolutionary way to tackle this project. Why not make a simple computer CPU on silicon? This could then be programmed to implement the calculator functions, and as time progressed these could be enhanced by developing this software. Besides giving the chip a longer and more profitable life, Intel was in the business of making memories—and computerlike architectures need lots of memory! Truly a brain wave. The Japanese company endorsed the Intel design for its simplicity and flexibility in late 1969, rather than the conventional implementation.

Federico Faggin joined Intel in spring 1970[9] and by the end of the year had produced working samples of the first chip set. This could only be sold to Busicom, but by the middle of 1971 they were in financial straits and in return for a payback of their $65,000 design costs, Intel was given the right to sell the chip set to anyone for non-calculator purposes. Intel was dubious about the market for this device, but went ahead and advertised the 4004 "Micro-Programmable Computer on a Chip" in the *Electronic News* of November 1971. The term **microprocessor unit** was not coined until 1972. The 4004 created a lot of interest as a means of introducing 'intelligence' into electronic products.

The 4004 MPU featured a von Neumann architecture using a four-bit data bus, with direct addressing of 512 bytes of memory. Clocked at 108 kHz, it was implemented with a transistor count of 2300.[10] Within a year the 8-bit 200 kHz 8008 appeared, addressing 16 Kbytes and needing a 3500 transistor implementation. Four bits is satisfactory for the BCD digits used in calculators, but eight bits is more appropriate for intelligent data terminals (like cash registers) which need to handle a wide range of alphanumeric characters. The 8008 was replaced by the 8080[11] in 1974, and then the slightly modified 8085 in 1976. The 8085 is still the current Intel 8-bit device.

The MPU concept was such a hit that many other electronic manufacturers clambered onto the bandwagon. In addition, many designers set up shop on their own, such as Zilog. By 1976 there were 54 different MPUs either available or announced. For example, one of the most successful families was based on the 6800 introduced

---

[8]Reputed to stand for INTELligence or INTegrated ELectronics.

[9]He was later to found Zilog (last word (Z) in Integrated LOGic) which became notable with the Z80 MPU, a rather superior Intel 8085.

[10]Compare with the Pentium Pro (also known as the P6 or 80,686) at around 5.5 million!

[11]Designed by Masatoshi Shima, who went on to design the 8080-compatible Z80 for Zilog.

by Motorola.[12] The Motorola 6800 had a clean and flexible von Neumann architecture, could be clocked at 2 MHz and address up to 64 Kbyte of memory. The 6802 (1977) even had 128 bytes of on-board memory and an internal clock oscillator. By 1979 the improved 6809 represented the last in the line of these 8-bit devices, competing mainly with the Intel 8085, Zilog Z80 and MOS Technology's 6502.

The MPU was not really designed to power conventional computers, but a small calculator company called MITS,[13] faced with bankruptcy, took a final desperate gamble in 1975 and decided to make and market a computer. This primitive machine, designed by Ed Roberts, was based on the 8080 MPU and interacted with the operator using front-panel toggle switches and lamps—no keyboard or VDU. The Altair[14] was advertised, and within a few weeks MITS had around 650 advance orders at about $400 each; going from $400,000 in the red to $250,000 in the black.

This first commercially successful[15] **personal computer (PC)** spawned a generation of computer hackers. Thus an unknown 19-year-old Harvard computer science student, Bill Gates, and a visiting friend, Paul Allen, in December 1975 noticed a picture of the Altair[16] on the front cover of *Popular Electronics* and decided to write software for this primordial PC. They called Ed Roberts with a bluff, telling him that they had just about finished a version of the BASIC programming language that would run on the Altair. Thus was born the Microsoft® Corporation.

In a parallel development, some two months later, 32 people in San Francisco set up the Home-brew Club, with initially one hard-to-get Altair between them. Two members were Steve Jobs and Steve Wozniak. As a club demonstration, they built a PC which they called the Apple.[17] By 1978 the Apple II made $700,000; in 1979 sales were $7 million, and then $48 million....

The Apple II was based on the low-cost 6502 MPU which was produced by a company called MOS Technology. It was designed by Chuck Peddle, who was also responsible for the 6800 MPU, and had subsequently left Motorola. The 6502 bore an uncanny resemblance to the Motorola 6800 family and indeed Motorola sued to prevent the related 6501 MPU being sold, as it even had the same pinout as the 6800. The 6502 was one of the main players in PC hardware by the end of the 1970s, being the computing engine of the BBC series and Commodore PETs amongst many others.

What really powered up Apple II sales was the VisiCalc spreadsheet package. When the business community discovered that the PC was not just a toy, but could

---

[12]Motorola was launched in the 1930s to manufacture motor car radios, hence the name 'motor' and 'ola', as in pianola. Motorola spunoff its microcontroller business in 2006 to Freescale Semiconductor. Also in 2006, Microchip overtook Motorola in achieving the largest share of the worldwide 8-bit microcontroller market in number and value.

[13]Located next door to a massage parlor in New Mexico.

[14]After a planet in *Star Trek*.

[15]There was an earlier design published in *Radio Electronics* of June 1974. The Mark 8 by Jonathan Titus was based on an Intel 8008 MPU.

[16]The picture was just a mock-up; they actually were not yet available; an early example of computer 'vaporware'!

[17]Jobs was a fruitarian and had previously worked in an apple orchard.

do 'real' tasks, sales took off. The same thing happened to the IBM PC. Reluctantly introduced by IBM in 1981, the PC was powered by an Intel 8088 MPU clocked at 4.77 MHz together with 128 Kbyte of RAM, a twin 360 Kbyte disk drive and a monochrome text-only VDU. The operating system was Microsoft's® PC/MS-DOS version 1.0. The spreadsheet package was Lotus 1-2-3.

By the end of the 1970s the technology of silicon VLSI fabrication had progressed to the stage that several tens of thousands of transistors could be integrated on a single chip. Microprocessor designers were quick to exploit this capability in one of two ways. The better known of these was to increase the size of the ALU and buses/memory capacity. Intel was the first with the 29,000-transistor 8086, introduced in 1978 as a 16-bit version of the 8085 MPU.[18] It was designed to be compatible with its 8-bit predecessor in both hardware and software aspects. This was wise commercially, in order to keep the 8085's extensive customer base from looking at (better?) competitor products, but technically dubious. It was such previous experience that led IBM to use the 8088 version, which had a reduced 8-bit data bus and 20-bit address bus[19] to save board space.

In 1979 Motorola brought out its 16-bit offering called the 68000 and its 8-bit data bus version, the 68008 MPU. However, internally it was 32-bit, and this has provided compatibility right up to the 68060 introduced in 1995 and the ColdFire RISC device launched in 1997. With a much smaller 8-bit customer base to worry about, the 68000 MPU was an entirely new design and technically much in advance of its 80X86 rivals.

The 68000 was adopted by Apple for its Macintosh series of PCs. However, the Apple Mac only accounted for less than 5% of PC sales. Motorola MPUs have been much more successful in the embedded microprocessor market, the area of smart instrumentation from egg timers to aircraft management systems. Of course, this is just the area which MPUs were developed for in the first place, and the number, if not the profile and value, of devices sold for this purpose exceeds those for computers by more than an order of magnitude.

In this applications area an MPU is 'buried' in the application circuit together with memory and various input and output interface circuits. The MPU with its program acts as the controller of the system by virtue of the software in program memory. Over 4 billion microprocessor and related devices are sold each year for embedded control, making up over 90% of the MPU market.

The second way of using the additional integrated circuit complexity that became available by the end of the 1970s was to keep a relatively simple CPU and use the extra silicon 'real estate' to implement on-board memory and input/output interface. In doing so, simple embedded control systems on a single chip became

---

[18] And the Intel 8086 architecture-based MPUs are by far the largest-selling MPU for computer-based circuitry.

[19] A 2^{20} address space is 1 Mbyte, and this is why for backwards compatibility MS-DOS was limited to 1 Mbyte of conventional memory, called real memory in a Microsoft® Windows environment.

possible and the overall chip count to implement a given function was thereby considerably reduced. The majority of control tasks require relatively little computing power, but the reduction in size (and therefore cost) is vital. A simple example of this is the intelligent smart card, which has a processor integrated into the card itself. Such microprocessor-based devices were called microcontrollers.[20] For instance, there are several hundred microcontrollers hidden in every home—in domestic appliances, entertainment units, PCs, communication devices, smart cards and in particular in the family's cars.

In terms of architecture, referring back to Figs. 3.1 and 3.2, the microprocessor is the central processor unit, whereas the microcontroller is the complete functioning computerlike system. As an example, consider the electronics of a car odometer monitoring system displaying total distance since manufacture and also a trip odometer. The main system input signal is a tachometer generating pulses on each rotation of the engine flywheel, which when totaled gives the number of engine revolutions—and the pulse-to-pulse duration could also give the road speed. Of course the actual road distance depends on the gearing of the engine, and thus we need to know which of the five gear ratios has been chosen by the driver at any time. This is shown as five lines G5,...,G1, (usually designated G[5:1]), originating from the gear box. One signal will be high for the appropriate forward gear, with reverse being ignored. Additional inputs are used to give a manufacturer's option of a mile or kilometer display, and a user input to reset the trip display to zero.

The display itself consists of seven 7-segment digits (see Fig. 6.8 on p. 173) to indicate up to (optimistically) 999999.9. As there are so many segments to control (49 in total), Fig. 3.8 shows the display data fed via a single digital line, shunted serially into a shift register; see Fig. 2.22 on p. 36. A second line provides clock pulses for the register with 49 clock pulses being needed to refresh the display.[21]

The trip odometer display comprises four digits, which will record up to 999.9.

Similarly two output lines are used to feed and clock the shift register, with 28 clock pulses needed to shift in a new 4-digit trip display.

The **resource budget** (list of subsystem functions) for this system is:

- An edge-triggered input for the tachometer pulse train, connected to a counter/ timer to add up engine revolutions.
- Seven static digital input lines to interface to the gear ratio, mi/km option and trip reset.
- Four output digital lines to clock the two shift registers and provide segment data.
- A microprocessor to do the calculations and to read/write to the input/ output ports, respectively.
- Program memory, usually ROM of some kind.
- Data memory for temporary storage of program variables, usually static RAM.
- Non-volatile storage for physical variables, such as total distance and distance since trip reset.

---

[20]The term *microcomputer* was an alternative term but was easily confused with early personal computers and has dropped into disuse.

[21]Many displays have this shift register built in as a complete subsystem.

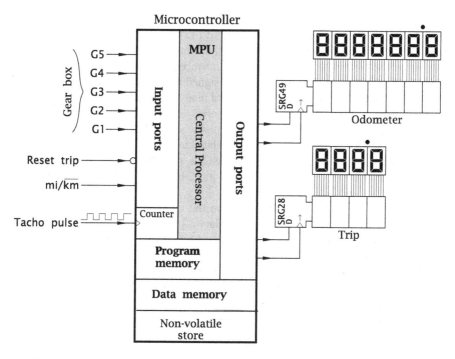

Fig. 3.8   An example of a system based on a microcontroller

This functionality could be implemented onto a single integrated circuit, and in this situation would be known as a microcontroller, that is, a microprocessor integrated with its support circuitry giving a complete microcomputer function. Of course the resource budget listed above is specific to our example. Although the core functions (microprocessor and memory) are common to a wide range of applications, the input/output (I/O) interface needs to be tailored to the task at hand. Some typical I/O functions are:

- I/O to interface to a serial bit stream of various synchronous and asynchronous protocols.
- Counter/timer functions to add up input events and to generate precision time-varying digital output signals.
- Analog-to-digital multiplex/conversion to be able to read and digitize analog inputs.
- Digital-to-analog conversion to output analog signals.
- Display ports to drive multidigit liquid crystal displays.

This alternative approach to using additional silicon resources led to the first MCUs in the late 1970s. For instance, the 35,000-transistor Motorola 6801, designed in response to a specific application from a car manufacturer, used the existing 6800 MPU as a core, with 2048 bytes of ROM program memory, 128 bytes of data RAM, 29 I/O lines and a 16-bit timer. With the viability of the MCU approach vindicated, a range of families, each based on a specific core but with individual family members having a different selection of I/O facilities, was introduced by the leading MPU manufacturers. For instance, the Motorola 68HC11 family (a development of the 6801 MCU) uses a slightly enhanced 6800 core. The 68HC12 and 68HC16 families use 16-bit cores but are designed to be upwardly compatible with the 8-bit 68HC11. It was quickly realized that many embedded applications did not even need the power of the (antique) 6800 core, and the 68HC05 family[22] had a severely reduced core and lower price. Actually, 4-bit MCUs, such as the Texas Instruments TMS1000 series, outsold all other kinds of processor until the early 1990s (and are still going strong) and 8-bit MCUs, now the most popular, are likely to continue in this role for the foreseeable future. Indeed the Motorola 14500 processor even uses one bit!

All these MPUs and MCUs are based on the von Neumann architecture used by mainframe computers. The alternative Harvard architecture first reappeared in the Signetics 8X300 MPU, and this was adapted by General Instruments in the mid 1970s for use as a **Peripheral Interface Controller** (**PIC**) which was designed to be a programmable I/O port for their 16-bit CP1600 MPU. When General Instruments sold off their microelectronics division in 1988 to a startup company called Arizona Microchip Technology, this device resurfaced as a stand-alone microcontroller. This family of microcontroller devices is the subject of the rest of our book.

Examples

Example 3.1 A greenhouse controller is to monitor an analog signal from a soil moisture probe and if below a certain value turn on a water valve for 5 seconds and off for 5 seconds. The source of the water is a tank with a float and if the level in the tank drops too low, a switch will be closed. In this event a buzzer is to be activated to sound the alarm.

Can you devise a system based on a microcontroller that will implement the system intelligence?

Solution The solution given in Fig. 3.9 is based on the car odometer of Fig. 3.8. The only new peripheral device is the analog port which can read and digitize the analog output of the soil moisture transducer. This works on the principle that the resistance of the soil between the two electrodes depends on the moisture content. This forms a potential divider and thus a varying voltage at the junction with the fixed resistor. The MCU can digitize this analog voltage, giving an internal digital equivalent byte, which is then compared in software with a predetermined value. Alternatively, the input port can simply be an analog comparator, giving a digital on/off response if the input voltage exceeds a value which can be set by the program.

On the basis of this diagram we can list the resource budget.

---

[22]The 68HC05 has found a niche as the computing engine of smart cards, where high-power computing is not a priority.

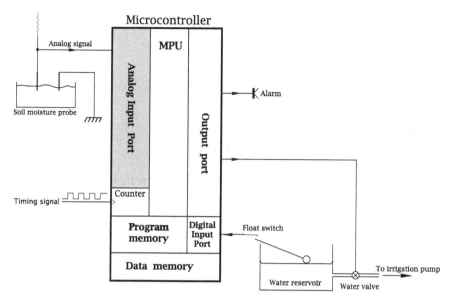

Fig. 3.9 A greenhouse environmental controller

- An input for an external oscillator, connected to a counter/timer to allow the MCU to calculate time. In practice the system clock can often be used by this internal timer to measure duration.
- A 1-input analog input line to measure the analog signal from the moisture detector.
- A 1-input digital line to check the level of the reservoir water tank.
- A 1-output digital line to open and close the water valve.
- A 1-output digital line to activate the buzzer alarm.
- A microprocessor to do the calculations and to read/write to the input/ output ports, respectively.
- Program memory, usually ROM of some kind.
- Data memory for temporary storage of program variables, usually static RAM.

Assuming that the software does not take up all the MCU's processing time, extra inputs may be used to monitor other environmental signals, such as temperature and light, to give a more comprehensive climate control.

Example 3.2 The most difficult problem that is being solved by a programmer is to define the problem that is being solved. This is the logical thought process that humans are (quite) good at and machines are not. The mark of a good programmer is one who has this ability of problem solving. It is a developed skill, coupled with some talent, and a good understanding of the problem being solved.

To illustrate this process, devise a sequence of simple steps that a MCU-controlled robot must perform to cross a busy road at a pedestrian-controlled crossing.

Solution

1. Walk up to the pedestrian crossing and stop.
2. Look at the traffic light.
3. Is it green for your direction of travel—make a decision?
4. IF the light is red THEN go to step 2 ELSE continue.
5. Look to the right.
6. Are there still cars still passing by?
7. IF yes THEN go to step 5 ELSE continue.
8. Look to the left.
9. Are there cars still passing by (there shouldn't be any by now, but you never know!)?
10. IF yes THEN goto step 5 ELSE continue.
11. Proceed across the street—carefully!

An alternative visual representation is illustrated in Fig. 3.10. This **flow chart** uses boxes to show statements, diamonds for decisions and ovals for entry and exit points. Lines with arrows give action paths, with annotations at decision points. Although this is not much of an advantage in this relatively simple case, for more complex situations with multiple decisions and flow paths, the visual presentation may have the edge in documenting system behavior. Neither the task list or flow chart is much use for very complex situations and here a hierarchy of descriptions starting with the more general and working down to the more particular must be implemented.

Now, this example may seem childish at first glance, but this is exactly what you should do every time you traverse a busy street with a pedestrian light-controlled crossing. This is also exactly what you need to tell a MCU-controlled robot to cross a street at a pedestrian crossing. This sequence of simple steps or instructions is called a **program**. Taken as a whole, the steps lead you across a busy road, which if a robot did it, would seem very intelligent. It is not intelligence; people are intelligent. The programmer that programmed these steps into the MCU would impart that intelligence to the robot.

Of course, the MCU-programmed robot would not know what to do when it got to the other side, since we did not tell it! In the case of the person, though, there has been some programming; it's called past experience!

Notice that the steps are numbered in the order they should be executed. The Program Counter, in this case the reader, starts with instruction 1 (the reset state) and ends with instruction 11. In a MCU the Program Counter automatically advances to the next step, after doing what the current step says, unless a *skip* or *goto* is encountered. A skip type of instruction directs the Program Counter to hop over the next instruction, usually based on some condition or outcome. A goto instruction directs the Program Counter to jump to any step in the program. Without these types of instructions, a program would be unable to make decisions or implement loops, where an action is repeated may times over; for instance, repetitively checking for a green light for as long as it takes.

Fig. 3.10 A flow chart
showing the robot how to
cross the road

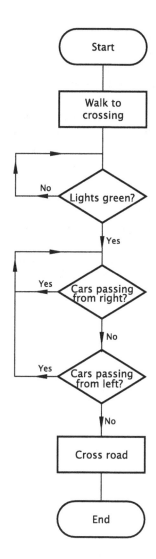

Self-Assessment Questions

3.1 Given the three instructions discussed in this chapter, can you devise a way of incrementing and also decrementing the contents of the Working register of Fig. 3.4?

3.2 If we add the new instructions addwf (ADD the byte in the Working register to that in the specified File) and movlw (MOVe a literal byte into the Working register) to our set of operations, devise a program to multiply the contents of File h'30' by two and place the resulting product in File h'20'. What would be the limitation of your program for an accurate outcome? What type of instruction would you need to overcome this?

3.3 Devise a program to enable the MCU-controlled robot of Example 3.2 to fill a glass of water from a tap/faucet.

3.4 The BASIC computer of Fig. 3.4 can fetch an instruction at the same time as it can execute an instruction. Discuss what features enable it to do these tasks in parallel.

3.5 Design a task list to program a robot to go to the nearest ATM and withdraw a specified amount of money, request a statement and return. Your consideration should include a request to print a statement and also what to do if your account does not have sufficient funds!

3.6 The gear inputs to the microcontroller system shown in Fig. 3.8 require five pins on the integrated circuit. MCU packages often have a small pin count; see, for example, Fig. 4.1 on p. 71. Can you think of any way to reduce the pin count and do you think this is economically viable? *Hint*: See Fig. 2.6 on p. 22.

3.7 In a similar attempt to reduce the pin count, can you think of a way to reduce the number of output pins driving the odometer and trip displays by one, and is this economically compatible?

Part II
The Software

In Part I we developed the concept of the Harvard architecture, ending up with our somewhat simplified BASIC computer. Although BASIC was entirely fictitious, it was designed with an eye toward the microcontroller that forms the basis for the rest of this book.

This part of the text looks mainly at the software aspects of our chosen micro-controller, the enhanced-range Microchip PIC® family. We will be covering:

- The internal structure of the enhanced-range PIC® microcontrollers.
- The instruction set.
- Instruction addressing.
- Software development using the MPLAB® integrated development environment.
- The assembly translation process.
- Subroutines and modular program design.
- Interrupt handling.
- The high-level language **C** and compilation.

16-bit core instruction set

16-bit Instruction	PIC18XXXX Mnemonic	W	F	N	V	Z	D	C	Operation summary
ADD Literal to W	addlw LL	√		√	√	√	√	√	w <- w + #LL
ADD W and F	addwf f,d,b	√	√	√	√	√	√	√	d <- w + f
ADD W and F with Carry	addwfc f,d,b	√	√	√	√	√	√	√	d <- w + f + C
AND Literal to W	andlw LL	√		√	•	√	•	•	w <- w · #LL
AND W to F	andwf f,d,b	√	√	•	√	•	•		d <- w · f
Bit Clear File bit n	bcf f,n,b		√	•	•	•	•	•	f_n <- 0
Bit Set File bit n	bsf f,n,b		√	•	•	•	•	•	f_n <- 1
Bit ToGgle File bit n	btg f,n,b		√	•	•	•	•	•	f_n <- $\bar{f_n}$
BRanch (relative) to <Label>	bra Offset			•	•	•	•	•	PC <- PC±Offset
Branch (relative) if Carry set to <Label>	bc offset			•	•	•	•	•	PC <- PC±offset IF C==1
Branch (relative) if Not Carry set to <Label>	bnc offset			•	•	•	•	•	PC <- PC±offset IF C==0
Branch (relative) if Zero to <Label>	bz offset			•	•	•	•	•	PC <- PC±offset IF Z==1
Branch (relative) if Not Zero to <Label>	bnz offset			•	•	•	•	•	PC <- PC±offset IF Z==0
Branch (relative) if Negative to <Label>	bn offset			•	•	•	•	•	PC <- PC±offset IF N==1
Branch (relative) if Not Negative to <Label>	bnn offset			•	•	•	•	•	PC <- PC±offset IF N==0
Branch (relative) if OVerflow to <Label>	bov offset			•	•	•	•	•	PC <- PC±offset IF V==1
Branch (relative) if Not OVerflow to <Label>	bnov offset			•	•	•	•	•	PC <- PC±offset IF V==0
Bit Test File bit n & Skip if Clear	btfsc f,n,b			•	•	•	•	•	PC++ IF f_n == 0
Bit Test File bit n & Skip if Set	btfss f,n,b			•	•	•	•	•	PC++ IF f_n == 1
CALL (jump to) subroutine[1]	call aaa			•	•	•	•	•	(TOS) <- PC, SP--, PC <- aaa
CALL and shadow context[1]	call aaa,FAST			•	•	•	•	•	plus save W, STATUS,BSR
CLeaR File	clrf f,b		√	•	•	√	•	•	f <- 00
CLeaR Watch Dog Timer	clrwdt			•	•	•	•	•	WDT <- 00
COMplement File	comf f,d,b	√	√	√	•	√	•	•	d <- $\bar{f}$
ComPare File, Skip if EQual	cpfseq f,b			•	•	•	•	•	PC++ IF f_n == W
ComPare File, Skip if Greater Than	cpfsgt f,b			•	•	•	•	•	PC++ IF f_n (unsigned) > W
ComPare File, Skip if Less Than	cpfslt f,b			•	•	•	•	•	PC++ IF f_n (unsigned) < W
Decimal Adjust W	daw	√		•	•	•	•	√	Correct sum of packed BCD digits
DECrement File	decf f,d,b	√	√	√	√	√	√	√	d <- f--
DECrement File & Skip if Zero	decfsz f,d,b	√	√	•	•	•	•	•	d <- f--, PC++ IF f == 0
DeCrement File & Skip if Not Zero	dcfsnz f,d,b	√	√	•	•	•	•	•	d <- f--, PC++ IF f != 0
GOTO (jump to) aaa[1]	goto aaa			•	•	•	•	•	PC <- aaa
INCrement File	incf f,d,b	√	√	√	√	√	√	√	d <- f++
INCrement File & Skip if Zero	incfsz f,d,b	√	√	•	•	•	•	•	d <- f++, PC++ IF f == 0
INcrement File & Skip if Not Zero	infsnz f,d,b	√	√	•	•	•	•	•	d <- f++, PC++ IF f != 0
Inclusive OR Literal to W	iorlw LL	√		√	•	√	•	•	w <- w + #LL
Inclusive OR W to F	iorwf f,d,b	√	√	√	•	√	•	•	d <- w + f
Load File Select Register n (0–2)[1]	lfsr n,LLL			•	•	•	•	•	FSR_n <- #LLL
MOV Literal into Bank SElect Register	movlb L			•	•	•	•	•	BSR <- #L
MOVe in File (load)	movf f,d,b	√	√	√	•	√	•	•	d <- f
MOVe source File to destination File[1]	movff f_s,f_d			•	•	•	•	•	$f_{destination}$ <- f_{source}
MOVe Literal into W	movlw LL	√		•	•	•	•	•	w <- #LL
MOVe W out to File (store)	movwf f,b		√	•	•	•	•	•	f <- w
MULtiply Literal by W	mullw LL			•	•	•	•	•	PRODH:PRODL <- W × #LL
MULtiply W by File	mulwf f,b			•	•	•	•	•	PRODH:PRODL <- W × f
NEGate (2's complement) File	negf f,b		√	√	√	√	√	√	f <- − f
No OPeration	nop			•	•	•	•	•	Do nothing

(continued of the next page)

(Continued)

16-bit Instruction	PIC18XXXX Mnemonic	W	F	N	V	Z	D	C	Operation summary
POP top of stack	pop			•	•	•	•	•	SP++
PUSH top of stack	push			•	•	•	•	•	(TOS) <- PC+2, SP--
Relative CALL subroutine	rcall	Offset		•	•	•	•	•	(TOS) <- PC, SP--, PC <- PC±Offset
RESET software MCLR	reset			•	•	•	•	•	All registers to MCLR Reset state
RETURN from subroutine	return			•	•	•	•	•	PC <- TOS
RETURN with shadowed context	return FAST			√	√	√	√	√	PC <- TOS; context
RETurn from subroutine with L in W	retlw	√		•	•	•	•	•	w <- #LL, PC <- TOS
RETurn From IntErrupt	retfie			•	•	•	•	•	GIEL|GIEH <- 1, PC <- TOS
RETurn From IntErrupt with context	retfie FAST			√	√	√	√	√	GIEL|GIEH <- 1, PC <- TOS, context
Rotate Left thru Carry File	rlcf f,d,b	√	√	√	•	√	•	b7	C <- [7 File 0] <- (rotate left through carry)
Rotate Left Not thru Carry File	rlncf f,d,b	√	√	√	•	√	•	•	[7 File 0] (rotate left)
Rotate Right thru Carry File	rrcf f,d,b	√	√	√	•	√	•	b0	[7 File 0] -> C (rotate right through carry)
Rotate Right Not thru Carry File	rrncf f,d,b	√	√	√	•	√	•	•	[7 File 0] (rotate right)
SET File to all 1s	setf f,b		√	•	•	•	•	•	f <- b'11111111'
SLEEP mode on	sleep			•	•	•	•	•	WDT <- 0, Clock off
SUB W from Literal	sublw LL	√		√	√	√	√	√	w <- #LL - w
SUBtract W from F	subwf f,d,b	√	√	√	√	√	√	√	d <- f - w
SUBtract W from F with Borrow	subwfb f,d,b	√	√	√	√	√	√	√	d <- f - (w + !C)
SUBtract F from W with Borrow	subfwb f,d,b	√	√	√	√	√	√	√	d <- w - (f + !C)
SWAP File nybbles	swapf f,d,b	√	√	•	•	•	•	•	d <- f[7:4] <-> f[3:0]
TaBLe ReaD/TaBLe WriTe from/to Program store as pointed to by TBLPTR[21:0] into/out of TABLAT									
Read	tablrd*			•	•	•	•	•	(TABLAT) <- <TBLPTR>
with TBLPTR post incremented	tablrd*+			•	•	•	•	•	(TABLAT) <- <TBLPTR++>
with TBLPTR post decremented	tablrd*-			•	•	•	•	•	(TABLAT) <- <TBLPTR->
with TBLPTR pre incremented	tablrd+*			•	•	•	•	•	(TABLAT) <- <++TBLPTR>
Write	tablwt*			•	•	•	•	•	<TBLPTR> <- (TABLAT)
with TBLPTR post incremented	tablwt*+			•	•	•	•	•	<TBLPTR++> <- (TABLAT)
with TBLPTR post decremented	tablwt*-			•	•	•	•	•	<TBLPTR-> <- (TABLAT)
with TBLPTR pre incremented	tablwt+*			•	•	•	•	•	<++TBLPTR> <- (TABLAT)
TeST File Skip on Zero	tstfsz f,b			•	•	•	•	•	PC++ IF f == 0
eXclusive OR Literal to W	xorlw LL	√		√	•	√	•	•	w <- w ⊕ #LL
eXclusive OR W to F	xorwf f,d,b	√	√	√	•	√	•	•	d <- w ⊕ f

√ : Flag operates normally	• : Not affected	L : 4-bit Literal (Bank no.)	
LL : 8-bit Literal	LLL : 12-bit Literal	b : Banking via the BSR	
f$_n$: File bit n	d : Destination; 0 = w, 1 = f	w : Working register	
WDT : Watch Dog Timer	PC : Program Counter	PC++ : Jump over next instruction	
== : Equivalent to	TOS : Top Of Stack	++ : Add one	
-- : Subtract one	( ) : Contents of	GIE : Global Interrupt Enable mask	
# : Constant	(TOS) : Contents of Top Of Stack	offset : ±128-word offset	
SP : Stack Pointer	context : W, STATUS, BSR	Offset : ±1024-word offset	
Note1 : Two-word instruction	FSR$_n$: File Select Register 0, 1, 2		
!= : Not equivalent to	WDT : Watch Dog Timer		
	< > : Pointed to by		

Chapter 4
The PIC18F1220 Microcontroller

Within a year of acquiring the intellectual rights to the General Instrument's Peripheral Interface Controller (PIC), as described on p. 60, Microchip had developed the
first of their range of Harvard architecture 8-bit microcontroller families. This low-
(or base-) range PIC16C5XX family, and currently the PIC10 and PIC12 families,
have a 33-instruction répertoire 12-bit Program store with parallel ports and an 8-bit
timer/counter. The execution unit processes all data as bytes, to match the 8-bit Data
store.

By 1992 the mid-range PIC16 family appeared. This has a 14-bit Program store;
the longer instruction word facilitating the access of larger Data stores. Two instructions were added to the original low-range set. The base set of interface devices was
expanded, with functions such as 16-bit timers, A/D converters, and serial ports;
together with an interrupt handling capability.

Known as the enhanced-range family, the PIC18 was introduced in 1999. Categorized along with all other previous introductions as an 8-bit MCU, as all data
processing is byte sized, the Program store is 16-bit, with 42 additional instructions
and many enhanced hardware features.

Microchip have since introduced ranges such as the digital signal processing-
oriented dsPIC30/33 (2004) and the PIC24 (2005) families. These 16-bit processors
are targeted to high-end applications, which require a faster processing throughput
with more powerful instructions. In 2007 Microchip announced the PIC32, with a
32-bit core. Nevertheless, the bulk of applications are expected to remain 8-bit for
the foreseeable future. For this reason, this book uses the enhanced-range PIC18
family

In this chapter we introduce the enhanced-range core from an architectural aspect. After completing this chapter you should:

- Understand the enhanced-range Harvard-based Microchip PIC microcontroller
  architecture;
- Appreciate the function, structure and memory map of the separate Program and
  Data stores;
- Appreciate the principle of banking in the Data store and its relationship to the
  Bank Select Register;

S. Katzen, *The Essential PIC18® Microcontroller,*
Computer Communications and Networks,
DOI 10.1007/978-1-84996-229-2_4, © Springer-Verlag London Limited 2010

- Be able to interpret the Status register bits **C**, **DC**, **Z**, **N** and **OV** flags;
- Know how to manipulate the contents of the Program Counter in conjunction with the PCLATU and PCLATH buffers;
- Recognize the interaction between the clock phases and the internal sequence of micro-operations;
- Be aware of the basic peripheral functions, using the PIC18F1220 as an exemplar.

From the point of view of software, all devices with the same core are identical. Indeed, there is considerable commonality across the entire range of 8-bit PIC MCU cores. At the time of writing (late 2007) there are over 150 members of the enhanced-range family. These mostly differ from each other in their memory capacity, mix and specification of peripherals and footprint (number of pins; ranging from 18 through 80—see p. 304). In this chapter we are mainly concerned with the processor core; that is the family CPU. For completeness we will briefly list the peripheral ports of our chosen exemplar, but we will leave a detailed discussion of these until Part III. Apart from interrupt handling, Part II generally is concerned with software issues, and these are common across the PIC18 family.

As our exemplar for Part II we have chosen the PIC18F1220/1320 (known generically as the PIC18F1X20). These are 18-pin devices and only differ in that the latter has double the Program store capacity at 4 kword. In Part III we use the PIC18F4420/4520 which have a larger footprint together with additional ports and peripheral devices—see Fig. 4.10.

The architecture of the PIC18F1220/1320 is shown in a simplified form in Fig. 4.1. Although initially this looks rather complex, it is little more than the architecture of our BASIC computer of Fig. 3.4 on p. 46 but with interface ports connected to the internal Data store's data bus. You should review this material now as background to our discussion. In essence the PIC MCU family is based on a Harvard structure with its separate Program and Data stores, and with peripheral interface ports mapped onto the Data Store's address space. That is, the various ports appear to the software to be in the Data store. The miscellaneous status and control registers, and even the Program Counter, also appears to the software to be in the Data store—see Fig. 4.10.

Fetch Unit
The fetch unit, shown close-up in Fig. 4.2, is primarily concerned in fetching instructions down into the Pipeline from the Program store. The location of each instruction is maintained by the Program Counter (PC). Copies of this 21-bit PC can be made into a local store called the stack, primarily to facilitate calls to subroutines and interrupt handling. A 21-bit Table Pointer can be used to point to any byte in the Program store, which can then be extracted into the 8-bit TABle LATch register.

Each instruction copied into the top of the Pipeline from the Program store, pushes the previous instruction down to the bottom which feeds the decoding circuitry. This in turn activates the appropriate logic in the execute unit in the correct sequence.

The Program Store
Central to the fetch unit is the Program store. Software in embedded systems is

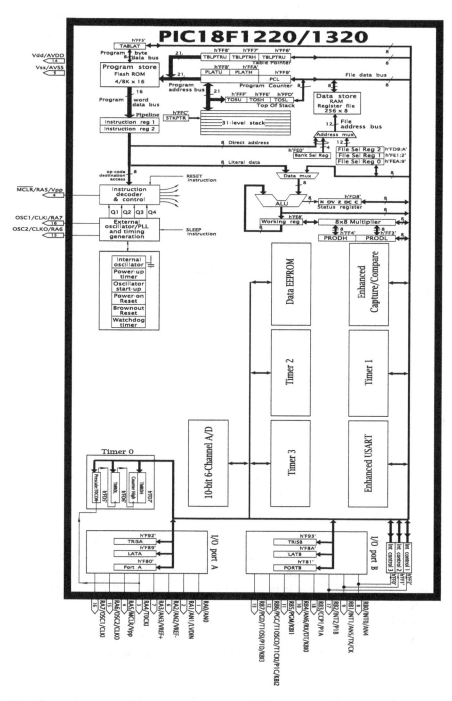

Fig. 4.1   Architecture of the PIC18F1220/1320 (PIC18F1X20) microcontrollers

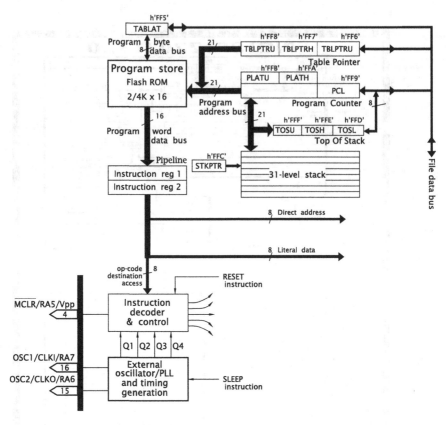

Fig. 4.2 A close-up look at the PIC18F1220/1320's fetch unit

invariably fixed, in that on power-on the microcontroller is expected to perform its duty without having to load in its operating program. This means that the Program memory will normally be ROM of some kind. Most PIC MCUs use some sort of electrically programmable technology; in the case of F parts this is Flash EEPROM. For the PIC18F1220, up to 1024 instructions can be stored, with each instruction comprising 16 binary bits; see Fig. 3.5 on p. 50. The PIC18F1320 holds up to 2048 words, but potentially the architecture of the expanded-range is such that family members can potentially store $2^{20} = 1$ Mword) of code and data. At the time of writing (late 2007) the largest implemented capacity is 64 k instructions; for instance the PIC18F8722.

From Fig. 4.3 we see that the Program store is actually organised as an array of bytes, even though instructions are stored as 16-bit words. The need to address individual bytes arises from the dual purpose nature of the code memory.

Software

The primary role of the Program store is to act as the repository of the instruction codes which define the program. We have already seen in Fig. 3.5

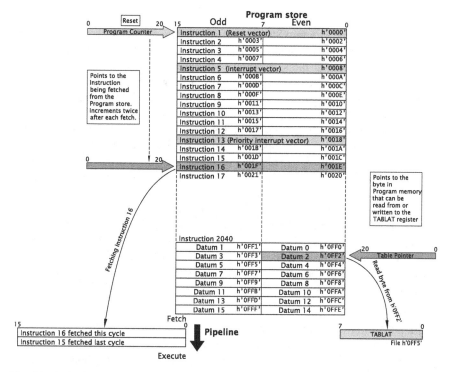

Fig. 4.3 A simplified look at the PIC18F1220's Program store

on p. 50 that these instructions are 16 bits wide.[1] The 21-bit Program Counter effectively increments twice when each instruction is fetched, and thus always holds an even address (PC[0] is 0). This is implemented by holding the least significant bit of PC permanently 0.

In Fig. 4.3 the program code is shown at the bottom addresses; starting with the first instruction at location h'0000'. This is where the PC resets to. Regardless of where the bulk of the program is located, there must always be an instruction at this so called **Reset vector**. This instruction is often a goto somewhere else in the store. Other special locations are h'00008', which is called the **Default Interrupt vector** and is where the PC is pointed-to whenever an enabled interrupt is responded to. This will be a goto instruction taking the execution focus to a piece of code called an interrupt service routine. Location h'00018' holds the **Low-priority Interrupt vector**. Interrupts are discussed in Chap. 7.

Apart from these three vectors, all store locations are equivalent; although usually executable code is in the lower end of the store. Figure 4.3 shows instruction 16 being extracted from location h'001E' into the top register of the Pipeline.

---

[1] Actually three instructions need two words; that is 32 bits.

Data

In this family the Program Store can also hold data of a semi-permanent nature as a table of bytes; for instance, strings of ASCII characters. Typically this data is programmed in at the same time as the executable code. In some devices (such as our exemplar) such data can even be written into the store by the program, while execution is temporarily halted. These TABle ReaD (`tblrd`) and TABle WriTt (`tblwt`) processes are the subject of Chap. 15.

When using the Program store in this way the 21-bit Table Pointer (actually made up of three separate Special-Purpose Registers, as shown in Fig. 4.10) point to the datum byte. This datum can then be copied out into the TaBle LATch (TABLAT) register. TABLAT has an analogous function for those processors that can write data bytes into this store.

The diagram depicts an array of data bytes in the topmost 16 bytes of memory. The 21-bit Table Pointer can address anywhere in stores of up to 2 Mbyte capacity. Data can be put anywhere in memory, apart from the three vectors, but care must be taken to keep away from executable code. Typically it will be located after the program code. Datum 2 at location h'0FF2' is shown being copied out to TABLAT.

Program Counter

The 21-bit PC can address a store of up to $2^{20} = 1$ M instructions. In our PIC18F1220 only $2^{11} = 2$ kword is implemented, and thus only the bottom 12 bits (including the superfluous bit 0) are implemented. Higher bits can be ignored. This 12-bit (13-bit for the PIC18F1320) register is normally incremented twice after each fetch, effectively acting as a binary counter. However, as we will see in the following chapter, there are a few instructions, such as `goto`, that will cause execution of the program to jump to another part of the Program store. Thus the Program Counter's normal up count can be overridden.

Although the PC is normally left to its own devices, it is possible for the program to 'get at' this register and override its normal progression. Typically this is useful when implementing look-up tables; for instance, see Program 6.6 on p. 175.

The low byte of the PC is directly accessible as a Special-Function Register in the Data store called PCL (Program Counter Low byte). The problem is that these SFRs are byte sized. Any changes to the PC must be made to all 21-bits simultaneously and thus the part(s) of the PC above these eight bits are not directly mapped as SFRs. They are indicated in Fig. 4.4 as 'buried'. Instead, two buffer SFRs are used to temporarily hold this data. These are called PCLATH (PC LATch High byte) and PCLATU (PC LATch Upper byte) corresponding to PC[15:8] and PC[21:16] respectively.

When PCL is written to (e.g. `movwf PCL`) the contents of pCLATH:PCLATU are synchronously transferred to their corresponding PC sector. Conversely, reading PCL (e.g. `movf PCL,w`) also copies the upper bits of the PC into their matching buffer registers.

To illustrate the process, consider that we want to add 24 to the PC; that is we want to branch forward 12 instructions (24 bytes). If we assume that there is a

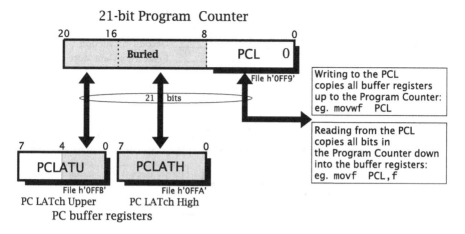

Fig. 4.4 Showing how all of the Program Counter is altered at the same time when accessing the PCL

General-Purpose Register File called `TEMP`, then the following code fragment will do the trick:

```
movf    PCL,w      ; Get the PC Low byte, other bits to buffers
movwf   TEMP       ; and copy it into memory
movlw   d'24'      ; Constant 24 (h'18') into Working register
addwf   TEMP,f     ; Add it effectively to the low byte of PC
movlw   0          ; Zero the Working register
addwfc  PCLATH,f   ; and add any carry-out to it
movlw   0          ; Zero again the Working register
addwfc  PCLATU     ; and add any carry-out to it
movwf   TEMP,w     ; Finally get the new value of PCL and
movwf   PCL        ; in writing to PCL update the rest of the PC
```

See also Example 6.5 on p. 195.

Pipeline

Two 16-bit registers implement the Pipeline. The top register holds the instruction that has just been fetched from the Program store at the location pointed to by the PC. The bottom register feeds the decoder circuits and is the instruction that is in the process of being executed. This allows an instruction to be fetched whilst at the same time the processor is executing the previously fetched instruction. This of course assumes that the instruction execution sequence is linear. For instructions that action a jump to another part of the Program store, the instruction sitting at the top of the Pipeline needs to be replaced by the far instruction. This process is known as *flushing* and adds an extra instruction cycle to the execution time. As you can see from Fig. 4.5, normally an instruction takes one instruction cycle to execute. Exceptionally, three instructions (out of 78) occupy two words in the Program store, and these require two cycles to execute.

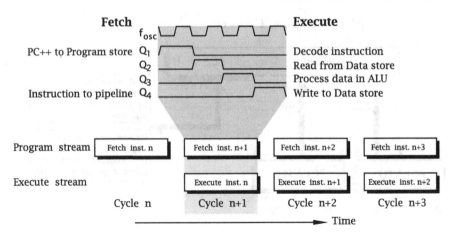

Fig. 4.5 Internal quadrature clock sequencing waveforms

Instruction Decoder

The Instruction decoder uses logic circuitry to decode each field of the 16-bit instruction and gate the appropriate addresses and data to the correct execution unit's circuitry and configure the ALU.

All PIC MCU families have an integral oscillator that generates the internal time-related sequences of micro-operations commanded by the Instruction decoder. The timing element is typically an external quartz crystal connected across pins OSC1 and OSC2 (Fig. 4.2), and this determines the clock frequency f_{osc}. More details are given in Chap. 10. Most enhanced-range devices have an upper frequency of 20 MHz.[2] There is no minimum frequency.

The oscillator is frequency divided as shown in Fig. 2.25 on p. 38, to give four internal non-overlapping quadrature clocks. These four pulses are used as part of the decoding logic to activate internal processes in time-dependent sequences. A consequence is that an **instruction cycle** takes four external clock frequency f_{osc} periods to complete; see Fig. 4.5. Thus with a 4 MHz crystal, the instruction cycle rate is $f_{osc}/4$, or one million per second, corresponding to a period of 1 µs.

The clock-related sequence of operations in the fetch unit are:

Q_1: Increment the Program Counter and copy onto the Program store address bus.

Q_4: Read the instruction code off the Program store's data bus into Instruction register 1 and at the same time move the previous instruction down the Pipeline into Instruction register 2, where it is presented to the Instruction decoder.

Stack

Eight 21-bit registers are stacked below and are connected to the Program Counter.

---

[2]There is a phase-locked loop mode where the crystal frequency is effectively multiplied by four. The maximum crystal frequency in this mode is 10 MHz, and this gives an effective clock frequency of 40 MHz, or instruction cycle time of 100 ns.

We will see in Chap. 6 that the Stack is used to hold past states of the Program Counter to 'remember' the jumping-off point when a subroutine is called up, and perform a similar function for interrupt handling—see Chap. 7.

The STKPTR (STacK PoinTeR) is a 5-bit up/down counter which increments every time a copy is made (the PC is said to be pushed into the stack), and decrements when a value is pulled (or popped) out of the stack.

The 21-bit datum to which the STKPTR points to is visible in the Top Of Stack register, and if desired this can be changed by writing to the three composite byte SFRs TOSU:TOSH:TOSL. In this manner, non-PC data can be pushed into the stack and pulled out at a later time. This gives a convenient and transparent method of passing data to and from subroutines—see Program 6.12 on p. 194 for an example.

Execute Unit

The 8-bit execute unit is responsible configured by the Instruction decoder; typically to read a datum from the Data store, processing it using the ALU and put the outcome back in the Data store. In accessing the Data store, various mechanisms are used to generate its 12-bit address. The ALU, which also includes an auxiliary multiplier, is set up according to the requirements of the instruction being executed and activates the five status flags. A single 8-bit Working register normally holds one of the ALU's operands, and may be selected as the destination of the resultant outcome. See Fig. 4.6.

Arithmetic Logic Unit

Central to the execution unit is the ALU (see Fig. 2.10 on p. 25) processing data from up to two sources. One of these is the 8-bit Working register. The other can be:

- A byte directly from a specified File in the Data store. For instance `addwf h'020',f` ADDs the contents of the Working register to the byte in File h'020'.
- A literal byte held as part of the instruction code; see Fig. 3.6 on p. 51. For instance, `addlw 5` ADDs the Literal 5 to the Working register.

The outcome in the former case can be directed either back into the Data store if the destination bit is 0 (see Fig. 3.5 on p. 50) or into the Working register if this bit is 1, e.g., `addwf h'020',w`.

As we shall see in the following chapter, the ALU can perform pass-through, logic inversion, AND, IOR, XOR, addition, subtraction, shift and multiplication operations on byte data. The multiplication circuitry is shown as a separate entity in the diagram. It can perform an unsigned 8×8 multiplication in one instruction cycle. In order to handle its 16-bit product, the two SFRs PRODH and PRODL are used to store this outcome.

Status Register

Associated with the ALU is the Status register, which holds five flag bits used to tell the software something about the outcome from an instruction. For instance, if there was a carry-out from an addition.

Carry Flag

Bit 0 of the Status register is the **C** flag. This primarily holds the carry out

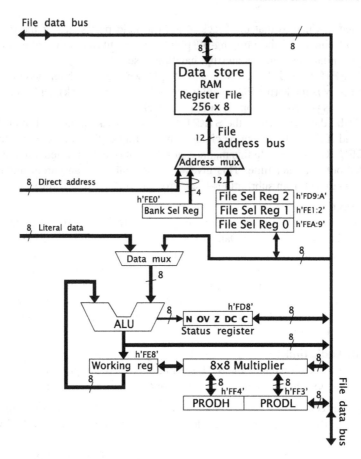

Fig. 4.6 A close-up view of the execute unit

from the last addition operation. Subtraction operations activate this bit as the *complement* of the borrow out—see Example 4.2. For instance, $24 - 12 = 12$: Borrow is 1 thus **C** is 0 and $12 - 24 = 88$: Borrow is 0 thus **C** is 1. **C** also functions as an input/output bit for some of the Rotate instructions, as shown in Fig. 5.14 on p. 132.

The label R/W ? in Fig. 4.7 indicates that this bit can be read from or written to and has an uncertain value on a Power-on Reset; its value does not alter on any other type of reset.

Digit Carry Flag

Bit 1 of the Status register is the **DC** flag. This operates in the same manner as the standard **C** flag but holds the carry out from the lower nybble to the upper nybble; that is, from bit 3 to bit 4. In the same manner **DC** holds the *complement* of the borrow out from bit 3 to bit 4.

Knowledge of the carry activity between the lower and upper halves of the byte is useful where binary coded decimal data is being manipulated; for

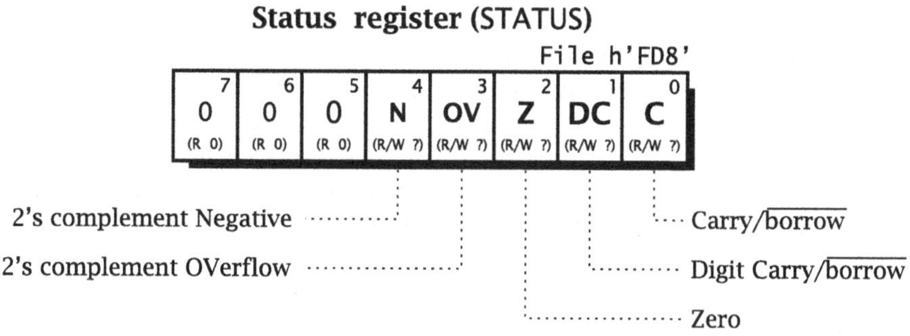

Status register (STATUS)

File h'FD8'

7	6	5	4	3	2	1	0
0	0	0	N	OV	Z	DC	C
(R 0)	(R 0)	(R 0)	(R/W ?)	(R/W ?)	(R/W ?)	(R/W ?)	(R/W ?)

2's complement Negative ··········   ···· Carry/$\overline{\text{borrow}}$

2's complement OVerflow ···············   ·········· Digit Carry/$\overline{\text{borrow}}$

·················· Zero

Fig. 4.7   The enhanced-range Status register

instance, see Example 4.5. Here each nybble holds a 4-bit representation of the decimal digits $0, \ldots, 9$ (see p. 6) and the half carry then indicates carries between decimal decades.

Zero Flag
Bit 2 of the Status register is the **Z** flag. This is set whenever the outcome of the instruction is zero, otherwise it is cleared.

OVerflow Flag
Bit 3 of the Status register is the **OV** flag. This is set whenever there is a 2's complement overflow from the 7-bit magnitude into bit 8; that is into the sign bit. The logic of this operation is shown in Fig. 1.5 on p. 14. This is only meaningful when the programmer is treating these quantities as 2's complements numbers.

Negative Flag
Bit 4 of the Status register is the **N** flag. This bit follows the logical value of bit 7 of the outcome. Where the programmer is treating the operands as signed 2's complement quantities, then a 0 indicates positive and a 1 negative—see p. 10.

———————————

Unlike most MCUs, there are no instructions to specifically clear or set a flag, such as sec for SEt Carry.[3] However, as we shall see in Fig. 4.10, the Status register is accessible as File h'FD8' in the Data store, and thus any instruction that can alter the contents of a File can potentially change the state of a flag. However, there is a problem in that many of these instructions inherently affect one or more flags (see for instance, Table 5.1 on p. 109) as part of their execution logic and this *overrides* any change that would result from the outcome of the instruction's execution. For instance, trying to use the Clear instruction to zero the flags clrf h'FD8 (see

———————————

[3]For instance, the Motorola 6800/5/11 families.

Table 5.2 on p. 113), actually sets the **Z** flag to 1 to indicate a zero outcome! The Bit Clear File and Bit Set File instructions (see Table 5.2) are recommended where an individual bit in the Status register needs to be altered, as these instructions do not inherently alter any flags. For instance, bsf h'FD8', 0 (Set Bit 0 in File h'FD8') is equivalent to sec and bcf h'FD8', 2 (Clear Bit 2 in File 3) will clear the **Z** flag.

All these bits are known as **flags**, or sometimes semaphores, as they signal an outcome of an instruction, such as a zero result. Bits 5, 6 & 7 are not implemented and always read as 0.

Data Store

Figure 4.8 shows the architecture of the enhanced-range Data store. This structure has a potential capacity of 4 kbytes, organised as 4096 × 8; each byte cell of which is called a **File** register. In fact the PIC18F1220 only implements 384 Files and the PIC18F4520 has 1664 Files. Just a few family members, such as the PIC18F8722, implement the full 4 kbyte complement.

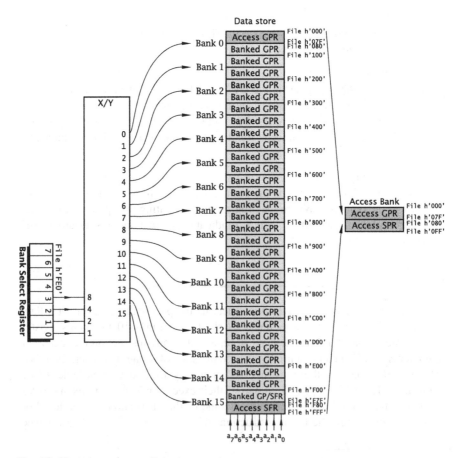

Fig. 4.8 The enhanced-range Data store structure

A 4 kbyte store needs a 12-bit address ($2^{12} = 4096$). As we have seen in Fig. 3.5 on p. 50, only eight out of the 16 bits defining an instruction are allocated to the operand address. There are four ways to extend the reach of this 8-bit address field.

Banking

The **Bank Select Register** BSR can hold four additional address bits, making up the required total of 12 bits as $(a_{11}a_{10}a_9a_8)a_7a_6a_5a_4a_3a_2a_1a_0$. As seen from this perspective, the Data store appears as 16 **banks**, each of which contains up to 256 Files. This organisation requires the Access bit in the instruction's machine code to be 1— as shown in Fig. 3.5.

In order to address, say, File h'20' in Bank 2, the programmer would have to:

- Load the constant 2 into the BSR.
- Ensure that the a-bit is 1.
- Specify File h'020' (or File h'220').

```
movlb 2          ; Move (copy) the constant 2 to the BSR
movf  h'220',w,1 ; Copy the datum in File h'220' into the
                 ; Working register using the Banking mode
```

As altering the BSR is so frequent an operation, the special instruction MOVe Literal to BSR (`movlb`) is provided (rather than the `movlw 2, movwf BSR` equivalent)—see Table 5.1 on p. 109.

Access RAM

When the a-bit is 0, the most-significant bit of the address a_7 is extended to make up a 12-bit address. In this situation the possible address range goes from (0000)0000-0000–(0000)01111111 and (1111)10000000–(1111)11111111; that is File h'000'–File h'07F' and File h'F80'–File h'FFF'. This is known as **Access RAM**, and is shown darkly shaded to the right of Fig. 4.8.

Figure 4.9 shows how the a-bit could be used by the Instruction decoder to switch in either the lower four bits of the BSR to make up the highest four address bits or else a replica of bit 7 of the instruction's address field.

The obvious question to ask at this point, is why waste a precious bit in the instruction code in switching between modes? The answer relates the dual use of the Data store. As well as holding variables that are stored and retrieved by the program, in what are called **General Purpose Registers** (GPRs), this RAM also holds the **Special Function Registers** (SFR). We have already met several of these; for instance, the Status register and PRODH:PRODL which hold the outcome from the Multiplier. We see from Fig. 4.8 that the top half of Access RAM holds all these SFRs—up to 128 of them.

SFRs relate to the configuration and control of both the core functions of the processor and the many peripheral modules. No matter which bank the Data store is configured for (that is the state of the BSR), the program can access these SFRs without the delay needed to change banks back and forth, by simply specifying Access RAM. For example, `movf PORTB,w, 0`, where PORTB is a SFR at File h'F81'. In

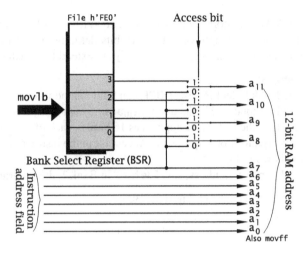

Fig. 4.9 Extending an 8-bit address to a 12-bit RAM address

the same manner the 128 Files at the bottom half of Access RAM can quickly be accessed no matter what the setting is of the BSR. Thus Access RAM can be thought of as a common zone across all banks.

The specific complement of the PIC18F1220's SFR locations are shown in Fig. 4.10. These tend to be in the same location across the family, although a different mix of peripheral devices will be supported by any particular family member. For example, the PIC18F4520 has a PORTC at File h'F82'. As we shall see in Table 8.1 on p. 243, fortunately the programmer does not need to remember absolute locations, as each device comes with a specific header file giving these addresses, which can be included at the start of the program.

No matter how little of the 4 kbyte Data store is implemented, all PIC18 devices have full Access RAM. Specifically, the PIC18F1220 has a full Bank 0 (that is File h'000' – h'0FF') as well as the SFR region File h'F80' – h'FFF'. Although this gives 384 Files, only the 256 GPR capacity is listed in the device description; the 128 byte SFR space is taken for granted. Similarly, the PIC18F4520 has Banks 0 though 5 fully populated, and is listed as having a 1536 (2^6)-byte capacity.

Pointers

The operand address of all our examples up to now have been constant, as they are an integral part of the binary code defining an instruction such as movwf h'020' and therefore a permanent fixture in the Program store. In many instances, this is too inflexible and a variable address is much more useful. Variable addresses, or **pointers**, are particularly useful in dealing with arrays or tables of data.

All processors use data pointers, although sometimes they are called index registers. The PIC18 family has three pointer registers, each of which can hold a full 12-bit Data store address. These are known as **File Select Registers**, named FSR0, FSR1 and FSR2. Each FSR is implemented as two SFRs; for instance, from Fig. 4.10 FSR0H:FSR0L at File h'FEA:9'. Both composite SFRs can be updated at the same time with a 12-bit constant, using the lfsr (Load File Select Register) instruction—see Table 5.1 on p. 109.

Addr	Register	Addr	Register	Addr	Register	Addr	Register
hFFF	TOSU (Top Of Stack Upper)	hFDF	INDF0 (INDirect File 2)[1]	hFBF	CCPR1H (Capture ComPaRe 1 High)	hF9F	IPR1 (Interrupt PRiority 1)
hFFE	TOSH (Top Of Stack Higher)	hFDE	POSTINC0 (POST INCrement 2)[1]	hFBE	CCPR1L (Capture ComPaRe 1 Low)	hF9E	PIR1 (Peripheral InteRrupt flag 1)
hFFD	TOSL (Top Of Stack Lower)	hFDD	POSTDEC0 (POST DECrement 2)[1]	hFBD	CCP1CON (Capture ComPare 1 CONtrol)	hF9D	PIE1 (Peripheral Interrupt Enable 1)
hFFC	STKPTR (STack PoInTeR)	hFDC	PREINC0 (PRE INCrement 2)[1]	hFBC	—[2]	hF9C	—[2]
hFFB	PCLATU (PC LATch Upper)	hFDB	PLUSW0 (PLUS Working register2)[1]	hFBB	—[2]	hF9B	OSCTUNE (OSCillator TUNE)
hFFA	PCLATH (PC LATch Higher)	hFDA	FSR2H (FIle SElect Register 2 High)	hFBA	—[2]	hF9A	—[2]
hFF9	PCL (PC Lower byte)	hFD9	FSR2L (FIle SElect Register 2 Low)	hFB9	—[2]	hF99	—[2]
hFF8	TBLPTRU (TaBLe PoInTeR Upper)	hFD8	STATUS (Status register)	hFB8	—[2]	hF98	—[2]
hFF7	TBLPTRH (TaBLe PoInTeR Higher)	hFD7	TMR0H (TIMer Register 0 High)	hFB7	PWM1CON (Pulse Width Mod1 CONtrol)	hF97	—[2]
hFF6	TBLPTRL (TaBLe PoInTeR Lower)	hFD6	TMR0L (TIMer Register 0 Low)	hFB6	ECCPAS (Enhanced CCP Auto Shutdown)	hF96	—[2]
hFF5	TBLAT (TaBLe LATch)	hFD5	T0CON (Timer 0 CONtrol)	hFB5	—[2]	hF95	—[2]
hFF4	PRODH (PRODuct High)	hFD4	—[2]	hFB4	—[2]	hF94	—[2]
hFF3	PRODL (PRODuct Low)	hFD3	OSCCON (OSCillator CONtrol)	hFB3	TMR3H (TIMer Register 3 High)	hF93	TRISB (TRIstate direction port B)
hFF2	INTCON (INTerrupt CONtrol)	hFD2	LVDCON (Low Voltage Detect CONtrol)	hFB2	TMR3L (TIMer Register 3 Low)	hF92	TRISA (TRIstate direction portA)
hFF1	INTCON 2 (INTerrupt CONtrol 2)	hFD1	WDTCON (WatchDog Timer CONtrol)	hFB1	T3CON (Timer 3 CONtrol)	hF91	—[2]
hFF0	INTCON 3 (INTerrupt CONtrol 3)	hFD0	RCON (Reset CONtrol)	hFB0	SPBRGH (Serial Port Baud Rate Gen High)	hF90	—[2]
hFEF	INDF0 (INDirect File 0)[1]	hFCF	TMR1H (TIMer Register 1 High)	hFAF	SPBRG (Serial Port Baud Rate Generator)	hF8F	—[2]
hFEE	POSTINC0 (POST INCrement 0)[1]	hFCE	TMR1L (TIMer Register 1 Low)	hFAE	RCREG (serial ReCeive Register)	hF8E	—[2]
hFED	POSTDEC0 (POST DECrement 0)[1]	hFCD	T1CON (Timer 1 CONtrol)	hFAD	TXREG (serial Transmit REGister)	hF8D	—[2]
hFEC	PREINC0 (PRE INCrement 0)[1]	hFCC	TMR2 (TIMer Register 2)	hFAC	TXSTA (serial Transmit STAtus)	hF8C	—[2]
hFEB	PLUSW0 (PLUS Working register 0)[1]	hFCB	T2CON (Timer 2 CONtrol)	hFAB	RCSTA (serial ReCeive STAtus)	hF8B	—[2]
hFEA	FSR0H (FIle SElect Register 0 High)	hFCA	PR2 (Prescaler timer 2)	hFAA	BAUDCTL (BAUD rate ConTroL)	hF8A	LATB (LATched port B)
hFE9	FSR0L (FIle SElect Register 0 Low)	hFC9	—[2]	hFA9	EEADR (EEprom ADdRess)	hF89	LATA (LATched port A)
hFE8	WREG (Working REGister)	hFC8	—[2]	hFA8	EEDATA (EEprom DATA)	hF88	—[2]
hFE7	INDF1 (INDirect File 1)[1]	hFC7	—[2]	hFA7	EECON2 (EEprom CONtrol 2)[1]	hF87	—[2]
hFE6	POSTINC1 (POST INCrement 1)[1]	hFC6	—[2]	hFA6	EECON1 (EEprom CONtrol 1)	hF86	—[2]
hFE5	POSTDEC1 (POST DECrement 1)[1]	hFC5	—[2]	hFA5	—[2]	hF85	—[2]
hFE4	PREINC1 (PRE INCrement 1)[1]	hFC4	—[2]	hFA4	—[2]	hF84	—[2]
hFE3	PLUSW1 (PLUS Working register 1)[1]	hFC3	—[2]	hFA3	—[2]	hF83	—[2]
hFE2	FSR1H (FIle SElect Register 1 High)	hFC2	ADCON0 (Analog Digital CONtrol 0)	hFA2	IPR2 (Interrupt Priority 2)	hF82	—[2]
hFE1	FSR1L (FIle SElect Register 1 Low)	hFC1	ADCON1 (Analog Digital CONtrol1)	hFA1	PIR2 (Peripheral InteRrupt flag 2)	hF81	PORTB (parallel i/o PORT B)
hFE0	BSR (Bank Select Register)	hFC0	ADCON2 (Analog Digital CONtrol2)	hFA0	PIE2 (Peripheral Interrupt Enable 2)	hF80	PORTA (parallel i/o PORT A)

Note 1: These registers do not physically exist (they are virtual)
Note 2: Unimplemented in this family member; always read as '0'

Fig. 4.10   The Special Function Registers in the PIC18F1220's Access bank

Once a pointer is set up, it may be *triggered* by referencing a phantom location linked to that FSR. For example, moving a datum from the Working register to INDF0 (INDirect File 0) will actually use the address in the linked FRS0 as the pointer into the destination location. If FSR0 was, say, h'220' (lfsr 0,h'220') then the instruction movwf INDF0 will actually copy the byte in WREG to File f'220'. The CPU's logic will pick up on the address of INDF0 (h'FEF' in Fig. 4.10) and gate the 12-bit address held in FSR0 to the Data store. No data will be copied into INDF0; indeed INDF0 does not actually exist; its address simply acts as a trigger for this form of **Indirect** addressing.

Although this seems an overly complicated way of accessing memory, the operand address is not fixed code, as it would be as part of an instruction, but is a *variable*. This means that the address of the operand can be altered as the program progresses. For instance, if the pointer were incremented after each write, complete tables or arrays can be processed. More details are given in Fig. 5.8 on p. 106.

MOVe File to File
The movff instruction can move a byte from anywhere in RAM to nearly anywhere else. movff is an example of only four instructions that occupy two words in the Program store (see p. 111) so that it can carry the full 12-bit address of both the source and destination Files in the one instruction. lfsr is also a double-word instruction for the same reason. For example, to copy the byte from File h'F81' (PORTB) to File h'220' we have movff h'F81',h'220'. There is no resort to banking or the Access RAM mechanism for this. However, because movff is a double-word instruction, it takes an extra instruction cycle to fetch it down from the Program store; that is two cycles to fetch and execute rather than one.

Peripheral Interface

Part III of this book concentrates mainly on the core peripheral interface capabilities of our MCU family, to enable them to interact to the outside world. Here it will be sufficient to list and briefly discuss the peripheral facilities available to our exemplar device.

Parallel Ports; Chap. 11
The ability to externally alter or monitor several digital lines at the same time is a virtually universal facility in microcontroller-based systems. Depending on the package size, enhanced-range PIC MCUs range from potentially 16 (e.g. the 18-pin PIC18F1220/1320) up to 70 (e.g. the 80-pin PIC18F8722) These figures are maximum, for as we shall see input/output (I/O) pins are often shared between several peripheral modules. For instance, we see from Fig. 4.1 that the digital port pin RA0 (Register (port) A, bit 0) is shared with the analog to digital converter's channel 0 input pin AN0; both of which share the MCU's pin 1. This problem is particularly severe for small pin-count devices.

The PIC18F1220 has 16 digital lines, divided up into two ports. Port A has eight lines mapped into Access RAM at PORTA (File h'F80'). Similarly, Port B has eight

I/O lines at PORTB (File h'F81'). Larger footprint devices have extra ports; for instance, PORTC at File h'F82' etc.

These ports can be thought of as a 'window' into the Data store, in that data written to, say, File h'F81' appear to the outside world on the corresponding pins RB7,..., RB0 (shorthand RB[7:0])—see Fig. 11.1 on p. 334. However, the electrical and logical behavior of these ports is more complex than that of a purely internal register File. This will be discussed in Chap. 11, but as an example, a port bit must be configurable as either an output (so that the CPU can control the state of the associated pin) or an input (so that the CPU can read the state of this pin). To do this, each parallel port register has an associated data direction register, which Microchip calls TRISA and TRISB, which map to File h'F92' and File h'F93' respectively. The term TRIS stands for TRIState—see Fig. 11.4 on p. 343 for its origin. Each bit n in a TRIS register controls the function of its linked pin as either an Output (bit $n = 0$) or Input (bit $n = 1$).[4] In the PIC18F1220/1320, pin RA5 is shared with the Master CLeaR input $\overline{\text{MCLR}}$. If the option is set to use this as a port pin (see Fig. 10.9 on p. 10.9), it can only ever be an input.

As an example, consider that we wish to make Port B pins RB[6:0] an input and pin RB7 an output. Then the set-up code would be:

```
        movlw h'7F'      ; Binary pattern 0111 1111 in W
        movwf h'F93',0   ; makes RB7 Output, RB[6:0] Input
                         ; in TRISB in Access RAM
```

Although this code fragment is correct and, with the aid of comments its function can be followed, the code is not very human readable. The alternative is more user friendly, but is identical as far as the assembler is concerned; see p. 52.

```
TRISB  equ   h'F93'  ; Data direction register @ File h'F93'
ACCESS equ   0       ; Specify Access RAM, a bit = 0

 movlw b'01111111'  ; Binary pattern 0111 1111 in W
 movwf TRISB,ACCESS ; makes RB7 Output, RB[6:0] Input
```

Obviously the latter is preferable. Although this might seem to be a cosmetic exercise, clarity reduces the chance of error and makes debugging and subsequent alteration easier. Realistic programs, rather than the code fragment illustrated here, use many variables and register bits, so lucidity is all the more important.

The two header lines of our program illustrate the means whereby the programmer tells the assembler translator program to substitute numbers for names. For instance, the line

---

[4]*Aid-mémoire*: A 0 configures for an Output pin, a 1 configures for an Input pin.

```
TRISB   equ   h'F93'
```

states that when the programmer uses the name TRISB as an operand, it is to be substituted by the number h'F93'[5] (that is, File h'F93'). The equ directive means "EQUivalent to." A **directive** is a pseudo instruction in that it does not usually produce actual machine code but rather is a means of passing information from the programmer to the assembler program. From now on we will give our Files and bits names for clarity. In practice the modifier ACCESS is not generally used, as the assembler will default to Access RAM for Files in the range h'000'–h'07F' and h'F80'–h'FFF' and BANKED for Files outside this range.

As an example, let us pulse pin RB7 High and then Low _/‾_ as follows:

```
bsf     PORTB,7       ; Pin RB7 High (set bit 7)
bcf     PORTB,7       ; then Low (clear bit 7)
```

where we are using the instruction bcf (Bit Clear File) and bsf (Bit Set File) to clear/set an individual bit in Port B to 0 or 1 respectively.

As we shall see in Table 5.2 on p. 113, these instructions work by first reading the target File into an internal holding register, modifying it by changing *one* bit and then writing back to the File. This type of instruction is described as **read-modify-write**. In our example, the target File is PORTB and the logic state of the bits are actually the voltage on the pins RB[7:0]. This does not normally lead to problems, but in some rare situations where pins are loaded beyond their specifications (usually ±25 mA) or being quickly changed into a capacitive load, the physical voltage on a pin may not match its primed logic state. In this circumstance, when being read, a nominally logic 1 bit may read as a 0, or vice versa. Thus when written back, more than one bit target bit may change!

To get around this problem, the PIC18F1220 has LATA (File h'F89') and LATB (File h'F8A') to shadow PORTA and PORTB respectively.[6] A read of a LATch register yields the logic state *before* it gets to the pins. This LATch register holds the logic state which is connected to the pins. Details of the circuitry are given in Fig. 11.4 on p. 343. At this point, all that it is necessary to observe, is that LATA and LATB can be interchanged with PORTA and PORTB respectively in nearly all cases, but provides isolation between the set logic state and the actual pin voltages.

Serial Ports; Chap. 12

Communication streams, ranging from intra integrated circuits through automobile networks up to trans-continental webs, are nearly always sent and received one bit at a time. Problems in doing this include bit rate, start and stop conditions, numbers

---

[5]Actually, we could equate TRISB equ h'F93',ACCESS in one go.

[6]Earlier PIC MCU families lack these Port LATch registers.

of bits and words in a frame, bit rate and format issues. Our exemplar MCU includes one Universal Synchronous-Asynchronous Receiver/Transmitter (USART) module which handles many of these issues for a range of common asynchronous (no clock) and synchronous (needs a clock signal) protocols.

Most members of the family feature a Synchronous Serial Port (SSP) to deal with serial links between integrated circuits. Some devices, such as the PIC18F8722 offer up to two USART and two SSP modules. Control Area Network (CAN; e.g. PIC18F4580) and Universal Serial Bus (USB; e.g. PIC18F4450) modules are also available in some family members.

Timers; Chap. 13

The ability to measure the duration of external events and generate timed wave-forms, is a frequent requirement in a digital processor. The first PIC MCU family (base-range) introduced in 1989, has an integrated 8-bit counter, originally called a Real-Time Counter-Clock (RTCC), which is able to totalize pulses, either exter-nally (usually from pin RA4) or internal instruction cycles. In either case, these can be optionally frequency divided by powers of two up to 2^8. Unfortunately, this 8-bit pre-divider (called a prescaler) flip-flop chain is shared with the Watchdog timer.

The mid-range PIC16 family renamed the RTCC **Timer 0** and introduced a 16-bit **Timer 1** and 8-bit **Timer 2**. These counters can be used with one or more **Com-pare/Capture/PWM** (CCP) modules to compare the timer state with settings in an internal register, capture the state when an external event occurs (pulses on pins) and generate pulse width modulated signals.

The enhanced-range family added a clone of Timer 1 called Timer 3, and some large-footprint members have an 8-bit clone of Timer 2, called Timer 4. In the PIC18F8722, Timer 3 and Timer 4 perform the same function as Timer 1 and Timer 2 for a second Master CCP module.

Timer 0 was expanded to 16 bits, but retained an 8-bit mode for backwards com-patibility. Additionally the prescaler frequency divider is now dedicated to the timer, with the Watchdog timer acquiring its own—see Fig. 13.1 on p. 455. We see from Fig. 4.1 that besides the two Timer 0 SFRs TMR0H and TMR0L at File h'FD7:6', which can synchronously be read from or written to, the T0CON (Timer 0 CONtrol at File h'FD5') allows the software to set various options and control the behavior of Timer 0—see Fig. 13.2 on p. 458.

- To turn on the timer.
- To select either an 8- or 16-bit count.
- To select either an external or internal source of counting pulses.
- If external, to pick a $\diagup$ or $\diagdown\!\!\_$ active edge.
- To switch in an 8-bit prescaler and select one of eight ratios from 2^1 though to 2^8 to frequency divide down the chosen pulse source.

As a general principle, all peripheral modules have one or more control registers, which allow the software to configure its operational mode and allow the program to monitor its state.

Analog-to-Digital Conversion; Chap. 14

In the real world the majority of physical quantities are analog in nature. Whilst it is possible to externally convert the analog signals to digital equivalents, to allow digital processing, many MCUs have an integrated **analog-to-digital converter** module for medium accuracy and conversion rates.

The first Microchip ADC module, introduced with the PIC16C71 (1994) had an 8-bit resolution. Later PIC16 devices used a 10-bit version and virtually all enhanced-range devices use a similar module. Unusually, the PIC18F4523 has a 12-bit resolution variant. In all cases an analog multiplexer is used to give several analog channels. Our exemplar device can have up to seven analog inputs, any one of which can be selected for conversion when required. The PIC18F8722 has 16 analog channels. These analog inputs are shared with digital I/O pins, and indeed on reset all shared pins are initialized as analog inputs. The ADC control registers can then be set to configure a subset of this shared resource as digital.

Maximum conversion rate for the PIC18F1220's ADC module is 30 ksps— 30×10^3 samples (conversions) per second. However, the majority of devices have a 100 ksps rate (e.g. the PIC18F4520) and some even go up to 200 ksps (e.g. the PIC18F2331); whilst the 12-bit module has a reduced rate of 80 ksps.

Most family members, but not our exemplar, have two or three analog comparators. This allows the software to determine if one or more analog signals exceed an internal or external analog voltage.

Data EEPROM; Chap. 15

The PIC16F84 (1994) was the first PIC MCU device to offer a block of **nonvolatile** memory (64 bytes), which could be used for long-term data storage. Most PIC18 devices have a 256-byte module, such as our exemplar. A few, such as the PIC18F8722, have a 1024-byte equivalent module. Although many enhanced-range devices can both read and write to the Flash EEPROM Program store, the standard EEPROM used for these modules have a greater endurance; typically 1 million writes compared with 100,000 for the Program store. In both cases a data retention of 40+ (some parts quote 100) years. This memory module is not part of the (volatile) Data store and is accessed through SFRs as a peripheral device. Any byte can be addressed and then read from or written to via the EE-DATA (EEprom DATA at File h'FA8') register, as addressed by the EEADR (EEprom ADdR at File h'FA9') register and controlled by the EECON1 (EEprom CONtrol 1 at File h'FA8') and EECON2 (EEprom CONtrol 2 at File h'FA7') SFRs. In particular, the phantom EECON2 register is used as part of an interlock sequence to avoid accidental writes to memory.

Interrupts; Chap. 7

All the various peripheral modules can **interrupt** the flow of execution (the background program) when one or more specified events occur; for instance, when a timer overflows. The three INTerrupt CONtrol registers INTCON, INTCON2 and INTCON3 registers give the software control over global enabling and prioritization of the interrupt system and handle interrupt requests that come from outside the processor, chiefly from the three pins INT0, INT1 and INT2 and also from Timer 0.

A peripheral module can set one or more flags in the two Peripheral Interrupt Registers PIR1 and PIR2, which potentially requests service. Each of these requests can be individually enabled by a corresponding bit in a Peripheral Interrupt Enable register PIE1 and PIE2. Finally, the Interrupt PRiority registers IPR1 and IPR2 set each of these interrupt requests as either normal priority (causes a leap to the instruction located at h'0008' in the Program store) or priority (can interrupt a normal interrupt service routine and go to the instruction at h'0018'). Once the processor recognises an interrupt request, it will switch to another program via either the Normal or High-Priority interrupt vectors—see Fig. 4.3) to service the peripheral device. This service routine is known as the foreground program.

Examples

Example 4.1 Discuss how the performance of the PIC MCU architecture is improved by incorporating pipelining into the design of the instruction-fetch unit. Do you foresee any problems associated with handling Jump instructions (such as Branch) in connection with the Pipeline's structure?

Solution The Pipeline is a precondition for the parallel operation of the fetch and execution units. That is, in order to allow the execution of instruction n whilst the next instruction $n + 1$ is being fetched from the Program store, internal storage must be provided to present the instruction code to the Instruction decoder. As all but four instructions are single 16 bit words, then the Pipeline's register structure and control is considerably simplified. Most conventional CISC processors have instructions that vary considerably in length. For instance, the 68HC11 MCU core has instructions that cover the range 1 through 4 bytes; that is, the fetch phase can take between 1 and 4 bus transactions. Some more sophisticated processors have multistage pipelines with each stage feeding part of the execution circuitry. Thus several streams of execution activity can occur simultaneously.

The problem with pipelines is that they presuppose that the program instructions will be executed sequentially as they are stored in memory. However, instructions that disrupt this smooth running and move on the Program Counter require that the Pipeline be emptied so that the destination instruction code travels down to the end of the pipe. For instance, if instruction k is bra nn (BRAnch nn places), then instruction $k + 1$ will be in the first stage of the Pipeline by the time the processor knows that the next step is actually to be the instruction nn words away. Thus a null instruction cycle needs to be executed which simply brings this destination instruction code into the Pipeline but does not execute instruction $k + 1$ whose code is at the end of the Pipeline. This is sometimes known as **flushing** the Pipeline. Instructions such as bra need two clock cycles to execute. Conditional Skip instructions (see Chap. 5) take two cycles when the skip is implemented and one otherwise. All other instructions always take one cycle, except the few double-word instructions which take an extra cycle.

Example 4.2 Can you determine why, after a subtraction or addition of a negative number, the setting of the **C** flag is the *complement* of the borrow-out. *Hint*: Look at 2's complement arithmetic on p. 9.

Solution Subtract instructions in all PIC MCU families work by 2's complementing the datum byte and then adding; as shown in Fig. 2.9 on p. 25. In this situation the resulting carry-out is 0 where a negative outcome is generated and 1 for a positive outcome. For instance:

1. $06 - 0A \rightsquigarrow 00000110 + 11110110 = (0)\ 11111100$ or -4 (no carry).
2. $0A - 06 \rightsquigarrow 00001010 + 11111010 = (1)\ 00000100$ or $+4$ (carry).

In both cases the Carry flag acts as an inverted borrow. This is in keeping with the RISC philosophy of the PIC MCU family, to keep the processor 'lean and mean'.

Exactly the same borrow inversion occurs if you specify a negative datum with an Add instruction, such as addlw -6. This is translated by the assembler to addlw h'FC', where h'FC is of course the 2's complement of 6.

Example 4.3 A smart alec programmer has decided to copy the contents of the Status register into File h'040' (Access bank) for safekeeping so that it can be returned later without alteration. However, bit 2 of the Status register sometimes changes state. Why is this?

Solution From p. 49 we see that the movf instruction will set the **Z** flag if the contents of the File in question is all zero, else it will clear **Z**. Thus the program fragment

```
movf   STATUS,w,0 ; Copy contents of File h'FD8' to W
movwf h'040',0    ; and to File h'040' in Access RAM
```

will indeed copy the contents of File h'FD8' into File h'040', but if the Status register bits are all zero, the **Z** flag will be set to 1, otherwise it will be cleared to 0.

The easiest way around this is to use the movff instruction (movff STA-TUS,h'040'), which we see from Table 5.1 on p. 109 does not influence the status flags.

Example 4.4 In the PIC18F4520 MCU banks 0 through 5 are fully populated with GPRs. Show how you could copy File h'520' and File h'521' to File h'020' and File h'021' respectively. Calculate how long this would take, assuming a clock frequency of 4 MHz, and how much Program storage would be necessary.

Solution Basically there are three ways of doing this.

Banking
We can use the BSR to switch between banks.

```
movlb  5              ; Pick Bank 5
movf   h'520',w,1     ; Get Datum 1 using Banking mode into W
movlb  2              ; Pick Bank 2
movwf  h'220',1       ; Store away using Banking mode
movlb  5              ; Pick Bank 5 again
movf   h'521',w,1     ; Get Datum 2 using Banking mode into W
movlb  2              ; Pick Bank 2 again
movwf  h'221',1       ; Store away using Banking mode
```

In total it will take eight instruction cycles to implement and occupy the same number of words in the Program store.

Pointers

We can use Indirect addressing to the moving, with two pointers (File Select Registers) for the two banks. Furthermore (but not explained until p. 106), if we use the two Indirect triggers POSTINC0 and POSTINC1, the two corresponding pointers FSR0 and FSR1 will be automatically incremented when used.

```
lfsr   0,h'220'    ; Pointer FSR0 (Pointer 0) set to h'220'
lfsr   1,h'550'    ; Similar-
ily FSR1 (Pointer 1) set to h'520'
movf   POSTINC1,w  ; Get Datum 1 from File h'520'& FSR1++
movwf  POSTINC0    ; Store away at File h'220' & FSR0++
movf   INDF1,w     ; Get Datum 1 from File h'521'
movwf  INDF0       ; and store away at File h'221'
```

This will also take eight instruction cycles, as each double-word lfsr instruction takes two cycles. However, this technique comes into its own for longer arrays of data.

Multiple Move

The movff double-word instruction allows us to copy any File directly (that is not via the Working register)from source to destination.

```
movff  h'520',h'220'   ; Copy Datum 1 to File h'220'
movff  h'521',h'221'   ; Copy Datum 2 to File h'221'
```

This takes four instruction cycles and the same number of instruction words.

For a clock rate of 4 MHz, an instruction cycle takes 1 μs—see Fig. 4.5. Thus the three techniques take 8 μs, 8 μs and 4 μs respectively to run.

Example 4.5 Write a program to increment a packed BCD quantity located in data memory at File h'020' in Access RAM. You can make use of two new instructions.

Program 4.1 Incrementing a packed BCD byte

```
;*****************************************************************
;* FUNCTION: Increments a BCD datum giving a BCD outcome        *
;* ENTRY    : BCD in File h'020'                                *
;* EXIT     : BCD+1 in File h'020'                              *
;* EXAMPLE  : 01111001 (79) + 1 = 10000000 (80)                *
;* *************************************************************** 
STATUS      equ    h'FD8'      ; The Status register
C           equ    0           ; Carry flag is bit 0
DC          equ    1           ; Digit Carry flag is bit 1
BCD         equ    h'020'      ; The BCD number is in File h'020'
; -------------------------------------------------------------
BCD_INC     incf   BCD,w,0     ; Binary inc'ed BCD number put in W
            addlw 6            ; Add six
            btfss STATUS,DC    ; Skip IF produced a half carry
            addlw -6           ; ELSE remove the correction of +6
; Now check the upper digit by adding 6 to it and checking carry
            addlw h'60'        ; Add h'60' (i.e. six to upper digit)
            btfss STATUS,C     ; Skip IF caused a carry
            addlw -h'60'       ; ELSE cancel the correction factor
; The incremented and corrected BCD number is now in W
            movwf BCD,0        ; Put it out in Access RAM
```

INCrement File (incf) adds one onto the contents of any File, and Bit Test File and Skip if Set (btfss) tests any bit in any File and skips the next instruction word it that bit is 1.

Solution Two binary-coded decimal (BCD) digits may be packed into a single byte to represent numbers up to 99. For instance, $\boxed{0100\ 1001}$ File h'020', represents BCD 49. Incrementing a number stored in this hybrid decimal-binary form using the normal binary addition rules may give an incorrect result. For instance, b'0100 1001 + 1' (49 + 1) gives b'0100 1010' (h'4A') after addition, but should give b'0101 0000' (h'50'). Similarly, b'1001 1001 + 1' (99 + 1) gives b'1001 1010' (h'9A') instead of b'0000 0000' plus a carry of 1 (h'1 00').

From these examples it can be seen that whenever any of the BCD decades equals ten after incrementation then it should be zeroed and one added to any higher decade. Based on this increment and add algorithm we can formulate the task list.

1. Increment the packed BCD byte using normal binary arithmetic.
2. IF the lower nybble of the outcome is ten then add six to the outcome.
3. IF the upper nybble of the outcome is ten then add six to it.

Program 4.1 gives an efficient implementation of this task list. After incrementing using normal binary rules, six is added to the previous outcome and the Decimal Carry (half carry) flag is checked for activity. The **DC** flag will only be set when the original nybble is ten (h'0A + 6 = 1 0'). In this case the add six operation is allowed to stand as the necessary correction, otherwise it is canceled by subtraction. The upper nybble (BCD digit) is checked and corrected in the same manner, but this time it is the full Carry flag that is tested. If this is set, then the addition of h'60' is allowed to stand, otherwise it is subtracted. This Carry flag could be used to set a hundreds digit if desired, to show overflow from 99 to 100. See p. 116 for a more efficient implementation.

An alternative approach would be to subtract nine *before* incrementation and if the **Z** flag is set then leave the digit at zero and increment the higher digit; otherwise add ten. Repeat for the upper digit.

Self-Assessment Questions

4.1 Where microprocessors are used in a general-purpose computing environment, the program is normally loaded into and run from read/write RAM memory. This means that the system can run a word-processor one minute and a spreadsheet program the next. Of course this means of operation is not applicable to embedded applications, where the program is stored in some variety of non-volatile read-only memory. Discuss why this is so and the virtues of ROM, EPROM and EEPROM implementations of non-volatile storage.

4.2 The goto instruction allows the software to jump to any part of the Program store. It does this by overwriting the Program Counter with a new 20-bit word address. Given that an instruction word is only 16 bits long, how do you think and instruction like goto handles this problem.

4.3 Given the effect of the movf instruction on the **Z** flag discussed in Example 4.3, how could you use this instruction to determine if the contents of any File is zero?

4.4 From Table 1.1 on p. 5 we see that the upper-case letters A through Z differ in coding from their lower-case siblings only in that bit 5 is 0 in the former instance and 1 in the latter. With the instructions we have de facto introduced in this chapter, how could you convert an ASCII character located in File h'020' from lower-case to upper-case?

4.5 Write a program to pulse pin RA0 High for 4 μs and then Low. You may assume a clock crystal of 4 MHz.

4.6 How could you bring pin RA1 High, then pulse RA0 four times and then RA1 is to go Low again? Your solution should include the setting for TRISA.

4.7 Given a 10 MHz crystal clock, how long would a PIC MCU take to execute the code fragment of Example 4.4?

4.8 Most digital watches use a 32.768 kHz crystal, commonly known as a watch crystal. Because of high production quantity, such crystals are low cost. Although this slows the processing rate, we shall see in Fig. 10.3 on p. 309 that the power dissipation is directly proportional to clocking frequency. Thus a watch crystal is an attractive low-cost proposition for many low-power applications.

Can you determine the instruction cycle time for such a system? What is the significance of the value 32,768 for timing circuits?

Chapter 5
The Instruction Set

Writing a program is somewhat akin to building a house. Given a known range of building materials, the builder simply puts these together in the right order. Of course, there are tremendous skills in all this; poor building techniques lead to houses that leak, are drafty and eventually may fall down!

It is possible to design a house at the same time as it is being built. Whilst this may be quite feasible for a log cabin, it is likely that the final result will not remain rainproof very long, nor will it be economical, maintainable, ergonomic or very pretty. It is rather better to employ an architect to design the edifice before building commences. Such a design is at an abstract level, although it is better if the designer is aware of the technical and economic properties of the available building materials.

Unfortunately, much programming is of the "off the cuff" variety, with little thought of any higher-level design. In the software arena, design means devising strategies and designing data structures in memory. Again, it is better if the design algorithms keep in mind the materials of which the program will be built, in our case the machine instructions.

At the level of our examples in this chapter, it will be this coding (building) task we will be mostly concerned with. Later chapters will cover more advanced structures which will help this process, and we will get more practice at devising strategies and data structures.

If you like to think of writing a program as analogous to preparing an elaborate meal, then for any given cooking appliance, such as a microwave oven or electric stove (the hardware) there are a range of processes. These processes—for instance, steaming, frying, boiling—are analogous to the **instruction set** which can be implemented by the CPU. The various ingredients that can be handled by a process are the instruction's data. Such data may lie in internal registers or out in memory. There are several different ways of specifying the **effective address** (**ea**) of an operand. These are known as **address modes**.

In keeping with the PIC microcontrollers' RISC-like philosophy, the enhanced-range core have a total of only 75 instructions; plus an optional eight which are primarily for handing software stacks used for **C** compiler operation; which we will ignore. Of these instructions, 71 are coded in a single 16-bit word. Four instructions need two words. An instruction word in the main holds the operation code;

S. Katzen, *The Essential PIC18® Microcontroller,*
Computer Communications and Networks,
DOI 10.1007/978-1-84996-229-2_5, © Springer-Verlag London Limited 2010

address, data or bit number, and destination and access bits. We covered a few of these instructions and address modes when discussing our BASIC computer back in Chap. 3 and also in Chap. 4; now would be a good time to review this material. Here we look at the various address modes and the core instruction set in some detail. Some of the more specialized instructions will be covered later; for instance, the subroutine-related instructions are covered in Chap. 6. A full instruction set is given on p. 66.

After reading this chapter you will:

- Know that an address mode is the way an instruction pin-points its data.
- Understand how Inherent, Literal, Absolute, File Direct, File Indirect and Bit address modes permit an instruction to target an operand for processing.
- Recognize how the binary structure of the instruction word impacts on the usage of instructions.
- Know that Movement instructions, copying data in-between the Working register and the Data store or between two Files, are the most used of the instruction categories.
- Appreciate that the processor can directly implement the common 2's complement arithmetic operations of addition, subtraction and negation, as well as multiplication, incrementation, decrementation and bit twiddling.
- Be able to compare or test data for differences and relative magnitude, and take appropriate action using the Conditional Branch instructions.
- Understand how the program flow can be diverted, based on the state of any bit or a zero overall value in a File.
- Know that a datum in the Data store can be rotate-shifted.
- Be able to use the four basic logic instructions to invert, set, clear, toggle, bit test and differentiate data.

Virtually all instructions act on data, either outside the CPU in its data or program memory space, or in internal registers. Thus the 16-bit instruction code must include bits which inform the CPU's instruction decoder *where* this data is being held. The exceptions to this are the few Inherent instructions, such as nop (No OPeration). Before looking at the instruction set we will discuss the various techniques used to specify the location of any operands.

The general symbolic form of an instruction is:

```
instruction mnemonic <operand A>,<operand B>
```

where operand A is the source datum or its location and operand B the destination. For instance, movf h'020',w (MOVe File) which copies a datum sourced from File h'020' in the Data store to its destination in the Working register.

There are some variations on this structure. $2\frac{1}{2}$-operand instructions are common. For instance, addwf [FILE],d adds the WREG register's contents to the specified File's contents and deposits the result either in WREG or back in the File register itself. Thus addwf h'020',f means "add the contents of WREG to that of File h'020' and put the outcome in File h'020'." This could be written in shorthand as [f020] <- W + [f020], where the brackets mean "contents of" and <- means "becomes". This notation is called **register transfer language (rtl)**.

Of course, this is not a true 3-operand instruction, as the destination must be one of the two source locations; that is WREG or File h'020'. A few instructions have only a destination specified; for example, `clrf h'020'`, and the Inherent instructions have no explicit operands.

Instructions can be classified by their address mode.

Inherent | 00000000 | XXXXXXXX |

Sixteen instructions do not need to explicitly specify an operand. At the binary code level, all these instructions are coded with the upper eight bits zero. For instance, `clrwdt` has a machine code of b'**000000000**0000100'. These instructions are:

clrwdt

`clrwdt` CLeaR WatchDog Timer resets this monostable and its pre-divider chain—see Fig. 13.1 on p. 455.

daw

Decimal Adjust Working register makes any required correction factor following a normal binary addition of two packed BCD formatted variables—see p. 116.

nop

No OPeration does nothing, apart from the resulting step of the Program Counter as a consequence of its fetch. Exceptionally, it has two binary codes; 0000000000000000 and 1111XXXXXXXXXXXX. Unprogrammed areas of the Program store are always logic 1 and so appear as blocks of `nop`s.

pop and push

This pair of instructions interact with the stack and are described on p. 181 in Chap. 6.

reset

This performs the software equivalent of an external reset as described on p. 329 in Chap. 10.

return and retfie

This pair of instructions action the completion of either a subroutine or interrupt service routine with a return to the caller or background program respectively.

sleep

Turns the main oscillator off and powers down the core to reduce power consumption—see p. 318 in Chap. 10.

Table Read and Table Write

There are four `tblrd` and four `tblwt` instruction variants which allow the processor to both read data bytes from the Program store and (in some family members) write data in blocks to the store. They are listed in Table 15.1 on p. 548.

Fig. 5.1 Machine code structure for the lfsr instruction

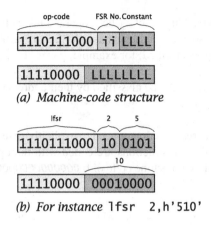

(a) Machine-code structure

(b) For instance lfsr 2,h'510'

Literal | 00001 | ??? | LLLLLLLL |

Literal instructions use the lower eight instruction word bits to specify a source operand which is a *constant* datum, rather than a byte in a File. For instance, addlw 06 is coded as b'11110000 00000110'. The destination of this type of instruction is *always* the Working register, and this is shown in the mnemonic. Thus in our example, the sum W + 6 is copied back into WREG. In rtl this is expressed as W <- W + #6, where the # (pound or hash) symbol denotes the following number as a constant or literal rather than a File address.

The exception to this structure is the lfsr (Load File Select Register) instruction, which we have already met on p. 82. This loads a 12-bit constant into one of three File Select Registers (FSRs). Each FSR carries a 12-bit address, and acts as a pointer into the Data store. As we see in Fig. 5.4, a FSR is made up of two 8-bit SFRs; for instance, FSR2 is actually a composite of FSR2H:FSR2L.

In order to code both the FSR number (0, 1 or 2) and the 12-bit literal, lfsr needs two words in the Program store. As shown in Fig. 5.1 the first word fetched from the Program store has a 10-bit op-code, a 2-bit FSR number and the top nybble of the literal. The lower eight bits of the second word holds the remaining lower byte of the literal. Rather unexpectedly, this second word also carries an auxiliary op-code. Any op-code beginning with b'1111' codes for nop! The reasoning behind this, which is common for all double-word instructions (e.g. see Fig. 5.2), is that if the Program Counter as a result of a Jump, Skip or Branch should point to the second word of a double-word instruction, it will be treated as a No OPeration.

From Fig. 5.1 we see that the assembler form of the instruction specifies the FSR number as operand A and the literal as operand B. For example, to put the constant h'456' in FSR1 we have lfsr 1,h'456'.

Absolute | 111011 | ?? | AAAAAAAA |

                                    plus | 1111 | AAAAAAAAAAAA |

Two instructions allow the program to jump to another instruction anywhere in the Program store. These are goto and two variations of call (CALL up or go to

Fig. 5.2 Machine code structure for the `goto` instruction

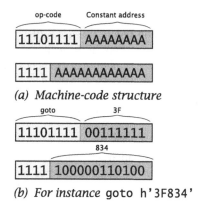

(a) *Machine-code structure*

(b) *For instance* `goto h'3F834'`

a subroutine—see Table 6.1 on p. 164. As this potentially requires a 20-bit word address[1] these instructions are both double-word.

As shown in Fig. 5.2, the first word carries an 8-bit op-code and the top byte of the absolute destination address. The second word has the op-code b'1111' for `nop` and then the remaining lower 12 bits of the 20-bit destination address.

Like all 2-word instructions, execution takes two instruction cycles.

Relative

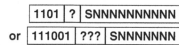

Like the Absolute address mode, Relative addressing is used to change the focus of the program to software in a different part of the Program store. However, it does this by adding an *offset* to the Program Counter, rather than overwriting its contents with a 20-bit absolute word address. For instance, `bra 08` will add eight to PC[21:1]; that is the program will effectively branch forward eight words up from the instruction *following* the `bra` instruction.

The instructions BRAnch and Relative CALL (`rcall` see Table 6.1 on p. 164) are coded with an 11-bit 2's complement signed offset field. As the PC has already incremented *before* the offset N is added, the maximum range of these instructions is ±1024 words forwards and backwards from the instruction. The offset code is shifted once left to give $2N$ before the addition, to effectively make the branch offset in words, rather than bytes.

The majority of Relative instructions are conditional Branches (see Table 5.4) in that the offset is only added if one of the status bits is set or clear. For instance, `bc -08` (Branch if Carry) will branch back eight words from the following instruction (seven from the `bc` instruction) if the **C** flag is 1. Conversely, `bnc` (Branch if No Carry set) will branch if the **C** flag is 0.

Relative Branch instructions have a signed 2's complement 8-bit offset, giving them a range of ±128 words. In practice, the programmer will rarely have to

---

[1]Don't confuse this with the File address in the Data store. In the Harvard structure the two stores are logically distinct with different address spaces.

calculate the offset for such Relative instructions; rather he/she will place a label at the desired landing point. The assembler will calculate the offset; for instance, bra LOOP. If LOOP is too far away, the assembler will output an error message. Where this is the case, conditional Branches can be augmented with either an unconditional Branch or a goto.

Although Relative addressing instructions are all single-word, like all instructions that modify the PC two instruction cycles are needed for execution.

File Direct | 0 | ????? | d | a | FFFFFFFF |

The majority of data which the program will process are located in the Data store. Instructions that specify that their source and/or destination operand lie in a File use this address mode. The bottom byte of the File address is carried in the lower eight bits of the instruction code—as shown in Fig. 3.5 on p. 50. This is extended to the necessary twelve bits to match the potential 4 kbyte Data store—as described in Fig. 4.9 on p. 82. In summary, if bit 8 of the instruction code (labeled a) is 0 then Files in the range h'000–07F' (GPRs) and h'F80–FFF' (SFRs) are accessed. To access anywhere else in RAM, this a-bit is set to 1 and the lower nybble of the Bank Select Register is used for the upper four bits of the address. This bank-oriented structure for the Data store is shown in Fig. 4.8 on p. 80.

Most instructions that use Direct addressing can dump the outcome either in the Working register or else back in RAM. Bit 9 of the instruction code, labeled d (see also Fig. 3.5 on p. 50), is used to specify the destination, as in the following example:

```
addwf  h'02C',w,0   ; Coded as 001001 0 0 00101100
addwf  h'02C',f,0   ; Coded as 001001 1 0 00101100
```

In both cases the byte contents of File h'02C' in Access RAM (a = 0) are added to the byte contents of the Working register. In the former instance, illustrated in Fig. 5.3(a), the outcome is put in WREG, leaving the File contents *unchanged* (d = 0), whilst in the latter, illustrated in Fig. 5.3(b), the original File data is *overwritten* (d = 1) with the sum.

In the base- and mid-range families, the Working register associated with the ALU is not mapped as a SFR in the Data store. It is accessible only in its role as a destination or as a result of an instruction explicitly referencing WREG; e.g. movlw. In the enhanced-range family, the Working register is also reachable as a SFR called WREG (File h'FE8' in Fig. 4.10 on p. 83). It therefore can take the place of the File address in Direct address mode instructions. This gives additional flexibility, in that all Direct addressing instructions can operate directly on the Working register, rather than the limited explicit subset previously allowed. For instance, setf sets up any File to b'11111111'. Thus setf WREG will make the contents of WREG all 1s.

The 8-bit address of the operand is *fixed* as an integral part of the instruction code, and thus cannot be changed as execution progresses. Although explicitly specifying its address may seem to be the obvious way to locate an object in the Data store, there are some situations where this restriction is rather onerous.

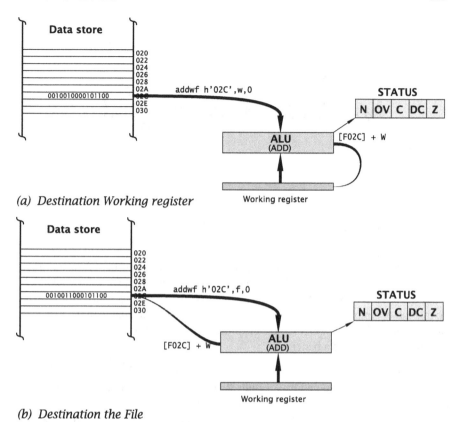

(a) *Destination Working register*

(b) *Destination the File*

Fig. 5.3 Selecting the destination for the instruction `addwf h'02C'`

Program 5.1 Clearing a block of Files the linear way

```
CLEAR_ARRAY clrf  h'020'  ; Clear File 32
            clrf  h'021'  ; and File 33
            clrf  h'022'  ; Each clrf
            clrf  h'023'  ; uses one instruction word
            clrf  h'024'  ; in the Program store
            clrf  h'025'  ; File 37 cleared
            clrf  h'026'  ; and so on
            ....  .....
            clrf  h'07E'  ; Clear File 126; nearly there
            clrf  h'07F'  ; Clear File 127; Phew!
```

As an example showing this lack of flexibility, suppose we wished to clear the contents of File registers in Bank 0 of a PIC18F1220 from File h'020'–File h'07F'. The obvious way to do this is to use the `clrf` instruction 96 times, as shown in Program 5.1.

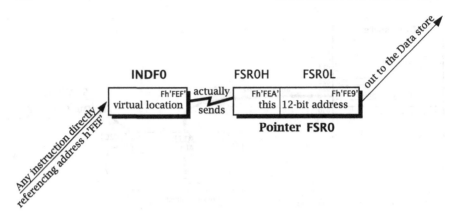

Fig. 5.4 The Indirect addressing mechanism shown for FSR0

Although this coding works, it is rather inefficient. Each of the 96 instructions does exactly the same thing, although on a different location. If we were to clear all 1536 GPRs in the PIC18F4520 then we would need 1536 clrf instructions, all to do this rather simple task. As there is only 16,384 locations available in the Program store for this device then this represents more than 9% of its entire capacity.

There has got to be a better way!

File Indirect   `0 ????? d a FFFFFFFF`

All processors feature some form of Indirect addressing, where one or more internal registers are used to hold the address of the operand in Data memory. Such address registers effectively are used as a **pointer** to the data. The key difference from Direct addressing is that the contents of a pointer register can be altered as execution progresses; that is, the address of the target datum is no longer fixed as bits in read-only (usually) Program memory but is now a *variable*. For instance, the array of data in Program 5.1 may be cleared by using a register to point to the target location and repeating in a loop while incrementing that pointer—see Fig. 5.6.

The principle of Indirect addressing as implemented in PIC MCUs is to 'trick' an unsuspecting Direct addressing instruction to substitute its operand address for the 12-bit address held in a **File Select Register**. It does this by using virtual SPRs, called **INDirect File registers**, to trigger the subterfuge. In the enhanced-range family, three[2] of these pointers are available. Figure 5.4 specifically shows FSR0, but FSR1 and FSR2 act in the same way. Each FSR is actually composed of two SRFs. In the case of FSR0, FSR0L at h'FE9' holds the bottom eight bits of the address and FSR0H at h'FEA' for the upper nybble. The composite register thus holds a full 12-bit address—see also p. 82.

Any instruction using Direct addressing can activate Indirect addressing by referencing a *trigger* address. For instance, to utilize the address in FSR0 to access the Data store, the pseudo address h'FEF', known as INDF0 (INDirect File 0), is used as

---

[2]In the base- and mid-range families only one 8-bit pointer is available, called FSR.

Fig. 5.5 Using a loop to clear an array of data

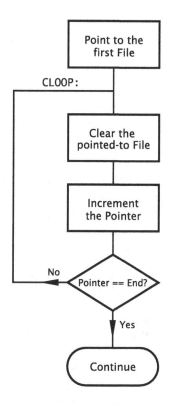

the target. Thus, if the content of FSR0 at some point of time were h'070', then the instruction `clrf INDF0` will actually clear File h'070' and not File h'FEF'. INDF0 is a virtual location, in that this SFR does not physically exist. Its contents can neither be read-from or written-to. Its sole use is to act as a trigger to switch through the contents of FSR0 as the actual operand address. Although this approach to Indirect addressing may seem rather convoluted, it requires very little additional logic in the processor and no extra clock cycles to execute; unlike the alternative techniques used by other MPU/MCUs.

As an example, let us repeat Program 5.1 but folding the linear structure into a **loop**, as shown in Fig. 5.5. A task list description of our program is now:

1. Set the FSR0 pointer to the initial array address.
2. Clear the pointed-to File by targeting the INDF0 File.
3. Increment the FSR0 pointer.
4. Check. Has the pointer gone over the top; in our case, has it reached h'080'? IF no THEN go to item 2.
5. Continue on to the next part of the program.

This process is perhaps more easily visualised in Fig. 5.6.

The coding for this scheme is shown in Program 5.2. The linear structure of the previous program has been folded into a loop, shown shaded. The execution path

Fig. 5.6 Walking through an array

keeps circulating around the `clrf` instruction, which is 'walked' through the array of Files from File h'020' upwards by incrementing the FSR0 pointer on each pass through the loop. Eventually the pointer advances beyond the desired range and the program then exits the loop and continues onto the next section of the code.

Program 5.2 has many new features, as we have yet to review the instruction set.

Program 5.2 Clearing a block of Files using a repeating loop

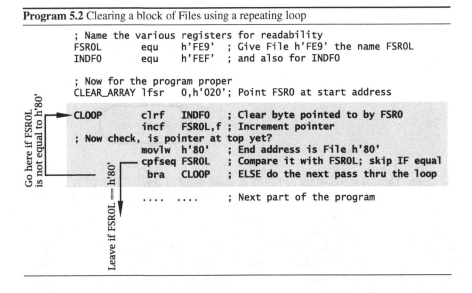

```
             ; Name the various registers for readability
   FSR0L         equ    h'FE9'  ; Give File h'FE9' the name FSR0L
   INDF0         equ    h'FEF'  ; and also for INDF0

             ; Now for the program proper
   CLEAR_ARRAY lfsr   0,h'020'; Point FSR0 at start address
   CLOOP         clrf   INDF0   ; Clear byte pointed to by FSR0
                 incf   FSR0L,f ; Increment pointer
             ; Now check, is pointer at top yet?
                 movlw  h'80'   ; End address is File h'80'
                 cpfseq FSR0L   ; Compare it with FSR0L; skip IF equal
                 bra    CLOOP   ; ELSE do the next pass thru the loop

             .... ....          ; Next part of the program
```

Task 1

The pointer is initialized to point to the first File to be cleared, by moving the 12-bit constant h'020' into FSR0 using the `lfsr` instruction—as described on p. 111. Nearly all loop routines involve some initialization before entry.

Task 2

The key clearing instruction uses the Indirect address mode by specifying the phantom File h'FEF' (INDF0) as the destination address—`clrf INDF0`. This line has a **label** associated with it called CLOOP. The assembler knows that this is a label and not an instruction, as it appears in the leftmost column of the source File. Lines without labels should begin with an indent of at least one space.

Task 3

Each pass around the loop involves advancing the pointer by one. This is done by incrementing FSR0. In this example, advancing FSR0L on its own will work, as the range of pointer values is limited to h'(0)20'–h'(0)80' in which FSR0H is always 0. Note that the destination is specified as the File and not the Working register.

Task 4

Unless you wish to go round the loop forever, you need a mechanism to eventually exit. In our case this is done by comparing the contents of the low byte of FSR0 with the constant h'80'; that is, with the first File after h'07F'. The instruction cpfseq (Compare File with W and Skip if EQual) will skip over the following instruction only if the contents of the File (that is FSR0L is equal to the byte in the Working register. In the program, the constant h'80' is copied into WREG prior with the comparison with FSR0L.

If there is not equality, then the following bra instruction actions a branch back to the instruction alongside the label CLOOP and the process is repeated, but with the pointer advanced one step. After 96 times around the loop, FSR0L will reach h'80' and with equality cpfseq will action a skip over the bra instruction and break out of the loop.

The result of executing Program 5.2 (and indeed Program 5.1) is shown in Fig. 5.7. Although the outcome is the same, our program now has only six instructions against the 96 of the linear equivalent, a 12:1 reduction. However, it takes six times as long to execute, due to the overhead of the various loop control instructions being executed 96 times! Normally the ratio of 'housekeeping' to core instructions in a loop is not as extreme as this particular example.

The requirements of scanning through an array of data is sufficiently common to warrant four additional Indirect modes of operation for each of the three pointers. Figure 5.8 lists the five distinct modes for each of the FSRs. Each mode is selected by invoking one of five different virtual trigger registers that exist for each of the three pointers.

File Registers																	
Address	00	01	02	03	04	05	06	07	08	09	0A	0B	0C	0D	0E	0F	
000	A1	90	31	72	F2	97	67	2B	FA	2D	DF	56	FD	D1	2B	0E	
010	40	06	CE	9B	79	FC	4E	39	84	6C	8A	1B	57	C4	4E	B0	
020	00	00	00	00	00	00	00	00	00	00	00	00	00	00	00	00	
030	00	00	00	00	00	00	00	00	00	00	00	00	00	00	00	00	
040	00	00	00	00	00	00	00	00	00	00	00	00	00	00	00	00	
050	00	00	00	00	00	00	00	00	00	00	00	00	00	00	00	00	
060	00	00	00	00	00	00	00	00	00	00	00	00	00	00	00	00	
070	00	00	00	00	00	00	00	00	00	00	00	00	00	00	00	00	
080	9D	06	92	0A	FC	04	84	AF	E0	25	FA	FE	16	50	4F	BB	

Hex Symbolic

Fig. 5.7 A view of the Data store after execution

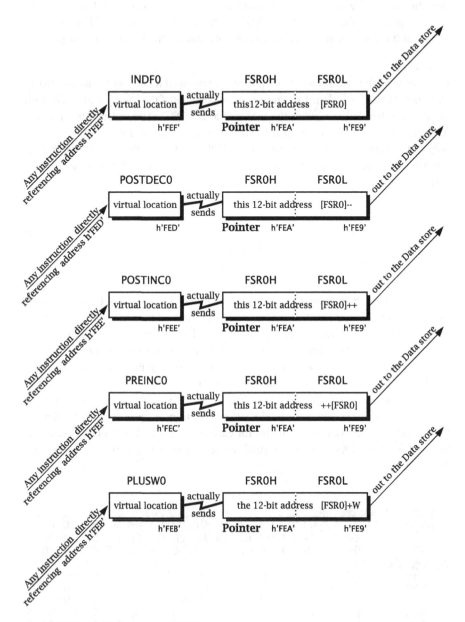

Fig. 5.8 Indirect addressing via FSR0

Program 5.3 Clearing memory using an auto-incrementing pointer

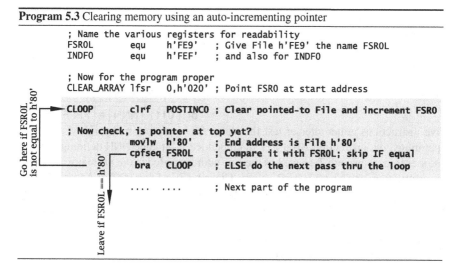

```
              ; Name the various registers for readability
              FSR0L      equ    h'FE9'  ; Give File h'FE9' the name FSR0L
              INDF0      equ    h'FEF'  ; and also for INDF0

              ; Now for the program proper
              CLEAR_ARRAY lfsr  0,h'020' ; Point FSR0 at start address

   CLOOP        clrf   POSTINC0 ; Clear pointed-to File and increment FSR0

              ; Now check, is pointer at top yet?
                   movlw  h'80'     ; End address is File h'80'
                   cpfseq FSR0L     ; Compare it with FSR0L; skip IF equal
                   bra    CLOOP     ; ELSE do the next pass thru the loop

                   ....  ....       ; Next part of the program
```

These are, where i indicates 0, 1, or 2:

INDirect*i*

For instance, `clrf INDF2` zeros the File pointed to by the 12-bit FSR2.

POSTDEC*i*

For instance, `clrf POSTDEC2` zeros the File pointed to by the 12-bit FSR2 and then decrements FSR2 *after* the action has been executed.

POSTINC*i*

For instance, `clrf POSTINC2` zeros the File pointed to by the 12-bit FSR2 and then increments FSR2 *after* the action has been executed.

PREINC*i*

For instance, `clrf PREINC2` *first* increments FSR2 and only then zeros the File pointed to by the 12-bit FSR2.

PLUSW*i*

For instance, if the contents of the Working register were h'06' then the instruction `clrf PLUSW2` would zero the File pointed to by FSR2 + 06. Neither the contents of FRS2 nor WREG would be altered. The contents of WREG are treated as a 2's complement signed number ranging from −128 through +127.

As a simple example of the use of these modes, Program 5.3 reworks our code, but using the Post-increment Indirect addressing mode. Replacing the target address INDF0 by POSTINC0 (File h'FEE') automatically increments FSR0 *after* it has been used to point into RAM. This has a two-fold advantage over our approach in Program 5.2.

1. We do not require the `incf` instruction. This not only reduces the size of code, but eliminating this housekeeping step saves 96 instruction cycle execution times.

2. The auto-incrementation is applied to the whole 12-bit address; that is FSR0H: FSR0L, whereas our `incf` instruction just operates on the low byte. For cross-bank arrays we would have to add two new instructions for a double-byte incrementation—see p. 119.

Bit Addressing

| 10 | ?? | NNN | a | FFFFFFFF |

or

| 0111 | NNN | a | FFFFFFFF |

Five instructions either alter or test the state of a *single* bit in any File. For these operations, the instruction word has an embedded 3-bit code NNN defining the bit number from 0 through 7, as well as the File address coded in the normal way. Thus the instruction `bcf h'020',7,0` (Bit Clear bit 7 in File h'020' in Access RAM) is coded as b'**1001 111** 00100000'. The other bit twiddling instructions are `bsf` (Bit Set in File) and `btg` (Bit ToGgle in File); the latter inverting the state of the target bit and leaving the other seven bits unchanged.

Two instructions allow the program to test (but not change) any bit in any File and optionally skip over the next instruction word. These are `btfsc` (Bit Test in File and Skip if Clear) and `btfss` (Bit Test in File and Skip if Set). We used the latter instruction in Program 4.1 on p. 92 to test bit 0 of the Status register (that is the **C** flag) and skip the next instruction if it was set.

So far we have classified instructions by the method they pin-point their operands. The alternative approach is to catalog the instruction set by function. Most of these functional groups will be discussed in this chapter. Those relevant to subroutines and interrupts are listed in Chap. 6, and control instructions pertaining to internal operation of the MCU hardware are left to Chap. 10.

In the instruction tables following; from left to right the instruction's mnemonic is listed, followed by the effect on the five status flags, with a • representing no change and √ normal operation. Finally a shortform description of the operation is given. The complete instruction set is given for reference on p. 66. If a more detailed reference is needed, any Microchip data sheet for the appropriate family (see the book's website) gives a detailed description for each instruction. However, because of the RISC nature of the PIC MCU architecture, instructions are rather minimalistic and details are easily consigned to memory.

Movement Instructions

About one in three instructions in any computer program, regardless of hardware or language, simply move data around without alteration between memory and internal registers. With this in mind the instructions in Table 5.1 are going to be the most used in the PIC MCU's repertoire.

All six Movement instructions *copy* byte data without alteration, either in between the Working register and a specified File, directly between Files or a constant byte (literal) into WREG, BSR or FSRi. Where the source is in a File its contents remain unaltered; it is simply copied into the destination. The `swap` instruction can also copy a datum from a File to WREG, but in the process interchanges the higher and lower nybbles.

Table 5.1 Move instructions

Operation		Mnemonic	Flags					Description
			N	**OV**	**Z**	**DC**	**C**	
Move								Copies a datum byte
	Literal to W	movlw kk	•	•	•	•	•	[W] <- #kk
	Literal to BSR	movlb kk	•	•	•	•	•	[BSR] <- #kk
	File	movf f,d,a	$\sqrt{}$	•	$\sqrt{}$	•	•	[d] <- [f]
1	File to File	movff f_s,f_d	•	•	•	•	•	$[f_d]$ <- $[f_s]$
	W to File	movwf f,a	•	•	•	•	•	[f] <- [W]
Load								Initializes a FSR pointer
1	Literal to FSRn	lfsr n,kkk	•	•	•	•	•	[FSRn] <- #kkk
Swap								Interchanges File nybbles
	File	swapf f,d,a	•	•	•	•	•	[d] <- [F(3:0)][F(7:4)]

•	Flag not affected	$\sqrt{}$	Flag operates in the normal way	
W	Working register	f	File register	
f_s	Source File	f_d	Destination File	
#kk	8-bit constant	#kkk	12-bit constant	
a	Access RAM	d	Destination, WREG or a File register	
BSR	Bank Select Register	<-	Becomes	
n	FSR 0, 1 or 2	1	2-word instruction	
[ ]	Contents of			

movlw | 10001101 | LLLLLLLL |

Copies the specified 8-bit constant (i.e. **literal**) into WREG. For instance, movlw h'80' initializes WREG to b'10000000'.

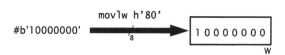

Note that by definition, the target is always the Working register, so another step is necessary to set up a File register to a constant value—see below.

movlb | 00000001 | LLLLLLLL |

We have already seen in Fig. 4.9 on p. 82 that treating the Data store as a banked structure, we need to be able to change the Bank Select Register to point to one of 16 banks. One way of doing this would be to initialize WREG and then copy the datum to BSR; e.g. movlw 4, movwf BSR enables entry into Bank 4 of RAM.

Where general-purpose data is spread over several banks, then this bank switching process will occur frequently and it is desirable that the overhead of changing BSR should be kept to a minimum. This is the *raison d'être* for including this instruction in the instruction set. As an example, consider copying the byte at File h'422' to Port B at File h'F81':

```
movlb              ; Point BSR to Bank 4
movf  h'22',w,1; Copy byte in File h'(4)22' in Banked RAM to W
movwf h'F81',0 ; and to PORTB in Access RAM
```

movwf | 0110111 | a | FFFFFFFF |

This instruction is used to copy (or store) the contents of WREG into a File. For instance, movwf h'023', 0 stores the byte in WREG to File h'023' in Access RAM.

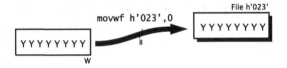

For example, to initialize File h'023' to, say, b'10000000':

```
movlw  h'80'     ; Set contents of W to b'10000000'
movwf  h'023',0 ; and copy to File h'023' in Access RAM
```

movf | 010100 | d | a | FFFFFFFF |

This instruction is normally used to copy (or load) the contents of any File into WREG. For instance, movf h'022',w,0 loads WREG with the contents of File h'022' in Access RAM.

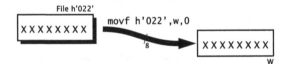

It is possible to specify the File itself as the destination; for instance, movf h'022',f,0. This circuitous command copies the byte contents of File h'022' back onto itself; that is there is no change in its contents! However, in the process the **Z** and **N** flags are activated as appropriate to the datum. Thus the instruction movf [File],f can be used in lieu of a TeST File for zero or negative instruction; i.e. tstf [File] that is commonly available to most other MPU/MCU families. Earlier PIC MCU families did not have a tstf instruction, but as we shall see in Table 5.4, the PIC18 family has a tstfsz (TeST File and Skip if Zero) instruction. As the btfss instruction can be used to test bit 7 (i.e. for negative) and skip, this use of movf is no longer as common as it was.

Given the property of most instructions acting on a File of specifying either the same File or the Working register as the destination, then a Move operation can be considered an implicit part of such instructions. As an example; for some situations to increment the contents of a File and then move it to WREG could be coded either as:

```
incf h'022',f,0 ; Increment File h'022's contents
movf h'022',w,0 ; and copy it into W
```

or

```
incf h'022',w,0 ; Copy File h'22's contents plus 1 to W
```

Of course the latter does not actually change the state of the File.

lfsr

1110111000	ii	LLLL

11110000	LLLLLLLL

The Load File Select Register double-word instruction is able to initialize any one of the three (ii) FSR pointer registers with a 12-bit literal—see Fig. 5.1. For example, `lfsr 1,h'422'` initializes **FSR1** to h'0422'; or FSR1H to h'04' and FSR1L to h'22'.

movff

1100	ssssssssssss

1111	dddddddddddd

The MOVe File to File instruction is unique in that it specifies two completely different Files, one holding the source datum and one the destination. A full 12-bit address is carried by this double-word instruction, and thus `movff` treats the Data store as a linear (that is non-banked) array—as described on p. 84.

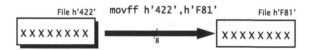

`movff` is especially useful for copying data in between SPRs and general-purpose storage. The diagram above shows a byte being relocated from File h'422' to Port B at File h'F81'. However, due to problems the interrupt service logic has with double-word instructions,the target should not be PCL nor any of the TOS (Top Of Stack) registers if interrupts are enabled.

swapf

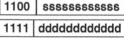

001110	d	a	FFFFFFFF

swapf interchanges the top and bottom 4-bit nybble in a File and places the outcome either back in the File or in the Working register. For instance, `swapf h'022',w,0`:

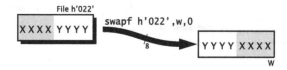

The swapf instruction is useful where the two nybbles are used to hold BCD digits. It is sometimes used as an alternative way of copying the contents of a File into WREG. Unlike movf, the Status flags are not altered with this instruction. The downside of this transparent equivalent is of course the two nybbles are swapped around in this process; but if required they can be unswapped by swapping a second time; i.e. swapf WREG, f—see p. 195.

Arithmetic Instructions

The base- and mid-range PIC MCU processors can do little more arithmetic than unsigned adding and subtracting byte operations. The enhanced-range family can do byte signed addition and subtraction, with and without carry/borrow and unsigned multiplication. Clearing, incremention and decrementation, negation setting and individual bit operations complete the repertoire. See Table 5.2.

Addition

Three addition instructions add two bytes together, generating a carry-out. Also in this category is an instruction to facilitate BCD addition.

addlw

This instruction allows the programmer to add an 8-bit *constant* (literal) to WREG; for instance, addlw b'10101010':

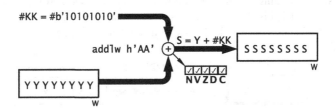

addwf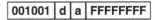

Adds a *variable* in the Data store to the byte constants of WREG. Unlike addlw, the destination of the outcome can be specified to be either the Working register or the original File. For instance, addwf h'026',f,0:

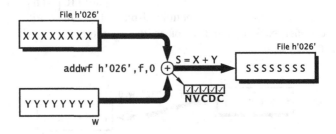

Table 5.2 Arithmetic

Operation		Mnemonic		Flags					Description
				N	**OV**	**Z**	**DC**	**C**	
Add									Binary addition
	Literal to W	addlw	k	✓	✓	✓	✓	✓	[W] <- [W] + #kk
	W to File	addwf	f,d,a	✓	✓	✓	✓	✓	[d] <- [W] + [f]
	W + C to File	addwfc	f,d,a	✓	✓	✓	✓	✓	[d] <- [W] + C + [f]
Bit twiddle									Alters a single bit
	Clear	bcf	f,n,a	•	•	•	•	•	[f_n] <- 0
	Set	bsf	f,n,a	•	•	•	•	•	[f_n] <- #1
	Toggle	btg	f,n,a	•	•	•	•	•	[f_n] <- $\overline{f_n}$
Clear									Zeroes destination byte
	File	clrf	f,a	•	•	✓	•	•	[f] <- #00
Decimal Adjust									Corrects addition of packed
	W	daw		•	•	•	✓	✓	BCD bytes
Decrement									Subtract one
	File	decf	f,d,a	✓	✓	✓	✓	✓	[f] <- [f] - #01
Increment									Add one
	File	incf	f,d,a	✓	✓	✓	✓	✓	[f] <- [f] + #01
Multiply									Unsigned 8 × 8 multiply
	Literal with W	mullw	k	•	•	•	•	•	[PRODH:L] <- [W] x #kk
	W with File	mulwf	f,a	•	•	•	•	•	[PRODH:L] <- [W] x [f]
Negate									2's complement sign change
	File	negf	f,a	✓	✓	✓	✓	✓	[f] <- −1[f]
Set									Sets all bits in a File to 1
	File	setf	f,a	•	•	•	•	•	[f] <- #h'FF'
Subtract									Binary subtraction
	W from literal	sublw	k	✓	✓	✓	✓	✓	[W] <- #kk - [W]
	W from File	subwf	f,d,a	✓	✓	✓	✓	✓	[d] <- [f] - [W]
	F − W − B	subwfb	f,d,a	✓	✓	✓	✓	✓	[d] <- [f] - [W] - $\overline{C}$
	W − F − B	subfwb	f,d,a	✓	✓	✓	✓	✓	[d] <- [W] - [f] - $\overline{C}$

#0	Single zero bit	#1	Single one bit
#00	Zero byte	#01	Byte h'01'
#kk	8-bit constant	n	3-bit specifier 0–7
f_n	Bit n of File		

addwfc | 001000 | d | a | FFFFFFFF |

Like all arithmetic operations, data are processed in 8-bit chunks. In order to facilitate multi-byte operations, instructions must be able to have regard to the carry/borrow-out generated by a previous instruction. The addwfc instruction is an extension to addwf but also adds the prior state of the **C** flag to the outcome. For instance, addwfc h'030',f,0:

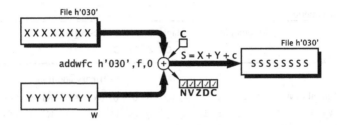

In order to illustrate addition of any length, we need to code a sequence of byte-sized additions which start from the least- and work their way up to the most-significant digit, with any carry from the nth digit being added into the $(n + 1)$th summation. The least significant addition has a presumed carry-in of 0 and carry-out from the most significant becomes the highest bit of the outcome. For instance, h'FF FF' + h'FF' = h'1 00 FE' ($65,535 + 255 = 65,790$).

To illustrate this process we will write a program to add an unsigned 16-bit number (the addend) to a 16-bit unsigned augend to give a 17-bit sum. The augend from Fig. 5.9 is located in Access RAM in the two locations File h'020' (low byte) and File h'021' (high byte). The addend is in File h'023:22' and the outcome is in File h'032:31:30'.

Given that we need to implement this process as a sequence of steps executable by byte instructions, then we need to produce a task listing.

1. Add the low byte of the augend to the low byte of the addend, generating the low byte of the sum and carry-out C1; Fig. 5.10(a).
2. Add with the carry C1 the high bytes of the augend and addend to give the high byte of the sum and a new carry-out C2; Fig. 5.10(b).
3. The upper byte of the sum is the last carry-out C2; either 0 or 1; Fig. 5.10(c).

As this is the first program of any substance in this chapter, a detailed visualization, such as shown in Fig. 5.10, has been given. For most instances, detail at this level is not helpful, and a more abstract flow chart can augment a task list.

In the listing of Program 5.4 the three tasks are identified by an appropriate comment.

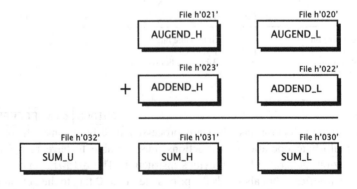

Fig. 5.9 The process

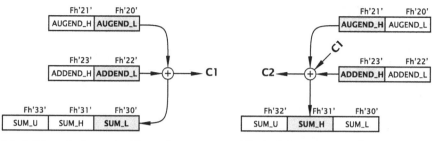

(a) Adding the least-significant bytes   (b) And the most-significant byte

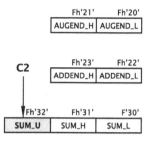

(c) The most-significant sum byte is the last carry-out

Fig. 5.10 Visualisation of the task

Preamble

All data is named using the equ directive. As discussed in p. 85, meaningful names rather than raw File addresses makes for a more readable program, with less chance for error and facilities debugging.

Task 1

The lower byte of the augend is loaded into WREG, added to the low byte of the addend and the outcome byte stored in memory as the lower byte of the sum. The **C** flag is set as appropriate by the addwf instruction and (fortunately) is not altered by the following movement instructions.

Task 2

The higher byte of the augend is fetched into WREG. It is then added with the previous (unaltered) state of the **C** flag to the higher byte of the Addend. The outcome is then copied into the high byte of the sum.

Task 3

Both the contents of the upper byte of the Sum and Working register are cleared; neither of which process affects the **C** flag. The instruction addwfc SUM_U, f, 0 thus effectively puts C_2 into the upper byte of Sum.

One point to reinforce in this program, is to ensure any intervening instructions between the carry-out/borrow-out and carry-in/borrow-in operations do not alter the **C** flag. In our case, movwf, movlw and clrf are **C**-flag neutral.

Program 5.4 The double-precision add program

```
AUGEND_L equ     h'020'        ; Name the two augend Files
AUGEND_H equ     h'021'
ADDEND_L equ     h'022'        ; Name the two addend Files
ADDEND_H equ     h'023'
SUM_L    equ     h'030'        ; Name the three sum Files
SUM_H    equ     h'031'
SUM_U    equ     h'032'


; Task1 ------------------------------------------------------
DP_ADD   movf    AUGEND_L,w,0 ; Get low byte of Augend into W
         addwf   ADDEND_L,w,0 ; Add low byte Addend, result in W
         movwf   SUM_L,0      ; and put away as low byte Sum

; Task 2 -----------------------------------------------------
         movf    AUGEND_H,w,0 ; Get high byte of Augend
         addwfc  ADDEND_H,w,0 ; Add high byte of Addend with Cin
         movwf   SUM_H,0      ; Put away as mid byte Sum
; Task 3 -----------------------------------------------------
         clrf    SUM_U,0      ; Zero upper sum byte (C unchanged)
         movlw   0            ; Zero Working reg. (C unchanged)
         addwfc  SUM_U,f,0    ; & add Cin, giving upper Sum byte

         .....   .....        ; Continue with next routine
```

---

daw   `0000000000001111`

An 8-bit File can be used to store two BCD digits, each decimal digit being encoded in a 4-bit nybble from b'0000' through b'1001' (0...9)—see p. 6. When normal binary addition is carried out on such a datum, a correction needs to be made to account for the six illegal combinations b'1010' through b'1111'. We have already designed a program to implement a BCD incrementation in Example 4.5 on p. 91. Essentially what we did in Program 4.1 was to check each nybble in turn and if it was b'1010' add six, to account for the unused combinations.

In the more general case of addition, the test before conditionally adding six needs to be expanded somewhat. The examination is if the nybble is any of the six unwanted patterns, or else a nybble carry-out (a half carry from the lower nybble or full carry from the higher nybble) has occurred. For instance, using one of the PIC MCU's Addition instructions to add h'18' + h'09' = h'21'. The low nybble digit 1 is perfectly legitimate, but the **DC** flag will be set to signify a half carry. We then need to add h'06' to give the corrected BCD value of h'27'.

The instruction daw automatically adjusts the contents of the Working register from an earlier addition of two packed BCD nybbles to give a correct packed BCD result. It implements the following algorithm.

- IF the lower nybble in WREG is greater than 9 or **DC** == 1 THEN add h'06' to the outcome.
- IF the higher nybble in WREG is greater than 9 or **C** == 1 THEN add h'60' to the outcome.

Both **DC** and **C** flags are altered as appropriate.

To illustrate the usage of this instruction, the following code fragment replaces that of Program 4.1.

```
BCD       equ    h'020'   ; The packed BCD datum is in File h'020'
BCD_INC incf   BCD,w,0  ; Binary incremented number put in W
          daw             ; Correct to BCD format
          movwf BCD,0    ; Put it back in Access RAM @ File h'020'
```

Subtraction

The four Subtract instructions mirror the Addition instructions, but with one important difference. Subtraction is not commutative; that is the order of the operands are important. For instance; 8–6 is not the same as 6–8.

sublw | 00001000 | LLLLLLLL |

This instruction subtracts the byte datum in WREG from a constant; that is $[L] - [W]$. This is a prime source of error as a common expectation is that this instruction should subtract the literal from WREG, rather than the other way around. For instance, if the contents of WREG were, say, h'64' (d'100'), then the instruction sublw 1 instead of subtracting 1 will give $1 - h'64' = h'9D'$, which is decimal 157 (actually in 2's complement form $-h'63'$). As an alternative, consider addlw h'FF'. This will give in our example $h'64 + FF = (1)63'$ (decimal 99). If we ignore the carry for the moment, the 8-bit outcome in WREG is one less than the original contents. Of course, knowing that h'FF' is the 2's complement of -1 then our instruction is really addlw -1, which makes more sense in our context.

Thus if the intention is to subtract a constant from the contents of WREG, then it is preferable to use the addlw instruction with the 2's complement of the literal. The assembler is happy to convert negative numbers to 2's complement equivalents; for instance, addlw -6 instead of addlw h'FA'.

subwf | 010111 | d | a | FFFFFFFF |

This SUBtract Working register from File($[F] - [W]$) instruction subtracts the byte contents of WREG from a variable in the Data store. Data may be either 2's complement signed or unsigned. As usual, the destination of the outcome can be specified to be either the Working register or the original File. For instance, subwf h'26',f,0.

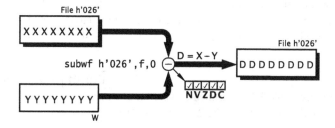

Program 5.5 The double-precision subtraction program

```
MINUEND_L      equ  h'020'    ; Name the two Minuend Files
MINUEND_H      equ  h'021'
SUBTRAHEND_L equ  h'022'     ; Name the two Subtrahend Files
SUBTRAHEND_H equ  h'023'
DIFFERENCE_L equ  h'030'     ; Name the two Difference Files
DIFFERENCE_H equ  h'031'

; Task1: First subtract the least-significant nybbles --------
DP_SUB movf   SUBTRAHEND_L,w,0; Get low byte of SUBTRAHEND
       subwf  MINUEND_L,w,0   ; subtract from MINUEND
       movwf  DIFFERENCE_L,0  ; & put away as lo byte Dif

; Task 2: Now subtract the highest nybble with borrow ---------
       movf   SUBTRAHEND_H,w,0; Get hi byte of SUBTRAHEND THEN
       subwfb MINUEND_H,w,0   ; sub from MINUEND with borrow-in
       movwf  DIFFERENCE_H,0  ; Put away as mid byte Difference

       .....  .....            ; Continue with next routine
```

As we discussed on p. 78 and Example 4.2 on p. 89, the **C** flag acts as the *complement* of the borrow-out after a Subtract instruction. Forgetting this complement is a fruitful source of programming errors!

subwfb | 010110 | d | a | FFFFFFFF |

This is an extension of subwf, which takes into account the state of the borrow-in state held in the **C** flag ($\overline{\text{borrow}}$). Here the outcome is the difference between the byte in a File and the sum of WREG and the complement of **C**; i.e. $[F] - ([W] + \overline{C})$.

In a similar manner to addwfc, this instruction is used for multiple-byte subtraction. As an example, consider subtracting an unsigned 16-bit Subtrahend from a like-sized Augend to give the Difference. As the quantities are magnitude only, any difference will be smaller than the originating quantities, so unlike Fig. 5.9, we only need two bytes to hold the outcome.

We see from Program 5.5 that the subtraction of the lower column is implemented using the plain subwf instruction. In using subwfb for the next significant nybbles, the borrow-out from the previous column is added in. As in the case of Program 5.4 any intermediate instructions do not alter the **C** flag.

The same code also works for 2's complement signed quantities. However, it is possible that the difference may be larger than either the Augend or Subtrahend. For instance, $+56 - (-27) = +83$. In this case overflow into the sign bit may occur and we will need a 3-byte Difference space. Program 5.9 shows the coding for this situation.

subfwb | 010101 | d | a | FFFFFFFF |

SUBtract File from WREG is a truncated version of subwfb, where the datum in a File is subtracted from that in WREG, i.e. $[W] - ([F] + \overline{C})$.

Single-Operand Instructions

The arithmetic operations of clearing, setting, incrementation and decrementation on a target File are covered in this sub-category.

`incf` | 001010 | d | a | FFFFFFFF |

The contents of the specified File plus one is placed either back in the File or else in WREG. In the latter case the original state of the File remains unchanged. For example, `incf h'26',0`:

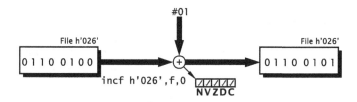

This instruction affects all the status flags in the same way as an Addition instruction[3] which makes it easy to implement multiple-precision increments. For example, if we wish to increment a 3-byte datum stored thus

Fh'22'	Fh'21'	Fh'20'
UPPER	HIGHER	LOWER

, then the low byte is incremented and if this generates a carry (or is zero) then the High byte is incremented, and so on. Thus:

```
      incf  LOWER,f,0   ; Add one
      bnc   NEXT        ; IF No Carry THEN branch to exit

      incf  HIGHER,f,0  ; ELSE increment next higher byte
      bnc   NEXT        ; IF No Carry THEN break out

      incf  UPPER,f,0   ; ELSE increment next significant byte

NEXT  ....  .....       ; Next code
```

`decf` | 000001 | d | a | FFFFFFFF |

This is the counterpart of `incf`, but subtracts one from the contents of the target File. Like `incf` all status flags are altered appropriately. Thus to decrement our 3-byte datum we have:

---

[3]In earlier families this instruction only activated the **Z** flag; see Chap. 5 of S.J. Katzen's *The Quintessential PIC® Microcontroller*, Springer.

```
      decf  LOWER,f,0   ; Subtract one
      bc    NEXT        ; IF No borrow THEN branch to exit

      decf  HIGHER,f,0  ; ELSE decrement next higher byte
      bc    NEXT        ; IF No borrow THEN break out

      decf  UPPER,f,0   ; ELSE decrement next significant byte

NEXT  ....  .....       ; Next code
```

Note that the breakout branch bc is taken when the **C** flag is 1. As the **C** flag is the complement of the borrow-out state, this indicates no borrow.

clrf | 0110101 | a | FFFFFFFF |

The contents of any File can be directly zeroed using this instruction—see Program 5.1 By targeting the Working register, we can also zero **WREG** (i.e. clrf WREG, 0) instead of using the specific instruction clrw available to earlier families.

setf | 0110100 | a | FFFFFFFF |

The contents of the specified File are set to b'11111111'; for example, setf h'026', 0:

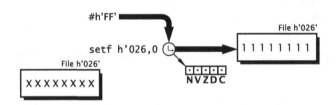

As an example of the use of this instruction, consider we have a double-byte 2's complement datum which we wish to extend to 3-bytes thus | EXTENSION | HIGHER | LOWER | (Fh'22', Fh'21', Fh'20'). From p. 11 we see that we need to extend the sign bit state to higher bytes. Thus if bit 7 of the Higher byte is 0 then the Extension byte is to be all zeros and otherwise all ones. For instance, if our datum is b'0,0000001 00101100' (d'300') then we want b'0,0000000 00000001 00101100'. If it was b'1,1111110 11010100' (−300) then the 24-bit version is b'1,1111111 11111110 11010100'. A simple routine to implement this would be:

```
      clrf   EXTENSION,0  ; Start by clearing it
      btfsc  HIGHER,7,0   ; IF bit 7 of High byte is 0 THEN skip
      setf   EXTENSION,0  ; ELSE make it all ones
```

Actually a `decf` `EXTENSION,f,0` instruction would do the same thing in this case ($00 - 1 = \text{h'FF'}$).

Bit Twiddling

Being able to either Clear, Set or Toggle any *individual* bit in any File is important, especially to manipulate the settings in the various SFRs controlling the processor and peripheral devices, including the physical state of a pin. The general machine structure of these bit twiddling (sometimes colloquially known as bit bashing or banging) instructions was discussed on p. 108.

All these instructions, along with others (such as `incf`) that appear to modify the contents of data in memory *in situ* actually transfer the byte into a temporary register, process the datum (e.g. set a single bit, `bsf`) using the ALU and then transfer the complete byte back to the Data store; all within a single instruction cycle. Sometimes this **read–modify–write** action can cause unintended side effects; see p. 344 for an example.

bcf | 1001 | NNN | a | FFFFFFFF |

This instruction enables the programmer to clear any *one* of the eight bits in the specified File. For example, to zero bit 0 of File h'FD8' without altering any other bit (actually this is the **C** flag in the Status register):

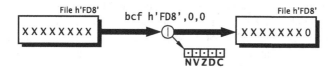

bsf | 1000 | NNN | a | FFFFFFFF |

This is similar to `bcf` but the targeted bit is set to 1. For instance, to set bit 5 of File h'026' we have `bsf h'026,5,0` (Bit Set 5 in File h'026' in Access RAM).

btg | 0111 | NNN | a | FFFFFFFF |

This instruction lets the program flip over the state of one bit in any File. Thus, to toggle bit 7 of File h'030' we have `btg h'030',7,0`.

Multiplication

Two Multiply instructions implement an 8×8 unsigned multiplication with the resulting 16-bit product being placed in the two SFRs PRODH and PRODL in a single instruction cycle. Multiplication of signed and larger data can be implemented as a series of 8×8 multiplications together with addition of partial products.

mullw | 00001101 | LLLLLLLL |

This instruction multiplies an unsigned byte in the Working register by an 8-bit constant. For example if the contents of WREG were h'E2' (d'114'), then the instruction `mullw h'0A'` will multiply it by ten and the outcome of h'AD08' (or d'1140') will appear in the SFR pair PRODH:PRODL after an execution time of one instruction cycle.

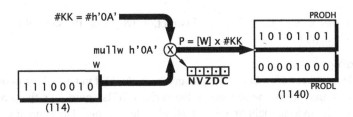

mulwf `0000001 a FFFFFFFF`

mulwf multiplies an unsigned byte in the Data store with a like datum in the Working register. Again the 16-bit outcome is placed in PRODH:PRODL. Neither the contents of the File or WREG are altered and the status flags remain unchanged by this instruction.

As an example showing the use of this instruction, consider that the 2's complement signed bytes in File h'020', called NUM_1, and File h'021', called NUM_2, are to be multiplied together to give a 16-bit signed product in PRODH:PRODL. If we simply multiply the two signed numbers together then we will get the wrong outcome. For instance, $+85 \times -5$ or 0, 1010110 × 1, 1111011 will give 0, 101010001010010, which appears to be +22,098 rather than −430. From this instance, we see that we will have to apply a correction factor if any of the operands are negative to give the outcome product the correct sign.

To determine this we have to look carefully at the form of a negative 2's complement byte number NUM; that is if its bit 7 is 1. From p. 9 we see that the 2's complement of −NUM is really 256 − NUM; or in general 2^n − NUM for an *n*-bit number. It turns out that inverting and adding one is an easy way to do this subtraction, but the former representation is more relevant to our problem here. On this basis if, say, NUM_1 is negative then when we use unsigned multiplication we have $(256 - \text{NUM\_1}) \times \text{NUM\_2} = (256 \times \text{NUM\_2}) - (\text{NUM\_1} \times \text{NUM\_2})$. If we thus take away $256 \times \text{NUM\_2}$ from the outcome, we will be left with $-\text{NUM\_1} \times \text{NUM\_2}$. However, a multiplication by 256 is the same as a shift left eight places, symbolized as <<8, and so subtracting NUM_2 from the high byte of the product will implement our correction factor. In a similar way we can check the sign of NUM_2 and if this shows negative, then subtract NUM_1 from PRODH. If both operands are negative then both corrections should be made.

Based on this approach we have as our task list.

- Multiply the two operands together as if unsigned quantities.
- Check sign of the first operand. IF negative subtract the *second* operand from the high byte of the product.
- Check sign of the second operand. IF negative subtract the *first* operand from the high byte of the product.

which is diagrammatically shown Fig. 5.11 with the example $+86 \times -5$, giving −430.

The implementation of this algorithm is shown in Program 5.6. mulwf is used to implement the first task. The btfsc instruction is then used to check the state of bit 7 of each of the two operands. If this bit is 0 (i.e. positive) then the following

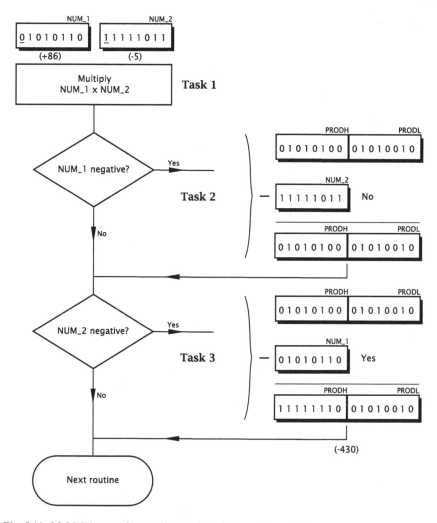

Fig. 5.11 Multiplying two 2's complement byte datums with `mulwf`

subtraction correction is skipped over, otherwise (negative) a correction is made. Thus between zero and two correction subtractions may be made. Irrespective of the sign of the operands, the program will always execute in six instruction cycles.

Logic and Shifting Instructions
All four basic logic operations of NOT, AND, Inclusive-OR and eXclusive-OR are provided as well as shifting, as shown in Table 5.3.

NOT
The NOT logic function of Fig. 1.1 on p. 12 inverts (1's complement) the logic state of the input. This differs from the Negate instruction which is invert plus one.

Program 5.6 Implementing a 8 × 8 2's complement multiplication

```
NUM_1     equ   h'020'    ; Signed Number 1
NUM_2     equ   h'021'    ; Signed Number 2
PRODL     equ   h'FF3'    ; Low byte of Product
PRODH     equ   h'FF4'    ; High byte of Product

; Task 1: Multiply the two operands ----------------------------
SIGN_MUL movf   NUM_1,w,0 ; Get Number 1 from memory into WREG
         mulwf  NUM_2,0   ; Multiply them

; Task 2: IF Number 2 is negative take away Number 1 >> 8 ----
         btfsc  NUM_2,7,0 ; Test NUM_2's sign bit, skip if +ve
         subwf  PRODH,f,0 ; ELSE take away NUM_1 from PRODH

; Task 3: IF Number 1 is negative take away Number 1 >> 8 ----
         movf   NUM_2,w,0 ; Now get Number 2 into WREG
         btfsc  NUM_1,7,0 ; Test NUM_1's sign bit, skip if +ve
         subwf  PRODH,f,0 ; ELSE take away NUM_2 from PRODH

NEXT     .....  ........  ; Next routine
```

comf | 000111 | d | a | FFFFFFFF |

The logic state of any specified File can be inverted. As an example, the instruction comf h'026',f,0 complements the contents of File h'026' in Access RAM: In the normal way the outcome can be placed either in the source File or in WREG, with the original contents being unchanged; for instance:

There is no specific instruction to complement the state of the Working register, rather use WREG as the target; i.e. comf WREG,f,0. As an alternative, the contents of WREG can be subtracted from b'11111111' to give the 1's complement; i.e., sublw h'FF'.

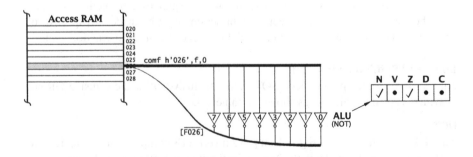

Table 5.3 Logic instructions

Operation		Mnemonic	Flags					Description	
			N	**OV**	**Z**	**DC**	**C**		
AND								Logic bitwise AND	
	Literal to W	andlw k	√	•	√	•	•	[W] <- [W] · #kk	
	W to File	andwf f,d,a	√	•	√	•	•	[d] <- [W] · [f]	
Complement								Invert or NOT	
	File	comf f,d,a	√	•	√	•	•	[d] <- $\overline{[f]}$	
Inclusive-OR								Logic bitwise Inclusive-OR	
	Literal to W	iorlw k	√	•	√	•	•	[W] <- [W] + #kk	
	W to File	iorwf f,d,a	√	•	√	•	•	[d] <- [W] + [f]	
eXclusive-OR								Logic bitwise eXclusive-OR	
	Literal to W	xorlw k	√	•	√	•	•	[W] <- [W] ⊕ #kk	
	W to File	xorwf f,d,a	√	•	√	•	•	[d] <- [W] ⊕ [f]	
Rotate File thru **C**								Circular shift via Carry	
	Left	rlcf f,d,a	√	•	√	•	b_7		
	Right	rrcf f,d,a	√	•	√	•	b_0		
Rotate File								Circular shift	
	Left	rlncf f,d,a	√	•	√	•	•		
	Right	rrncf f,d,a	√	•	√	•	•		

·	Boolean bitwise AND	+	Boolean bitwise Inclusive-OR
⊕	Boolean bitwise eXclusive-OR	$\overline{[f]}$	Bitwise inverse of the File contents

AND

From Fig. 1.2 on p. 13 you will recall the following relationship:

- ANDing a bit variable with 0 *always* gives a 0 output.
- ANDing a bit variable with 1 yields an unchanged logic state.

On this basis we can zero a selected group of bits in a datum byte by ANDing with the appropriate pattern.

In the same manner, ANDing a datum with a test pattern to clear all unwanted bits can also be used to check if a selected group of bits in the datum is zero. If this is true, the overall result will be zero and the **Z** flag will be set accordingly.

andwf

000101	d	a	FFFFFFFF

The andwf instruction bitwise ANDs the contents of WREG together with the contents of any File, with the outcome being placed either in that same File or in WREG. For example, to AND each bit of WREG with each corresponding bit in File h'026', with the outcome being put back in File h'026' we have:

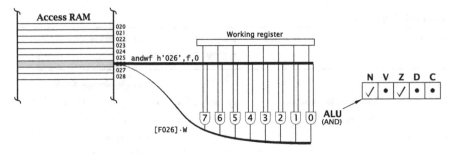

For instance, to clear the upper six bits of File h'026':

```
movlw  b'00000011'  ; Mask pattern in Working register
andwf  h'026',f,0   ; AND it with contents of File h'026'
```

The alternative would be to use the bcf instruction six times.

To see how the AND function can be used to test for a set of zero bits, consider a controller for a washing machine where the eight switches in the control panel are read via Port B, that is, File h'F81'. It is desired to jump to a routine to implement a Fast Wash if the switches connected to bits 7 and 6 are both zero; that is, the GO and FAST switches are closed (logic 0). Here is how it could be done:

```
movlw  b'11000000' ; The test mask
andwf  PORTB,w     ; ANDed with PORTB
bz     FAST_WASH   ; Branch if Zero to routine FAST_WASH
.....  .....       ; Next routine
```

By ANDing the contents of File h'F81' with b'11000000', the lower six bits will be cleared. The overall outcome in WREG will be all zero only if *both* bits 7 and 6 of Port B are 0 before this action. In this case the **Z** flag will be set and the following Branch if Zero instruction will be taken and the program will transfer to the instruction located at the label FAST_WASH. If a *single* bit in a File is being tested for zero then it is more efficient to use btfsc to directly check that bit—see p. 138.

andlw   00001011 | LLLLLLLL

The contents of WREG can be bitwise ANDed with a byte literal. For instance:

$$10001110\;\big|_W \quad \overset{\text{andlw h'0F'}}{\sim\!\sim} \quad \mathbf{0000}1110\;\big|_W$$

where the high nybble of WREG is zeroed and the low nybble is left untouched.

Inclusive-OR

From Fig. 1.3 on p. 13 you will recall the following relationship:

- IORing a bit variable with 0 yields an unchanged logic state.
- IORing a bit variable with 1 *always* gives a 1 output.

On this basis we can set a selected group of bits in a datum byte to 1 by IORing with a suitable pattern.

iorwf | 000100 | d | a | FFFFFFFF |

In a similar manner to andwf, the contents of any File can be bitwise Inclusive-ORed with the contents of WREG. Thus to IOR each bit in WREG with its corresponding bit in File h'026', with the outcome being put back in the File, we have:

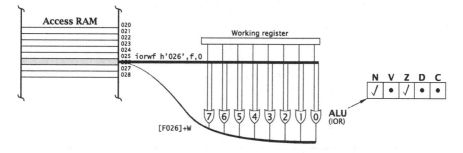

For instance, to set to 1 the top seven bits in File h'096' in Banked RAM we have:

```
movlw  b'11111110'; The mask byte
iorwf  h'096',f,1 ; Set top 7 bits to 1, lowest bit unchanged
```

where we are assuming that the BSR is set to h'00' (its Power-on Reset state) for Bank 0.

iorlw | 00001001 | LLLLLLLL |

The contents of WREG can be bitwise IORed with a byte literal. For instance, to set the lower two bits of the Working register to 1:

$$\boxed{1000\,1110}_{\text{W}} \quad \overset{\text{iorlw 03}}{\leadsto} \quad \boxed{1000\,11\mathbf{11}}_{\text{W}}$$

eXclusive-OR

From Fig. 1.4 on p. 14 you will recall the following relationship.

- XORing a bit variable with 0 yields an *unchanged* logic level.
- XORing a bit variable with 1 inverts or *toggles* the state of the input logic level.

Another useful property of the XOR logic operator is as a logic differentiator. A close inspection of the truth table of Fig. 1.4(a) on p. 14 shows that the output of an XOR gate is 1 if the two input logic levels are different and 0 if they are the same. Thus bitwise XORing two bytes together will produce a byte output with 0 in locations where the two input bits are the *same* and a 1 where they *differ*.

xorwf | 000110 | d | a | FFFFFFFF |

The contents of any File can be bitwise eXclusively-ORed with the contents of WREG. Thus to XOR each bit in WREG with its corresponding bit in File h'026' with the outcome being put back in the File:

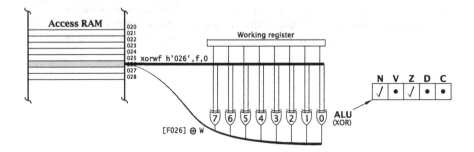

For example, to toggle the top two bits of File h'036' in Access RAM we have:

```
movlw  b'11000000'    ; The mask byte
xorwf  h'36',f,0      ; Toggle top two bits only of File h'036'
```

As an example showing the use of XOR to isolate *changes* between two bit patterns, consider a program routine that continually monitors Port B, to which has been connected eight switches as part of the control panel of a washing machine. The routine waits until a switch is moved.

```
START   movf   PORTB,w,0   ; Get initial state of switches
        movwf  h'020',0    ; Put away at File h'020'

S_LOOP  movf   PORTB,w,0   ; Sample switches
        xorwf  h'020',w,0  ; Check for alterations from original
        bnz    NEXT        ; Skip out IF Z flag clear (non zero)
        bra    S_LOOP      ; ELSE Branch back and check again

NEXT    .....  ......      ; Next routine
```

Two possible scenarios are:

| 10011110 | | xorwf h'020',w,0 | | | | | |
| File h'020' | | | 10011110 | w = | 00000000 | w | Z = 1 |

| 10011110 | | xorwf h'020',w,0 | | | | | |
| File h'020' | | | 10001110 | w = | 00010000 | w | Z = 0 |

The outcome in WREG reflects any changes. In the first case there are no differences between the latest sample and the original switch settings put away in File h'020'. In the second situation, Switch 4 has just been thrown from 1 to 0.

You can determine which switch changed by shifting the outcome (the change byte) right, counting until the residue is zero; see Fig. 5.13. You can also determine the type of change (0 → 1 or 1 → 0) by ANDing the change byte to the original switch settings in File h'020', i.e., andwf h'020',w,0. If the outcome at bit 4 is a 0, then the original state must have been 0 and therefore the change must have been 0 → 1, and vice versa. In our example, bit 4 has gone 1 → 0.

$$\boxed{10011110}_{\text{File h'020'}} \xrightarrow{\text{andwf h'020',w,0}} \boxed{00010000}_{\text{W}} = \boxed{00010000}_{\text{W}} \; Z = 0$$

xorlw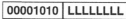

The contents of WREG can be bitwise XORed with a byte literal. For instance, to invert the lower nybble in WREG:

$$\boxed{10001110}_{\text{W}} \xrightarrow{\text{xorlw h'0F'}} \boxed{1000\mathit{0001}}_{\text{W}}$$

Shifting

Shifting data left or right is a fundamental operation found in all digital systems. We saw in Fig. 2.22 on p. 36 how this could be done in hardware. Without exception, all MPU/MCU devices have ALUs that allow various combinations of Shift Right and Shift Left operations to be performed.

The enhanced-range PIC MCU family has two pairs of instructions in this category. Each pair shifts the contents of any File one place either right (>>, from most to least-significant) or left (<<). They differ in their interaction with the **C** flag.

rrcf $\boxed{001100 \mid d \mid a \mid \text{FFFFFFFF}}$

Rotate Right File through Carry shifts the byte contents of the specified File once right, with the incoming bit coming from the **C** flag, which is simultaneously replenished with the outgoing bit. This circular action is emphasised in Fig. 5.12. In essence, this is a Shift Right function but with the Carry flag acting as a sort of bit 8 buffer.

With this diagram in mind the programmer can do a plain shift right with a zero coming in (as in Fig. 2.22) by first clearing the Carry bit before rotating; for instance, for File h'030':

```
bcf   STATUS,C,0 ; Zero the Carry bit in the Status register
rrcf  h'030',f,0 ; Now rotate right \verb#»#
```

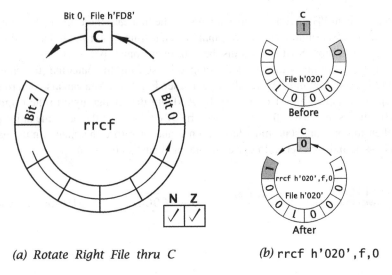

(a) Rotate Right File thru C   (b) rrcf h'020',f,0

Fig. 5.12 Rotating the contents of a File once right

One use of the shifting operation is to bitwise examine a datum. For example, assume that the state of an array of eight switches from a mobile phone has been copied into File h'026'. You are required to find the *leftmost* open switch, where you can assume that an open switch reads as 1 and a closed switch as logic 0. For instance, if the reading was:

0 SW8	0 SW7	*1* SW6	0 SW5	1 SW4	1 SW3	1 SW2	1 SW1

then the outcome in WREG should be 6 (b'00000110') for SW6.

The coding given in Program 5.7 uses the Working register as a counter. As the Carry flag is cleared each time *before* the shift, logic 0s are brought in from the left.[4] Eventually the residue will become all zeros and the process should then terminate. Thus **00010111** (1) ↝ **00001011** (2) ↝ **00000101** (3) ↝ **00000010** (4) ↝ **00000001** (5) ↝ **00000000** (6).

A task list for this problem, also shown diagrammatically in Fig. 5.13, would be:

1. Zero KEY_COUNT
2. DO shift right and increment WHILE SWITCH_PATTERN is not zero
   a. IF residue is zero THEN break
   b. Shift left SWITCH_PATTERN once
   c. Increment KEY_COUNT
3. KEY_COUNT holds the position of the leftmost open switch

---

[4]MPU/MCUs that have Logic Shift instructions always shift in 0s irrespective of the state of the **C** flag; for instance, Motorola's 1sr (Logic Shift Left).

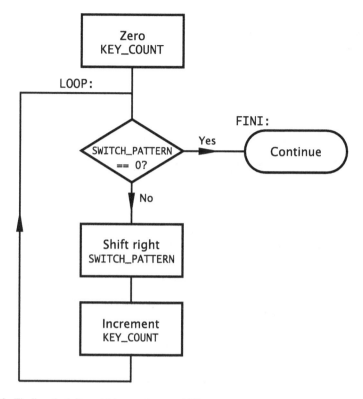

Fig. 5.13 Finding the leftmost 1 by continuous shifting

Shifting right pops out the rightmost bit into the Carry flag. Replacing bz (Branch if Zero) by bc (Branch if Carry) would determine the position of the *rightmost* bit. In many situations repetitively shifting into the Carry flag can be used to examine the data on a bit-by-bit basis. For instance, we could modify our program to total the number of set bits in the byte, as in Program 5.10.

Notice that Program 5.7 returned a zero outcome if no switch was open. If the test for zero had been done after the Shift operation, then it would not be possible to distinguish between this situation and Switch 1 alone being open. It is important to design your software for robustness, so that limiting conditions, such as this, are dealt with. Finally note that movf is used to test the File for zero by copying its contents back onto itself, as described on p. 110.

rlcf | 001101 | d | a | FFFFFFFF |

Rotate Left File through Carry is similar to rrcf but, as shown in Fig. 5.14, the shift direction is from the low to the high bit position.

As an example of the use of rlcf we note from p. 11 that we can use shifting to the left to multiply by powers of two. For instance:

```
00000110   (6)   <<
00001100   (12)  <<
00011000   (24)  <<
00110000   (48)  <<
          etc.
```

where the **C** Shift-Left operator << is used to indicate a shift left.

Program 5.7 Scanning the File looking for the highest 1

```
PATTERN    equ    h'026'    ; Pattern is in File h'026'
STATUS     equ    h'FD8'    ; The Status reg is File h'FD8'
C          equ    0         ; in which bit 0 is the C flag
WREG       equ    h'FE8'    ; The Working reg, in Access RAM

; Task 1 --------------------------------------------------------
HIGH_BIT   clrf   WREG,0    ; Zero the count

; Task 2: DO right shift & inc count, WHILE datum isn't zero
; Task 2a -------------------------------------------------------
LOOP       movf   PATTERN,f,0 ; Test for residue zero?
           bz     FINI        ; If zero THEN Branch to exit

; Task 2b -------------------------------------------------------
           bcf    STATUS,C,0  ; Carry flag (carry-in) cleared
           rrcf   PATTERN,f,0 ; Shift datum right

; Task 2C -------------------------------------------------------
           incf   WREG,f,0    ; Continue by adding one to count
           bra    LOOP        ; and do another shift

; Task 3 --------------------------------------------------------
FINI              .....  ...... ; KEY_COUNT is in W
```

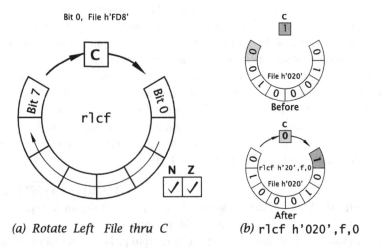

(a) *Rotate Left File thru C*   (b) rlcf h'020',f,0

Fig. 5.14 Rotating the contents of a File once left

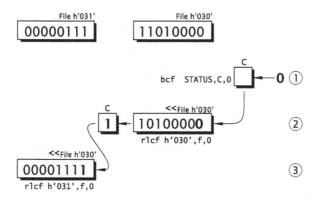

Fig. 5.15 Shifting a double-byte datum once to the left to multiply by two

To illustrate the process, assume that if we have the 16-bit number b'00000111 11010000' (which is decimal $1024 + 512 + 256 + 128 + 64 + 16 = 2000$), then this will be stored in two Files; for example:

00000111		11010000	
File h'031'		File h'030'	

After shifting once left we have:

00001111		10100000	
File h'031'		File h'030'	

which is decimal 4000 ($2048 + 1024 + 512 + 256 + 128 + 32 = 4000$).

The problem is that our `rlcf` instruction can only shift a single byte at a time. Thus we need to break this down to three steps, as illustrated in Fig. 5.15.

1. Clear the Carry flag so that we will rotate in a 0.
2. Rotate the low byte left, with b_7 being popped out into the Carry flag.
3. Rotate in the carry-out of the previous Rotate operation into the high byte.

From the diagram we see that the process is straightforward, with the carry-out from the previous File becoming the carry-in for the second File. The routine is thus:

```
bcf STATUS,C,0  ; Clear C flag; will hold the incoming bit
rlcf h'030',f,0 ; Rotate into the lo byte, MSB pops out into C
rlcf h'031',f,0 ; Rotate into the high byte
```

rrncf

010000	d	a	FFFFFFFF

rlncf

010001	d	a	FFFFFFFF

Rotate Right/Left File Not through the Carry flag is similar to `rrcf`/`rlcf` pair except that the **C** flag is not used as an intermediary—see Fig. 5.16.

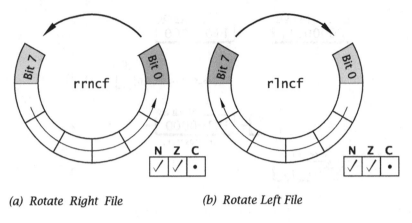

(a) Rotate Right File   (b) Rotate Left File

Fig. 5.16 The two plain Rotate instructions which do not pass through the **C** flag

Ancestor families only provided the Rotate through Carry instructions, using the mnemonics `rrf` and `rlf`, but these instructions distort the datum in that the state of the **C** flag is injected into the File on shifting. These plain Rotates illustrated here are non destructive, in that rotating n places and then back n places preserves the original datum. This can be useful if the 8-bit field is used to store several datum of smaller bit groups. For example, if File h'050' has three bit fields; arg1 = File h'050'[7:6]', arg2 = File h'050'[5:2] and arg3 = File h'050'[1:0]. The packed datum thus is organised as $\boxed{\text{arg}_1\text{arg}_1 \mid \text{arg}_2\text{arg}_2\text{arg}_2 \mid \text{arg}_3\text{arg}_3}$. It is desired to extract arg2 into the Working register right-aligned but leave the original data unchanged. The following code is one possibility.

```
rrncf   h'050',f,0    ; Rotate the packed datum right once
rrncf   h'050',w,0    ; and again aligning arg2 right in W
andlw   b'00000111'   ; Mask out arg1 and arg3 from W
rlncf   h'050',f,0    ; Rotate the packed datum left to restore
```

Program Counter Instructions

Anything other than trivial software requires an interaction with events occurring either internally (such as a timer overflowing) or externally (such as a temperature rising above a preset level)—see Fig. 3.8 on p. 59. In order to respond to environmental events, the program must be able to jump to an appropriate service routine. This entails modification of the state of the Program counter such as adding two to the PC, causing it to **skip** over one program word. Other possibilities are adding on a signed number, or offset, and thus causing a **branch** forward or backwards, or overwriting the entire contents resulting in the focus of execution to **go to** a new location in the Program store.

Whatever the mechanism of the jump, this may be **unconditional**; that is always taken. Alternatively, this jump may depend on the state of some bit in a File or input pin, or frequently the setting of a flag in the Status register.

Table 5.4 Program Counter and decision instructions

Operation	Mnemonic		Flags N OV Z DC C	Description
Absolute jump				Goto a fixed instruction
[1] Goto an instruction	goto	aaaaa	• • • • •	[PC]<-aaaaa
Branch unconditionally	bra	Offset	• • • • •	[PC]<-[PC]+Offset
No operation				Do nothing
	nop		• • • • •	[PC]<-[PC]+2
Branch conditionally				IF condition true THEN branch
Branch if Carry	bc	offset	• • • • •	IF C==1: [PC]<-[PC]+offset
Branch if No Carry	bnc	offset	• • • • •	IF C==0: [PC]<-[PC]+offset
Branch if Negative	bn	offset	• • • • •	IF N==1: [PC]<-[PC]+offset
Branch if No Carry	bnc	offset	• • • • •	IF N==0: [PC]<-[PC]+offset
Branch if OVerflow	bov	offset	• • • • •	IF OV==1: [PC]<-[PC]+offset
Branch if No OVerflow	bnov	offset	• • • • •	IF OV==0: [PC]<-[PC]+offset
Branch if Zero	bz	offset	• • • • •	IF Z==1: [PC]<-[PC]+offset
Branch if No Zero	bnz	offset	• • • • •	IF Z==0: [PC]<-[PC]+offset
Bit test and skip				Check bit in File and skip if true
Bit Test in File, Skip if 0	btfsc	f,n,a	• • • • •	PC = PC+2 IF f_n == 0
Bit Test in File, Skip if 1	btfss	f,n,a	• • • • •	PC = PC+2 IF f_n == 1
Decrement and skip				Decrement & skip if result is #00
File	decfsz f,d,a		• • • • •	d <- f--, PC-- IF [f] == #00
File	dcfsnz f,d,a		• • • • •	d <- f--, PC-- IF [f] != #00
Increment and skip				Increment & skip if result is #00
File	incfsz f,d,a		• • • • •	d <- f++, PC++ IF [f] == #00
File	infsnz f,d,a		• • • • •	d <- f++,PC++ IF [f] != #00

++	Increment contents	--	Decrement contents
==	is equivalent to	!=	Is not equivalent to
aaaaa	Absolute 20-bit instruction address	[1]	2-word instruction
Offset	11-bit signed word offset	offset	8-bit signed word offset

Unconditional Jumps

These three instructions *always* alter the state of the PC. However, used in conjunction with the Conditional instructions listed later on, they are an indispensable part of the decision making mechanism.

bra | 11010 | SNNNNNNNNNN |

BRanch Always adds on a signed 11-bit word offset S,NNNNNNNNNNN to the state of the whole PC, which then shifts execution to a new setting *relative* to the bra instruction—as described on p. 99. Actually, the PC is already pointing to the following instruction, due to pipelining—see Fig. 4.3 on p. 73. Thus the instruction bra .+8, shown to the left of Fig. 5.17, will end up nine words beyond. Notice how the assembler uses the . operator to indicate current setting of the PC.

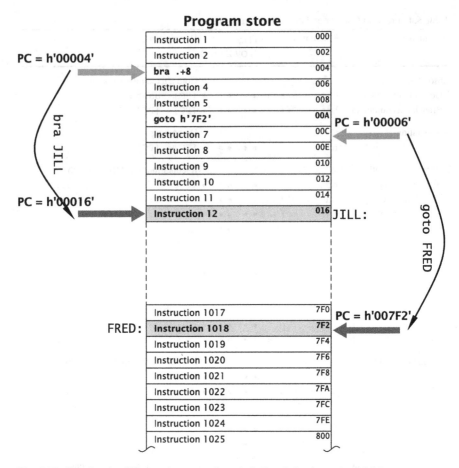

Fig. 5.17 Showing the difference between a branch (*left*) and absolute go to (*right*)

In practice the programmer will normally place a label at the destination instruction; in our example the instruction at h'00016' is labeled JILL. This will enable the assembler to work out the offset. The total range of bra is ±1024 words.

In the normal course of events the Program Counter has been incremented to h'00006' and the Instruction located here has already been fetched into the top of the Pipeline, ready to be executed in the next instruction cycle. However, when the bra .+8 instruction at the bottom of the Pipeline is executed, the computed Program store address (h'00006 + 00010 = 00016') is placed in the PC (note that 8 words is h'010' or 16 bytes). This means that the next instruction to be executed is the one at this location. To permit this to happen, Instruction 12 must be fetched down into the Pipeline, overwriting the now unwanted code. This process is known as *flushing*, and takes an extra instruction cycle to implement. Thus bra takes two instruction cycles to execute.

goto

11101111	AAAAAAAA

1111	AAAAAAAAAAAA

This instruction allows the programmer to jump to a specified instruction *anywhere* in the Program store. Unlike the `bra` instruction, a full 21-bit address is superimposed over the original contents of the Program Counter. In order to hold this address, `goto` is a 2-word instruction with the first word holding the bottom eight address bits and the second word the top 12 bits. As usual, A_0 is forced to 0. The second word is also prefixed with the op-code for a `nop` instruction, in case a jump is performed in the middle of the instruction. An example of this is shown in Fig. 5.19. In the example shown to the left of Fig. 5.17, the instruction `goto h'3F9'` is located in the Program store at location h'0000A'.

In the diagram, location h'007F2' is labeled `FRED` (the following colon is optional). Using labels rather than absolute locations is strongly recommended (see also p. 104) as the programmer does not easily know where an instruction is located in the Program store, and in any case, this location may change as the program develops.

`goto` is sometimes known as a long jump, as compared to the short-jump `bra` alternative. Compared to the latter, `goto` takes an extra word of storage. However, despite its double-word structure, it still only takes one additional instruction cycle to execute.

nop

0000000000000000

or

1111	XXXXXXXXXXXX

No OPeration does not alter the state of the system in any way, but the PC will increment as a consequence of the instruction code being fetched from the Instruction store. Thus, its sole outcome is `[PC] <- [PC] + 2`. This takes one instruction cycle, so its main use is to implement a short delay, 1 µs for a 4 MHz clock rate. For instance, to pulse Port A's pin 2 low for 2 µs and then high we have:

```
bcf    PORTA,2,0   ; Pin RA2 low
nop                ; for 2 us
nop
bsf    PORTA,2,0   ; and now high
```

with the assumption that bit 2 of Port A has been set up as an output (see p. 85) and that pin `RA0` was high before entering the routine.

The second binary coding of this instruction is found in the second word of a 2-word instruction. This is to avoid erratic behaviour if execution jumps directly into the second word of such an instruction—see Fig. 5.19.

Conditional Skips

The majority of instructions in Table 5.4 skip or branch only if the outcome of a process meets a specified criterion. This is normally signaled by the setting of a bit in a register or pin; typically a status flag. Decisions are binary, as indicated as an

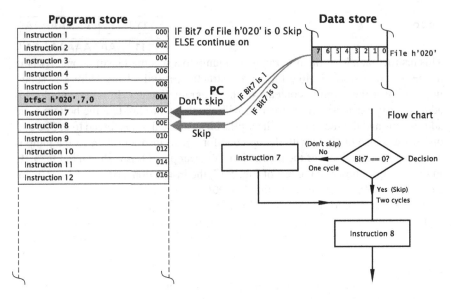

Fig. 5.18 Skipping over the next instruction whenever bit 7 of File h'020' is clear

IF-ELSE statement or ◇ symbol in a flow chart. More complex multiple pathways are coded as a series of binary decisions; for example as shown in Fig. 5.25.

btfsc | 1011 | NNN | a | FFFFFFFF |

btfss | 1010 | NNN | a | FFFFFFFF |

All binary decisions fundamentally turn on the state of a single binary bit in a File. As such, Bit Test in File and Skip if Clear/Set are all that are necessary to implement any binary decision and all the other listed Conditional instructions are simply convenient derivatives.

Figure 5.18 illustrates the situation where Instruction 6 in the Program store is btfsc h'020',7,0. The execution of this instruction checks bit 7 of File h'020' and on the basis of its state implements one of two outcomes:

1. IF bit 7 is 0 THEN skip over Instruction 7 and execute Instruction 8.
2. IF bit 7 is 1 THEN continue on as normal to Instruction 7.

Often Instruction 7 is a bra, so the program can react to the state of any bit in the Data store by branching to an appropriate routine.

When a skip occurs, the Pipeline requires to be flushed, as do all instructions that disrupt the orderly progression of the Program Counter. This means that a btfsc or btfss instruction takes one instruction cycle to execute if no skip takes place and two instruction cycles if a skip is executed. However, there are circumstances where it takes three instruction cycles if a skip is performed. As an example, if consider that we want to sample the state of pin RB7 and if high jump to a routine labeled CONTINUE. This could be implemented by using btfsc PORTB,7,0 to

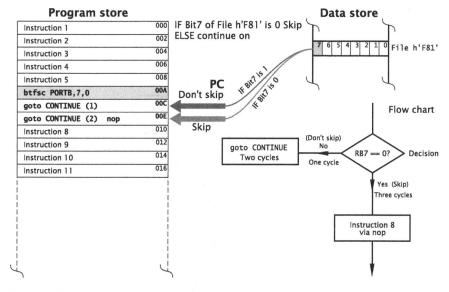

Fig. 5.19 Skipping into the middle of a double-word instruction

skip over a following `goto` instruction if RB7 were 0 otherwise with no skip the program could go to CONTINUE.

```
SAMPLE btfsc PORTB,7,0 ; Check pin RB7. Skip IF == 0 (low)
       goto CONTINUE   ; ELSE jump out to CONTINUE
       bra  SAMPLE     ; Try again waiting for high
```

As shown in Fig. 5.19, if the skip was taken (the pin being low) then the PC would land in the middle of the 2-byte `goto` instruction. As we have already seen, the second word of all double-byte instructions is coded as a pseudo `nop` instruction, so all that will happens is for the program to progress to Instruction 8, but incurring an extra instruction cycle delay. Thus the execution time for all Jump instructions is listed as 1, 2 or 3 instruction cycles. In our case, if the instruction labeled CONTINUE was closer than 1024 words away, `goto CONTINUE` should be replaced by `bra CONTINUE`.

As a matter of style, the instruction following a Skip instruction is often indented, to show that it is to be hopped over.

decfsz | 001011 | d | a | FFFFFFFF |

dcfsnz | 010011 | d | a | FFFFFFFF |

DECrement File and Skip if Zero represents an alternative way of making a decision. As a combination of the instruction pair `decf` followed by `btfss STATUS,Z`, this instruction allows the programmer to decrement the contents of any File, and if the outcome is zero, then skip over the next instruction.

Fig. 5.20 Pulse pin RA0
repeating 20 times

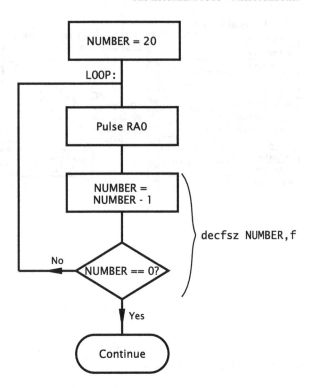

A typical use of this instruction is to count the number of passes through a loop. For example, suppose it is necessary to pulse Port A pin RA0 low 20 times. To implement this task, shown in Fig. 5.20, we have the code, assuming a 4 MHz crystal:

```
      movlw  d'20'     ; Put decimal 20 into W and copy into
      movwf  h'09F',1  ; File h'09F' (Bank 0) as a loop counter
;------------------------------------------
LOOP bcf     PORTA,0,0 ; Pin RA0 low
      nop              ; One extra cycle delay
      bsf     PORTA,0,0 ; and now high
;------------------------------------------
      decfsz h'9F',f,1 ; Count down and skip out IF zero
       bra    LOOP      ; ELSE repeat loop if not zero
      .....  ......    ; Continue
```

The original code shown between dashed lines is cocooned by the decrementing test which skips out of the loop whenever the contents of File h'09F' in Bank 0 reach zero. We are assuming that the BSR has been cleared; as it would be on Reset. Notice the assembler notation d'20' for *decimal* 20; see p. 266. This is equivalent to h'14' but more readily understood by the programmer. Incidentally, only one nop is used, as the bsf and bcf between them add an extra cycle delay to the total.

DECrement File and Skip if Not Zero is the counterpart which skips over the next instruction word if the contents of the target File is not zero. That is decf followed by btfsc STATUS,Z.

incfsz | 001111 | d | a | FFFFFFFF |

infsnz | 010010 | d | a | FFFFFFFF |

INCrement File and Skip if Zero increments rather than decrements the contents of the specified File. If this causes the contents to roll over to zero, e.g., h'FC → FD → FE → FF → 00' then the following instruction will be skipped over. It thus behaves as the instruction pair incf followed by btfss STATUS,Z. In the case of our example of Fig. 5.20, if we were to preload File h'03F' with −20 (or h'EC') and replace decfsz h'03F',f,0 by incfsz h'03F',f,0 then this will give the same outcome by counting up rather than counting down.

INCrement File and Skip if Not Zero is the counterpart which skips over the next instruction word if the contents of the target File are not zero. That is incf followed by btfsc STATUS,Z.

As an example of the use of both these instructions, consider that we wish to decrement the triplet File | ARG_U | ARG_H | ARG_L | and when the array overflows to h'00 00 00' go to the routine labeled OVERFLOW. The task list would be:

1. Increment the low byte and IF no overflow THEN break.
2. Increment the high byte and IF no overflow THEN break.
3. Increment the upper byte and IF no overflow THEN break.
4. ELSE go to OVERFLOW.

A suitable routine based on this would be:

```
        incfsz  ARG_L,f,1 ; Increment low byte & skip IF zero
        bra     NEXT      ; ELSE break out
        incfsz  ARG_H,f,1 ; Increment high byte & skip IF zero
        bra     NEXT      ; ELSE break out
        infsnz  NEXT,f,1  ; Inc. upper byte & skip IF not zero
        goto    OVERFLOW  ; ELSE go off to the specified routine

NEXT ......  ......       ; Continue after incrementation
```

tstfsz | 0110011 | a | FFFFFFFF |

TeST File and Skip if Zero behaves like a movf [FILE],f followed by btfss STATUS,Z. It enables the program to side-step if the contents of the target File is zero.

As an example, consider a bank of eight switches that are connected to Port B (see p. 85); as shown to the left of Fig. 5.21. An open switch gives a high voltage

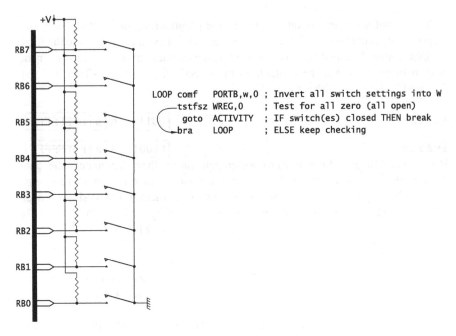

```
LOOP comf    PORTB,w,0 ; Invert all switch settings into W
     tstfsz  WREG,0    ; Test for all zero (all open)
        goto ACTIVITY  ; IF switch(es) closed THEN break
     bra     LOOP      ; ELSE keep checking
```

Fig. 5.21 Testing a bank of switches for any switch closed

(logic 1) when open, due to the pull-up resistor, and low (logic 0) when closed. We want to continually monitor the switch array until one or more switch closes, then go to a routine labeled ACTIVITY.

When all switches are open, the voltage pattern presented to the port will be b'11111111'. The code to the right of the diagram first inverts the pattern with a target of the Working register. Thus an all-open array will give the pattern b'00000000' in WREG. If this is the situation then tstfsz will skip over (actually into the middle of) the goto ACTIVITY instruction and the test will be repeated in an endless loop. Once one or more switch closes, the datum will be non-zero and the skip will not be taken. The program will then jump to the required routine.

cpfseq (ComPare File & Skip if Equal) | 0110001 | a | FFFFFFFF |

cpfsgt (ComPare File & Skip if Greater Than) | 0110010 | a | FFFFFFFF |

cpfslt (ComPare File & Skip if Less Than) | 0110000 | a | FFFFFFFF |

One of the more important operations is the *comparison* of the magnitude of two numbers. Mathematically this can be done by *subtracting* the two quantities. If we are comparing the datum in a File with the byte in the Working register then the outcome of [W]−[f] gives the actual magnitude difference between the operands. However, in most cases it is sufficient to determine the relative magnitude of the quantities, e.g., is the datum in the File greater than the datum in WREG? For unsigned numbers, this is determined by checking the state of the **C** and **Z** flags in the Status register.

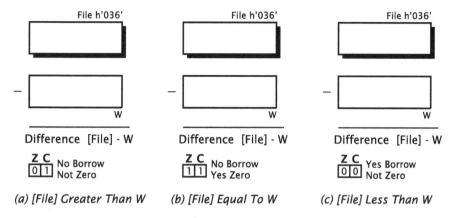

Fig. 5.22 Unsigned comparison of the contents of File h'036' with WREG

Datum [f] *greater than* Working register .... No borrow, non-zero
Datum [f] *equal to* Working register .... Zero
Datum [f] *less than* Working register .... Borrow, non-zero

In terms of our processor, the **C** flag represents the *complement* of the borrow after subtraction and the **Z** flag is set on a zero outcome. Thus:

[f] *Greater than or equal* [W] : [f]–[W] gives no borrow; (C = 1).
[f] *Equal to*   [W] : [f]–[W] gives Zero;   (Z = 1).
[f] *Less than*   [W] : [f]–[W] gives a borrow; (C = 0).

Figure 5.22 illustrates this, where the byte in File h'036' is to be compared to that in the Working register. The instruction subwf h'036',w,0' generates the difference and alters the **C** and **Z** flags as shown, giving the three magnitude outcomes. The actual difference in WREG is irrelevant, but overwrites the original contents, which may have to be saved before the comparison.

The three Compare instructions listed here have several advantages over the Subtract–Bit Test process used in earlier PIC MCU families. These instructions also do the subtraction [f]–[W] but throw away the difference, which does not then overwrite the contents of WREG. Non of the flags are altered. A single instruction replaces the two or more needed with the more basic technique and the logic is clearer.

As an example, consider a series of comparisons with fixed values for a 255-litre fuel tank. A sensor at the bottom of the tank indicates the remaining volume of fuel as a linear function of pressure. Assume that the sensor represents the capacity as a byte that can be accessed at Port B (see p. 85), which we give the name FUEL. We wish to write a routine that will light an 'empty' light (at bit 0 at Port A) if the capacity is below 20 liters and ring an alarm buzzer (bit 1 at Port A) if below 5 liters. Both output peripherals are active on logic 1. See Fig. 5.23.

In Program 5.8 the constant 4 is loaded into WREG and the fuel reading in Port B is compared with this value for greater than. If true then the following instruction,

Fig. 5.23 Comparisons made
in the fuel warning system

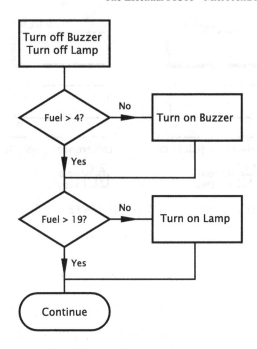

Program 5.8 Coding the fuel tank warning system

```
DISPLAY equ   h'F80'                ; Port A
LAMP    equ   0                     ; in which RA0 is the Lamp
BUZZER  equ   1                     ; and RA1 is the Buzzer

ALARM   bcf   DISPLAY,BUZZ,0        ; Turn off the Buzzer
        bcf   DISPLAY,LAMP,0        ; Turn off the Lamp

        movlw 4                     ; Set up to compare with 4 liters
        cpfsgt FUEL,0               ; Skip IF Fuel is greater than 4
         bsf  DISPLAY,BUZZ,0        ; ELSE sound buzzer

        movlw d'19'                 ; Set up to compare with 19 liters
        cpfsgt FUEL,0               ; Skip IF Fuel is greater than 19
         bsf  DISPLAY,LAMP,0        ; ELSE turn on Lamp

NEXT    ..... ......                ; Continue with the next routine
```

---

which turns on the buzzer, is skipped over. Similarly, the fuel reading is checked for
>19 and if true the lamp turn-on is skipped. In this manner a series of tests can be
made, the outcome of each of these taking an appropriate action.

Note the use of the bsf (Bit Set in File) instruction to set the appropriate pin in
Port A, which we assume to have been initialized as an output. In the same manner
the bcf instruction is used to turn off the lamp and buzzer at the beginning of the
routine.

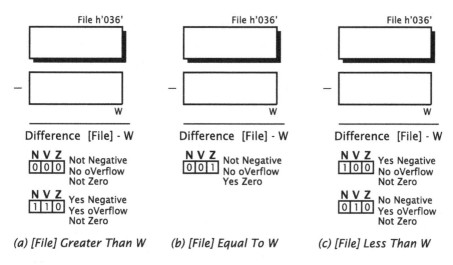

Fig. 5.24 Signed comparison of the contents of File h'036' with WREG

The Compare instructions and the flag tests outlined in Fig. 5.22 only apply to unsigned quantities. If the data is signed then subtraction is used, but this time the **N**, **OV** and **Z** flags need to be checked, as shown in Fig. 5.24.

Conditional Branches

The PIC18F instruction set has four pairs of Conditional Branch instructions. Unlike the Unconditional Branch `bra` which always adds an 11-bit signed offset to the PC (see p. 135) a Conditional Branch instruction adds on an 8-bit signed offset *only* if one of the four principle status flags (not **DC**) is either set or clear. The potential branch range is ±128 program words (back or forwards). The four pairs are:

bc   (Branch if Carry) | 11100010 | SNNNNNNN |

bnc (Branch if No Carry) | 11100011 | SNNNNNNN |

bn   (Branch if Negative) | 11100110 | SNNNNNNN |

bnn (Branch if Not Negative) | 11100111 | SNNNNNNN |

bov (Branch if OVerflow) | 11100100 | SNNNNNNN |

bnov (Branch if No OVerflow) | 11100101 | SNNNNNNN |

bz   (Branch if Zero) | 11100000 | SNNNNNNN |

bnz (Branch if Not Zero) | 11100001 | SNNNNNNN |

We have already used some of these instructions; for example `bc` on p. 119. For a more extensive illustration, consider the double-precision subtraction of Program 5.5. Here we coded for a 2-byte difference on the basis that its value would be

Program 5.9 A double-precision signed subtraction routine

```
MINUEND_L      equ  h'020'   ; Name the two Minuend Files
MINUEND_H      equ  h'021'
SUBTRAHEND_L   equ  h'022'    ; Name the two Subtrahend Files
SUBTRAHEND_H   equ  h'023'
DIFFERENCE_L   equ  h'030'    ; Name the three Difference Files
DIFFERENCE_H   equ  h'031'
DIFFERENCE_U   equ  h'032'

; Double-precision subtraction as Program 5.5 -----------------
; Task1 ------------------------------------------------------
DP_SUB movf    SUBTRAHEND_L,w,0; Get low byte of SUBTRAHEND
       subwf   MINUEND_L,w,0   ; subtract from MINUEND
       movwf   DIFFERENCE_L,0  ; and put away as lo byte Diff


; Task 2 -----------------------------------------------------
       movf    SUBTRAHEND_H,w,0; Get hi byte of SUBTRAHEND THEN
       subwfb  MINUEND_H,w,0   ; sub from MINUEND with borrow-in
       movwf   DIFFERENCE_H,0  ; Put away as mid byte Difference
; Extend sign into Upper byte --------------------------------
       clrf    DIFFERENCE_U,0  ; Zero the Upper byte
       bnn     NEXT            ; Branch if last subt. wasn't -ve
       setf    DIFFERENCE_U,0  ; ELSE make it 11111111
NEXT   bnov    CONTINUE        ; IF No OVerflow THEN finished
       comf    DIFFERENCE_U,f,0; ELSE invert the Upper byte
CONTINUE   ... .......        ; Continue on
```

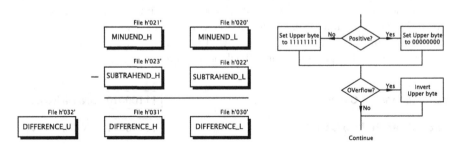

(a) Minuend - subtrahend = difference   (b) Generating the Upper byte

Fig. 5.25 Double-precision signed subtraction

smaller than either of the minuend or subtrahend. However, if these quantities are signed, then this is no longer necessarily true. For instance, $+30,000 - (-30,000) = +60,000$. The maximum positive quantity that can be represented in signed 16-bit format is $+32,680$ (b'0111 1111 1111 1111'). What will happen in our instance, is that there will be overflow into the sign bit; $+60,000$ is b'1110 1010 0110 0000' which is -5536.

In order to cope with this overflow, we need to provide space to extend the sign bit—see p. 120. In Fig. 5.25(a) this is shown as DIFFERENCE_U. If there is no overflow then this is simply an 8-bit expansion of the sign bit in DIFFERENCE_H; that is b'00000000' if positive or b'11111111'. On the other hand if there has been an overflow then bit 7 of DIFFERENCE_H is incorrect—see Fig. 1.5 on p. 14. As shown in the flowchart of Fig. 5.25(b), the previous expansion is simply inverted to give the true signage. In our instance above this will give b'0000 0000 1110 1010 0110 0000' or d'60,000'.

The subtraction routine of Program 5.9 is identical to that of Program 5.5. The sign extension makes use of the bnn and bnov instructions to test if the outcome of the last subtraction was positive or negative and if a sign overflow occurred. This sequential series of tests is possible as Conditional Branch instructions do not change the state of the status flags. The coding follows the flow chart except that the Upper byte is initially cleared before checking for negative. Neither clrf nor setf affect the flags.

Examples

Example 5.1 Some early computers used a bi-quinary code to represent BCD digits. This is a 7-bit code with only two bits set to one for any combination:

01	00001	0
01	00010	1
01	00100	2
01	01000	3
01	10000	4
10	00001	5
10	00010	6
10	00100	7
10	01000	8
10	10000	9

Although this is highly inefficient (with only ten out of a possible 128 code combinations being used) it does have the advantage that it is very easy to decide when an error has occurred. Design an error-detection routine to check the bi-quinary byte in File h'020' in Access RAM. Assume that the most-significant bit is zero. If an error occurs then the Working register is to be set to h'FF', otherwise zero.

Solution All we need to do here is to determine when there are more or less than two bits set to one. Based on this approach we have the task list:

1. Count the number of ones in the bi-quinary byte.
2. Zero WREG.
3. IF the count is not two THEN finish with WREG set to h'FF' to signal an error.

Program 5.10 shows a possible coding implementing this algorithm. Here the loop continually shifts the bi-quinary byte left until the residue is zero. After each shift, when the Carry flag is set, the bit count is incremented. On exit from the

Program 5.10 Bi-quinary error detection

```
WREG         equ    h'FE8'      ; The Working register
STATUS       equ    h'FD8'      ; Status register is File h'FD8'
C            equ    0           ; Carry flag is bit0
Z            equ    2           ; Zero flag is bit2
BI_QUIN      equ    h'020'      ; Bi-quinary byte is in File h'020'
COUNT        equ    h'021'      ; The bit count is put here

BI_QUINARY clrf   COUNT,       ; Bit count is cleared
; Task 1 ----------------------------------------------------
LOOP         bcf    STATUS,C    ; Clear carry flag
             rlcf   BI_QUIN,f,0 ; Rotate code left
             bnc    NEXT        ; No Carry THEN next?
             incf   COUNT,f,0   ; ELSE increment the count
NEXT         tstfsz BI_QUIN,0   ; IF zero THEN skip out of loop
             bra    LOOP        ; ELSE repeat shift and count
; Tasks 2 & 3 ----------------------------------------------
             movf   COUNT,w     ; Get count
             addlw  -2          ; Compare with two (W - 2)
             bz     FINI        ; IF ZERO finished with W == 00
             setf   WREG,0      ; ELSE put h'FF' (-1) in W

FINI         ....   .....       ; Next routine
```

loop, two is subtracted from the bit tally after moving into WREG. If it is zero, then the routine is completed and the h'00' setting of WREG shows a correct outcome. Otherwise h'FF' is placed in WREG to show a fault. This is equivalent to decimal -1 and is traditionally used to note an error situation. There are 20 code combinations in all which have two ones, of which only 10 are legitimate. Can you think of a simple extension to the routine to weed out these additional double-one code patterns?

Example 5.2 Eight-bit PIC MCUs do not have instructions to divide. However, division can be implemented by continual subtraction. For instance, to divide a number by ten you can count how many times ten can be subtracted before a borrow-out is generated. The count is then the quotient and the residue is the remainder. Using this technique, write a routine to convert a binary number byte of magnitude no greater than h'63' (decimal 99) in File h'020' to two Binary Coded Digits to be placed in File h'021:22' ordered as Tens:Units—see p. 6.

Solution Dividing the binary number by ten generates a quotient between 0 and 9 (remember the maximum value is 99) and a remainder. The quotient is the number of tens and the remainder is the number of units.

The simplest way of doing this, illustrated in Fig. 5.26, is to keep subtracting ten (addlw -d'10' or addlw -h'0A'). Keeping a count in the TENS File register gives the number of subtractions until a borrow is generated. The required number of tens is one less than this tally; that is, the number of successful subtractions. Adding that one extra ten back again to the residue gives the remainder, which is the units tally.

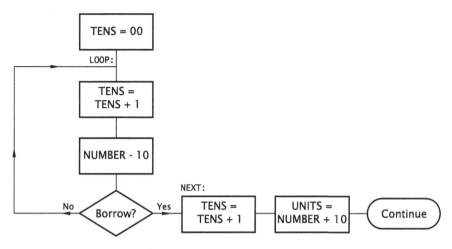

Fig. 5.26 Conversion of a byte up to 99 to BCD

Program 5.11 Binary to 2-digit BCD conversion

```
BINARY    equ   h'020' ; Binary byte is in File h'020'
TENS      equ   h'021' ; The quotient is put here
UNITS     equ   h'022' ; The remainder is put here

; First divide by ten
BIN_2_BCD clrf  TENS,0     ; Zero the loop count
          movf  BINARY,w,0 ; Get binary byte into W
; DO subtract ten and counting WHILE no borrow is generated
LOOP      addlw -d'10'     ; Subtract decimal ten
          bnc   NEXT       ; IF a borrow (C == 0) THEN exit loop
          incf  TENS,f,0   ; ELSE record one more ten subtract
          bra   LOOP       ; and do it again
; Correct for one ten too many, and hence determine the Units
NEXT      addlw d'10'      ; Add on a ten
          movwf UNITS,0    ; and copy remainder into memory
          ..... ......     ; Next routine
```

Example 5.3 Another approach to division is to express the divisor as the sum of fractional powers of two. For instance, the binary approximation to the fraction $\frac{1}{3}$ is:

$$\frac{1}{3} = \frac{1}{2} - \frac{1}{4} + \frac{1}{8} - \frac{1}{16} + \frac{1}{32} - \frac{1}{64} + \frac{1}{128} \cdots .$$

Using this series, write a program that will divide a byte N in the Working register by three, with the quotient being in the same register at the end. You can use File h'020' temporary storage for the quotient.

Solution The coding shown in Program 5.12 simply zeros the Quotient byte and then continually shifts the datum in the Working register to get the various fractions.

Program 5.12 Dividing by three

```
QUOTIENT equ h'020'        ; Put the final Quotient here
STATUS   equ h'FD8'        ; The Status register
C        equ 0             ; in which bit 0 is the Carry flag
WREG     equ h'FE8'        ; The Working register

DIV_3 clrf  QUOTIENT,0     ; Zero the outcome
      bcf   STATUS,C,0     ; Carry = 0
      rrcf  WREG,f,0       ; Shift right once to give N/2
      movwf QUOTIENT,0     ; and copy into Quotient = N/2

      bcf   STATUS,C,0     ; Carry = 0
      rrcf  WREG,f,0       ; Shift again once right to give N/4
      subwf QUOTIENT,f,0   ; Subtract to give Q = N*(1/2-1/4)

      bcf   STATUS,C,0     ; Carry = 0
      rrcf  WREG,f,0       ; Shift again to give N/8
      addwf QUOTIENT,f,0   ; Add to give Q = N*(1/2-1/4+1/8)

      bcf   STATUS,C,0     ; Carry = 0
      rrcf  WREG,f,0       ; Shift again to give N/16
      subwf QUOTIENT,f,0   ; Sub to give Q = N*(1/2-1/4+1/8-1/16)

      bcf   STATUS,C,0     ; Carry = 0
      rrcf  WREG,f,0       ; Shift again to give N/32
      addwf QUOTIENT,f,0   ; Add: Q = N*(1/2-1/4+1/8-1/16+1/32)

      bcf   STATUS,C,0     ; Carry = 0
      rrcf  WREG,f,0       ; N/64
      subwf QUOTIENT,f,0   ; Q = N*(1/2-1/4+1/8-1/16+1/32-1/64)

      bcf   STATUS,C,0     ; Carry = 0
      rrcf  WREG,f,0       ; N/128
      addwf QUOTIENT,w,0   ; N*(1/2-1/4+1/8-1/16+1/32-1/64+1/128)
                           ; Outcome now in the Working register
      .....  .......       ; Next
```

These are then either added to or subtracted from this Quotient, which tends towards the final value. This value is then copied down into WREG as specified.

The outcome up to $\frac{1}{128}$ is 0.3359375, which is within 0.78% of the exact value. With an 8-bit datum there is no point in including any further elements in the series.

If greater accuracy is desired, then the original number can be extended to a 16-bit datum, by adding a zero lower byte. The series can then be extended to give a resolution down to one part in 32,768, with double-precision shifting and arithmetic operations.

Example 5.4 A certain temperature logging system samples every hour and at the end of a day the 24 samples are to be found *in situ* in the Data store between File h'030' and File h'047'. Write a program to scan through this array and evaluate the average daily temperature. This average is to be located in Bank 0 at File h'080'.

Program 5.13 Average daily temperature

```
FSRL0      equ   h'FE9'      ; Low byte of Pointer 0
POSTINC0   equ   h'FEE'      ; Post-increment Indirect trigger
TEMP_0     equ   h'030'      ; Array starts @ File h'30'
SUM        equ   h'048'      ; Grand total to be in File h'048:49'
AVERAGE    equ   h'080'      ; Average byte is to be here in Bank 1

; Task1: Clear grand total and Average --------------------------
AV_DAILY   clrf  SUM,0       ; LSbyte sum zeroed
           clrf  SUM+1,0     ; MSbyte sum zeroed

; Task2: Point to Temp[0] ---------------------------------------
           lfsr  0,TEMP_0    ; Put address of 1st byte in pointer 0

; Task3:  DO ----------------------------------------------------
; Tasks3A&B : Add Temp[i] to the double-byte sum and inc pointer
LOOP       movf  POSTINC0,w,0 ; Get Temp[i]: FSR0++
           addwf SUM,f,0     ; Add LSB sum to it and put away
           bnc   NEXT1       ; IF no carry, don't increment MSB
           incf  SUM+1,f,0   ; ELSE pass carry on

; Task3C: REPEAT WHILE i < 24 -----------------------------------
NEXT1      movlw TEMP_0+d'23'; Put the end address in W (TEMP[23])
           cpfsgt FSRL0,0    ; IF pointer FSRL0 is > than THEN skip
           bra   LOOP        ; ELSE repeat

; Task4: Divide by 24 to give the average -----------------------
           clrf  AVERAGE,1   ; Zero the average (in Bank 0)
; Keep subtracting 24 and keep a count until a borrow-out
DIV_24     movlw d'24'       ; Put the constant 24 in W
           subwf SUM,f,0     ; Take away 24 from the sum LSB
           bc    NEXT2       ; IF no borrow out (C==1), THEN skip
           decf  SUM+1,f,0   ; ELSE decrement high byte
           bnc   FINI        ; IF this generates a borrow THEN fini
NEXT2      incf  AVERAGE,f,1; ELSE record one more subtract 24
           bra   DIV_24      ; and do next subtract

FINI       .....  ......     ; Next routine
```

Solution Finding the average involves walking through the array, in the manner of Fig. 5.6, adding each element to a 2-byte grand total. On completion this total is divided by 24 to give the average function:

$$\frac{\sum_{i=0}^{23} \text{Temp}[i]}{24}$$

Based on this approach we have as a task list:

1. Clear Average.
2. Point to Temp[0] ($i = 0$).
3. DO
   (a) Add Temp[i] to the 2-byte grand total.
   (b) Increment i.
   (c) Repeat WHILE $i < 24$.
4. Divide by 24.

Program 5.13 directly implements the task list, summing each datum byte by adding to the double-byte location File h'048:49', which has been cleared before

Program 5.14 Converting Celsius to Fahrenheit

```
PRODL      equ      h'FF3'     ; Low byte of Product
PRODH      equ      h'FF4'     ; High byte of Product
CELSIUS    equ      h'080'     ; The Celsius temperature byte
FAHRENL    equ      h'081'     ; Low byte of Fahrenheit equivalent
FAHRENH    equ      h'082'     ; High byte in Bank 0

; Task1: Multiply by nine ------------------------------------------
C_TO_F     movlw    9              ; Nine
           mulwf    CELSIUS,1  ; Multiply Celsius

; Task2: Divide by five by subtracting until borrow from high byte
           clrf     FAHRENL,1  ; First zero the 2-byte Fahrenheit
           clrf     FAHRENH,1  ; equivalent temperature
DIV_5      movlw    5              ; Five
           subwf    PRODL,f,0  ; Take away from low byte of x9
           bc       NEXT           ; IF no borrow (C==1) skip
           decf     PRODH,f,0  ; ELSE decrement high byte
           bnc      ADD_32         ; IF a borrow THEN continue
; DO a double-byte increment of Fahrenheit every successful subtract
NEXT       incf     FAHRENL,f,1 ; Add one onto the 2-byte Fahrenheit
           bnc      DIV_5          ; IF no Carry do another subtraction
           incf     FAHRENH,f,1 ; add one onto high byte
           bra      DIV_5          ; and do another subtraction

; Task3:  Add 32 to Fahrenheit -----------------------------------
ADD_32     movlw    d'32'      ; Thirty two
           addwf    FAHRENL,f,1 ; Add it to the low byte
           bnc      FINI           ; IF no carry THEN thats it
           incf     FAHRENH,f,1 ; ELSE pass carry on

FINI       .....    ......     ; Next routine
```

entry to the loop. Division is accomplished by repetitively subtracting 24 from the final total. This is similar to the $\div 10$ routine of Program 5.11 but this time the single-byte constant is taken off the double-byte dividend. The number of successful subtracts is the quotient, which in this case is the truncated Average. Of course it would be more accurate to round up if the remainder is more than half of the divisor.

Example 5.5 In the last example we evaluated the average temperature over a 24-hour period. On the basis that this an unsigned byte representing a Celsius value, design a routine to convert this to its Fahrenheit equivalent. As the range 0°C though 255°C maps to 32°F through 491°F we need two bytes to store the outcome. This 16-bit product is to be located at $\boxed{\text{FAHRENH}}$ F h'082' $\boxed{\text{FAHRENL}}$ F h'081'.

Solution The relationship between the two tasks is given by:

$$\text{Fahrenheit} = (\text{Celsius} \times 9)/5 + 32.$$

Based on this approach we have as a task list:

1. Multiply datum by nine.
2. Repetitively subtract five from the double-byte product.
3. Add 32 to the double-byte outcome.

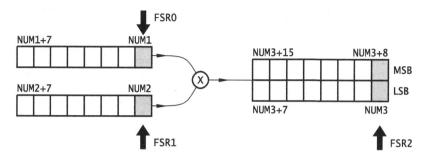

Fig. 5.27 Multiplying for array multiplication

The task list implemented in Program 5.14 uses the `mulwf` instruction to multiply Celsius byte by the preloaded constant nine. The double-byte product in `PRODH:PRODL` is then continually decremented by five with the double byte `FAHRENH:FAHRENH` being incremented every time this is done. When a borrow-out is generated from the decrementation of `PRODH`, then the quotient in `FAHREHH:FAHRENL` is augmented by 32 to give the final outcome.

Although this routine illustrates unsigned Celsius values, using the signed multiplication routine of Program 5.6 to replace the `mulwf` instruction would allow both positive and negative Celsius temperatures to be converted.

Example 5.6 Consider two arrays of eight unsigned numbers labeled in Fig. 5.27 as NUM1 and NUM2. All arrays are to be placed in Bank 0, with NUM1[] located at File h'080' through File h'087' and NUM2[] in File h'088' through File h'08F'. We wish to multiply each element i to give an array of eight unsigned 16-bit numbers NUM3[].

$$\left|\sum_{i=0}^{7}\right. NUM3[i] = NUM1[i] \times NUM2[i]$$

The resulting product array is assigned to File h'090' through File h'09F' with the MSB offset from the LSB of each element of NUM3[i] by eight Files.

Solution Program 5.15 uses FSR0 to walk through NUM1[] and FSR1 to point to the corresponding NUM2[] array. As NUM3[] is really two byte arrays displaced by eight, FSR2 is used to point to the LS Byte of the Product array. These three pointers are initialized to the location of element 0 of each array, shown shaded in Fig. 5.27, using the `lfsr` instruction.

The core of the program is the `mulwf` instruction. This generates a 16-bit product into the SFRs PRODH:PRODL from data in WREG and the designated File. By using the Post-increment Indirect addressing mode to move the first datum into WREG with FSR0 and also for FSR1 which points to the second array byte, the operation is walked through the two arrays.

In order to copy the contents of each byte from PRODL into the LS Byte array NUM3[] and PRODH into the MS Byte array, the program uses the `movff` instruction. This 2-word instruction (see p. 111) uses two 12-bit File addresses to locate

Program 5.15 Coding for the array multiplication

```
FSR0L       equ    h'FE9'        ; Low byte of first pointer
POSTINC0    equ    h'FEE'        ; Pointer 1's post increment mode
POSTINC1    equ    h'FE6'        ; Pointer 2's post-increment mode
POSTINC2    equ    h'FDE'        ; Pointer 3's post-increment mode
PLUSW2      equ    h'FDB'        ; Pointer 3's offset mode
PRODL       equ    h'FF3'        ; Low byte of product
PRODH       equ    h'FF4'        ; High byte of product

NUM1        equ    h'080'        ; Start location of Array 1
NUM2        equ    h'088'        ; Start location of Array 2
NUM3        equ    h'090'        ; Start location of Array 3

ARRAY_M lfsr   0,NUM1            ; Point to LSByte of Number 1
        lfsr   1,NUM2            ; Point to MSByte of Number 2
        lfsr   2,NUM3            ; Point to LSWord of Number 3

M_LOOP  movf   POSTINC0,w,0      ; Get NUM1[n] & advance pointer 1
        mulwf  POSTINC1,0        ; Multiply by NUM2[n] & advance ptr2

        movlw  8                 ; Offset for NUM3 LSB:MSB
        movff  PRODH,PLUSW2      ; Copy High byte Product into MSB
        movff  PRODL,POSTINC2;   ; Copy lo byte Prod into LSB; ptr3++

        movlw  NUM1+8            ; Check pointer 1 has not overrun
        cpfseq FSR0L,0           ; IF reached h'088' THEN finished
        bra    M_LOOP
        ...... ........          ; Next routine
```

source and destination data anywhere in the Data store without the need to use the
banking mechanism. The contents of PRODH are first copied into the MS Byte by
using the Plus W Indirect mode. As WREG has been preset to 8, effectively this
datum is copied to a location eight Files above that pointed to by FSR2. Finally,
PRODL is copied into the LS Byte array location pointed to by FSR2. By using the
Post-increment Indirect mode for this destination, this pointer too will walk through
the Product array. The process terminates whenever the FSR0 pointer reaches h'088'.

Self-Assessment Questions

5.1 Can you deduce what function the following code fragment performs on the
data byte in the Working register?

```
addwf   FILE,w
subwf   FILE,w
```

5.2 How could you extend Example 5.2 to give an outcome packed into a single-
byte TENS:UNITS in File h'021'? This is known as packed BCD where each

byte holds two decade nybbles rather than one digit per byte. *Hint*: Consider making use of the swapf instruction.

5.3 Develop Example 5.2 to give a 3-digit BCD outcome, removing the restriction that the original binary byte should be limited to decimal 99. The outcome is to be in File h'023:22:21' as HUNDS : TENS : UNITS respectively.

5.4 As part of a Data memory testing procedure, each File in Bank 0 (that is File h'000' through File h'0FF') is to be set to the pattern b'01010101' (h'55'). If an error is detected (datum not stored correctly) the routine is to exit with −1 in the Working register and FSR0 pointing to one beyond to the erroneous location, otherwise WREG should be zero. Using Program 5.2 as a model, design a suitable coding to implement this task.

5.5 Data from an array of data memory between File h'030' and File h'04F' is to be transmitted byte-by-byte to a distant computer over the Internet. In order to allow the receiver to examine the data and check for transmission errors it is proposed to append a single byte, which is the 2's complement of the *8-bit sum* of all the data bytes together. If all the received data bytes plus this **checksum** byte are similarly added then the outcome should be zero if no error has occurred. Code a routine to scan through this data, placing this checksum in File h'020'.

5.6 Based on the data logger specified Example 5.4, write a program to evaluate the *maximum* daily temperature. By the end of the routine this is to be in File h'07F'.

5.7 Example 5.4 evaluated the average of an array of hourly temperature samples by summing all bytes and then subtracting 24 until the residue dropped below zero. Write an extension to this program to round the average to the nearest integer; that is, if the remainder is more than 12 then round up.

5.8 One simple way of encrypting a data byte is to reverse the order of bits. For example b'10111100' $\longrightarrow$ b'00111101'. Write a routine to implement this reversal on a data byte in File h'020'. The encrypted outcome is to be in the Working register. You can use location File h'021' as a temporary workspace and WREG as a loop counter. *Hint*: Use the Rotate Right through Carry and Rotate Left through Carry instruction eight times.

5.9 Reverse encryption is a somewhat weak coding as once the code is broken all subsequent messages can be unscrambled! Teleprinter traffic used by the German high command during the Second World war was based on the Vernam cipher. Essentially this eXclusive-ORs each plain text code pattern with a key.[5] This key is usually a pseudo-random number sequence, such as that generated in SAQ 6.6.10 on p. 203. However, the key could be simply a daily crib sheet. Figure 5.28 shows an example based on 8-bit ASCII code groups, although originally a 5-bit Baudot/Murray code was used.[6]

---

[5] Actually a patchboard was used on the Lorentz teleprinter to transpose bits within each code group to further complicate attempts to break the code.

[6] The ten Colossi 25,000 valve/tube digital processor based in Bletchley park were designed to eXclusive-OR monitored text with various keys at a rate of 25,000 characters per second. For more details see http://www.codesandciphers.org.uk.

	H	e	l	l	o	
Transmitter	01001000	01100101	01101100	01101100	01101111	Plain string
⊕	11110000	11101101	00111000	00000011	01001111	Key
	10111000	10001000	01010100	01101111	00100000	Encrypted text
Receiver	10111000	10001000	01010100	01101111	00100000	Encrypted text
⊕	11110000	11101101	00111000	00000011	01001111	Key
	01001000	01100101	01101100	01101100	01101111	Plain string
	H	e	l	l	o	

Fig. 5.28 Illustrating the Vernam cipher used to encrypt teleprinter traffic

Design a routine to encipher a string of ASCII-coded characters terminated with a NUL character (h'00') located in the range File h'080' – h'08F' with a key string *in situ* in File h'090' – h'09F'. The resulting enciphered text is to be placed in File h'0A0' – h'0AF' for subsequent transmission.

5.10 A simple digital low-pass filter can be implemented using the algorithm:

$$\texttt{Array[i]} = \frac{S_n}{4} + \frac{S_{n-1}}{2} + \frac{S_{n-2}}{4}$$

where S_n is the nth sample from an eight-bit analog to digital converter located at Port B.

Write a routine assuming that the three byte memory locations to store S_n, S_{n-1} and S_{n-2} are located at File h'022:21:20' respectively. The outcome Array[i] is to be located at File h'048'.

5.11 Consider a 24-bit word stored in the Data store at the three locations

24	File h'032'	16	15	File h'031'	8	7	File h'030'	0

Design a routine to count the number of bits set to 1 in this triple-byte datum.

5.12 A certain television show has eight contestants who are evenly divided into Team A and Team B. Each member has a switch, giving logic 1 when pressed, which may all be read simultaneously by the microcontroller at Port B. Team A switches appear on the lower four bits of the port.

Write a routine that will:

- Decide when a response to the question has been made, any switch closed (non zero at Port B).
- Determine the team identity that has responded by either clearing File h'020' for Team A, otherwise setting it to any non-zero value to signify Team B.
- Ascertain which team member pressed the switch by putting the member number 0–3 in File h'021'.

5.13 Parity is a simple technique to protect digital data from corruption by noise. Odd parity adds a single bit to a word in such a way as to ensure the overall packet has an odd number of 1s. Write a routine that takes an 8-bit byte stored

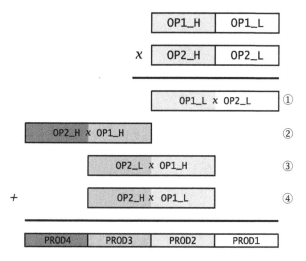

Fig. 5.29 Multiplying two double-byte numbers to give a 4-byte product

at File h'020' and alters its most significant bit to comply with this specification. You can assume that bit 7 is always 0 before the routine begins. *Hint*: Determine if a binary number is odd or even by counting the number of bits as in Example 5.1 and then examining its least significant bit. All powers of two are even except $2^0 = 1$. Thus without exception all odd numbers must have this bit set to 1.

5.14 Write a routine to multiply two 16-bit operands to give a 32-bit product. Operand 1 is located in File h'021:20' as OP1_H:OP1_L, Operand 2 is stored in File h'023:22' as OP2_H:OP2_L and File h'083:82:81:80' holds the 4-byte Product PROD4:PROD3:PROD2:PROD1.

Figure 5.29 shows how the Product can be built up with four 8 × 8 multiplications, suitably aligned and added. Shifting one byte eight places left is effectively multiplying by 2^8, so overall the cross product is mathematically defined as:

$$PROD_{16} = (OP1\_H \times OP2\_H)2^{16} + (OP1\_H \times OP2\_L$$
$$+ OP1\_L \times OP2\_H)2^8 + (OP1\_L \times OP2\_L).$$

Your target should be 24 instructions in all; taking 28 instruction cycles to execute (2.8 μs at a clock speed of 40 MHz).

Chapter 6
Subroutines and Modules

Good software should be configured as a set of interacting modules rather than one large program working straight through from beginning to end. There are many advantages to modular programming, which is almost mandatory when code lengths exceed a few hundred lines or when a project is being developed by a team.

What form should such modules take? In order to answer this question we will look at the use of program structures designed to facilitate this modular approach and the instructions associated with it.

After completing this chapter you will:

- Appreciate the need for modular programming.
- Have an understanding of the structure of the Hardware stack and its use in the call–return subroutine mechanism.
- Understand the term 'nested subroutine'.
- See how parameters can be passed to a subroutine and returned to the caller.
- Be able to write a transparent subroutine, having a minimal impact on its environment, using either the Hardware or a Software stack to pass parameters and provide a temporary workspace.
- Know how to use the optional Extended instruction set to more efficiently implement Software stacks.

Take a look at the inside of your personal computer. It will probably look something like the photograph in Fig. 6.1, with a motherboard hosting the MPU, assorted memory and other support circuitry, and a variable number of expansion sockets. Into this will be plugged a disk controller card and a video card. There may be others, such as a soundboard, USB or network card. Each of these plug-in cards has a distinct and separate logical task and they interact via the services supplied by the main board, the motherboard.

Advantages of this **modular** construction are:

- Flexibility; that is, it is relatively easy to upgrade or reconfigure by adding or replacing plug-in cards.
- Can reuse from previous systems.
- Can buy standard boards or design specialist boards in-house.
- Easier to maintain.

S. Katzen, *The Essential PIC18® Microcontroller,*
Computer Communications and Networks,
DOI 10.1007/978-1-84996-229-2_6, © Springer-Verlag London Limited 2010

Fig. 6.1 Modular hardware implementing a PC

Of course there are a few disadvantages. A fully integrated motherboard (such as commonly found in a laptop computer) is smaller and potentially cheaper than an equivalent mother-daughterboard configuration. It is also likely to be more reliable, as input and output signals do not have to traverse so many sockets and plugs. However, when they do occur, faults are often more difficult to track down and rectify.

Modular programming uses the same principle to construct 'software circuits', i.e. programs. A formal definition of modular programming is:

> An approach to programming in which separate logical tasks are programmed separately and joined later.[1]

Thus, to write a program in a modular fashion we need to decompose the specification into a number of stand-alone routines, each implementing a well-defined task. Such a module should be relatively short, be well documented and be easy for a human, not necessarily the original programmer, to understand.

The advantages of a modular program are similar to those for modular hardware, but even more compelling:

- Modules can be tested, debugged and maintained on a stand-alone basis; this makes for overall reliability.
- Can be reused from previous projects or bought in from outside.
- Easier to update by changing modules.

Deciding how to segment a program into individual stand-alone tasks requires expertise. The actual coding of such tasks as sub-programs is no different than the

---

[1]From *Chambers Science and Technology Dictionary*, Cambridge University Press, 1988.

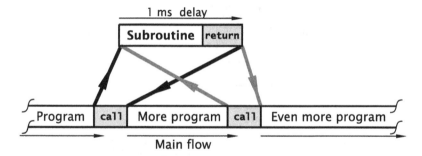

Fig. 6.2 Subroutine calling

examples we have given in previous chapters, such as that shown in Program 5.5 on p. 152. There are a few additional instructions associated with such subprograms, and these are listed in Table 6.1. We will look at these and some useful techniques in constructing software in the remainder of the chapter.

Program modules may be entered by calling from other software or by a hardware event external to the CPU. The latter may be a voltage at one of the processor pins or an internal peripheral interface wanting service, such as a Timer module overflowing. In the former case modules at assembly level are universally known as **subroutines**; as they are in some high-level languages such as FORTRAN and BASIC.[2] In the latter they are classified as **interrupt service routines**. The techniques for writing these interrupt modules and their entry and exit techniques are sufficiently different to warrant a separate treatment in Chap. 7. Here we will look at subroutines.

Subroutines are the analog of hardware plug-in cards. Consider the situation where a 1 ms delay task is to be implemented. This may be needed to generate a 500 Hz tone to alert an aircraft pilot to look at the control panel warning lights for various scenarios, such as low fuel or overheating. In a modular program, this delay would be implemented by coding a 1 ms delay subroutine, which would be *called* by the main program as necessary to, say, continually force a port pin high and low for 1 ms durations. This is represented diagrammatically in Fig. 6.2.

In essence, calling up a subroutine involves nothing more than placing the address of the first subroutine instruction in the Program Counter (PC); that is, doing a goto. Thus, if our initial instruction was located at, say, h'0400', then goto h'0400' would seem to do the trick. Assuming the programmer has labeled the subroutine entry point instruction DELAY_1MS, as in Program 6.1, we have goto DELAY_1MS.

The problem really is how to get back again! Somehow the MCU has to remember from where in the caller program the subroutine was entered, so that it can return to the *next* instruction in the caller's sequence. This can be seen in Fig. 6.2, where the jumping-off point can be from either of two *points* in the main program. Indeed it can even be called from another subroutine—see Fig. 6.4.

---

[2]Other high-level languages use the terms *function* (**C** and Pascal) or *procedure* (Pascal).

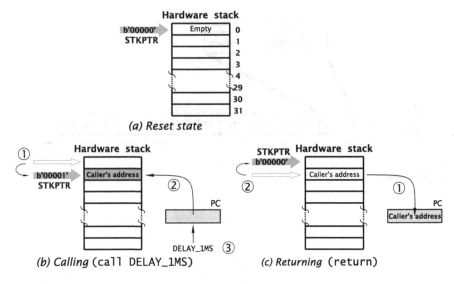

(a) Reset state

(b) Calling (call DELAY_1MS)   *(c) Returning* (return)

Fig. 6.3 Using the Hardware stack to hold return addresses

One possibility is to place this address in a designated Address register or memory location prior to jumping off. As the return mechanism, this can then be moved back into the Program Counter at the end of the subroutine. This approach breaks down whenever one subroutine wishes to call another. Then the secondary subroutine will overwrite the return address of the first, and the main program can never be regained. To get around this problem, more than one register or memory location can be used to hold a stack of return addresses. This last-in first-out stack structure is shown in Fig. 6.3(a).

The PIC18 MCU core has a stack of 32 21-bit registers, which are exclusively used to hold subroutine and interrupt (collectively known as function) return addresses.[3] Shown in Fig. 6.3, this structure is known as a **Hardware stack**. This stack is outside the PIC MCU's Data store memory map and normally its contents are left alone. However, unlike earlier family architectures, there are ways of seeing and altering the contents of this stack—see p. 179. For the moment, we will concentrate on the automatic operation of the stack.

Associated with this stack is a 5-bit dead-end counter known as the **Stack Pointer** (**SP**), which is located in the **STKPTR** register—see Fig. 6.12. Whenever a call instruction is executed the Stack Pointer is *automatically* incremented and the Program Counter state is copied into the pointed-to stack cell. This state is the address of the instruction *after* the call instruction, as the PC has already been incremented and the following instruction is being fetched into the pipeline. After the PC has been **pushed** into the stack it is overwritten by the destination instruction address.

---

[3]The 12-bit base-range core devices have only two 11-bit stack registers and the mid-range core has an 8-deep 13-bit Hardware stack.

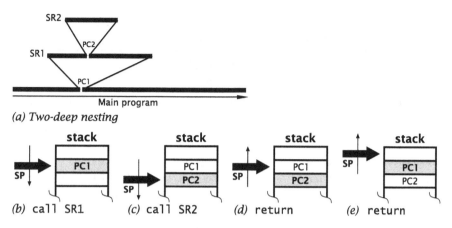

Fig. 6.4 Nested subroutines

As in the analogous `goto` instruction, `call` is a double-word instruction and so the first subroutine instruction can be anywhere in the Program store.

Figure 6.3(a) shows the stack configuration after a Power-on Reset with the SP zeroed. In Fig. 6.3(b) the situation is shown after a call to a subroutine labeled `DELAY_1MS`. The execution sequence of this `call  DELAY_1MS` is:

1. The Stack Pointer is incremented.
2. Copy the 21-bit contents of the PC into the stack at the location pointed to by the Stack Pointer. Effectively this stored datum will be the address of the instruction following the `call` instruction.
3. The destination address `DELAY_1MS`, that is the location of the entry point instruction of the subroutine, overwrites the original state of the PC. This causes the program execution to transfer to the subroutine.

The exit point from the subroutine should be a `return` instruction. This reverses the push action of the preceding `call` and pulls the return address back from the stack into the PC, as shown in Fig. 6.3(c). The execution sequence for `return` is:

1. Copy the 21-bit address in the stack pointed to by the Stack Pointer into the Program Counter.
2. Decrement the Stack Pointer.

Thus no matter where the subroutine was called from, it will return to the instruction just past the original `call` instruction when the subroutine has been completed.

The beauty of the stack mechanism is its handling of **nested** subroutines. Consider the situation in Fig. 6.4, where the main program calls the first-level subroutine SR1 which in turn calls the second-level subroutine SR2. In order eventually to get back to the main program, the outward progression sequence must be precisely matched by the inward path. This pattern is reflected in the **last-in first-out** (**LIFO**) structure of the stack mechanism, which can automatically handle any arbitrary nesting sequence up to the depth of the stack. Actually cell 0 is never used (see Fig. 6.3(c)) as the SP is always incremented *before* the PC is pushed onto

Table 6.1 Subroutine and interrupt handling instructions

Operation	Mnemonic	Description
Call		Transfer to subroutine
Call subroutine	`call   aaa`	Push PC on to stack, `PC <- <aaa>`
Fast call	`call   aaa,1`	As above but saves W, STATUS, BSR in Fast stack
Relative Call	`rcall  ±offset`$_{11}$	Push PC on to stack, `PC <- PC + offset`
Return		Transfer back to caller
from subroutine	`return`	Pull original PC back from Stack
Fast return	`return 1`	As above but recovers W, STATUS, BSR
with literal in W	`retlw`	Put literal in W and return as above
from interrupt	`retfie`	Return with the GIE flag in INTCON[7] set
Fast from interrupt	`retfie 1`	As above but recovers W, STATUS, BSR
Stack		Stack control
	`pop`	Pull data from stack into TOS
	`push`	Push PC into stack

the stack, so the effective depth of the stack is 31 and not 32. As we shall see in Chap. 7, the stack mechanism is also used to handle interrupts. Thus, in a system using both subroutines and interrupts, the nesting depth will be somewhat less. This LIFO mechanism can even handle the (painful) situation where a subroutine calls itself! Such a subroutine is known as **recursive**. This structure is so useful that virtually all MPU/MCUs support subroutines in this manner.

As the stack-Stack Pointer mechanism is part of the PIC MCU's hardware and requires no initialization, from the programmer's perspective only the following points are relevant:

- The subroutine should be invoked using the `call` instruction.
- The entry point to a subroutine should be labeled, and this label is then the name of that subroutine.
- The exit point from the subroutine should be a Return instruction.

Instructions that are associated with coding functions are listed in Table 6.1. For reference these are:

`call`

1110110	F	AAAAAAAA

1111	AAAAAAAAAAAA

This double-word instruction pushes the address of the *following* instruction into the stack and then transfers to the target instruction.

If the F (Fast) bit is 1 then copies are made of WREG, STATUS and BSR into a set of shadow registers prior to the transfer—see Program 6.2.

`rcall`

11011	SNNNNNNNNNN

Relative CALL is a 1-word version of `call` which allows the program to switch to a subroutine which is located not more than ±1024 words away. However, like the double-word `call` it takes two instruction cycles to execute.

return   `000000000001001` `F`

Pops out the last return address from the stack, effectively returning control to the caller. If the F-bit is 1 then the state of the WREG, STATUS and BSR registers are restored from their shadow counterparts—see Program 6.2.

retlw   `00001100` `LLLLLLLL`

Similar to a plain `return`, this Literal instruction copies the specified 8-bit literal into the Working register before going back to the caller—see Program 6.6.

retfie   `000000000001000` `F`

RETurn From an Interrupt and Enable is the counterpart to `return` used to go back from a function entered via an interrupt event, as will be described in the following chapter. As well as popping out the stacked address, it re-enables the appropriate interrupt system. Like `return`, it can also restore the WREG, STATUS and BSR states prior to the interrupt.

pop   `0000000000000110`

Decrements the Stack Pointer and allows the software to change the contents of the newly pointed-to stack cell—see p. 179.

push   `0000000000000101`

Increments the Stack Pointer and copies the current value of PC into the newly pointed-to stack cell—see p. 179.

For our first example, let us code the 1 ms delay subroutine of Fig. 6.2. Creating a delay in software is simply a matter of doing nothing for the relevant duration. A common way of doing this is to count down an initial constant to zero, as shown in Fig. 6.5. By choosing an appropriate constant, the delay can be tailored to the desired value. Obviously, this delay will depend on the PIC MCU's oscillator rate. For this example we will assume a clock rate of 4 MHz, giving an instruction cycle of 1 μs—see Fig. 4.5 on p. 76. Program 12.8 on p. 407 gives an example which can cope with a range of clock rates.

Consider the subroutine shown in Program 6.1. Here a constant N is placed in the Working register, and this value is decremented down to zero inside a 3-instruction loop. The subroutine then exits using a `return` instruction.

In order to calculate the total number of instruction cycles the program takes, and thus determine a value for N, we need to evaluate an execution cycle budget.

1. The `call DELAY_1MS` instruction used by the caller to jump to the subroutine takes two instruction cycles (2~) to execute.
2. The `movlw` instruction preceding entry into the loop takes one cycle.
3. The 1~ `addlw` instruction decrementing the contents of WREG takes in total N cycles (N times round the loop).

Fig. 6.5 Delaying by
counting N times

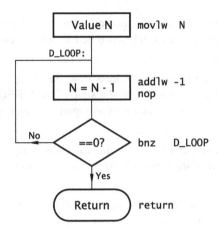

Program 6.1 A 1 ms delay subroutine

```
;   *************************************************************
;   * FUNCTION: Delays for nominally 1ms with a 4MHz crystal   *
;   * ENTRY    : None                                          *
;   * EXIT     : Flags and W altered                           *
;   *************************************************************
N           equ     d'249'   ; Delay parameter computed in the text

DELAY_1MS movlw   N          ; Set up loop                          1~

; LOOP ---------------------------------------------------------------
D_LOOP      addlw   -1        ; Decrement count                      N~
            nop               ; Put in extra cycle                   N~
            bnz     D_LOOP    ; Repeat unless zero          2*(N-1)+1~
; ------------------------------------------------------------------
            return            ;                                      2~
```

4. The 1~ nop instruction is added to boost the number of loop cycles inside the
   loop. It takes one instruction cycle and does not alter the status flags; necessary
   for the following test for zero. In total N cycles are added to the budget.
5. The bnz instruction branching back to the top of the loop if the **Z** flag is not set
   (is WREG zero after the last decrement?) is also executed N times. Each branch
   taken takes two instruction cycles. However, the very last time when the loop exits
   takes only one cycle. Thus the total delay is $2 \times (N - 1) + 1$.
6. The final return takes two cycles.

The total number of cycles is then:

$$2 \,(\texttt{call}) + 1 (\texttt{movlw}) + N (\texttt{addlw}) + N (\texttt{nop}) + [2 \times (N - 1) + 1]\,(\texttt{bnz})$$

$$+ \,2 \,(\texttt{return})$$

Equating this to 1000 cycles gives:

$$2 + 1 + N + N + [2 \times (N - 1) + 1] + 2 = 1000,$$
$$4 + (4 \times N) = 1000,$$
$$4 \times N = 996,$$
$$N = 249.$$

Our delay subroutine is pretty limited in that the Working register, like all data registers, is only eight bits wide and thus the maximum value of N is b'11111111' or decimal 255. Actually, in the case of our subroutine in Program 6.1, a value $N = 0$ gives the longest delay! This is because WREG is decremented *before* being tested for zero. So the sequence would actually go h'00 $\rightarrow$ FF $\rightarrow$ FE $\rightarrow \cdots \rightarrow$ 01 $\rightarrow$ 00'. Thus effectively N acts as if it were d'256', giving a maximum delay of $4 + (4 \times 256) = 1028$ cycles, or 1.028 ms with a 4 MHz crystal.

Simple as it is, our exemplar program illustrates an important characteristic of functions; it alters its environment. In this instance, on return to the caller, the contents of both the Working and Status registers have been altered. In both cases the writer of the code will take this into account and there will be no problem. However, if a team is working on the project or if the function is being reused from another project, then there is ample opportunity for confusion. At the very least, a subroutine should be well documented in its header comments, narrating what Files are altered and what resources are being used and assumptions that are made. For instance, the clock frequency.

One way around this problem is to allocate GPRs to store vulnerable data on entry to the subroutine and retrieve it just prior to exit. Of course, care is needs to be taken that these data will not be overwritten by some other process, such as another subroutine called from within the current function. In all but the most elementary subroutine, the Status flags and Working register will be changed, and in many cases the state of the Bank Select register will be altered as data is fetched from various banks of RAM. To speed up the save and retrieve process for these core resources, the PIC18 instruction set provides **Fast** versions of the call and return instructions—see p. 164. A Fast call automatically stashes away WREG, STATUS and BSR in three shadow registers; known as the **Fast stack**. Conversely, a Fast return restores these states from the Fast stack and only then pops out the caller's return address into the PC. Confusingly, the Microchip assembler treats these alternative instructions as variants, and they are designated as such by a ,1 in the operand field; e.g. call DELAY_1MS,1 and return 1. Strictly a ,0 should be used for an ordinary (slow) call, but this is the default and is usually omitted—see also p. 241.

As an example, let us repeat our 1 ms delay subroutine, but this time for an 8 MHz clock frequency. For this situation, we are going to have to double the number of cycles, as each instruction cycle now only takes 0.5 µs. In Program 6.2 we do this by adding four additional nop instructions inside the loop, and as our following calculation shows, we require an additional four cycles outside the loop. The total number of cycles is now:

$$2\,(\texttt{call}) + 1(\texttt{movlw}) + N\,(\texttt{addlw}) + 5 \times N(5, \texttt{nop})$$
$$+ [2 \times (N - 1) + 1]\,(\texttt{bnz}) + 4(4, \texttt{nop}) + 2(\texttt{return})$$

Program 6.2 A 1 ms delay subroutine for an 8 MHz clock

```
;   ***********************************************************
;   * FUNCTION     : Delays for nominally 1ms with a 8MHz crystal *
;   * ENTRY        : None (uses Fast stack)                       *
;   * EXIT         : Delays 1ms @ 8MHz                            *
;   * ENVIRONMENT  : No change                                    *
;   ***********************************************************
N           equ    d'249'   ; Delay parameter computed in the text

DELAY_1MS   movlw  N        ; Set up loop                        1~

; LOOP ---------------------------------------------------------
D_LOOP      addlw  -1       ; Decrement count                    N~
            nop             ; Put in one extra cycle             N~
            nop             ; Put in four extra cycles           N~
            nop             ;                                    N~
            nop             ;                                    N~
            nop             ;                                    N~
            bnz    D_LOOP   ; Repeat unless zero         2*(N-1)+1~
; Fine tune total cycles --------------------------------------
            nop             ;                                    1~
            nop             ;                                    1~
            nop             ;                                    1~
            nop             ;                                    1~
            return 1        ; Fast return                        2~
```

Equating this to 2000 cycles gives:

$$2 + 1 + N + N + 4N + [2 \times (N - 1) + 1] + 4 + 2 = 2000,$$
$$8 + (8 \times N) = 2000,$$
$$8 \times N = 1992,$$
$$N = 249.$$

Adding nops in this manner can be used to design delay routines that can cope with different clock frequencies. Thus adding an appropriate number of nops will allow the programmer to 'tweak' our subroutine to cope with crystals between 4 and 40 MHz; see also Program 12.8 on p. 407. How many nops would you need to give a 1 ms delay with a 20 MHz crystal?

This technique isn't much use if you need a substantially longer delay. This can be achieved by using a GPR as a second decrementing counter, effectively encapsulating a kernel comprising our 1 ms delay loop; as shown shaded in Fig. 6.6. If we execute this kernel 100 times, then we will have a 100 ms delay.

The coding of our 100 ms delay subroutine is shown in Program 6.3. The File, named COUNT1, is initialized to d'100' on entry and thereafter the inner 1 ms loop is executed. When WREG reaches zero and the inner loop exits, the File count is decremented *in situ* using the instruction decfsz COUNT1, f. The outer loop only exits when COUNT reaches zero, that is, after 100 inner loops. As long as this outer count remains non-zero, the inner 1 ms delay is re-executed.

Fig. 6.6 A nested loop delay
algorithm

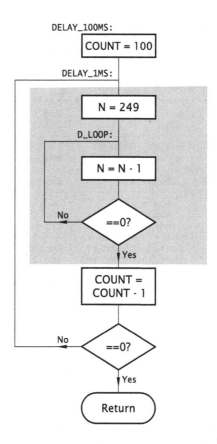

Our simplified treatment of the timing of Program 6.3 is of course not completely accurate, as we have ignored the time taken by the instructions in the outer loop, such as decfsz. However, to compensate somewhat, the number of cycles in the inner loop has dropped four cycles to $4 \times N$, giving an overall time reduction of 100×4 cycles, as the entry call and exit return instruction now belongs to the outer loop. The actual delay turns out to be 99,905 ms; that is accurate to better than 0.1%. Adding a single nop in the outer loop will give an extra delay of 100 cycles, giving 100.005 ms, or 5 µs too long in 100,000 µs.

The maximum delay possible with this program is 256,000 cycles, which could give us our 100 ms delay even up to a 10 MHz crystal, or up to 256 ms delay with a 4 MHz crystal. For even longer delays, we could use a triple-loop structure, potentially giving more than a minute delay. For instance, see Example 6.2.

Our 100 ms delay program is an example of a double-void subroutine, in that no parameters (cf. signals in our hardware plug-in card analog) are sent to it and nothing is returned—just the side effect of a delay (and the alteration of a File). Most subroutines process parameters made available at entry time and provide data at return time.

Program 6.3 A 100 ms delay subroutine

```
; *********************************************************
; * FUNCTION    : Delays for nominally 100ms with a 4MHz xtal *
; * ENTRY       : None                                       *
; * EXIT        : Delay 100ms                                *
; * ENVIRONMENT : File h'030' zero (Fast call)               *
; *********************************************************
COUNT1       equ    h'030'    ; Use File h'30' as a loop counter
N            equ    d'249'    ; Delay parameter computed in text
; -----------------------------------------------------------
DELAY_100MS  movlw  d'100'    ; Initialize outer loop count to 100
             movwf  COUNT1    ; held in File h'030'
; Outer loop -----------------------------------------------
DELAY_1MS    movlw N          ; Set up loop                    -
; Inner loop -----------------------------------------------  -  -
D_LOOP       addlw -1         ; Decrement count            -    -
             nop              ; Put in four extra cycles   -    -
             bnz   D_LOOP     ; Repeat unless zero         -    -
; -----------------------------------------------------------
             decfsz COUNT1,f  ; Decrement outer loop count     -
             bra    DELAY_1MS ; and repeat until zero          -
; -----------------------------------------------------------
             return 1         ; Fast return (restores registers)
```

Fig. 6.7 System view of
$K \times 100$ ms delay subroutine

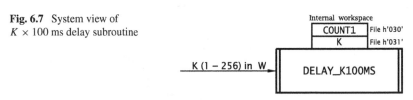

As a simple example, consider the extension of Program 6.3 to give a delay of
$K \times 100$ ms, where K is a byte parameter 'sent' by the caller. The system view of
this function is shown in Fig. 6.7 as a single input signal of range 1–256, with no
output signal; that is, with a void output. This diagram also documents the location
of all **local variables** used internally by the subroutine. This latter attribute is useful
in checking for multiple usage of a File register between different subroutines and
callers. Notice the double line vertical borders commonly used in flow diagrams to
denote functions.

As there is only one input byte-sized parameter, the most convenient place to
place K in the calling program is in the Working register. Thus to call up a 5 s
delay, the caller could use the sequence:

```
movlw  d'50'       ; 50 x 0.1s gives 5 seconds
call   DELAY_K100MS ; Go to it!
```

The subroutine itself in Program 6.4 implements the task list:
1. DO:
    (a) Delay 100 ms.
    (b) Decrement K.
2. WHILE ($K > 0$).
3. End.

Program 6.4 A $K \times 100$ ms delay subroutine

```
; ***********************************************************
; * FUNCTION    : Delays for around K x 100ms @ 4MHz       *
; * EXAMPLE     : K = 100, delays 10 seconds              *
; * ENTRY       : K in W, range 1 - 256                    *
; * EXIT        : Delay K x 100ms                          *
; * ENVIRONMENT: Flags and W altered. Files h'030:31' zero *
; ***********************************************************
COUNT1      equ      h'030'     ; 100ms loop counter
K           equ      h'031'     ; Temporary storage for K
            #define  N  d'249'  ; Delay parameter

DELAY_K100MS
            movwf    K          ; Put K away in a File

; DO 100ms delay --------------------------------------------
DELAY_100MS movlw    d'100'     ; Setup outer loop cnt to 100-
            movwf    COUNT1     ;                           -
; DO 1ms delay ---------------------------------------------- -
DELAY_1MS   movlw    N          ; Set up loop             - -
; ------------------------------------------------------- -
D_LOOP      addlw    -1         ; Decrement count      -  -  -
            nop                 ; One-cycle delay      -  -  -
            bnz      D_LOOP     ; IF not THEN repeat   -  -  -
; ------------------------------------------------------- -  -
            decfsz COUNT1,f     ; Dec 100's loop count -  -
            bra     DELAY_1MS   ; & repeat until zero  -  -
; Decrement K ---------------------------------------------- -
            decfsz   K,f        ;                          -
            bra      DELAY_100MS; Rept 100ms delay WHILE K>0 -
; ----------------------------------------------------------
FINI    return
```

The actual coding simply copies the parameter from the Working register into File h'031' before entering the following delineated coding, which is identical to Program 6.3 and gives a single 100 ms delay. On completion of this fixed delay, K is decremented *in situ* and the delay block repeated until K reaches zero. Thus the 100 ms block is repeated K times.

Because K is tested for zero *after* the 100 ms delay is executed[4] an initial value of $K = 0$ will be treated as $K = 256$, giving a delay range of 0.1–25.6 s. Testing *before* the loop[5] would give a range 0–25.5 s. Again the actual time calculation is approximate, as we have ignored instructions in the outer loops.

As WREG is needed to set up COUNT1 and time the inner 1 ms loop, it cannot be used directly to hold K during the subroutine. In fact, if the caller had known that File h'031' was used by the subroutine to hold K then it could have been passed

---

[4]Known to **C** programmers as a DO-WHILE loop.

[5]Known to **C** programmers as a WHILE loop.

Program 6.5 An alternative $K \times 100$ ms delay subroutine

```
;  ****************************************************************
;  * FUNCTION     : Delays for around K x 100 ms @ 4MHz         *
;  * EXAMPLE      : K = 100, delays 10 seconds                  *
;  * RESOURCE     : DELAY_100MS called                          *
;  * ENTRY        : K in W, range 1 - 256                       *
;  * EXIT         : Delay                                       *
;  * ENVIRONMENT  : W, Status flags and File h'030' altered     *
;  ****************************************************************
K           equ     h'030'      ; Temporary storage for K

DELAY_K100MS
            movwf   K           ; Put K away in a File

; Task 1: DO 100 ms delay------------------------------------
DK_LOOP call    DELAY_100MS

; Task 2: Decrement K----------------------------------------

            decfsz K,f          ; Decrement K

; Task 3: WHILE K > 0----------------------------------------
            bra     DK_LOOP     ; REPEAT WHILE K > 0

            return
```

directly through this File. However, the less the caller has to know about the 'innards' of its subroutines the better it will be, on the basis that a subroutine should disturb its environment as little as possible. DELAY_K100MS is not very good in this respect, using two Files for its internal use and altering both **WREG** and Status flags.

Notice that the directive #define is used in this program to give N its substitute value, rather than equ. Either works to inform the assembler that the defined symbol is to be replaced by d'249'. However, #define is more suggestive of the symbol substitution than equ which is normally used to name a File or bit in a File—see also p. 267.

As an example of what could go wrong, Program 6.5 shows an implementation of the task list but calling the 100 ms block as the existing Program 6.3 subroutine; that is, a nested subroutine. Here File h'030' is used as a store for K oblivious to the fact that this File is also used by subroutine DELAY_100MS as a loop counter. The effect of this interaction is to make K zero on return from DELAY_100MS, which when decremented will always give a non-zero outcome. Thus the delay is infinite and the system locks up! Simply changing K equ h'030' to K equ h'031' fixes the problem; but if another member of the team with responsibility for the DELAY_100MS subroutine alters its internal storage map without communicating this to other team members, then catastrophe may occur! Thus even though each subroutine would have been passed when tested on its own, certain combinations of calling sequences could cause failure. We will return to this problem later.

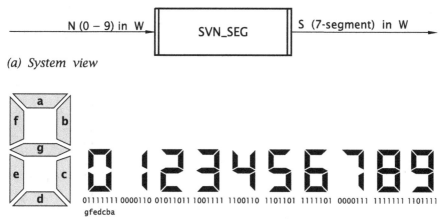

(a) System view

(b) The 7-segment font

Fig. 6.8 The 7-segment display

Incidentally, the `call` instruction in this program could be replaced by `rcall`, assuming that `DELAY_100MS` is nearby. However, note that the Fast option is not available to this instruction.

Program 6.4 is still void, in that no data was returned to the caller on exit. For our next example we will code a non-void subroutine that will activate a decimal readout. Many numeric electronic displays are based on a selective activation of seven segments in the manner shown in Fig. 6.8. These segments are typically implemented using light-emitting diodes (see Fig. 11.19 on p. 371) or electrodes in a liquid-crystal cell.

The system view of our subroutine is shown in Fig. 6.8(a). Here the input signal is a 4-bit binary code representing the ten decimal digits as b'0000–1001' in the Working register. The output, also in WREG, is the corresponding 7-segment code to activate the digit as listed in Table 6.2. This code assumes that a segment is lit/opaque on a binary 1 and is dark/clear on a binary 0. Depending on the physical connections used, the opposite polarity is possible.

Most MPU/MCUs deal with **look-up tables** by storing the codes as part of the program memory and copying the Nth byte out of the table as the mapping function. In the base- and most of the mid-range PIC MCU families, the Harvard structure makes code in the Program store inaccessible as data.[6] In these cases look-up tables are implemented as an array of `retlw` instructions; each returning a constant byte in the Working register. This approach is shown in Table 6.2. In this example bit 7 has arbitrarily made logic 0; but in practice this could be used for other purposes; such as to activate the decimal point.

In developing a coding based on this table structure, the mechanism for element N extraction is to execute the Nth `retlw` instruction. This will place the

---

[6]But see Program 15.5 in *The Quintessential PIC® Microcontroller*, Springer, 2nd ed. 2005 for an exception.

174 The Essential PIC18® Microcontroller

Table 6.2 The 7-segment look-up table showing byte [*N*] being extracted

PC	Table [*i*]	Display	
+0	retlw b'00111111'	; 0	
+2	retlw b'00000110'	; 1	
+4	retlw b'01011011'	; 2	
+6	retlw b'01001111'	; 3	
+8	retlw b'01100110'	; 4	
+10	retlw b'01101101'	; 5	= Table[6]
N ⟹ +12	retlw b'01111101'	; 6	
+14	retlw b'00000111'	; 7	
+16	retlw b'01111111'	; 8	
+18	retlw b'01101111'	; 9	

instruction literal in the Working register and then do a normal return from subroutine back to the caller. As each instruction occupies two bytes in the Program store, then the value of N will need to be doubled to act as an offset to the entry value of the PC. In the example shown, if N is six, then the sixth retlw, located at PC $+ (2 \times 6)$, is executed; returning with the code b'01111000' for 6 in WREG.

The coding shown in Program 6.6 implements this selection mechanism by simply doubling N (by adding the contents of WREG to itself) and then adding this to the lower byte of the Program Counter; that is, PCL in File h'FF9'. As the PC is already pointing to the first retlw instruction, after the addition it then points to the Nth retlw, and this is the exit point from the subroutine.

The code in Program 6.6 takes no account of the possibility that the datum in WREG is greater than h'09'. Of course it shouldn't be, but robust code should cope with all contingencies, even if they technically cannot occur. This is especially true if the code module is to be reusable for general-purpose applications. What would happen if this situation arose and how could you add to the code to gracefully return an error code, say -1, in this eventuality?

This approach of adding a byte number in WREG to the *low* byte of the Program Counter (PCL) to select one of N Return instructions is deceptively simple. Although it works in most situations where the table is small, for the unwary programmer, it can cause seemingly unpredictable system crashes. The problem arises, as altering PCL with the instruction addwf PCL, w only alters the lower eight bits of the 21-bit Program Counter. If the addition should cause overflow, then the net effect is to effectively move the Program Counter proper backwards! For instance, if the subroutine of Program 6.6 happened to be located at h'001F8' (that is, the label SVN_SEG was h'001F8') and if the contents of WREG happened to be h'04'

Program 6.6 The software 7-segment decoder

```
; ****************************************************************
; * FUNCTION    : Returns byte[N] in table                     *
; * FUNCTION    : where N is the contents of WREG              *
; * EXAMPLE     : IF WREG = 06 THEN returns code b'01111101'*
; * ENTRY       : N range 00 - 09 in WREG                      *
; * EXIT        : Table entry N in WREG                        *
; * ENVIRONMENT: WREG = 2N, Status flags altered               *
; ****************************************************************

SVN_SEG addwf  WREG,w         ; Adds WREG to itself to give 2N
        addwf  PCL,f          ; Add WREG to PCL, giving PC + 2N
;              -gfedcba
        retlw  b'00111111'  ; Code for 0; Returned if N = 0
        retlw  b'00000110'  ; Code for 1; Returned if N = 1
        retlw  b'01011011'  ; Code for 2; Returned if N = 2
        retlw  b'01001111'  ; Code for 3; Returned if N = 3
        retlw  b'01100110'  ; Code for 4; Returned if N = 4
        retlw  b'01101101'  ; Code for 5; Returned if N = 5
        retlw  b'01111101'  ; Code for 6; Returned if N = 6
        retlw  b'00000111'  ; Code for 7; Returned if N = 7
        retlw  b'01111111'  ; Code for 8; Returned if N = 8
        retlw  b'01101111'  ; Code for 9; Returned if N = 9
```

on entry, then the outcome of the instruction addwf PCL,f would be to leave the Program Counter at h'(001)F8 + 08 = (001)00' rather than h'00200'. The instruction located at h'00108' (if any) is unlikely to be a Return instruction and so we have exited a subroutine illegally and left the state of the Stack unbalanced. The exact position of a subroutine in the Program store is not easy to predict, as the programmer is unlikely to know in advance where the subroutine is located in memory; that is, what value the PC will have at the beginning of the subroutine. Even if he/she checks the assembler listing file (see Table on p. 247) for the value of SVN_SEG, this can change if subsequent alterations are made to other parts of the program. It is possible to devise code to allow this address boundary to be crossed, but at the expense of complexity—see Example 6.5.

Storing data using a series of retlw instructions is rather inefficient in that a 16-bit instruction is being used to store each 8-bit datum. All enhanced-range family members have implemented a technique of being able to read a single byte datum directly from the Program store using the tblrd instruction—see Program 15.6 on p. 551.

Using the Working register to transfer information to and from a subroutine is limited to a single byte datum each way. Where several pieces of information of byte or greater sizes are to be passed, then GPRs must be pressed into service for this conduit. An example of this is shown in Program 6.7 where a 2-byte datum labeled DIVIDEND_H:DIVIDEND_L is to be divided by a byte DIVISOR The outcome is to be a 2-byte Quotient QUOTIENT_H:QUOTIENT_L and a single byte REMAINDER; as described in Fig. 6.9.

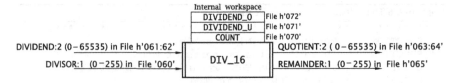

Fig. 6.9 System diagram for the 16 ÷ 8 division subroutine

It is possible to divide by repetitively subtracting the divisor from the dividend/residue and counting until the residue drops below the divisor. Whilst this is practical for single-byte dividends, such as in Program 5.11 on p. 149; for larger dividends the execution time becomes very lengthy. For instance, $65,535 \div 3$ would require 21,845 subtractions.

A rather more efficient technique is based on an algorithm which uses a combination of shifting and subtraction; as briefly illustrated on p. 11. For our 2-byte example this only requires a fixed 16 passes through the loop. The task list in this instance is:

1. Extend Dividend internally to four bytes.
2. DO
   (a) Shift 4-byte Dividend left once.
   (b) Subtract (Divisor `<<16`) from Dividend/residue.
   (c) Shift borrow (Carry bit) left into Quotient (Quotient × 2).
   (d) IF no borrow (`C == 1`) THEN update the Dividend/residue with the difference.
   (e) Repeat 16 times.
3. Remainder is the residue Dividend shifted right 16 times.

As shifting the Divisor left 16 times (`<<16`) essentially means aligning it beyond the 2-byte Dividend, we need to extend the latter by adding an upper byte. Indeed, as the Dividend will be shifting left during this process, a further overflow byte needs to be added to give a 4-byte Dividend array. Both these extension bytes will need to be initialized as zero. Our memory map for our coding with this in mind is illustrated in Fig. 6.10.

Program 6.7 declares the variables that are passed to and from the subroutine at the beginning of the main program. Keeping all these **global** declarations in one part of the program and using a different File for each overall global variable reduces the possibility of interaction but at the expense of rather extravagant use of scarce Data memory resources. Temporary local storage is declared within each subroutine, as its need will be 'thrown away' after the subroutine is terminated. However, interaction can still occur in local storage where nested subroutine structures are used.

The coding follows the task list closely.

DIV_16: Instructions 1–4
The two **local** (internal) variables used to extend the Dividend; that is DIVIDEND_O and DIVIDEND_U, are zeroed. The initialization is wrapped up by setting the local variable COUNT to 16, to act as the loop counter.

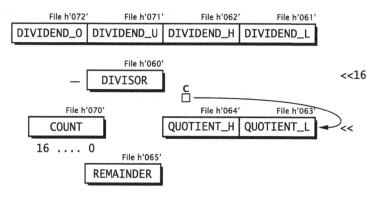

Fig. 6.10 Memory map for the shift and subtract division subroutine

DIV_LOOP: Instructions 5–9
Inside the shift and subtract loop, the 4-byte Dividend chain is shifted left one place, as described in Fig. 5.15 on p. 133.

Instructions 10–13
The Divisor byte is subtracted from DIVIDEND_O:DIVIDEND_U; effectively −(DIVISOR <<16).

Instructions 14 & 15
The 2-byte Quotient is shifted left once with the Carry flag (not borrow) from the previous subtraction coming in from the right as the new bit 0.

Instructions 16–18
Based on this new bit 0 of the Quotient, if there was no borrow-out ($Q_0 = 1$) then the difference from the subtraction overwrites the extension bytes of the Dividend; giving the new residue Dividend.

Instructions 19 & 20
The loop count is decremented and the process repeated a total of 16 times before breaking out.

Instructions 21 & 22
The final residue Dividend shifted right 16 places is the Remainder. Copying the Upper byte of the residue chain to REMAINDER effectively implements this >>16 operation.

In order to use this subroutine, the caller puts the Divisor into File h'060' and the two Dividend bytes DIVIDEND_H and DIVIDEND_L respectively into File h'062:61'. On return, the 16-bit QUOTIENT:H and QUOTIENT_L can be read at File h'064:63' respectively. The Remainder is accessed at File h'065'.

As an example, consider that the bytes located at File h'043:2' are to be divided by the byte at File h'046'.

```
movff  h'042',DIVIDEND_L  ; Get low byte of Dividend
movff  h'043',DIVIDEND_H  ; and high byte in situ
movff  h'046',DIVISOR     ; Get Divisor in place
call   DIV_16             ; Go to it!
; On return the Quotient now in File h'064:63'
; and Remainder in File h'065'
```

Program 6.7 The 16 ÷ 8 division subroutine

```
; Global declarations
DIVISOR     equ  h'060'     ; On entry the Divisor is here
DIVIDEND_L equ  h'061'      ; and the Dividend low byte is here
DIVIDEND_H equ  h'062'      ; Dividend high byte
QUOTIENT_L equ  h'063'      ; The Quotient is built up here
QUOTIENT_H equ  h'064'
REMAINDER  equ  h'065'      ; The Remainder on exit

; ****************************************************************
; * FUNCTION   : Divides a 2-byte Dividend by a 1-byte Divisor    *
; * FUNCTION   : Giving a 2-byte Quotient and 1-byte Remainder    *
; * EXAMPLE    : Dividend = h'FFFF' (65,535); Divisor = h'0A' (10) *
; * EXAMPLE    : Quotient <- h'1999' (6553); Remainder <- h'05'   *
; * ENTRY      : DIVISOR, File h'060'; DIVIDEND_L, File h'061'     *
; * ENTRY      : DIVIDEND_H, File h'062'                           *
; * EXIT       : QUOTIENT_L, File h'063'; QUOTIENT_H, File h'064'  *
; * EXIT       : REMAINDER, File h'065'                            *
; * ENVIR'MENT: COUNT, DIVIDEND_U, DIVIDEND_O (Files h'070 -- 72')*
; * ENVIR'MENT: and W, STATUS altered                             *
; ****************************************************************
; Local declarations
COUNT        equ  h'070'     ; Holds the Quotient bit count
DIVIDEND_U equ  h'071'       ; Overflow byte for shifted Dividend
DIVIDEND_O equ  h'072'       ; Holds any overflow from Dividend

DIV_16      movlw  d'16'     ; Sixteen times around the loop
            movwf  COUNT     ; Zero the Bit count
            clrf   DIVIDEND_U ; Zero the Upper Dividend byte
            clrf   DIVIDEND_O ; Zero the Overflow byte
; Shift 4-byte Dividend left one place --------------------------
DIV_LOOP    bcf    STATUS,C  ; In any case shift Dividend left
            rlcf   DIVIDEND_L,f
            rlcf   DIVIDEND_H,f; All 32 bits
            rlcf   DIVIDEND_U,f
            rlcf   DIVIDEND_O,f
; Subtract Divisor << 16 from Upper byte of Dividend --------------
            movf   DIVISOR,w  ; Get Subtrahend
            subwf  DIVIDEND_U,w; Dividend High - Subtrahend
            btfss  STATUS,C   ; Skip if no Borrow
              decf DIVIDEND_O,f; ELSE take away 1 from Overflow byte
; Shift the borrow into the 2-byte Quotient ----------------------
            rlcf   QUOTIENT_L,f; Shift Borrow into Quotient
            rlcf   QUOTIENT_H,f
; Update the Dividend/residue if no borrow ----------------------
            btfsc  QUOTIENT_L,0; Skip IF this bit is a 1
              movwf DIVIDEND_U  ; ELSE update Dividend Upper byte in W
            clrf   DIVIDEND_O  ; Always zero the overflow Dividend
; Loop housekeeping --------------------------------------------
            decfsz COUNT,f    ; Record one more bit
            bra    DIV_LOOP   ; Repeat 16 times in all
; The remainder is the High byte in the Dividend -----------------
REMAIN      movff  DIVIDEND_U,REMAINDER
            return            ; Return to caller
```

Hardware stack

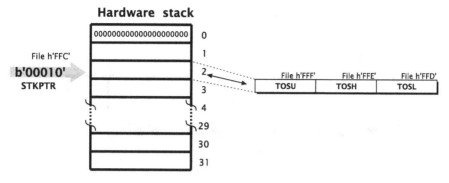

Fig. 6.11 A more detailed view of the Hardware stack; peeking into cell 2

For more advanced programming structures, such as multi-tasking and real-time operating systems (RTOS), the Hardware stack in the PIC18 core has been significantly enhanced compared to previous families. Figure 6.11 shows how the Hardware stack integrates with the STKPTR (STacK PoinTeR) and triplet Top Of Stack TOSU:TOSH:TOSL SFRs; collectively known as TOS. At any instant the contents of the TOS registers reflect the 21-bit datum in the cell pointed-to by the Stack Pointer. Changing the TOS will effectively alter the contents of this stack cell, allowing the software both to examine and optionally change the value of a return address or other data previously pushed into the Hardware stack. By moving the StacK Pointer up and down, a range of stack-located data can be accessed; for instance, to set up a table of subroutine (or interrupt) return addresses.

This increased flexibility comes at a price. The possibility of the Hardware stack overflowing (full) or underflowing in certain circumstances is increased. In earlier families an over/underflowing stack causes the Stack Pointer to wrap around, and overwrite previously stored return addresses, causing catastrophic failure. In the PIC18 Hardware stack, this situation can be trapped.

Figure 6.12 shows that as well as the five Stack Pointer bits **SP[4:0]**, STKPTR also holds two status bits which signal this error situation. **STKFUL** (STacK FULl) will be set to 1 whenever the Stack Pointer (SP) reaches 31. In a similar manner, **STKUNF** (StacK UNderFlow) will be set whenever SP decrements to cell 0. The precise action that takes place whenever a stack over/underflow occurs depends on the state of the **STVREN**[7] (STack oVer/underflow Reset ENable) Configuration fuse bit—see Appendix B. Essentially these Configuration fuses are set or cleared when the software is blasted into the Program store and enable the designer to setup a series of options, such as what should happen when there is a stack malfunction.

STVREN ON (= 1)—default
In this situation:

---

[7]Confusingly, in some devices, including the PIC18F1220, this is called STVR.

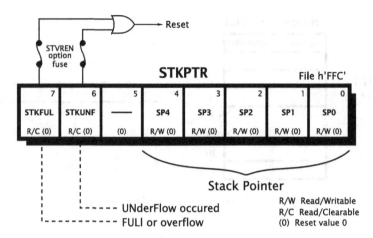

Fig. 6.12 The enhanced-range STacK PoinTeR register

- A push (e.g. from a Call type instruction) into cell 31 will set STKFUL and copy the Program Counter into the stack and then will reset the MCU. This is equivalent to bringing the Master CLeaR ($\overline{\text{MCLR}}$) pin low, as described on p. 323, and the SFRs will be initialized into their Reset state. In the case of STKPTR, bits SP[4:0] will be cleared; effectively resetting the stack. However, STKFUL will remain set to 1.
- When the Stack Pointer is pointing to cell 1, any further pop (e.g. from a Return type instruction) will set STKUNF and reset the MCU as above.

STVREN OFF (= 0)

- When the stack overflows, STKFUL will be set. Any subsequent pushes of the PC will not overwrite this data, and thus newer data will be lost. The Stack Pointer will remain at 31.
- When the stack underflows, STKUNF will be set and the next pop will return a value of zero to the PC. Effectively, this will cause the program to restart from its Reset vector. This is not the same as a reset, as the state of the various SFRs will not be changed.

In all cases the STKUNF and STKFUL bits can only be subsequently cleared by software; e.g. `bcf STKPTR, 7`. A Power-On Reset (POR) will also clear the Stack status bits.

As the Stack Pointer component of STKPTR is readable and writeable (R/W) it can be moved up or down using normal instructions. For instance, `movlw 5`, `addwf STKPTR, f` will move the Stack Pointer up five places. The contents of the newly targeted cell can then be examined through the TOS registers, and changed if desired. The disadvantage of this approach is that it doesn't activate the over/underflow detection mechanism. Also in some cases, such as multi-precision arithmetic, altering the Status flags can be an issue.

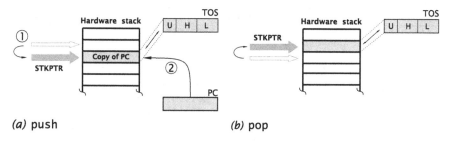

Fig. 6.13 Illustrating the push and pop instructions

Apart from the Call and Return instructions, the PIC18 instruction set includes the two instructions push and pull, that are specifically designed to alter the Stack Pointer without causing a jump to or return from a subroutine.

push   `0000000000000101`

This instruction, shown in Fig. 6.13(a) increments the Stack Pointer and then copies the 21-bit contents of the PC (which is already pointing to the instruction *following* push) into the stack, which is then accessible through TOS and can be changed. This is similar to call, but without overwriting the PC with a destination address. Of course, the original contents of this Hardware cell will be destroyed by the Push action, and if this is a problem incf STKPTR, f can be used as a substitute (but this latter alter Status flags).

pop   `0000000000000110`

As shown in Fig. 6.13(b), pop decrements the SP, and the TOS value now changes to reflect the value previously pushed into the stack. A series of pop instructions can be used to move the SP back up if desired, and clrf STKPTR will effectively reset it to the bottom of the stack.

Neither push nor pull instruction affects the STKFUL and STKOVF flags nor any Status bits.

As an example, consider that we wish to code a subroutine to convert a 16-bit natural binary word in File h'061:60 to five BCD digits, located in the array:

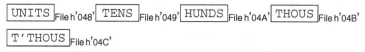

For instance; h'FFFF' ⤳ 06 05 05 03 05.

The simplest way to convert binary to decimal is to repetitively divide by ten, with the series of remainders giving the BCD digits; units first (Horner's method). In our instance $65,535 \div 10 \rightsquigarrow 6553r5$; $\div 10 \rightsquigarrow 655r3$; $\div 10 \rightsquigarrow 65r5$; $\div 10 \rightsquigarrow 6r5$; $\div 10 \rightsquigarrow 0r6$.

Now we already have a $16 \div 8$ subroutine in Program 6.7 which our new subroutine can call. If we wish to use, say, FSR0 as a pointer to our BCD array, then all we

Program 6.8 16-bit binary to BCD conversion

```
; ******************************************************************
; * FUNCTION   : Converts a 16-bit binary word to 5-digit BCD    *
; * EXAMPLE    : b'1111 1111 1111 1111' -> 06 05 05 03 05        *
; * ENTRY      : 2-byte binary number in DIVIDEND_H:DIVIDEND_L   *
; * EXIT       : 5 BCD digits in File h'048 -- 04C' Units first  *
; * RESOURCES  : Subroutine DIV_16.                              *
; * ENVIR'MENT : DIVISOR, DIVIDEND:4, QUOTIENT:2, W, S'S altered *
; ******************************************************************

; First save FSR0 in stack ------------------------------------
BIN_2_BCD push                 ; Move Stack Pointer up one
          movf   FSR0L,w       ; Push FSR0 into Stack
          movwf  TOSL
          movf   FSR0H,w
          movwf  TOSH

; Divide binary word by ten five times; giving decimal digits --
          lfsr   0,h'048'      ; Point to a BCD array
M_LOOP    call   DIV_16        ; Divide by ten
          movff  REMAINDER,POSTINC0 ; Put the Remainder in the array
          movff  QUOTIENT_L,DIVIDEND_L ; Set up the next Dividend
          movff  QUOTIENT_H,DIVIDEND_H
          movlw  h'4C'         ; Check for h'4D'
          cpfsgt FSR0L         ; Compare with Low byte of FSR0
          bra    M_LOOP        ; IF not yet there THEN repeat

; Restore entry state of FSR0 ---------------------------------
          movff  TOSH,FSR0H    ; Get back out the entry value
          movff  TOSL,FSR0L
          pop                  ; Backup to the caller's return address

          return               ; ELSE done
```

need to do is initialize this to File h'048', and in a loop call our ÷10 subroutine five times, putting the return remainder in our array with automatically incrementation, and using the return quotient as the new dividend.

One problem remains. Although the enhanced-range family is richly endowed with File Pointer registers compared to previous families (three 12-bit as compared to one 8-bit FSR); nevertheless they are in great demand and unless enough care is used, our subroutine may be called from another routine which also uses FSR0. Using a Fast Call/Return will automatically save and retrieve WREG, STATUS and BSR, but any other common resources need to be 'manually' stashed away if necessary. In our case, we could reserve two fixed GPR Files for this task. However, this is not foolproof, as a nested subroutine or interrupt service routine may inadvertently overwrite this when also saving its local value of FSR0. A better way of making transparent use of FSR0 is to push a copy of it into the Hardware stack *before* the call to BIN_2_BCD and then pop it out of the stack after return.

On entry to our subroutine listed in Program 6.8, instruction 1 moves the Stack Pointer up one cell (and also superfluously copies PC + 2 into this cell). The following four instructions simply copy the low- and high-bytes of FSR0 into TOSL

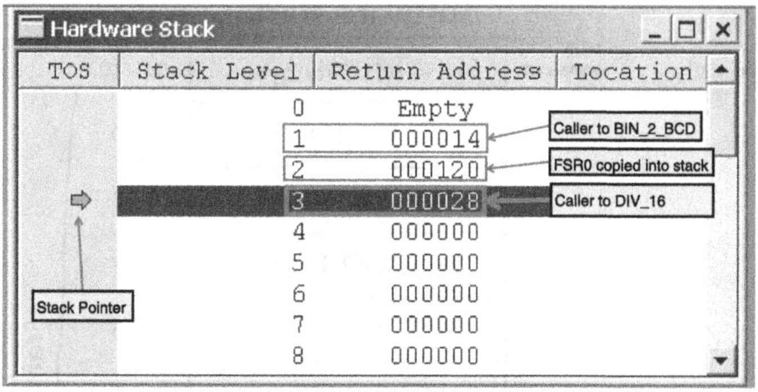

Fig. 6.14 A view of the stack with MPLAB SIM on arrival at subroutine `DIV_16`

and TOSH respectively via WREG.[8] This overwrites the contents of this cell. In Fig. 6.14 cell 2 is shown holding this original value, which in this instance is h'120'. Of course, should other nested subroutines also use FSR0 and save it in the same way, then the new copy will be further up the stack each time this is done. Thus, within the capacity of the stack, there will not be a problem with corruption.

Our program is now free to initialize FSR0 to File h'048' and then call `DIV_16` with a Divisor of ten. On return, the two Quotient bytes are copied as the new Dividend and the Remainder copied into memory with the FSR0 used as a Post-Incrementing pointer. When this pointer exceeds h'04C' the process is complete. On return from the subroutine `DIV_16` the Stack pointer will be pointing to the cell holding the original value of FSR0 (cell 2 in the diagram) and these two bytes are then copied back into FSR0H:FSR0L. Finally the Stack Pointer is popped back down a cell to point to the caller's address before returning from subroutine `BIN_2_BCD` (cell 1 in the screenshot holds the caller's address h'00014'.

Although this example only saved FSR0, the principle can be extended for other Files; GPRs as well as SFRs—see Example 6.7. Where the Fast Call/Return mechanism cannot be used; for instance with interrupts active or more than one nested subroutine, the Working register, Status and Bank Select Registers can be pushed and popped to/from the stack.

Although the Hardware stack can be used as described to preserve system information (or **context**), as well as passing parameters back and forth between caller and subroutine (see also Example 6.7), its use in this regard is really rather limited. This is only partly to do with its restricted size, but also the overhead of continually moving the Stack pointer up and down to access these data.

As an alternative it is possible to simulate a stack-like structure in normal Data memory, using a spare File Select Register as a **Pseudo Stack Pointer** (**PSP**). In the examples given in this chapter, FSR2 is used in this regard. If we allocate the block of Files h'0A0–0FF' for our **Software stack**, then that gives us 96 bytes of storage.

---

[8]The instruction `movff` should not be used with the TOS or PC registers as *destination* as corruption may occur if interrupts are in use, but they can be used as *source* data.

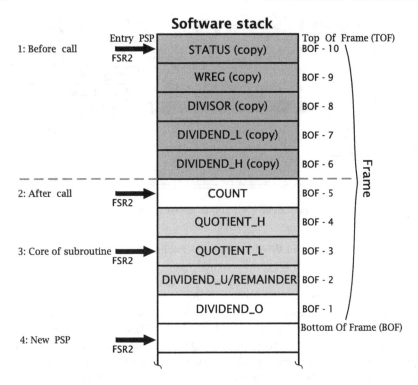

Fig. 6.15 The Software stack from the perspective of subroutine DIV_16S

To set up this stack, all that needs to be done is to initialize FSR2 at the start of the program code following the Reset vector; lfsr 2,h'0A0' in the code fragment of p. 185.

To illustrate this concept, consider a stack-oriented version of the division subroutine of Program 6.7. We are going to use the stack structure, shown in Fig. 6.15, to both pass the Divisor and Dividend to the subroutine and also preserve the Working and Status register states. On return the Quotient and Remainder will be available to the caller in memory on the stack. The code fragment following shows how this is done. For the purposes of this example we have assumed that prior to the call the Divisor is located in File h'020' and Dividend in File h'022:21'. On return the Quotient is to be put into File h'029:28' and Remainder in File h'02A'.

(a) Make a copy of the Status and WREG registers using the movff instruction, with the destination pointing into the Software stack, using FSR2 with Post-increment Indirect addressing. The movff instruction does not change any Status flags. Effectively this moves the PSP down the stack (up in address).

(b) In an identical manner, copy the Divisor and Dividend (two bytes) into the stack frame, automatically advancing FSR2.

(c) Call the subroutine, with the two variables passed in the frame.

(d) On return the PSP will be restored to the Top Of Frame (TOF). On this basis, us-ing the Working register as an offset to FSR2 (see p. 107) both Quotient bytes at

TOF + 7:6 and Remainder at TOF + 8 are copied to their destination. Finally the two system variables at the Top Of Frame (TOF and TOF + 1 are copied across.

```
; Global variables

DIVISOR     equ    h'020'          ; Divisor found here
DIVIDEND_L  equ    h'021'          ; 2-byte Dividend
DIVIDEND_H  equ    h'022'
QUOTIENT_L  equ    h'028'          ; 2-byte Quotient
QUOTIENT_H  equ    h'029'          ; to be put here
REMAINDER   equ    h'02A'          ; & Remainder to put here

; In the beginning set up Top Of Stack
MAIN    lfsr  2,h'0A0'             ; Reset value of PSP (FSR2)
;
;
; Some time later when ready to call subroutine
; (a) ------------------------------------
        movff STATUS,POSTINC2      ; Copy STATUS into S'stack
        movff WREG,POSTINC2        ; Copy WREG into S'stack

; (b) ------------------------------------
        movff DIVISOR,POSTINC2     ; Copy Divisor into S'stack
        movff DIVIDEND_L,POSTINC2; Copy the 2-byte Dividend
        movff DIVIDEND_H,POSTINC2; into Software stack

; (c) ------------------------------------
        call  DIV_16S              ; Go and DO the division

; (d) ------------------------------------
        movlw 6                    ; Point to QUOTIENT_H (TOF+6)
        movff PLUSW2,QUOTIENT_H    ; Copy byte out of frame
        incf  WREG,f               ; Point to QUOTIENT_L
        movff PLUSW2,QUOTIENT_L    ; Copy byte out of frame
        incf  WREG,f               ; Point to REMAINDER
        movff PLUSW2,REMAINDER

; Restore WREG & STATUS without affecting any Status flags --
        movf  PREINC2,f            ; Move PSP up one (TOF+1)
        movff POSTDEC2,WREG        ; Copy Working reg out; PSP-
        movff INDF2,STATUS         ; & lastly return Status (TOF)
```

Coding of the DIV_16 subroutine is given in Program 6.9. This uses the same processes as Program 6.7, but the variables are accessed relative to the PSP. On entry, FSR2 is incrementally moved towards the Bottom Of Frame all the while initializing the two short-term variables and the three local variables (shown shaded) that will eventually be returned to caller.

When the intialization has been completed, the PSP will be pointing to the first new location beyond the frame, labeled in Fig. 6.15 as BOF. In this diagram, all variables are labeled relative to this point. For instance, to decrement the temporary

Program 6.9 16 ÷ 8 division subroutine using a Software stack

```
; ***************************************************************
; * FUNCTION   : Divides a 2-byte Dividend by a 1-byte Divisor  *
; * FUNCTION   : Giving a 2-byte Quotient and 1-byte Remainder  *
; * EXAMPLE    : Dividend = h'FFFF' (65,535); Divisor = h'0A' (10)*
; * EXAMPLE    : Quotient <- h'1999' (6553); Remainder <- h'05'  *
; * ENTRY      : DIVISOR @ BOF-8, DIVIDEND_L @ BOF-7             *
; * ENTRY      : DIVIDEND_H @ BOF-6                              *
; * EXIT       : QUOTIENT_L @ BOF-3; QUOTIENT_H @ BOF-4          *
; * EXIT       : REMAINDER @ BOF-2                               *
; * ENVIR'MENT: PSP is FSR2                                      *
; ***************************************************************
DIV_16S   movlw   d'16'         ; Set up loop count to 16
          movwf   POSTINC2      ; As PSP-5
          clrf    POSTINC2      ; QUOTIENT_H = 00 (PSP-4)
          clrf    POSTINC2      ; QUOTIENT_L = 00 (PSP-3)
          clrf    POSTINC2      ; DIVIDEND_U = 00 (PSP-2)
          clrf    POSTINC2      ; DIVIDEND_O = 00 (PSP-1)

; Shift 3-byte Dividend left one place ---------------------------
DIV_LOOP  bcf     STATUS,C      ; In any case shift Dividend left
          movlw   -7            ; Prepare to point to DIVIDEND_L
          rlcf    PLUSW2,f      ; All four bytes
          movlw   -6
          rlcf    PLUSW2,f      ; Making sure not to disturb
          movlw   -2            ; the Carry flag
          rlcf    PLUSW2,f
          movlw   -1
          rlcf    PLUSW2,f

; Subtract Divisor << 16 from Upper byte of Dividend --------------
          movlw   -8            ; Prepare to get Divisor
          movf    PLUSW2,w      ; into W (Subtrahend)
          movf    POSTDEC2,f    ; Move PSP up one
          movf    POSTDEC2,f    ; and to point to DIVIDEND_U
          subwf   INDF2,w       ; Dividend Upper - Subtrahend
          movf    PREINC2,f     ; Point to Overflow Dividend byte
          btfss   STATUS,C      ; Skip if no Borrow
          decf    INDF2,f       ; ELSE subtract 1 from Overflow byte
; Shift the borrow into the 2-byte Quotient ----------------------
          movf    POSTDEC2,f    ; Move down two places
          movf    POSTDEC2,f    ; to get to low byte of Quotient
          rlcf    POSTDEC2,f    ; Shift Borrow into Quotient Low
          rlcf    POSTINC2,f    ; and then into Quotient High
; Update the Dividend/residue if no borrow -----------------------
          btfsc   POSTINC2,0    ; Skip IF bit Q0 is a 0 (Borrow-in)
          movwf   INDF2         ; ELSE update Dividend Upper byte
          clrf    PREINC2       ; Always zero the overflow Dividend
          movf    PREINC2,f     ; Return to BOF location
; Loop housekeeping ---------------------------------------------
          movlw   -5            ; Prepare to point to Loop Count
          decfsz  PLUSW2,f      ; Record one more bit
          bra     DIV_LOOP      ; Repeat 16 times in all
; The remainder is the Upper byte in the Dividend ----------------
REMAIN    movf    FSRL2,w       ; Move down ten places
          addlw   -d'10'        ; Add ten to low byte of FSR2
          movwf   FSR2
          btfss   STATUS,C      ; Skip IF no Borrow
          decf    FSRH2,f       ; ELSE return borrow to High FSR2

          return                ; Return with data in S'stack
```

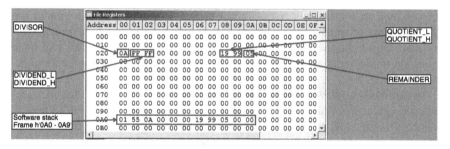

Fig. 6.16 A view into the Data store after a run with h'FFFF' ÷ 0A' (65,535 ÷ 10)

variable COUNT, an address of BOF + 5 will have to be accessed. There are three ways of doing this, all of which are used in the coding.

- Put an offset in WREG and use this with the Plus W Indirect address mode—see p. 107. In our example, movlw 5; incf PLUSW2,f. This technique is also used to access the four Dividend bytes when shifting the chain left once. The Carry flag is not altered between accesses.
- Use Automatic Increment and Decrement Indirect address modes to move FSR2 up and down. If it is not intended to alter any values, just increment or decrement FSR2 then the instruction movf PREINC2,f or similar will simply copy a da-tum onto itself, but change FSR2 according to the address mode used. This latter technique is used in the coding prior to shifting the borrow-out into the Quotient.

  As this approach involves actually changing the PSR, the programmer needs to keep track of its value and problems can occur when Branch and Skip instructions are used, as these can conditionally jump over instructions modifying PSP.
- A constant offset can be added to or subtracted from FSR2, remembering that this register is really two single-byte SFRs; FSRH2 and FSRL2.

  In our coding this technique is used to subtract ten from the value of FSR2 to move the PSP back to the Top Of Frame position, prior to returning to the caller. This rollup effectively cleans up the frame, allowing the same GPRs to be reused later whenever a new subroutine is called which also makes use of the Software stack. Of course before this, any data required by the caller must be collected and set into its final resting place, as previously described.

Figure 6.16 shows the Data store after execution with an initial value for DIVIDEND:2 of h'FFFF' and for DIVISOR of h'0A' (65,535 ÷ 10). The outcome h'1999' (6553) is in the Software stack along with the remainder of h'05'. Both vari-ables have also been copied into QUOTIENT:2 and REMAINDER.

Program 6.9 requires 40 instructions as compared to 22 in Program 6.7. Its exe-cution time of 478 cycles also compares unfavorably with 282 cycles. However, its transparency gives superior reusability and robustness. Furthermore, a fully trans-parent function is re-entrant; that is can call itself. These attributes are valued for high-level language compilers, which nearly always use stack-oriented implemen-tations at assembly-level. In larger assembly-coded systems, a Software stack model is likely to be more economical in its use of scarce data memory, as after a frame has been cleaned up, such memory can safely be reused. However, programs running on PIC MCUs are often not very complex and the comparatively small Program memory may restrict the use of this relatively extravagant technique. Where real-time execution time is critical the additional burden of stack handling is unlikely to be worthwhile.

To facilitate high-level compiler design (see Fig. 9.1 on p. 277) and to increase the code efficiency of high-level language script, many of the newer PIC18 family members (not the PIC18F1220) provide an optional extension, adding eight instructions and a new address mode to the core instruction set. This option is controlled by the XINST fuse, which can be set when the code is blasted into the Program store—as described on p. 316. Normally the processor defaults to disabling this extension.

Although the Extended instruction set are designed primarily for optimizing high-level language compilers, it is possible to use them for general assembly-level coding. For reference they are briefly reviewed here.

addfsr | 11101000 | ii | LLLLLL |

Augments the 2-byte File Select Register i with a 6-bit literal (0–31); for instance, addfsr 0,6 (add six to FSR0). Unlike the similar process at the end of Program 6.9, neither status flags nor the contents of WREG are altered. This makes this instruction useful when moving the PSP in the middle of a multi-byte arithmetic process.

subfsr | 11101001 | ii | LLLLLL |

The subtractive counterpart of addfsr. For instance, subfsr 2,d'10' subtracts ten from the 2-byte FSR2; potentially replacing the five penultimate instructions in Program 6.9.

addulnk | 1110100011 | LLLLLL |

At the end of a stack-based function, the PSP has to be restored to its entry value and then execution returned to the caller. This process is called **unlinking**. The ADD UNLinK instruction adds a 6-bit literal to FSR2 and then executes a Return; that is it combines the two processes into one for stacks that grow downwards from the entry PSP—see Fig. 6.20.

subulnk | 1110100111 | LLLLLL |

This subtractive counterpart to addulnk is used for stacks that grow upwards; as in Fig. 6.15. For instance, to unlink Program 6.9 we could have subulnk d'10'; replacing the last six instructions!

movsf | 110010110 | LLLLLLL$_s$ |

| 1111 | ddddddddddddd |

This variant of movff specifies the source File as a 7-bit literal offset to the pointer FSR2. Thus, movsf [8],WREG copies the byte at the effective address FSR2 + 8 (i.e. eight higher than the cell pointed to by FSR2) to the Working register. This means that FSR2 does not have to be altered to access data in the stack. Like movff, the destination should not be any of the PC, TOS or INTCON registers—see Program 6.16 for an example.

movss | 111010111 | LLLLLLLₛ |

                                             | 1111 | XXXXX | LLLLLLLₐ |

This instruction is similar to `movsf` but *both* source and destination are pin-pointed as an offset from FSR2. Thus `movss [6],[2]` copies the byte from cell six above FSR2 into cell two above FSR2.

pushl | 11111010 | LLLLLLLL |

Copies the specified 8-bit literal into wherever FSR2 is pointing to and then decrements FSR2. For instance, `pushl 6` sets the pointed to cell to six and then moves the pointer down one.

callw | 0000000000010100 |

This version of `call` generates the destination address of the target subroutine as PCLATU:PCLATH:WREG. By storing an array of `goto` instructions, up to 64 subroutines (remember that a `goto` instruction occupies four bytes in the Program store) can be accessed as an indexed array, with the Working register holding the index. For instance, if PCLATU = h'00', PCLATH = h'20' and WREG = h'08', then `callw` will save the PC on the Hardware stack in the normal way and then jump to the instruction at h'002008', which will normally (but not necessarily so) be a go to the start of an associated subroutine. This subroutine is terminated by a Return instruction in the normal way, but note that there is no Fast option.

As well as adding the eight listed instructions, this Extended mode also enables a new kind of Indirect address mode, known as **Indexed Literal Offset**. This re-interprets any address in Access RAM (see p. 81) in the range h'000–05F' as a literal to be offset from FSR2. For instance, if FSR2 = h'040', then `clrf [h'12']` will actually clear File h'052', where the [] brackets indicate Indirect—see Program 6.16. The instruction `clrf h'12'` is no longer permitted if the Extended instruction set is enabled. However, if the programmer really wants to clear the absolute location File h'012' then there are two ways to do this. This high-jacking of Absolute to Indexed Literal Offset address modes is only implemented if the a-bit is 0 (see p. 81). Normally the programmer does not have to explicitly specify Access as against Banking addressing, as the assembler will take care of the setting of this bit. However, if the BSR is h'00' (as it is on a Power-on Reset) then the instruction `clrf h'012',1` will use the Banking mode to clear File h'(0)12'. Alternatively, if FSR2 is zeroed then File h'012' is the target. This re-interpretation does not apply to locations above File h'05F' in the Access bank.

Examples

Example 6.1 Write a subroutine to give a fixed 208 μs delay. Assume a 4 MHz processor clock rate.

Solution For a short time period like this, the code outlined in Program 6.1 provides adequate delay. With a 4 MHz clock, one cycle is 1 μs and thus we require 208 cycles. From p. 166 we have:

$$4 + 4 \times N = 208 \text{ cycles,}$$

$$N = 51.$$

Thus replacing N by d'51' will give the required delay. What value of N would you use if a 20 MHz crystal were used?

Example 6.2 At the other end of the spectrum write a subroutine to give a delay of one minute.

Solution Sixty seconds can be implemented as 240×250 ms. Our solution Program 6.10 closely follows the coding in the triple-loop Program 6.4 which carries out a $K \times 100$ ms delay. The maximum value of K is 255, which would only give 25.5 s, but we can increase the middle loop to 250 ms and thus give increments of $\frac{1}{4}$ s. If we now use an outer loop with a 240 count, we have our 60 s delay.

Comments in the listing give the full delay calculation, which totals 59.941206 s, accurate to approximately 0.1%. Once again the routine can be padded out with nop

---

Program 6.10 A 1-min delay program

```
; ***********************************************************
; * FUNCTION: Delays for approx a minute for a 4 MHz XTAL    *
; * ENTRY   : None                                          *
; * EXIT    : Status & W altered, Files h'030:31' zero      *
; ***********************************************************
; Local variables
COUNT1        equ     h'030'      ; Counter at File h'030'
COUNT2        equ     h'031'      ; and File h'031'

DELAY_1_MIN movlw   d'240'      ; Put 240 as the MS count, 1~
            movwf   COUNT2      ;                          1~

; Outer loop (250ms ----------------------------------------------
DELAY_250MS movlw   d'250'      ; Put 250 as mid count, 240*1~   -
            movwf   COUNT1      ; for a 250ms delay,      240*1~ -
;                                                                -
; Mid loop (1ms) -------------------------------------------------  -
DELAY_1MS   movlw   d'249'      ;                       250*240~    -   -
;                                                                -   -
; Inner loop -----------------------------------------------   -   -
D_LOOP        addlw   -1        ;                   249*250*240~  -   -   -
              nop               ;                   249*250*240~  -   -   -
              bnz     D_LOOP    ;[2*(249-1)+1*250*240)~  -   -   -
; -----------------------------------------------------------  -   -
              decfsz  COUNT1,f  ;               (250+1)*240~        -   -
              goto    DELAY_1MS ;           2*(250-1)+1*240~        -   -
; --------------------------------------------------------   -
              decfsz  COUNT2,f  ;                         240+1~        -
              goto    DELAY_250MS;         2*(240-1)+1~               -
; --------------------------------------------------------   -
              return            ;                             2~
```

instructions. Each nop after the first decfsz adds $250 \times 240 = 60,000$ cycles; so one will change the shortfall to an excess of 1206 cycles in 60,000,000 cycles, or better than $+0.002\%$.

Example 6.3 Write a subroutine to evaluate the square root of a 16-bit integer located in File h'027:26' and return the 8-bit outcome in the Working register.

Solution The crudest way of doing this is to try every possible integer k from 1 upwards, generating k^2 by multiplication and checking that the outcome is no more than Number. An equivalent but (perhaps) slightly more sophisticated approach is based on subtracting the series 1, 3, 5, 7, 9, 11,... from Number until underflow occurs. Counting the number of subtractions gives the nearest square root. This series comes from the relationship:

$$k^2 = \sum_{I=0}^{k} (2 \times I) + 1.$$

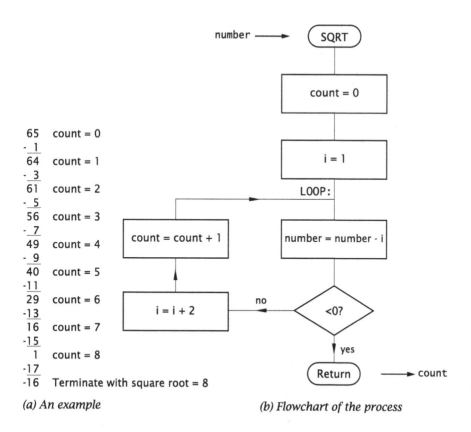

(a) An example   (b) Flowchart of the process

Fig. 6.17 Finding the square root of an integer by repetitive subtraction

On this basis a possible structure for this function is:

1. Zero the loop count.
2. Set variable I (the magic number) to 1.
3. DO forever:
   (a) Take I from Number.
   (b) IF the outcome drops below zero THEN BREAK out.
   (c) ELSE increment the loop count.
   (d) Add 2 to I.
4. Return loop count as $\sqrt{\text{Number}}$.

Program 6.11 Coding the square root subroutine

```
; Global declarations -------------------------------------
WREG          equ    h'FE8'  ; Working register
NUM_L         equ    h'026'  ; Number low byte
NUM_H         equ    h'027'  ; Number high byte

; ******************************************************************
; * FUNCTION  : Calculates the square root of a 16-bit integer   *
; * EXAMPLE   : Number = h'FFFF' (65,535), Root = h'FF' (d'255') *
; * ENTRY     : Number in File h'027:26'                         *
; * EXIT      : Root in W and in COUNT.                          *
; * ENVIR'MENT: Files h'037:35' and Status register altered      *
; ******************************************************************

; Local declarations --------------------------------------
COUNT        equ    h'035'     ; The loop count
I_L          equ    h'036'     ; Magic number low
I_H          equ    h'037'     ; Magic number high

; Task 1: Zero loop count ---------------------------------
SQR_ROOT clrf   COUNT

; Task 2: Set magic number I to one -----------------------
         clrf   I_L
         clrf   I_H
         incf   I_L,f

; Task 3: DO ----------------------------------------------
     ; Task 3(a): Number - I -------------------------------
SQR_LOOP movf   I_L,w     ; Get Low byte magic number
         subwf  NUM_L,f   ; Subtract from Low byte Number
         movf   I_H,w     ; Get Hi byte magic number and
         subwfb NUM_H,f   ; subtract with borrow from Hi byte

     ; Task 3(b): IF underflow THEN exit -------------------
         bnc    SQR_END   ; No Carry is Borrow. IF true terminate

     ; Task 3(c): ELSE increment loop Count ----------------
         incf   COUNT,f

     ; Task 3(d): Add two to the magic number I:2 ----------
         movlw  2         ; Add two to Low byte of I
         addwf  I_L,f
         clrf   WREG      ; Zero Working register
         addwfc I_H       ; and add Carry bit to upper byte I
         bra    SQR_LOOP  ; and do another subtract and test

; Task 4: Return loop count as the square root ------------
SQR_END  movf   COUNT,w   ; Copy into WREG
         return           ; and return to caller
```

An example giving $\sqrt{65} = 8$ is given in Fig. 6.17(a) using this series approach. A flowchart visualizing the task list is also given in Fig. 6.17(b). The coding in Program 6.11 follows the task list closely. The maximum value of the loop count is h'FF', as $\sqrt{65535} \approx 255$. Thus a single byte at File h'035' is reserved for this local variable. Similarly the maximum possible value of the magic number is 511 (h'1FF') and so the two registers File h'037:36' are reserved for this local variable. This of course means that Task 3(a) entails a double-byte subtraction. By zeroing the Working register (which does not affect the **C** flag) and using the subwfb instruction, any borrow-out will be subtracted from the high byte of the ever decreasing Number. If a borrow is generated from this high-byte subtraction then the outcome is under zero and the loop is exited using the Conditional bnc instruction (a borrow-out is signaled by no Carry). Otherwise COUNT is incremented and I augmented by two and the process repeated.

Actually the loop Count is always half of (I less one), so COUNT is not really required. Instead, on return the 16-bit value I can be shifted once right. This divides by 2 and by throwing away the one that pops out into the Carry flag, effectively subtracts one—I is always odd and so its least significant bit is always 1. Try coding this alternative arrangement.

Example 6.4 Example 6.3 used several GPR Files to hold local variables. Repeat the coding using the Hardware stack to give an equivalent transparent subroutine with a minimal environmental impact.

Solution In implementing a function of this nature, the Hardware stack has to make room for both variables passed to the subroutine and for temporary storage of local variables. All this is in addition to its normal automatic storage of the return address. The organization of the Hardware stack used to support these three roles is shown in Fig. 6.18. Pre call, both bytes of the Number to be rooted are pushed into a single stack cell—which can hold up to 21 bits. The precall sequence will be something like:

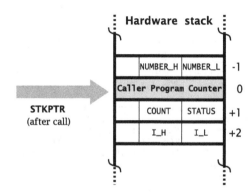

Fig. 6.18 Organizing the Hardware stack to implement a transparent square-root subroutine

Program 6.12 Coding the transparent square-root subroutine

```
; ***********************************************************
; * FUNCTION   : Calculates the square root of a 16-bit integer    *
; * EXAMPLE    : Number = h'2710' (10,000), Root = h'64' (d'100')  *
; * ENTRY      : Number in STKPTR-1                                 *
; * EXIT       : Root in W                                          *
; * ENVIR'MENT: TOS altered. Four additional Hardware stack cells* 
; ***********************************************************
; Task 1: Zero loop count ------------------------------------------
SQR_ROOT push             ; Move STKPTR up one to cell +1
         swapf   STATUS,w  ; Use swapf twice to
         swapf   WREG,w    ; get entry Status unchanged into W
         movwf   TOSL      ; and into the stack
         clrf    TOSH      ; Zero COUNT

; Task 2: Set magic number I:2 to 0001 -----------------------------
         push             ; STKPTR++ up one to cell +2
         clrf    TOSL      ; I:2 <- 0001
         incf    TOSL,f
         clrf    TOSH

; Task 3: DO -------------------------------------------------------
   ; Task 3(a): NUMBER:2 - I:2 -------------------------------------
SQR_LOOP movf    TOSL,w    ; Get Low byte magic number I_L
         pop              ; Move down to point to NUMBER:2
         pop              ; by moving STKPTR
         pop              ; down three to cell -1
         subwf   TOSL,f    ; Subtract from NUMBER_L
         bc      SQR_NEXT  ; Skip IF no Borrow
          decf   TOSH,f    ; ELSE take one away from NUMBER_H and
         bnc     SQR_END   ; IF borrow (underflows) THEN terminate
SQR_NEXT incf    STKPTR,f  ; Now go and get Hi byte I_H
         incf    STKPTR,f  ; STKPTR++ to cell +1
         incf    STKPTR,f  ; STKPTR++ to cell +2
         movf    TOSH,w    ; Get I_H into WREG & go back down to
         pop              ; point to NUMBER:2
         pop              ; STKPTR-- to cell 0
         pop              ; STKPTR-- to cell -1
         subwf   TOSH,f    ; and subtract I_H from NUMBER_H

   ; Task 3(b): IF underflow THEN exit ----------------------------
         bnc     SQR_END   ; No Carry is Borrow-out.

   ; Task 3(c): ELSE increment loop count -------------------------
         incf    STKPTR,f  ; Now go up to point to loop Count
         incf    STKPTR,f  ; STKPTR++ to cell +1
         incf    TOSH,f    ; and increment COUNT

   ; Task 3(d): Add two to the magic number -----------------------
         incf    STKPTR,f  ; Now point to I:2
         movlw   2         ; Add two to low byte
         addwf   TOSL,f
         clrf    WREG      ; Zero Working reg
         addwfc  TOSH      ; and add Carry bit to upper byte I
         bra     SQR_LOOP  ; and do another subtract

; Task 4: Return loop count as the square root ---------------------
SQR_END  movlw   2         ; Move STKPTR up two
         addwf   STKPTR,f  ; to point to loop Count in cell +1
         movf    TOSH,w    ; Copy into W (is answer)
         movff   TOSL,STATUS ; Get Status register back
         pop              ; Back to the caller's address (cell 0)
         return           ; and return to caller
```

```
push                 ; Ensure STKPTR is pointing to valid cell
movf    NUM_L,w      ; Copy low byte Number into bits cell[7:0]
movwf   TOSL         ; via TOSL
movf    NUM_H,w      ; and high byte into bits cell[15:8]
movwf   TOSH         ; via TOSH
call    SQR_ROOT     ; Go and do square root of Number
pop                  ; Clean up stack & the square root is in W
```

The listing of our modified subroutine shown in Program 6.12 is based on the listing of Program 6.11. In particular the Task list items line up. The only difference is the movement of STKPTR up and down to gain admission to the data in the appropriate stack cell. The pop instruction is used to adjust STKPTR towards the bottom of the stack (upwards in the diagram). Whilst push will move the Stack Pointer towards the top of the stack (downwards in the diagram) it will overwrite the cell's contents with the 21-bit Program Counter. Where this is a problem, either the incf instruction is used, or sometimes a constant is added to STKPTR. With these latter techniques, care must be taken, as the Status flags and in the last case the Working register, will be altered. In particular, once the lower bytes I_L and NUMBER_L have been added, any borrow-out is handled immediately, as the C flag will be altered in the process of accessing I_H. All these problems make the use of a Software stack a more attractive proposition for anything beyond saving registers!

In our listing the entry state of STATUS is also saved in the stack, as part of Task 1. This is not as straightforward as might be expected. Copying it to the Working register using the movf instruction will alter the Z flag. movff cannot be used with a TOS register as destination. Instead, the rather obscure use of swapf will copy a nybble switched over version into the Working register without any status change. Swapping it again rights the switch and it can then be saved into the stack.

The final Task 4 moves COUNT into the Working register to pass back as the square root. The STKPTR is then manoeuvred into position to allow a return to the caller.

Example 6.5 The 7-segment decoder subroutine of Program 6.6 has several flaws, the major of which being the possibility that when the offset in the Working register ($\times2$) is added to PCL, an overflow may occur, and the final value of the 21-bit Program Counter will not reflect one of the tabulated retlw instructions. This is a fatal error in that the program execution will navigate away from the subroutine with little chance of ever legitimately returning to the caller.

It is possible to make this type of look-up table more robust by adding the offset (doubled to reflect the 2-byte nature of the retlw instruction) to not only PCL but also PCLATH and PCLATU. Show how this could be done for a look-up table to display the hexadecimal digits 0–9 and A, b, C, d, E and F.

Solution A working solution is shown in Program 6.13. The structure of the actual table of retlw instructions is similar, but with six additional entries and the label TABLE_7 attached to the first of these instructions. The objective of the preliminary

Program 6.13 A robust hexadecimal 7-segment look-up table

```
; *******************************************************
; * FUNCTION    : Returns byte[N] in table            *
; * FUNCTION    : where N is the contents of W         *
; * EXAMPLE     : IF W = 06 THEN returns code b'01111101' *
; * ENTRY       : N range 00 - 09 in W                 *
; * EXIT        : Table entry N in W                   *
; * ENVIRONMENT: W = 2N, Status flags altered          *
; *******************************************************

SVN_SEG andlw b'00001111' ; Remove any erroneous upper nybble
        addwf  WREG,w       ; Adds W to itself to give 2N
        addlw  LOW TABLE_7  ; Add onto the lo byte Table address
        movwf  TEMP         ; and store away
        movlw  HIGH TABLE_7 ; The High byte of the Table address
        movwf  PCLATH       ; Into PCLATH
        clrf   WREG         ; Add zero plus
        addwfc PCLATH,f     ; any carry
        movlw  UPPER TABLE_7 ; Now the Upper Table address
        movwf  PCLATU       ; into PCLATU
        clrf   WREG         ; Zero WREG
        addwfc PCLATU,f     ; and add any carry to PCLATH
        movf   TEMP,w       ; Get the saved PCL from above
        movwf  PCL          ; and update the whole array

; -------------- xgfedcba -----------------------------------
TABLE_7 retlw b'00111111' ; Code for 0; Returned if N = 0
        retlw b'00000110' ; Code for 1; Returned if N = 1
        retlw b'01011011' ; Code for 2; Returned if N = 2
        retlw b'01001111' ; Code for 3; Returned if N = 3
        retlw b'01100110' ; Code for 4; Returned if N = 4
        retlw b'01101101' ; Code for 5; Returned if N = 5
        retlw b'01111101' ; Code for 6; Returned if N = 6
        retlw b'00000111' ; Code for 7; Returned if N = 7
        retlw b'01111111' ; Code for 8; Returned if N = 8
        retlw b'01101111' ; Code for 9; Returned if N = 9
        retlw b'01110111' ; Code for A; Returned if N = 10
        retlw b'01111100' ; Code for b; Returned if N = 11
        retlw b'00111001' ; Code for C; Returned if N = 12
        retlw b'01011110' ; Code for d; Returned if N = 13
        retlw b'01111001' ; Code for E; Returned if N = 14
        retlw b'01110001' ; Code for F; Returned if N = 15
; -------------------------------------------------------------
```

routine is to double the entry offset in the Working register and then add this on to the address TABLE_7, putting the outcome in the complete 21-bit PC. As there is exactly 16 entries in the table, ANDing WREG with h'0F' (b'00001111') removes any incorrect value above 15, to increase the security of the subroutine.

To dismember the 3-byte address TABLE_7, the Microchip assembler has the three directives **upper**, **high** and **low**. If TABLE_7 where, say, h'0001F8', then the instruction addlw low TABLE_7 will add the *low* byte (h'F8' in our instance) to WREG. Similarly, addlw high TABLE_7 adds bits 15–8 of TABLE_7 (h'01' in our case) to the Working register, whilst addlw upper TABLE_7 adds bits 20–16.

Once the offset × 2 in WREG is added to TABLE_7[7:0], it is stored away in a Temporary File. PCLATH is then updated with TABLE_7[15:8] and any carry-out from this first addition added. Finally, PCLATU is overwritten with TABLE_7[20:16] plus any previous carry-out. To get all this into the program counter, the routine copies the low byte of this addition from TEMP and then the act of copying this into PCL will at the same time copy PCLATH:PCLATU into the entire PC—as described in Fig. 4.4 on p. 75.[9] At this point in the program, the PC is now set to TABLE_7 + WREG × 2. The next instruction to be executed will be the appropriate retlw.

Although our resultant program is robust and can cope with tables up to 256 entries (Program 6.6 can only handle 128 entries), nevertheless there are much more efficient techniques of accessing tables of constants in the Program store—see Program 15.6 on p. 551. How might you extend the program to cope with up to 65,536 entries—assuming a Program store of sufficient capacity?

Example 6.6 A certain vending machine has to display the message "insert coin for cola" using a single 7-segment display, with a 400 ms delay between characters. Show how this could be done, assuming that the display is connected to Port B, which has been configured as an output, and the unit is clocked at 4 MHz.

Solution There are several solutions to this problem. The one adopted in Program 6.14 is based on a table giving the code for each character in turn as it appears in the message; duplicating when necessary and assuming no Space character.

If the last table entry is a special character; in our case all zeros or NUL, then the driving subroutine can interpret this as End Of String.

The driving subroutine PUT_STR simply uses a Count, passed to WREG prior to calling the look-up table STR_INSERT_COIN. Depending on the Count, element *N* will be extracted from the table. On return, the byte in the Working register is tested for zero. If this is true, then COUNT is zeroed and the process repeated from character 0. Otherwise, the returned character code is copied into Port B and subroutine DELAY_K100MS (see Program 6.4) is called with a parameter of 4 to give the required delay.

Using an End-Of-String character is a common technique employed in string handling, as it allows the same driving function to be used irrespective of the string

---

[9]Note that movff cannot be used to do this copy operation into the PCL in one, as the PC, TOS and INTCON registers are forbidden as destination for this instruction.

Program 6.14 Displaying a 7-segment coded character string

```
; *******************************************************************
; * FUNCTION  : Sends out all characters from string table    *
; * FUNCTION  : terminated by NUL, to Port B at 400ms rate     *
; * ENTRY     : None                                           *
; * EXIT      : String displayed                               *
; * RESOURCE  : Subroutines STR_INSERT_COIN, DELAY_K100MS       *
; * ENVIR'MENT: Byte COUNT, COUNT1, K, WREG & Status altered    *
; *******************************************************************
PUT_STR   clrf    COUNT             ; Start character number at zero

PUT_LOOP  movf    COUNT,w           ; Test number
          call    STR_INSERT_COIN   ; Convert it
          tstfsz  WREG              ; Skip IF returns with NUL char
           bra    NEXT              ; ELSE display digit
          bra     PUT_STR           ; Otherwise start all over again

NEXT      movwf   PORTB             ; Send it out
          movlw   4                 ; Do 400ms delay
          call    DELAY_K100MS      ; Delay
          incf    COUNT,f           ; Next number
          bra     PUT_LOOP          ; ELSE start again

          return

; *******************************************************************
; * FUNCTION   : Returns byte[N] in table                      *
; * FUNCTION   : where N is the contents of W                  *
; * EXAMPLE    : IF W = 06 THEN returns code b'00111001'       *
; * ENTRY      : N range 00 - 17 in W                          *
; * EXIT       : Table entry N in W                            *
; * ENVIRONMENT: Status flags altered                          *
; *******************************************************************
STR_INSERT_COIN
          addwf   WREG,w            ; Double offset
          addwf   PCL,f             ; and add to PC

;                 xgfedcba
STRING    retlw   b'00000110' ; Code for I   ; Returned if N = 0
          retlw   b'00110111' ; Code for N   ; Returned if N = 1
          retlw   b'01101101' ; Code for S   ; Returned if N = 2
          retlw   b'01111001' ; Code for E   ; Returned if N = 3
          retlw   b'00110001' ; Code for r   ; Returned if N = 4
          retlw   b'01110000' ; Code for t   ; Returned if N = 5
          retlw   b'00111001' ; Code for C   ; Returned if N = 6
          retlw   b'00111111' ; Code for O   ; Returned if N = 7
          retlw   b'00000110' ; Code for I   ; Returned if N = 8
          retlw   b'00110111' ; Code for N   ; Returned if N = 9
          retlw   b'01110001' ; Code for F   ; Returned if N = 10
          retlw   b'00111111' ; Code for O   ; Returned if N = 11
          retlw   b'00110001' ; Code for r   ; Returned if N = 12
          retlw   b'00111001' ; Code for C   ; Returned if N = 13
          retlw   b'00111111' ; Code for O   ; Returned if N = 14
          retlw   b'00111000' ; Code for L   ; Returned if N = 15
          retlw   b'01110111' ; Code for A   ; Returned if N = 16
          retlw   b'00000000' ; Code for NUL ; Returned if N = 17
```

length—up to 255 characters in our case. However, there are other ways of dealing with problems of this nature; can you think of any alternatives? Of course, we could improve the robustness of the look-up table as described in Program 6.13.

Example 6.7 Design and code a transparent subroutine based on a Software stack, that will accept a 7-bit datum in the Working register and return an 8-bit odd one's parity version also in the Working register. Assume that FSR2 has been initialized to act as the Pseudo Stack Pointer.

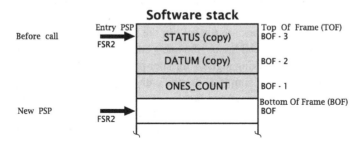

Fig. 6.19 Software stack for the odd one's parity subroutine

Program 6.15 Coding a transparent odd one's parity subroutine

```
; *******************************************************************
; * FUNCTION    : Modifies a 7-bit datum to give 1's parity        *
; * EXAMPLE     : Datum = 01010101 -> 11010101                     *
; * ENTRY       : 7-bit Datum in WREG                              *
; * EXIT        : 8-bit 1's parity datum in WREG                   *
; * ENVIR'MENT  : FSR2 used as PSP. Fully transparent              *
; *******************************************************************
          movff    WREG,POSTINC2  ; Copy Datum on stack
          clrf     POSTINC2       ; 1's Count zeroed

PARITY_LOOP
          bcf      STATUS,C       ; Shift Datum
          rlcf     WREG,f         ; once left into Carry
          bnc      PARITY_NEXT    ; Skip IF zero popped out
          movf     POSTDEC2,f     ; PSP-- points to 1's Count
          incf     POSTINC2,f     ; Increment 1's Count; PSP++

PARITY_NEXT
          tstfsz   WREG           ; Check residue from shifted datum
          bra      PARITY_LOOP    ; IF not zero THEN DO another shift

; Check one's count and if even (bit 0 == 0) THEN make bit 7 = 1
          movf     POSTDEC2,f     ; PSP-- points to 1's Count
          btfsc    INDF2,0        ; Test bit 0 of 1's Count
          bra      PARITY_FINI    ; IF == 1 THEN finished
          movf     POSTDEC2,f     ; PSP-- points to copy Datum
          bsf      POSTINC2,7     ; Set bit 7 to give odd 1's parity
                                  ; & PSP++
PARITY_FINI
          movf     POSTDEC2,f     ; PSP-- points to Datum
          movff    POSTDEC2,WREG  ; Copy into WREG and PSP-- moves

          return                  ; back to return value
```

Solution Our subroutine needs to implement the following tasklist:

1. Count the number of ones in the initial 7-bit Datum.
2. If even (that is bit 0 of the Count is 0) then set bit 7 to one.

The Software stack structure for this subroutine is shown in Fig 6.19. As the Datum is to be passed back and forth through the Working register, nothing needs to be copied into the stack prior to the call. In the subroutine three cells are opened as a Frame to hold the pre-call Status, the one's count initialized to zero and a copy of the Datum from the Working register for later possible modification.

The body of the routine simply shifts the entry Datum in the Working register inside a loop, counting any ones and breaking out when the residue is zero. Based on the state of bit 0 of this tally, bit 7 of the Datum in the Stack is set and then this byte copied back into the Working register. Finally the stack is unlinked by positioning the Pseudo Stack Pointer, in the guise of FSR2, to its entry point and returning.

Example 6.8 Using the Extended Instruction set detailed on p. 188, show how Program 6.15 can be more efficiently coded.

Solution Most of the new instructions and the Indexed Literal Offset address mode work better with a grow-down Software structure. That is the PSP is decremented as data is pushed out into the stack. This paradigm is shown in Fig. 6.20, which is basically an 'upside down' version of Fig. 6.19.

Based on the illustrated structure, the subroutine listed in Program 6.16 is similar to Program 6.15 but FSR2 at the low address of the Frame after the initialization phase. Cell data is accessed using Indexed Literal addressing. Thus bsf [2], 7 will set bit 7 of the byte two cells *above* the Frame bottom; that is FSR2 + 2. As the Frame is three bytes deep, the final addulnk 03 instruction rolls up (cleans) the Frame and returns to the caller.

In processors with the option of an Extended instruction set, the XINST fuse must be set and the assembler notified that this set is to be used. Thus in the preamble we need something like this:

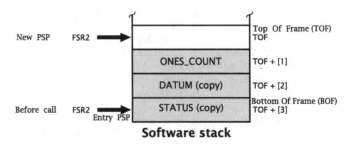

Software stack

Fig. 6.20 Grow-down Software stack for the odd one's parity subroutine

Program 6.16 Using the Extended instruction set to code a transparent odd one's parity subroutine

```
; ********************************************************************
; * FUNCTION    : Modifies a 7-bit datum to give 1's parity        *
; * EXAMPLE     : Datum = 01010101 -> 11010101                     *
; * ENTRY       : 7-bit Datum in WREG                              *
; * EXIT        : 8-bit 1's parity datum in WREG                   *
; * RESOURCES   : Uses Extended instruction set                    *
; * ENVIR'MENT  : FSR2 used as PSP. Fully transparent              *
; ********************************************************************
PARITY     movff   STATUS,POSTDEC2 ; Save Status on S'ware stack
           movff   WREG,POSTDEC2   ; Copy Datum on stack
           pushl   0               ; 1's Count zeroed

PARITY_LOOP
           bcf     STATUS,C        ; Shift Datum
           rlcf    WREG,f          ; once left into Carry
           bnc     PARITY_NEXT     ; Skip IF zero popped out
           incf    [1],f           ; Increment 1's Count

PARITY_NEXT
           tstfsz  WREG            ; Check residue from shifted datum
           bra     PARITY_LOOP     ; IF not zero THEN DO another shift

; Check one's count and if even (bit 0 == 0) THEN make bit 7 = 1
           btfsc   [1],0           ; Test bit 0 of 1's Count
           bra     PARITY_FINI     ; IF == 1 THEN finished
           bsf     [2],7           ; Set bit 7 to give odd 1's parity

PARITY_FINI
           movsf   [2],WREG        ; Returned value in WREG
           movsf   [3],STATUS      ; Copy Status into STATUS
           addulnk 3               ; Unlink stack
```

```
       config XINST = ON       ; XINST option active
       list   ep = PIC18F4520  ; Use 18F4520 Extended Processor
       org    0.               ; Start main code at h'00000'
MAIN   lfsr 2,h'05F'           ; Initial PSP is File h'05F'
       .... .......            ; More code
```

More details of these assembler directives are given in Chap. 8.

Self-Assessment Questions

6.1 A certain student has coded his 1 ms delay subroutine of Program 6.1 thus:

```
DELAY_1MS movlw  d'249'   ; Set up loop
D_LOOP    addlw  -1       ; Decrement count
          nop             ; Put in extra cycle
          bnz    DELAY_1MS ; Repeat unless zero
          return
```

What will be the outcome?

6.2 Create a subroutine that will read Port B every hour and copy its value into File h'044'. You can base it on a 60-minute version of Program 6.10. Say why this may not be a good use of the PIC MCU's resources.

6.3 In the same manner as Program 6.14 write a subroutine to display once when called, the string "calibrate", using upper- and lower-case glyphs as necessary, at a rate of 250 ms per character. You may assume a clock rate of 8 MHz.

6.4 Program 6.7 showed how a 16-bit Dividend could be divided by an 8-bit Divisor to give a 16-bit integer Quotient and an 8-bit Remainder. Extend this coding to give a 24-bit Quotient comprizing 16-bit integer and 8-bit fraction. For instance $100 \div 40 = 2.5$ or in binary $0000000001100100 \div 00101000 = 0000000000000010.10000000$. You will need to deal with a variable QUOTIENT_F to hold the fractional byte and will of course do 24 shift and subtract operations.

6.5 Repeat the division example of Program 6.9, but this time using the Extended instruction set available to the PIC18F4520 processor. For an efficient implementation replace the diagram of Fig. 6.15 by the push-down stack of Fig. 6.21.

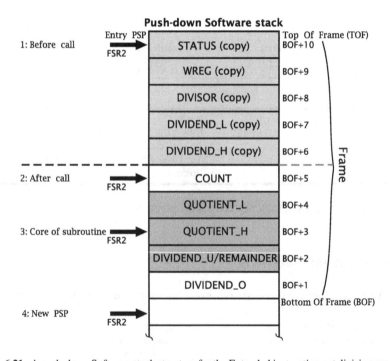

Fig. 6.21 A push-down Software stack structure for the Extended instruction-set division program

6.6 Readings of the state of a mechanical switch can be erratic, as the contacts will bounce for several milliseconds when closed; thus giving a series of 1s and 0s. Similar considerations apply to electronic devices such as phototransistors when passing through a shadow. Although this problem can be fixed with hardware, it is usually more cost effective to use a software solution.

Devise a subroutine that will return with the stable state of a switch connected to Port B pin RB7 as bit 7 of the Working register. Stability is defined as 5000 (h'1388') reads all giving the same value. The other bits of WREG on return are undefined.

6.7 An analog-to-digital converter is connected to Port B. Repeat SAQ 6.6, but this time defining stability as 1000 identical reads, and returning with the stable digitized analog voltage in WREG.

6.8 The subroutine in SAQ 6.7 returns the stable value of a noisy digitized signal, assuming 1000 identical values. Using this subroutine, code a main routine that will generate how this stable reading differs from a preceding value previously stored in location File h'040'. Each bit that differs is to be logic 1. Generate the position of the rightmost change bit in File h'041'; with a zero denoting no difference and 1 through 8 for bits 0 through 7 respectively.

6.9 The subroutine of SAQ 6.7 will not return a value when relatively high-frequency noise is present on the analog signal, as the resulting digital jitter will ensure that 1000 identical readings rarely occur. As an alternative, noise reduction can be obtained by taking the average of multiple readings. If the noise is random then n readings will give a noise improvement of $\sqrt{n}$. Devise a subroutine that will read Port B 256 times and return the 8-bit rounded average in WREG; which will give an increase in signal to noise ratio of 16.

6.10 The circuit diagram of Fig. 6.22 shows a 7-bit pseudo-random number generator (PRNG) based on a shift register with an Exclusive-OR gate feedback. Devise a subroutine to send these 127 binary random numbers to Port B as a single burst. An initial non-zero value for the PRN is to be passed to the subroutine in the Working register. For instance, if this seed value were 01, then the first 32 7-bit hexadecimal values (with bit 7 = 0) are:

```
02 04 08 10 20 41 03 06 0C 18 30 61 42 05 0A 14
28 51 23 47 0F 1E 3C 79 72 64 48 11 22 45 0B 16 ...
```

What would happen if the initial value of the random number was zero?

Fig. 6.22 A 7-bit pseudo-random number generator

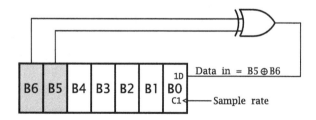

6.11 Mathematically to convert from Celsius to Fahrenheit we have:

$$F = C \times \frac{9}{5} + 32.$$

Write a transparent subroutine where the temperature is passed via the Working register, ranging from 0°C through 100°C, with the Fahrenheit equivalent being returned in the same location. Make use of the DIV_16S subroutine of Program 6.9 for your transparent division.

Chapter 7
Interrupt Handling

The subroutines discussed in Chap. 6 are predictable events in that they are called up whenever the program dictates. Real-time situations, defined as where the processor interacts in concert with external physical events, are not as simple as this. Very often something happens beyond the core CPU which necessitates prompt action from the processor. The vast majority of controllers have the capability to deal with a range of such events that disrupt their smooth running. In the case of a microcontroller, requests for service may come from an internal peripheral device, such as a timer overflowing, or from the outside world from a source entirely external to the device. At the very least, on an external reset (a type of outside hardware event) the MCU must be able to get (vector) to the first instruction of the main program. In the same manner an external service request, or interrupt, when answered must lead to the start of the special subroutine, known as an interrupt service routine or interrupt handler.

Although PIC MCU devices have different mixes of internal peripherals; such as analog, serial and timer ports, the response to interrupt requests are handled in the same manner. In this chapter we will concentrate on how the enhanced-range family deals with such requests, and in particular to those external to the device. The particular circumstances of how and why various internal peripheral modules generate interrupts will be discussed as appropriate in Part III.

After reading this chapter you will:

- Be aware of the need for interrupt handling.
- Appreciate the concept of a vector table as a jumping-off point for Reset, Compatible, Low- and High-priority interrupt events.
- Follow the sequence of events when the PIC microcontroller recognizes an interrupt request in both Compatible and Priority modes.
- Understand the principle of latency.
- Recognize the function of the global interrupt enable masks in empowering groups of requests.
- Understand the operation of the local interrupt mask, flag and priority triplet bits as appropriate to the various sources of interrupts.
- Be able to write a simple interrupt handler according to the principles:

S. Katzen, *The Essential PIC18® Microcontroller,*
Computer Communications and Networks,
DOI 10.1007/978-1-84996-229-2_7, © Springer-Verlag London Limited 2010

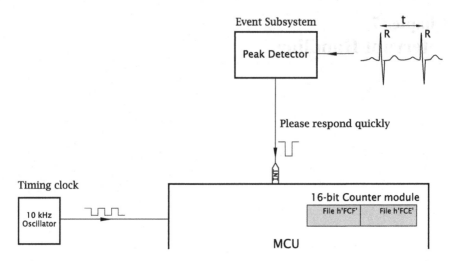

Fig. 7.1 Detecting and measuring an external event

- Context switching.
- Determination of interrupt source.
- Core process.
- Return via the retfie instruction.

A simple example of a time-sensitive situation is shown in Fig. 7.1. Here we wish to measure the elapsed time between the R-points of an electrocardiogram (EKG)[1] signal, which by definition is an external **real-time** event. The time resolution is to be 0.1 ms and the maximum peak-to-peak duration is likely to be no more than 1.5 seconds. In order to measure this time, a free-running 16-bit counter clocked at 10 kHz can be used as the time base. As we shall see in Chap. 13, all enhanced-range PIC MCUs have several internal 16-bit counters. Here we will assume that the state of the count can be read at any time from the appropriate Files. If the count at the last R-point is stored in two spare GPRs, then subtraction of the count at the current R-point will give the required beat-to-beat duration.

The next problem is how to detect the signal peak, as by definition the patient's heart is not synchronized to the MCU! One technique is to continually read the ECG signal and perform a peak-detection algorithm to determine the R-point. Now this **polling** technique will have to be carried out 10,000 times each second in order to keep to the specified resolution. Taking a nominal human heart rate of 60 beats per minute, 99.99% of the time no peak will be detected. Essentially, this means that the processor will spend the vast majority of its processing power just looking out for one event in 10,000.

The alternative approach is to use external hardware whose sole task is to find the peak signal. That peak-picking hardware could be an analog circuit or even a MCU

---

[1]From the German *Electrocardiogram*; ECG in UK.

with an analog-to-digital converter dedicated to this one task—see Example 14.2 on p. 527. Whatever the implementation, the peak-picker sends a signal to the main processor when an R-point has been detected. This signal **interrupts** the MCU, which must drop whatever it is doing and read the counter within 100 μs, if a counter tick is not to be missed.

In the situation where external processes happen in their own good time and are in no way synchronized to the processor, there has to be some way for certain events to interrupt the background process and direct it to attend to their immediate need. Polling a series of external events is adequate where nothing much happens quickly outside and/or there are few parameters to monitor and little processing to do. The possibility of missing anything important can be reduced by increasing the polling rate, but there comes a time when the processor does little else but sample peripheral data. This resource burnout is especially a problem when there are many signals to poll in a short period of time.

The downside of interrupt-driven real-time monitoring is additional hardware complexity and the greater intricacy of the hardware—software interaction. If you are confused, consider the telephone system. It would be possible to have a telephone network where the subscriber would pick up the phone every, say, 5 minutes and ask "Is there anyone there?" Apart from the bother (processing overhead) of doing this,[2] the caller might get bored and hang up. You could reduce the chance of this happening by increasing the polling rate to, say, once per minute. But you would then end up spending all your time on the phone and, depending on how popular you are, getting only a few hits a day. That is, 99% of your effort would be wasted.

This is obviously ridiculous, and in practice an interrupt-driven technique is used so that you only respond when the buzzer sounds. This is highly efficient, but at the cost of a lot more complexity for the phone company, as the signaling side of the system can be more demanding than the speech side. There is another problem too, in that you (cf. the processor) have no idea when the phone will ring. And it surely will be at the most inconvenient time. Thus you have to (unless you have an iron will) break off what you are doing at the drop of a hat. For instance, if you happen to be in the middle of solving a problem in your head you should save your partial results before responding, so, when finished, you can return to where you left off.

Microcontrollers can respond to interrupt requests from a wide range of sources, either physically outside the chip or from the various internal ports and peripheral devices supported by that particular family member. For instance, our PIC18F1220 has 12 separate sources of interrupts originating from these modules, plus three from outside via pins labeled INT0, INT1 and INT2—pins 8, 9 and 17 in Fig. 4.1 on p. 71. These are shared with bits 0, 1 and 2 of Port B; that is RB0, RB1 and RB2 respectively. The programmer can choose to disable the complete interrupt system (which is the Power-on Reset default) or enable/disable each of these sources individually, as well as prioritize most of them. Because the response to an interrupt

---

[2]It would of course make it easier just to ignore the phone!

request is essentially the same, no matter whence it arises, we will in this chapter refer mainly to these external, or **Hardware**, interrupts.

Keeping in mind the randomness of an external event, the response of any processor to an **interrupt request** will normally be something like:

1. Finishing the current instruction.
2. Automatically saving, at the very least, the state of the Program Counter (PC)—which is needed to get back. Some processors, such as the enhanced-range PIC MCUs, also automatically save the Status register and other internal registers at this point.
3. Entering the appropriate interrupt service routine.
4. Executing the defined task.
5. Restoring the processor state and returning to the point in the background program from where control was first transferred.

Essentially, signaling an interrupt causes the PIC MCU to drop whatever it is doing, save its position in the interrupted **background program** on the Hardware stack and go to a special subroutine known as an **interrupt service routine** (ISR). This **foreground program** is basically just a subroutine entered at the behest of an external happening.

The minutiae of the response to an interrupt request varies somewhat from processor to processor. In the case of the PIC18 family, this depends slightly on which one of two interrupt modes the processor is configured for. These modes are:

Compatible mode; IPEN = 0

This is the default Power-on Reset state where bit 7 of the RCON (Reset CONtrol) register is 0—see Fig. 10.14 on p. 325. All sources of interrupt service requests have an individual local mask bit, but cannot be prioritized. Provided that the processor is not already in an interrupt service routine, execution will be transferred to the instruction located at h'00008'; known as the **Default interrupt vector**.

Priority mode; IPEN = 1

By setting RCON[7] to 1, all sources of interrupt requests, apart from that from the INT0 pin, can be configured as either High- or Low-priority.

If an enabled source requests service, then the response of the processor depends on its priority designation.

- If the request is High-priority, then provided that the processor is not already executing a High-priority interrupt service routine, execution will be transferred to the instruction at the **High-Priority interrupt vector** at h'00008'. This is the same location used by the default Compatible mode for its interrupt vector.
- If the request is tagged as Low-priority, then provided that the processor is not already executing *either* a High- or Low-priority interrupt service routine; execution will be transferred to the instruction located at the **Low-Priority interrupt vector** at h'00018'.

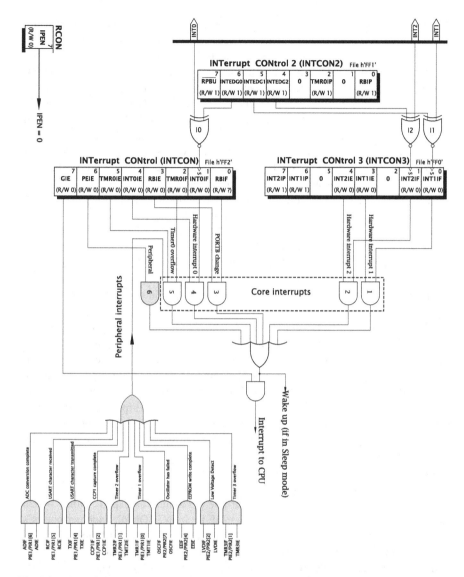

Fig. 7.2   The PIC18F1220 MCU's interrupt logic in the default Compatible mode

In looking at these modes in more detail, we will begin with the default Compatible mode. This name comes from its compatibility with the mid-range PIC16 family interrupt architecture. To understand the logic behind this structure, shown in Fig. 7.2, we need to briefly consider the evolution of interrupt handling in 8-bit PIC MCU families.

The base-range family has no interrupt handling facilities and thus these devices are restricted to polling in their interaction with the outside world. The rapid inter-

action of a processor to external events happening outside its control is a measure
of its caliber. With this in mind, when Microchip introduced the mid-range PIC16
family, they included an interrupt system capable of handling service requests from
up to four sources; namely an external voltage edge on the INT pin, a change of
state on the top four Port B pins, an overflow from the 8-bit Timer 0 and from one
device-specific module. For instance, the now obsolete PIC16C71 could generate
a service request when its Digital to Analog Converter (DAC) module completed
its conversion and similarly, the superseded PIC16C84 originated an interrupt when
the write-to action on its non-volatile Data EEPROM was complete. Each source of
interrupt set a bit in the INTerrupt CONtrol register (INTCON) to signal (or flag) the
various requests. For instance, a rising edge on the INT pin set bit INTF. Each of
these flag bits had an associated mask bit; INTE in our instance, which allowed the
programmer to control what sources were permitted to request an interrupt. Finally,
the General Interrupt Enable (GIE) bit when 0 disabled the complete interrupt logic,
which was the situation after a Power-on Reset.

All these mask (enable) and flag bits were located in the INTCON register. However, this 8-bit register had no room for the one additional peripheral device's interrupt flag, which had to be placed elsewhere; usually in the peripheral's control
register.

Later and current members of the family added additional peripheral modules, all
of which can generate one or more interrupts. For instance, the 18-pin PIC16F627A
has seven additional peripheral modules, besides the three core services. As an evolution of the early structure, these additional flag and mask bits are all located in one
or more Peripheral Interrupt (PIR) and Peripheral Interrupt Enable (PIE) registers.
These form their own logic group, with a separate overall PEripheral (group) Interrupt Enable (PEIE) bit in INTCON[6]; replacing the specific single peripheral enable
bit in the early family members. This PIC16 logic structure is shown in Fig. 7.6 in
my *The Quintessential PIC® Microcontroller*, and it would be useful to compare
this Fig. 7.2.

The Compatible mode interrupt request logic shows the ten sources of interrupt
from the PIC18F1220's complement of peripheral modules as a single group, which
can enabled/disabled as an entity with the PEIE bit in INTCON[6]. Each peripheral's
service request sets its interrupt flag in one of the two Peripheral Interrupt Flag (PIR)
registers and this is ANDed with a corresponding local mask bit in the like Peripheral Interrupt Enable (PIE) register—see Fig. 7.5. For instance, when the Timer 2
module overflows, it sets its interrupt flag TMR2IF in PIR1[1]. If the TMR2IE bit in
PIE1[1] has been previously set, then this request will cause the output of the group
OR gate to go logic 1. If the group mask bit PEIE in INTCON[6] is also 1 then this
event will initiate an interrupt response—as described in Fig. 7.3.

All the Hardware interrupts, from pins INT0, INT1 and INT2 and that from
Port B and Timer 0 are handled as core interrupts, with interrupt and mask bits in
the three (increased from one) INTerrupt CONtrol registers. As well as holding the
global, group and five local core mask bits and interrupt flags, these control registers allow the software to set which edge at the INTerrupt pins set their associated
interrupt flag. For instance, if INTEDG1 is 1 (its Power-on Reset state) then a rising

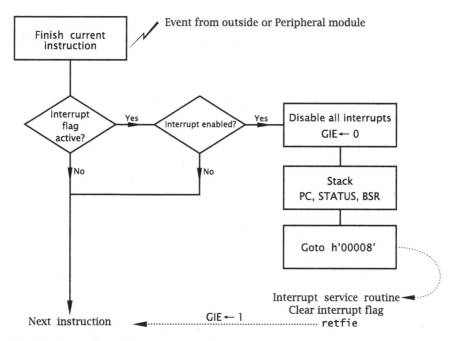

Fig. 7.3   Responding to an interrupt request in Compatible mode

edge on pin INT1 will set the INT1IF flag in INTCON3[0]. These registers also hold the Priority bits for the core interrupt sources. Like the Peripheral interrupt group, unmasked requests are OR'ed for onward transmission to the interrupt logic.

All mask bits are logic 0 after a Power-on Reset; that is all sources of interrupt are disabled. However, the software is still able to poll the relevant interrupt flags, as they continue to be set as appropriate to their monitor function.

What happens when one (or more) of the fifteen unmasked interrupt flags are set? The Compatible mode response is illustrated in Fig. 7.3 and in more detail here:

1. The processor checks once during each instruction (Q_4 in Fig. 4.5 on p. 76) for an interrupt request from an enabled source. Even if this request is active, the instruction continues to completion; that is, execution does not break part way through the instruction, even in a multi-cycle or 2-word instruction.
2. If there is no valid request, the PIC MCU simply continues on to the next instruction and the process is repeated.
3. If there is a valid enabled request, then the next three instruction cycles are involved in moving execution to the instruction located in the Interrupt vector h'00008'. This comprises a cycle to allow the instruction at the top of the pipeline to be executed[3] plus two more dummy cycles to flush the Pipeline; after which the instruction in the Interrupt vector is ready to be executed. This 3- to 4-cycle

---

[3]Alternatively, this could be the final cycle of a 2-cycle and/or 2-word instruction.

delay from the instant of the external signal to the INTn pin and beginning the execution of the first instruction of the ISR is known as **latency**. It is impossible to be more precise due to the time-random nature of the external request signal, which can occur anywhere in the instruction cycle. Requests from internal peripheral modules have a similar 3-cycle latency.

4. During this latency period the PIC MCU does four things:

   (a) The complete interrupt system is disabled, to ensure that once an interrupt response is in train, any further interrupt requests are locked out. This is done by clearing bit 7 of the INTerrupt CONtrol register INTCON, which is labeled in Fig. 7.2 as General Interrupt Enable (GIE). GIE is an example of an **interrupt mask**, as it is able to mask out interrupt activity. After a Power-on Reset, GIE is clear; so by default interrupt activity is disabled.

   (b) The state of the 21-bit Program Counter is pushed into the Hardware stack, in exactly the same manner as for a call instruction—see Fig. 6.3 on p. 162. As for subroutines, this is to allow the processor to return to the interrupted background program after the interrupt service routine has terminated. As the enhanced-range PIC MCUs have a 31-deep Hardware stack, subroutines nested to depth of 30 can be called from an ISR—assuming that this stack is not used for anything else.

   (c) Simultaneously, a copy of the Working, Status and Bank Select registers are saved on the Fast stack—as described on p. 167. This **context** can be retrieved if the Fast bit in the retfie instruction is 1, as described on p. 165; i.e. retfie 1.

      As the 3-cell Fast stack is (unfortunately) identical to that used for Fast subroutines and it is *always* overwritten when an interrupt response is underway. Unlike subroutine Fast calls, this is *not* optional. Thus Fast subroutine calls should not be made at any point in the background program in an interrupt-enabled system. They can be called from an ISR, provided that a Fast retfie is not used.

   (d) The first instruction of the ISR is *always* in location h'00008' in the Program store (see Fig. 4.3 on p. 73) for Compatible mode interrupts. This final step of the sequence is to change the PC to this instruction address, known as the **Default interrupt vector**. The first instruction of the ISR is here, but usually this will be a goto, or sometimes a bra, to the interrupt handling software elsewhere in Program memory—see Program 7.1.

5. Like a subroutine, an ISR must be terminated by a Return instruction. However, in this case not only has the PC to be pulled back out of the Hardware stack to move execution back to the interrupted program but the GIE bit in the INTCON register must be set to re-enable the interrupt capability. This counteracts the clearing of this bit in 4(a) above, on entry to the ISR. The Return instruction relevant to this situation is retfie (RETurn From Interrupt and Enable)—see Table 6.1 on p. 164. Thus on re-entry to the background program, any pending or future interrupts can be serviced. As pointed out above, like the ordinary return instruction, a Fast option is available to restore the entry WREG, STATUS and BSR register states; provided that no Fast calls to subroutines are made in the body of the ISR.

Apart from the two additional Hardware interrupt pins, the major enhancement in the PIC18 interrupt capability from the predecessor PIC16 family, is the optional provision to prioritize interrupts. This option is exercised by setting the IPEN bit in RCON[7] to 1—see Fig. 10.14 on p. 325. In the Compatible mode, once the processor enters an interrupt service routine any subsequent interrupt from another source will be locked out, until the terminating `retfie` instruction sets the GIE switch bit—see Fig. 7.3. Whilst this is fine in most cases, there are situations where some sources requesting service need particularly urgent attention compared to others. For instance, in our ECG monitoring system of Fig. 7.1, if the processor does not respond within 100 µs then when it does get round to reading the timer, this will not reflect the proper R-point instant. If one of the other interrupt services deals with, say, sending a message to a remote terminal via a slow serial link; then this relatively inconsequential process should not be allowed to hog the attention of the core at the expense of a time-critical process. In such situations the more critical request should be allowed to interrupt another lesser-important service.

In the Priority mode, the program can set the priority of each interrupt *individually* as either High or Low. We see from Fig. 7.4 that as before, each source of service request sets its own interrupt flag, which is ANDed with its local enable bit. Compared with the Compatible mode of Fig. 7.2, this logic is duplicated, with one array of AND gates generating a composite High-priority request and the other a Low-priority request. Each gate in the former case ANDs with an associated priority bit, and the latter with the inverse priority bit. In the case of the Peripheral group, these priority bits are held in the two Interrupt Priority registers, IPR1 and IPR2, as shown in Fig. 7.5. These correspond to the PIR1:PIR2 and PIE1:PIE2 registers, which hold the interrupt flags and enable bits respectively.

As an example, shown in Fig. 7.5, are the two AND gates corresponding to Timer 1. When overflow occurs in this peripheral module, the TMR1IF in PIR1[0] is set. If the software has previously set the linked TMR1IE bit in PIE1[0], then a potential service request is generated. Depending on the state of TMP1IP in IPR1[0], this will be either Low-priority (TMR1IP = 0) or High-priority (TMR1IP = 1). On a Power-on Reset all priority bits are 0 and thus interrupts default to Low-priority.

High-priority interrupts are enabled as a group with enable bit GIEH, which shared with GIE as bit INTCON[7]—as shown in Fig. 7.2. Likewise, GIEL can mask out Low-Priority interrupts, and this is shared with PEIE in INTCON[6]. Masking out High-priority interrupts (GIEH = 0) also disables all Low-priority interrupts.

Core interrupts have their priority bits in INTCON2 and INTCON3, as shown in Fig. 7.2, but otherwise function in the same way. The one exception is the Hardware interrupt requested from pin INT0, which is always High-priority.

Comparing the flow charts of Fig. 7.6 and Fig. 7.3, we see that the response to an enabled interrupt request is similar to the Compatible mode; with the following major differences:

1. High-priority interrupts picking up their first ISR instruction at the High-Priority vector at h'00008' (the same as the Default vector in the Compatible mode) can interrupt a Low-Priority ISR already in progress. No interrupt can muscle in on a High-Priority ISR.

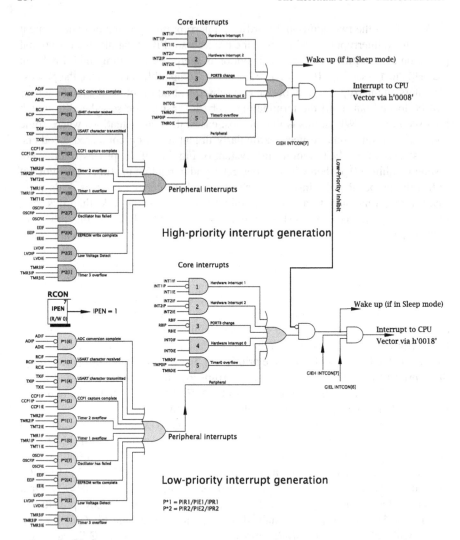

Fig. 7.4 The PIC18F1220 MCU's interrupt logic in Priority mode

2. Low-Priority interrupts pick up their first ISR instruction at the Low-Priority
   vector at h'00018'. A Low-Priority interrupt will be locked out if a High-Priority
   ISR is in progress, as the GIEH bit will be zero in this instance.

Both High- and Low-Priority interrupts automatically save the 21-bit PC in the
Hardware stack; so an interrupted Low-Priority ISR can be resumed when the High-
Priority ISR has been completed. However, the WREG, STATUS and BSR registers
are saved in the same Fast stack, and thus if a mixture of High- and Low-Priority
interrupts are used in a system, only the former should use a fast `retfie` to re-

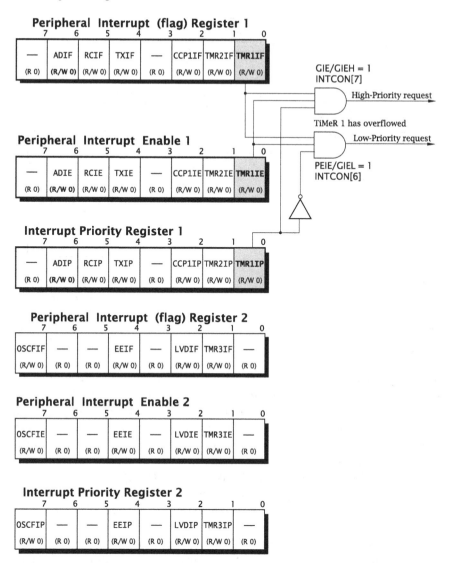

Fig. 7.5 Peripheral group Interrupt, Enable and Priority registers for the PIC18F1220; showing the logic for the Timer 1 module as an example

store these registers. Using a `retfie 1` to terminate a Low-Priority ISR will risk corruption from a possible High-Priority interrupt.

In all interrupt modes, the raw unglobally masked request signal can be used to awaken the processor if it is in a power-down or **Sleep** state. As we will see in Chap. 10, the current consumption of the device can be considerably reduced to typically less than 1 µA if processing is stopped and the PIC MCU is put in a state of suspended animation. For instance, monitoring the temperature profile at the bottom

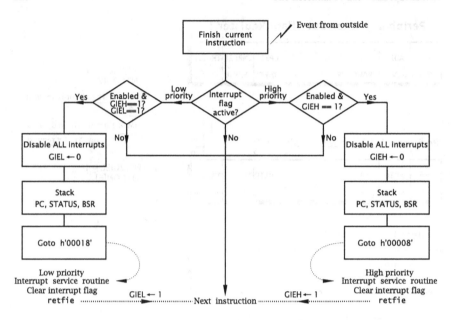

Fig. 7.6 Responding to an interrupt request in Priority mode

of a lake over a period of a year at one-hour intervals using a battery-powered data logger requires processing for a tiny proportion of the time. Placing the PIC MCU in this power-down mode after each sample has been taken and stored will reduce the necessary battery capacity. The sleep instruction initiates this mode. An interrupt from an outside source, in this case a low-power hourly oscillator, is used to wake the PIC MCU up. As we see from Figs. 7.2 and 7.4, this awakening is independent of the setting of any of the Global mask bits located in INTCON[7:6].

In all cases, an interrupt request is initiated when a peripheral module or external device causes its associated interrupt flag to be set. For instance, if Timer 3 overflows, it will *automatically* set TMR3IF—that is PIR2[1]. This will happen, even if the TMR3IE mask bit is 0. In virtually all cases, once an interrupt flag is set, it will remain 1 until cleared by the program. For this reason, it is *vital* that before exiting an interrupt service routine that the initiating interrupt flag is cleared before returning; e.g. bcf PIR2,TMR3IF. Failure to do so would mean that on return from the ISR another request will immediately be made ... *ad infinitum*!

To illustrate the software aspects of interrupt handling, consider an absolutely bare bones example. In order to count customers coming into a small shop, a low-power laser light beam/photocell is placed across the entry door; as shown in Fig. 7.7. When the shopper breaks the beam, the resulting pulse _/‾_ requests service from the monitoring microcontroller, which is away in its main background

Fig. 7.7 Monitoring
customers entering the shop

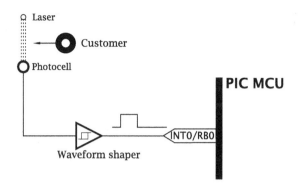

routine doing something else; maybe handling the communication link between the point of sale (POS) terminal and the main inventory computer.

From the software point of view we specify that each time the customer enters the shop a GPR called TALLY is incremented. Of course customers will also be leaving, but if the entrance is relatively narrow we can divide the number of breaks by two to get the actual number of bodies. This will limit the customer count, but we can easily increase the total by using extra GPR Files. We assume that the system is reset at the beginning of business each day, so we are not expecting a customer base greater than 126.

In keeping with the minimalistic character of this example, we will assume that the interrupt system remains in the Power-on Reset default Compatible mode, so we can use the Default interrupt vector h'00008' and ignore prioritization issues.

Program 7.1 shows the two vectors at the top of the listing. As specified by the directive org 00000 (see p. 245) at location h'00000', which is where the PIC MCU resets, is the instruction goto MAIN. Similarly, the instruction goto PERS_COUNT actions a switch to the routine located at PERS_COUNT, should the Program Counter alight at location h'00008'. Thus if the PIC MCU responds to an interrupt, we have the sequence *interrupt* ⤳ h'00008' ⤳ PERS_COUNT.

The Main program itself simply zeroes the customer Tally and sets both the INT0IF and GIE mask bits to enable external Hardware interrupts via pin INT0. As we can see from Fig. 4.1 on p. 71, INT0 is shared with bit 0 of Port B (pin RB0) which also doubles as an analog input to the analog to digital converter module (pin AN4). As all PIC MCUs default to analog input, the ADC's control register ADCON1 is set to configure this pin as a digital input—see Fig. 14.12 on p. 510.

The following endless loop represents a skeleton of the processor's background tasks. Normally an embedded system's software will be structured as a DO-FOREVER loop, with calls outside to subroutines and dislocations to interrupt service routines.

Interrupts happen randomly as viewed by the software and thus, unless masked out, may happen at any part of the background software, including in the middle of a subroutine or even with a High-priority request when executing a Low-priority ISR. An ISR foreground routine uses the internal SFR registers in the same way as any other software, so conflict over such resources will exist. For instance, the

Program 7.1 People counting

```
ADCON1  equ   h'FC1'
INTCON  equ   h'FF2'    ; The INTerrupt CONtrol reg in which
INT0IF  equ   1         ; bit1 is the Hardware interrupt0 flag
INT0IE  equ   4         ; and bit4 is the associated mask bit
GIE     equ   7         ; and bit7 is the Global mask bit

TALLY   equ   h'020'    ; Keeps tally of passing customers
; Reset vector ---------------------------------------------
        org   00000     ; Resets to h'00000' in Program store
        goto  MAIN      ; Go to start of background routine
; Interrupt vector -----------------------------------------
        org   h'00008'  ; Goes to h'00008' if interrupt accepted
        goto  PERS_COUNT ; Go to start of foreground ISR

; Background program starts by initialization --------------
MAIN    movlw b'11111111' ; Make ports all digital
        movwf ADCON1    ; rather than default analog!

        bsf   INTCON,INT0IE; Enable Hardware interrupts
        bsf   INTCON,GIE  ; Enable interrupt system overall
        clrf  TALLY       ; Zero the customer count
; Main endless loop ----------------------------------------
M_LOOP                    ; Do this
                          ; Do that
                          ; Do the other
        bra   M_LOOP      ; and repeat

; ----------------------------------------------------------

; **********************************************************
; * FUNCTION  : ISR increments TALLY on entry            *
; * ENTRY     : None                                     *
; * EXIT      : TALLY incremented                        *
; * ENVIR'MENT: Transparent if Compatibility or High-priority *
; **********************************************************
PERS_COUNT
        bcf   INTCON,INT0IF; Clear the INT0 Interrupt flag
; =========================================================
        incf  TALLY,f     ; Record one more event
; =========================================================
        retfie 1          ; Restore WREG, STATUS & BSR & return
```

background program could be testing the contents of a File when an interrupt occurs. The Skip or Conditional Branch instruction which follows the test could be dependent on, say, the state of the Zero flag in the Status register. However, the ISR will in all probability alter **Z** and thus on return the background program will execute this instruction, oblivious of the fact that execution has been transferred in the interregnum. Any change to **Z** would cause an erroneous branch in the background program. Trying to debug this sort of problem is virtually impossible because the effect of such an interrupt is sporadic, as the particular bug depends on the interrupt occurring at just this wrong time and wrong place—something it may do perhaps once a week—and thus is difficult to reproduce.

Apart from Status register problems, the Working register is almost certain to be altered in an ISR. Where an ISR alters the current Data store bank the consequences can be even more serious. For instance, if BSR is set to h'04' when an interrupt occurs, and is changed to, say, h'01' in the ISR, then any further Banked access by the unsuspecting background program will be in Bank 1 rather than Bank 4. Again this aftermath will appear to happen sporadically; depending where exactly the interrupt occurs. For this reason, the Status, Working and Bank Select registers are *always* automatically saved in the Fast stack before entry to the ISR—as shown in Figs. 7.3 and 7.6 and on p. 167.

If SFRs besides STATUS, WREG and BSR are altered in the foreground routine, then they can be manually saved. Typically this will be done by pushing into the Hardware or Software stack, as illustrated in Program 6.8 on p. 182 and the code fragment on p. 185 or by using fixed locations in memory.[4] However this is done, this process is known as **saving the context**. Due to the capricious nature of an interrupt, all ISRs need to be super transparent; in the same manner as transparent subroutines discussed in the last chapter.

With this in mind, in general an ISR is normally divided into three phases.

Context Switching

A copy is made of the contents of any SFR that the ISR is going to alter, unless only used by this ISR. If necessary, the same can be done for shared GPRs not being used to pass data between the background software.

In our example, no SFRs are altered beyond the three automatically saved, and the ISR is not servicing a Low-priority request in the Priority mode. The GPR TALLY is altered (incremented) deliberately to record a person breaking the beam. TALLY is an example of a **global** variable, which can be used by both the background and foreground programs, and is used to pass information back from the ISR.

Our example's simplicity is partly because we have assumed that service requests can only come from one source; that is from a Hardware interrupt at pin INT0. Where requests can originate from several origins, the ISR will firstly have to determine whence it came from before saving the appropriate context. This is done by testing each of the appropriate interrupt flags in turn. Thus in a system where an interrupt can be requested from Timer 1 as well as from pin INT0 we would have something like:

```
btfsc INTCON,1   ; Check INT0IF for Hardware interrupt
 goto EXTERNAL_0 ; IF set THEN go to INT0 handler
btfsc PIR1,0     ; Check TMR1IF Timer1 interrupt flag
 goto TIMER1_ISR ; IF PIR1[0] set, go to TMR1 handler
....  ......
```

[4]See Program 7.1 in my *Quintessential PIC® Microcontroller* for an example of the latter technique.

Core Function

This is where the processing is done. In our example, all that is done is to increment TALLY. After which the active interrupt flag INT0IF in INTCON[1] must be cleared. If this is not done, then on return to the background program another interrupt request will immediately be set in train.

In the more general case with several request origins, there will be an interrupt handler for each source. Each will follow on from its specific context switching routine. Only the interrupt flag specific to the handler will need to be cleared, so that on return any pending requests can then be serviced.

As a general rule an absolute minimum of processing should be done in any handler routine. This reduces the possibility of interaction with process resources and especially when there are multiple sources of interrupts, facilitates a real-time response. It also helps in debugging which is notoriously difficult in this random-like environment.

Restoring the Context

The exit process first pulls out any saved registers. This reinstates the state of these Files to their entry value.

Finally, the exit instruction retfie automatically sets the appropriate Global Interrupt Enable mask bits in INTCON[7:6] to re-enable the interrupt system. Execution then returns to the background software. In our example, GIE is set and as we are using the Fast version retfie 1, the context is concurrently restored.

Where multiple interrupt handlers are coded, it is usual to exit via a single retfie rather then using multiple exit points. Generally both subroutines and interrupt functions should be structured with only one entry and one exit point

Even with transparency, issues can arise with shared data. In particular, in dealing with events where multiple-precision data are monitored and changed by both background and foreground routines. Consider as an example a real-time clock (RTC) which updates four Files holding time in the 4-byte multiple-precision format HOURS:MINUTES:SECONDS:JIFFY, where the JIFFY byte holds tenths of seconds; see Example 7.3. We assume an external 10 Hz oscillator interrupts the PIC MCU ten times per second, and the ISR updates the time-array.

Consider now that this RTC is part of a central heating controller. At 09:00:00:00 hours the water pump is to be toggled from on to off by the background program. One day this has been done and the time is now 09:59:59:09. The background program, which spends most of its time just looking at the time, reads the hours as 09. Getting interested, it is just going to read the Minutes variable when the Jiffy oscillator 'ticks'. The MCU is interrupted and the RTC now is updated to 10:00:00:00. On return the background program now reads in succession 00, 00, 00. Thinking that it is now 09:00:00:00 it toggles the pump on and thereafter the on and off periods are interchanged indefinitely!

Of course it is bad design practice to use a toggle action; instead the pump should be switched off at 9 am rather than toggled. At least in this latter case the harm done

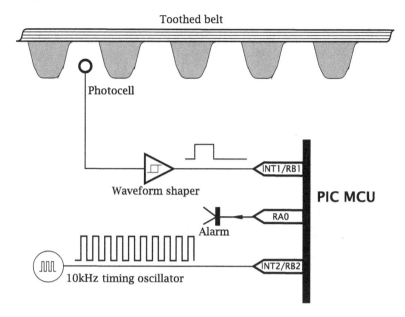

Fig. 7.8 Monitoring the teeth integrity of a belt drive

would be time limited. In general, the interrupt handler should be disabled by clearing the appropriate mask where such multiple-precision data manipulation routines are being executed in the background. Any interrupts occurring during this time will be acknowledged when the mask is subsequently set, although events could be missed if the masked-out period is too long.

For our final example, let us consider an interrupt-driven system which makes use of the Priority-mode interrupt configuration. Mechanical power is frequently transmitted using a drive belt. This is typically structured as an integral body and teeth urethane cast. This is reinforced with spiraled cord and a yarn of carbon fiber.

In order to give warning of a missing tooth, it is proposed that a light source/photocell generate a pulse with each passing tooth. A 10 kHz oscillator is to be used to allow a processor to calculate the inter-tooth period with a resolution of 0.1 ms. As can be seen from Fig. 7.8, it is proposed to use the conditioned tooth signal to create an interrupt at the INT1 pin; whilst in parallel, the timing oscillator creates an interrupt at INT2 every 100 µs. A buzzer will be connected to Port A's pin RA0 to sound the alarm. In practice there will likely be a serial communication link to send a status report to a remote location. Some typical real-world signals can be seen in Fig. 7.9. The rather noisy output from the photocell is shown in the bottom trace, with the cleaned up tooth signal of the top waveform being input to the INT1 pin.

The system software is to be partitioned into three routines. The Main or background routine and two ISRs implementing the foreground code.

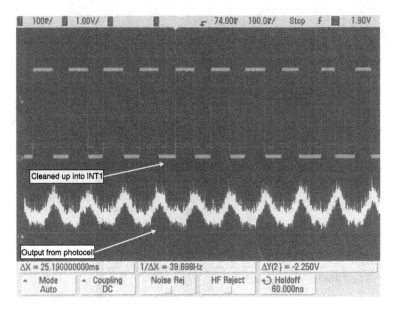

Fig. 7.9 An oscillogram showing the raw and conditioned tooth pulse train

Routine MAIN sets up the parallel ports and interrupt system. This is followed by an endless loop, looking for an absent tooth. The algorithm used in the implemention of this loop is to set a variable ALARM_COUNT to 255 whenever the comparison is true; that is the measured period is too long. Each time the Alarm routine is entered with this variable non-zero then the sounder is activated for 1 ms. Thus after an over-long period is detected, the sounder will activate. Should no more incidents be detected, then the alarm will cease after this variable decrements to zero; for not less than 255 ms.

High-priority function JIFFY is entered on each 'tick' of the 10 kHz timing oscillator connected to INT2. Each instance of this ISR results in an increment of the 2-byte variable TIMER_H:TIMER_L. That is TIMER:2 is incremented every 100 μs.

Low-priority function TOOTH is entered on each passing tooth.

This routine reads the 2-byte Timer count and uses this to generate a new moving average. TIMER:2 is then zeroed. Both this average and current period count are used by the background routine to conditionally activate an alarm situation.

The routine below shows the vector table actioning a jump to these three functions. The PIC MCU will reset to the instruction at h'00000' in the Program store on a Power-on Reset, and as shown, execution will then go to the beginning of the routine labeled MAIN. Likewise, a High-priority interrupt (the oscillator connected to pin INT2) will vector via h'00008' and hence to the ISR JIFFY. The Low-priority ISR TOOTH (belt photocell connected to pin INT1) is entered via a Branch from the vector at h'00018'.

```
; The three vectors ----------------------------
        org     h'00000'  ; Reset vector
        bra     MAIN      ; Go to the background software


        org     h'00008'  ; Interrupt on oscillator
        bra     JIFFY     ; Go and execute the High-
priority ISR


        org     h'00018'  ; Interrupt on passing tooth
        bra     TOOTH     ; Go and execute the Low-
priority ISR
```

The executable code (less the vector table and various equates) is shown in Program 7.2. Looking at each routine in more detail, we have:

The Background Routine
The task list for the MAIN routine is:

1. Ensure that the parallel port pins are set-up as digital (MCU defaults to analog inputs on Power-on Reset) so that pins RB1, RB2 can be used for the interrupt inputs and pin RA0 can be used as an output to drive the digital alarm device.
2. As we are going to use the Priority interrupt mode, bit IPEN in RCON[7] needs to be set to 1.
3. Both the INT1 and INT2 Hardware interrupts need to be locally enabled by setting the INT1IE and INT2IE Enable mask bits in INTCON3[3:4] respectively.
4. The former interrupt should be set as Low-priority by clearing its INT1IP bit in INTCON3[6] (its default Power-on Reset value anyway) and setting INT2IP in INTCON3[7].
5. Globally both High- and Low-priority interrupts will need to be enabled by setting bits GIEH and GIEL in INTCON[7:6] respectively.
6. With the interrupt system set-up to respond to Low-priority INT1 and High-priority INT2 interrupts, the remainder of the background software will be an endless loop comparing the latest Period reading with the running average.
   a. FOREVER DO:
   b. Compare $0.75 \times$ Period > EMA?
   c. IF TRUE THEN sound alarm for nominally 0.25 s,

where EMA represents the Exponential Moving Average of all the antecedent inter-tooth period timings and Period is the last recorded tooth-tooth period. If a tooth is missing, then this Period reading will be somewhat larger than the running average. Using a $\frac{3}{4}$ factor gives some margin for a de-accelerating belt and timing jitter.

The coding itself copies the Period data $\boxed{\text{PERIOD\_H}}_{\text{F h'073'}}\boxed{\text{PERIOD\_L}}_{\text{F h'072'}}$ into two Temporary locations $\boxed{\text{TEMP\_H}}_{\text{F h'021'}}\boxed{\text{TEMP\_L}}_{\text{F h'020'}}$ and then generates $\frac{1}{4}$ of the Period by shifting this copy twice right—lines 27 through 34. Subtracting this from the original PERIOD:2 data gives the $\frac{3}{4}$ fraction in TEMP:2.

Subtracting this modified Period from the top two Exponential Moving Average bytes will not give a borrow-out (carry is 1) if EMA:2 $\geq$ TEMP:2 and this actions

Program 7.2 Belt-drive monitor software

```
; ****************************************************************
; * This is the start of the background Main routine           *
; ****************************************************************
; First initialize the hardware --------------------------------
MAIN      movlw    b'11111111'    ; Make ports all digital
          movwf    ADCON1         ; rather than analog
          bcf      TRISA,0        ; With RA0 an Output
          bcf      PORTA,0        ; Starting with the alarm off

; Now initialize the interrupt system --------------------------
          bsf      RCON,IPEN      ; Select Priority mode
          bsf      INTCON3,INT1IE ; Enable Hardware interrupt 1
          bsf      INTCON3,INT2IE ; Enable Hardware interrupt 2
          bcf      INTCON3,INT1IP ; INT1 interrupts Low-priority
          bsf      INTCON3,INT2IP ; INT2 interrupts High-priority
          bsf      INTCON,GIEH    ; Enable Hi-priority int
          bsf      INTCON,GIEL    ; and Lo-priority int system

          clrf     PERIOD_L       ; Zero the initial PERIOD:2
          clrf     PERIOD_H
          clrf     TIMER_L        ; and Time count
          clrf     TIMER_H
          clrf     EMA_DEC        ; and Exponential Moving Average
          clrf     EMA_L
          clrf     EMA_H

; This is the endless loop checking the Timer period isn't less
; than 0.75 of the tooth period running average ----------------
; First generate 0.75 x PERIOD ---------------------------------
MAIN_LOOP movff    PERIOD_H,TEMP_H; Copy PERIOD:2 to TEMP:2
          movff    PERIOD_L,TEMP_L

          bcf      STATUS,C       ; Now >> 2 to give
          rrcf     TEMP_H,f
          rrcf     TEMP_L,f
          bcf      STATUS,C
          rrcf     TEMP_H,f
          rrcf     TEMP_L,f       ; PERIOD/4

          movf     TEMP_L,w       ; Subtract from PERIOD
          subwf    PERIOD_L,w
          movwf    TEMP_L
          movf     TEMP_H,w
          subwfb   PERIOD_H,w
          movwf    TEMP_H         ; to give PERIOD*3/4 in TEMP:2

; Now compare (EMA:2 - 3/4*PERIOD:2) < ? -----------------------
          movf     TEMP_L,w       ; EMA:2 - (0.75*PERIOD:2)
          subwf    EMA_L,w        ; First subtract low bytes
          movf     TEMP_H,w       ; THEN high bytes
          subwfb   EMA_H,w
          bc       ALARM          ; IF no Borrow THEN DO Alarm ELSE
          setf     ALARM_COUNT    ; reset Alarm loop C'nt to 255

ALARM     tstfsz   ALARM_COUNT    ; IF Alarm C'nt 0, turn off sound
          bra      SOUND          ; ELSE sound the alarm
          bcf      PORTA,0        ; Turn off the sounder
          bra      MAIN_LOOP      ; and repeat the loop
```

(continued on the next page)

Program 7.2 (*Continued*)

```
SOUND     bsf       PORTA,0          ; Turn on the sounder
          decf      ALARM_COUNT,f    ; One more beep
          call      DELAY_1MS        ; for a millisecond
          bra       MAIN_LOOP        ; and repeat the loop
;  ****************************************************************
;  * FUNCTION : INT2 ISR increments TIMER:2 on entry            *
;  * ENTRY    : None                                            *
;  * EXIT     : TIMER:2  incremented                            *
;  * ENVIR'MENT: Transparent High-priority                      *
;  ****************************************************************

JIFFY infsnz TIMER_L,f      ; Increment TIMER:2
      incf   TIMER_H,f
      bcf    INTCON3,INT2IF  ; and reset interrupt flag
      retfie 1


;  ****************************************************************
;  * FUNCTION  : INT1 ISR updates Exponential Moving Average    *
;  * ENTRY     : TIMER:2, EMA:3                                 *
;  * EXIT      : EMA:3, PERIOD:2 updated.  TIMER:2 cleared      *
;  * ENVIR'MENT: Transparent Low-priority                       *
;  ****************************************************************
TOOTH push
      movwf  TOSL            ; Save WREG in the Hardware stack
      swapf  STATUS,w        ; Copy Status not changing any flag
      movwf  TOSH            ; and push into stack

; Now clear update PERIOD:2 and clear the TIMER:2 counter -----
      movff  TIMER_L,PERIOD_L; Make a copy
      movff  TIMER_H,PERIOD_H
      clrf   TIMER_L         ; Before clearing the Timer count
      clrf   TIMER_H

; Now update the 3-byte EMA ------------------------------------
      movf   EMA_L,w         ; Prepare to subtract
      subwf  EMA_DEC,f       ; EMA/256 from EMA
      movf   EMA_H,w         ; to give
      subwfb EMA_L,f         ; EMA = (255/256)*EMA
      clrf   WREG
      subwfb EMA_H,f
      movf   PERIOD_L,w      ; Now add Y/256
      addwf  EMA_DEC,f       ; to give
      movf   PERIOD_H,w
      addwfc EMA_L,f
      clrf   WREG
      addwfc EMA_H      ; EMA(n) = (255/256)EMA(n-1) + Y(n)/256

; Now reset the interrupt flag and restore the context ========
      bcf    INTCON3,INT1IF
      swapf  TOSH,w          ; Retrieve entry value of STATUS
      movwf  STATUS
      movff  TOSL,WREG       ; and entry Working register
      pop
      retfie
```

a jump into the Alarm routine. If there is a borrow-out then the loop count variable ALARM_COUNT is set to h'FF' before entering the Alarm routine—lines 41 through 47.

The Alarm routine itself tests ALARM_COUNT. If this is zero, then pin RA0 is set low, which turns the sounder off and the endless main routine repeated. Otherwise, ALARM_COUNT is decremented and pin RA0 set high to activate the sounder. A 1 ms delay (see Program 6.1 on p. 166) ensures that where ALARM_COUNT is not set to h'FF' on entry, then the alarm will continue to be sounded for 256×1 ms (actually a little longer, due to the surrounding code and interrupts) as the loop count decrements.

The Jiffy Interrupt Service Routine

This INT2 ISR entered on each cycle of the 10 kHz is rather elementary. All that is done is to increment the Timer $\boxed{\text{TIMER\_H}}_{\text{F h'071'}} \boxed{\text{TIMER\_L}}_{\text{F h'070'}}$. As this variable is read and then zeroed in the TOOTH ISR, its reading at that point will be the inter-tooth period in 100 µs jiffies.

The High-priority status of this handler means that the Fast stack retains its integrity, as it can't be interrupted by a Low-priority interrupt. Thus returning with a Fast retfie is all that has to be done to preserve the context. As in all ISRs, the interrupt flag (here INT2IF) must be cleared before this return.

The Tooth Interrupt Service Routine

The core task for the TOOTH ISR is to use the inter-tooth tally originating from the JIFFY ISR to update a running average, which will in turn be used by the background routine to compare with the inter-tooth period and inform the Alarm code. As the belt speed will vary with time, an average that will take into account historical data whose weight diminishes with age and with a low RAM storage and processing overhead is required.

In this example an Exponential Moving Average (EMA) is computed to give the baseline period. This is based on the relationship:

$$\text{EMA}_n = \alpha \text{EMA}_{n-1} + (1 - \alpha)S_n$$

where EMA_n is the new computed moving average, EMA_{n-1} the value on entry and S_n the new Timer reading (i.e. entry inter-tooth period) and α a small constant smoothing factor.

In our coding of Program 7.2 we have used $\alpha = \frac{1}{256}$, giving

$$\text{EMA}_n = \frac{255}{256}\text{EMA}_{n-1} + \frac{1}{256}S_n.$$

That is, each new Period sample S_n contributes $\frac{1}{256}$ of its value to the grand ensemble; the latter having lost the same fraction of its value.

This type of evolving or moving average is termed exponential, as an event occurring in the past will have an exponentially decaying influence on the overall outcome. In our coding, a sample taken 256 readings ago will contribute about 37% ($\frac{1}{\exp}$) of its initial effect on the outcome. As the belt speeds up or slows down, the EMA will track such changes with a time constant of 256 inter-tooth spans. Figure 7.10 shows an actual experiment based on a step Period reading of 16.384 ms. After four time constants (1024 samples) the EMA has stabilized to its steady-state value.

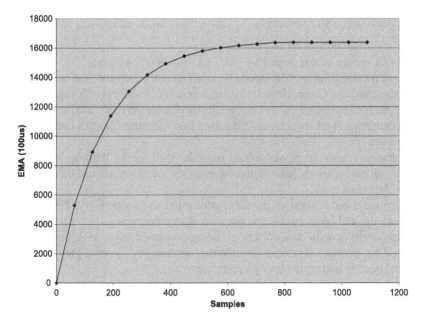

Fig. 7.10 Response to a step change of 16,384 in the tooth-tooth period

Turning to the ISR code itself. As the High-priority JIFFY ISR can interrupt this routine and alter the Fast stack, the context needs to be explicitly saved. This is done in lines 7 through 10 by first saving the Working and then Status registers in TOSH:TOSL in the pre-Pushed Hardware stack. Actually a swapf is used instead of movf to avoid altering the **Z** flag—see p. 195. When the context is restored at the end of the routine in line 31, swapf is again used to retrieve **STATUS**. The two Swaps in series cancel, leaving the exit status unchanged.

After the context has been saved, the setting of the TIMER_H:TIMER_L pair is copied into PERIOD_H:PERIOD_L. This is the Period data used to update the running average and also (multiplied by $\frac{3}{4}$) by the background routine in its comparison process. After this, TIMER:2 is cleared (lines 14 & 15), ready for on-going incrementation in the JIFFY ISR, which as High-priority, can interrupt this Low-priority handler.

The Exponential Moving Average is stored as a 3-byte array, with the LS byte EMA_DEC being treated as fractional (right of the decimal, or more correctly, binary point) $\boxed{\text{EMA\_H}}_{\text{F h'062'}} \boxed{\text{EMA\_L}}_{\text{F h'061'}} \boxed{\text{EMA\_DEC}}_{\text{F h'60'}}$. In order to create the fraction $\frac{255}{256}$ EMA, the low byte EMA_L is subtracted from EMA_DEC and the high byte EMA_H from the resulting EMA_L byte, with the borrow-out being taken from EMA_H. Effectively this subtracts EMA:3 slid right a byte ($\div256$) from the original value, giving $\frac{255}{256}$ EMA:3. By adding PERIOD_L to EMA_DEC and PERIOD_H to EMA_L with a carry-out to EMA_H; effectively PERIOD:2 slid right one byte (i.e. $\frac{\text{PERIOD:2}}{256}$) has been added to the EMA. This process gives the new EMA, and clearing INT1IF followed by restoration of the context completes the routine.

In conclusion, ISRs are similar to subroutines, but keep in mind the following points:

- The ISR should be terminated by retfie instead of return.
- Any SPRs altered in the ISR should be saved on entry and retrieved on exit if they are also used elsewhere. If the ISR is High-priority, or there is only one ISR in the code, then a Fast retfie will automatically preserve WREG, STATUS and BSR. The exception is where such an ISR calls a Fast subroutine, as this overwrites the Fast stack.
- Parameters cannot be passed to and from the ISR via the Working register. Instead, global variables (data in known memory locations) should be used as required or via a stack.
- ISRs should be as short as possible, with minimal functionality. This helps in debugging, and helps ensure that other events are not missed.
- Where multiple-byte data objects are being processed by an ISR, consideration should be given to disabling the interrupt system (by clearing the appropriate global masks) during any background access.

Examples

Example 7.1 Consider a conveyor belt in a pea-canning factory. As part of the automatic packing system, a photocell generates a single short pulse for each passing can, in the manner of Fig. 7.7. After each batch of 24 cans, a nominal 1 ms pulse _／‾＼_ is to be generated using Port A's pin 0 (RA0) and this triggers the packing mechanism's electronics.

Solution The software is shown in Program 7.3. The Reset vector at h'00000' actions a transfer to the Main background routine. The MCU powers up in the default Compatible mode with its interrupt vector at h'00008' and this causes the code to jump to the foreground ISR labeled CAN_COUNT.

As interrupts are automatically disabled on a Power-on Reset, the various Files and ports are normally set to their initial value at the beginning of the background program before interrupts are enabled. This eliminates the possibility of servicing an interrupt before the initialization code has been completed. The initialization schedule is:

1. Clearing bit 0 of Port A will ensure that pin RA0 starts low after Reset.
2. All parallel port pins are configured as analog inputs on Reset. To change Port A bit 0 to an output, the associated bit in the TRISA SFR must be cleared. In addition, setting the ADC module's CONtrol register 1 to all 1s makes all parallel port pins digital—see Fig. 14.12 on p. 510.
3. GPR File EVENT recording the photocell pulse count and BATCH; which is set to non-zero in the ISR whenever a batch of 24 cans has passed, are both zeroed.
4. Setting the Global Interrupt Enable mask bit now enables the interrupt system and specifically setting INT0IE enables interrupts from the INT0 pin.

Program 7.3 Program for the pea-canning packer

```
EVENT     equ    h'060'          ; Keeps count of cans of peas
BATCH     equ    h'061'          ; Signals when a 24 can lot passes
; ----------------------------------------------------------------
          org    h'00000'        ; Power-on Resets here
          bra    MAIN            ; Go to start of background routine
; ----------------------------------------------------------------
          org    h'00008'        ; The Default Interrupt vector
          bra    CAN_COUNT       ; Go to start of foreground ISR
; ----------------------------------------------------------------

; ****************************************************************
; * This is the start of the executable code,                  *
; * beginning with the main or background routine              *
; ****************************************************************
; First initialize the hardware, Variables and Interrupt INT0 --
MAIN      movlw  b'11111111'     ; Make ports all digital
          movwf  ADCON1          ; rather than analog
          bcf    TRISA,0         ; With RA0 an Output
          bcf    PORTA,0         ; Starting with the alarm off
          clrf   BATCH           ; Zero the Batch signal
          clrf   EVENT           ; and the can count
          bsf    INTCON,GIE      ; Enable ALL interrupts
          bsf    INTCON,INT0IE   ; Enable external INT0 interrupts

; WHILE Batch signal is zero DO nothing ----------------------
M_LOOP tstfsz   BATCH           ; Check BATCH == 0?
          bra    M_GO            ; Skip out IF not
          bra    M_LOOP          ; ELSE try again

; Pulse on the 24th can (BATCH set to non-zero ----------------
M_GO      clrf   BATCH           ; Zero the Batch signal
          bsf    PORTA,0         ; Bring line RA0 high
          call   DELAY_1MS       ; Wait for one millisecond
          bcf    PORTA,0         ; and go low again
          bra    M_LOOP          ; DO forever
; ****************************************************************
; * FUNCTION : INT0 ISR increments EVENT and IF >= 24 THEN     *
; * FUNCTION : zeroes EVENT and makes BATCH non zero           *
; * ENTRY    : EVENT:1, BATCH:1                                *
; * EXIT     : EVENT:1 and BATCH:1 updated                     *
; * ENVIR'MENT: Transparent Compatible mode                    *
; ****************************************************************
CAN_COUNT
          incf   EVENT,f         ; Record one more event
          movlw  d'23'           ; Check for 24 events (a batch)
          cpfsgt EVENT           ; > 23? Yes THEN skip
          bra    CAN_EXIT        ; ELSE finished
          clrf   EVENT           ; Yes, so zero can count and tell
          incf   BATCH,f         ; the world that 24 cans have passed
; ================================================================
CAN_EXIT
          bcf    INTCON,INT0IF   ; Clear the Hardware Interrupt flag
          retfie 1               ; & return with context to background
```

The core of the main background routine simply repetitively checks the contents of BATCH. This is normally zero, but the foreground ISR sets this whenever each batch of 24 cans have passed. When this is the case BATCH is zeroed and RA0 is brought high and a 1 ms delay subroutine called.[5] RA0 is dropped low and the loop is then repeated.

When an interrupt occurs, as triggered by a can breaking a beam, then execution will be transferred to the ISR, that is, Interrupt ↝ h'0008' → CAN_COUNT. The functional sector of this handler simply increments the datum EVENT. By comparing this with the constant 23, the program determines when the Event tally rises above this value. When this occurs, BATCH is incremented to signal the background program that 24 cans have passed and EVENT is zeroed to give a modulo-24 count.

Example 7.2 In a food processing factory, cans of baked beans on a conveyer belt continually pass through a tunnel oven, as shown at the top of Fig. 7.11, where the contents are sterilized. Photocell detectors are used to sense cans, both entering and leaving the oven. The output of the sensors are logic 1 when the beam is broken.

You are asked to design an interrupt-driven interface for this system, combining the two signals to activate the PIC MCU's *one* INT0 input. You may assume that the INT1 and INT2 pins are in use for other non-interrupt duties. A buzzer connected to Port B's pin RB7 is to be sounded if the number of tins in the oven exceeds four, indicating that a jam has occurred.

Solution The hardware aspect of this example presents two problems. The first of these involves distinguishing which cell, IN or OUT, generates a request. In Fig. 7.11 each cell clocks an associated D flip flop when the beam is broken. As the D input is permanently tied to logic 1, the clocked flip flop output goes to logic 1. ORing both of these interrupt flags together generates a falling edge at the INT0 pin if *any* beam is broken.

Both the IN and OUT external flags can be read at Port A pins RA0 and RA1, and this allows the ISR software to distinguish between the two events (can-in and can-out). The appropriate flag can then be reset by toggling the appropriate flip flop reset, using two further port lines RA2 and RA3 for Cancel_in and Cancel_out respectively.

To show how this operates, consider a can has just broken the Out cell beam, as shown in the diagram. The following sequence occurs.

1. The resulting pulse clocks the OUT flag.
2. The flip flop goes high, which in turn brings pin RA1 high and via the OR gate pin INT0/RB0. This requests a Hardware interrupt.
3. When the PIC MCU transfers to the interrupt handler it checks the state of both flip flops by testing pins RA1 and RA2. In this case it finds RA1 high and in software pulses pin RA3 low.
4. This resets the OUT flip flop and hence cancels the interrupt request from this source.

---

[5] Of course the delay subroutine can be interrupted, which will randomly slightly lengthen the delay. In time-critical situations interrupts should be disabled before calling the delay subroutine and re-enabled on return.

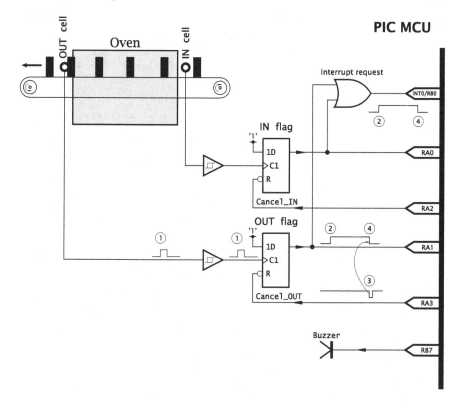

Fig. 7.11 Oven safety hardware

One problem remains: If one event follows another before the ISR software has time to reset the appropriate external flip flop, that second event will be missed, as the OR gate will hold `INT0` low. In this situation no further edge can occur and the interrupt system will be permanently disabled! This can be circumvented in software by polling both external flags before exiting the ISR and taking the appropriate action if either pin is still high.

The interrupt service routine for this hardware configuration is given in Program 7.4. The meat of the code simply checks each of the external flip flops in turn. Depending on the state of these flip flops, one of three pathways through the code is followed:

1. If pin `RA0` is high then a can has broken the IN beam and one is added to the Event counter, kept in a GPR File labeled `EVENT`. The external IN flip flop is reset. If the total is greater than four, the buzzer is turned on by bringing `RB0` low, otherwise it is turned off. Repeat check.
2. If pin `RA1` is high then a can has broken the OUT beam and one is taken away from the Event counter. This time the external OUT flip flop is reset. Again the total is checked against the boundary of four and the buzzer set to its appropriate state. Repeat check.

Program 7.4 Foreground ISR for oven safety

```
; ****************************************************************
; * FUNCTION  : INT0 ISR increments EVENT for an IN can        *
; * FUNCTION  : Decrements for an OUT can & alarms IF > 4       *
; * ENTRY     : EVENT:1                                         *
; * EXIT      : EVENT:1                                         *
; * ENVIR'MENT: Transparent Compatible mode                    *
; ****************************************************************
OVEN  btfsc  PORTA,0       ; Check, IN signal?
      bra    IN            ; IF non zero, a can has just gone in
      btfsc  PORTA,1       ; Check for OUT signal
      bra    OUT           ; IF non zero, a can has just gone out
; ============================================================
; The exit point
      bcf    INTCON,INT0IF ; Clear the Hardware interrupt flag
      retfie 1             ; and return to interrupted background
; ============================================================
; The ISR core
IN    incf   EVENT,f       ; Record a can gone in (count up)
      bcf    PORTA,2       ; Clear external IN flag
      bsf    PORTA,2       ; by pulsing its reset
      bra    ALARM         ; and check for alarm situation

OUT   decf   EVENT,f       ; Record a can gone out (count down)
      bcf    PORTA,3       ; Clear external OUT flag
      bsf    PORTA,3       ; by pulsing its reset

ALARM movlw  4             ; Is Can count > 4?
      cpfsgt EVENT         ; Skip IF less than
      bra    BUZ_OFF       ; ELSE OK, turn the buzzer off
      bsf    PORTB,7       ; Turn buzzer alarm on
      bra    OVEN          ; and repeat poll of cells flags
BUZ_OFF
      bcf    PORTB,7       ; Turn buzzer off
      bra    OVEN          ; and repeat poll of cell flags
```

3. If neither flip flop is set then the ISR exits after resetting the internal INT0IF flag and doing a Fast return to restore the context.

This sequence is repeated whenever actions 1 or 2 have been completed. This ensures that the situation where both beams are broken simultaneously or within a short time frame, will be properly serviced.

The main background program is not shown here. It will be similar to that of Program 7.3 in that the various ports will be set up, the Event counter File cleared and interrupts enabled. It is likely that this background program will be in charge of sounding the alarm and other consequential tasks rather than implementing this as part of the ISR, in keeping with the philosophy of reducing the size of the foreground code. In a practical system the background program would probably drive a numeric display showing the aggregate of cans (four was a ridiculous value, chosen for illustrative purposes only) in the oven. Also some means of resetting to a non-zero value after a jam and some sign in the (erroneous) event of a subzero count being computed must be facilitated.

Example 7.3 On p. 213 a central heating real-time clock was discussed. Write an ISR to add one onto the array of Files holding the four time bytes in a 24-hour time representation, on each 0.1 s interrupt. Each byte location is to hold two binary-coded decimal (BCD) digits; for instance BCD 40 in the File labeled MINUTES is represented as b'0100 0000'. This packed binary-coded decimal format is described on p. 116.

Solution Each time the PIC MCU enters the ISR, one Jiffy must be added to the array of bytes HOURS:MINUTES:SECONDS:JIFFY. The base of each byte count differs in that JIFFY rolls over at a count of ten (i.e., modulo-10), SECONDS and MINUTES have a modulo-60 count and HOURS is modulo-24. Based on this scenario we have as a task list

1. Add one onto the JIFFY count.
2. IF this gives 10 THEN zero JIFFY and add one onto the SECONDS count; ELSE goto EXIT.
3. IF this gives 60 THEN zero SECONDS and add one onto the MINUTES count; ELSE goto EXIT.
4. IF this gives 60 THEN zero MINUTES and add one onto the HOURS count; ELSE goto EXIT.
5. IF this gives 24 THEN zero HOURS.
6. EXIT

The example specified that the datum format should be packed BCD. Thus, 59 minutes should be stored as b'0101 1001' or h'59'. This means that the incrementation process has to preserve this BCD format. Subroutine BCD_INC shown in Program 7.5 uses the daw instruction to correct the standard binary incrementation. Also the FSR0 pointer is decremented ready to deal with the next datum in the chain.

Based on this subroutine, coding for this task list is given in Program 7.6. As File Select Register 0 is used in the ISR, both of the bytes FSR0H:FSR0L need to be pushed out into the Hardware stack on entry and popped back on exit. The Fast stack still holds the basic context.

Program 7.5 Incrementing a packed-BCD byte with maximum value of 99

```
; ********************************************************
; * FUNCTION: Adds onto  packed BCD byte, maximum value 98  *
; * ENTRY   : FSR0 points to byte                          *
; * EXIT    : BCD byte incremented; W and STATUS altered   *
; * EXIT    : FSR0 decremented                             *
; ********************************************************
BCD_INC   incf    INDF0,w    ; BCD + 1 in W
          daw                ; Correct it to BCD
          movwf   POSTDEC0   ; Restored and FSR0 decremented
BCD_EXIT  return
```

Program 7.6 Coding the real-time clock ISR

```
HOURS     equ     h'020'     ; Space for the 2-digit Hour count
MINUTES   equ     h'021'     ; Space for the 2-digit Minute count
SECONDS   equ     h'022'     ; Space for the 2-digit Seconds count
JIFFY     equ     h'023'     ; Space for the 0.1s predivision


; ****************************************************************
; * FUNCTION : INT0 ISR BCD increments 0.1s time chain         *
; * ENTRY    : JIFFY:SECONDS:MINUTES:HOURS bytes               *
; * EXIT     : JIFFY:SECONDS:MINUTES:HOURS plus one            *
; * RESOURCE : Subroutine BCD_INC                              *
; * ENVIR'MENT: Transparent Compatible mode                    *
; ****************************************************************
; First save context ===========================================
RTC       push
          movf    FSR0L,w    ; Copy FSR0 into Hardware stack
          movwf   TOSL
          movf    FSR0H,w
          movwf   TOSH

; The core code ================================================
          lfsr    0,SECONDS  ; Initialize FSR0 to Seconds count

; Task1 --------------------------------------------------------
          incf    JIFFY,f    ; Add one onto Jiffy count

; Task2 Jiffy handling -----------------------------------------
          movlw   d'9'       ; Compare to ten (>9)
          cpfsgt  JIFFY      ; Skip IF Yes
          bra     EXIT       ; ELSE finished
          clrf    JIFFY      ; Clear Jiffy count

; Task 3 Seconds handling --------------------------------------
          call    BCD_INC    ; BCD Increment pointed-to byte
          movlw   h'59'      ; Compare with 0101 1001 (59 BCD)
          cpfsgt  SECONDS    ; Skip IF > 59
          bra     EXIT       ; ELSE finished
          clrf    SECONDS    ; Clear Seconds count

; Task 4 Minutes handling --------------------------------------
          call    BCD_INC    ; BCD Increment pointed-to byte
          movlw   h'59'      ; Compare with 0101 1001 (59 BCD)
          cpfsgt  MINUTES    ; Skip IF > 59
          bra     EXIT       ; ELSE finished
          clrf    MINUTES    ; Clear Minutes count

; Task 5 Hours handling ----------------------------------------
          call    BCD_INC    ; BCD Increment pointed-to byte
          movlw   h'23'      ; Compare with 0010 0011 (23 BCD)
          cpfsgt  HOURS      ; Skip if > 23
          bra     EXIT       ; ELSE finished
          clrf    HOURS      ; ELSE zero Hours count
```

(continued on the next page)

Program 7.6 (*Continued*)

```
; Task6 Restore context and exit ================================
EXIT      bcf      INTCON,INT0IF ; Clear the Hardware INT0 flag
          movff    TOSL,FSR0L    ; Retrieve FRS0
          movff    TOSH,FSR0H
          pop                    ; Restore SP and restore
          retfie   1             ; core regs & return from interrupt
```

The core of the ISR is sectioned as shown to follow the task list. After each incrementation, the datum is compared with the base literal. If greater than, then the datum is zeroed and the next datum incremented.

Example 7.4 A certain vending machine channels coins of various denominations past one of six microswitches connected to Port B. Any coin will close one switch and pull the appropriate pin low, as shown in Fig. 7.12.

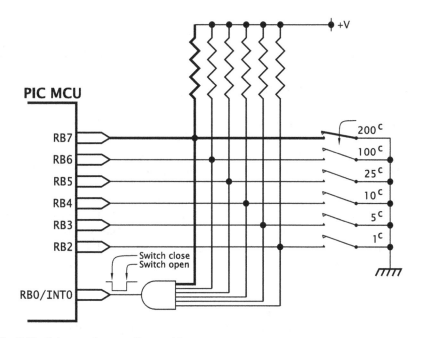

Fig. 7.12 Coin entry for a vending machine

Write the foreground ISR so that the appropriate quantity is added to a GPR File called MONEY. You can assume that the background routine has set up the INTCON register to enable Hardware interrupts via the RB0/INT0 pin.

Solution As shown in Program 7.7 each switch is tested in turn. Any pin which is low reflects a logic 0 in the corresponding Port B bit. With the coin mechanism outlined, only one switch will be closed at any time, so the scanning need not exit after a successful find.

Program 7.7 Interrupt handler for the vending machine

```
VEND   movf    MONEY,w       ; Get current money tally

       btfss   PORTB,7       ; Check for $2
        addlw  d'200'        ; IF 0 THEN add 200
       btfss   PORTB,6       ; Check for $1
        addlw  d'100'        ; IF 0 THEN add 100
       btfss   PORTB,5       ; Check for 25c
        addlw  d'25'         ; IF 0 THEN add 25
       btfss   PORTB,4       ; Check for 10c
        addlw  d'10'         ; IF 0 THEN add 10
       btfss   PORTB,3       ; Check for 5c
        addlw  5             ; IF 0 THEN add 5
       btfss   PORTB,2       ; Check for 1c
        addlw  1             ; IF 0 THEN add 1

       movwf   MONEY         ; Return sum to File MONEY

; The exit point ================================================
       bcf     INTCON,INT0IF ; Clear the Hardware interrupt flag
       retfie 1              ; & return with context to background
```

Self-Assessment Questions

7.1 Rewrite Program 7.3 to deal with a packing quantity of one gross (144). The can count is to be kept in packed BCD (Hundreds and Tens:Units) which can be used by the background software to display the can tally.

7.2 What changes to Example 7.2 would you have to make to allow for a maximum value in the oven of 1000?

7.3 Based on Fig. 7.1, adapt the software of Program 7.2 to compute a 2-byte Difference variable, as well as a 256-sample time constant EMA, on each cardiac beat. As well as updating these variables, the R-point ISR is to set a GPR labeled EVENT to a non-zero value, to signal the background routine that a new reading is available. The background endless loop is to call a subroutine labeled VAR only on each *new* value of DIFFERENCE:2 and zero EVENT. The alarm is to be sounded for the duration of each Event in which the EMA period is less than $\frac{3}{4}$ of the new Period reading.

7.4 The speed of a rotating shaft can be measured by using a coded disk to generate a pulse on each angular advance of 10°, which can be used to interrupt a PIC MCU. If the top speed is 20,000 revolutions per minute, what is the absolute maximum duration of the ISR in this worst-case situation to avoid missing pulses? You may assume a crystal frequency of 4 MHz.

7.5 An electronic tape measure determines distance by pulsing an ultrasonic transmitter and detecting the time it takes for the echo to return. The hardware for this echo sounder is shown in Fig. 7.13 and makes use of the INT1 and INT2 Hardware interrupts.

   The maximum range is specified to be 2.5 m with a reading resolution of 1 cm. The speed of sound in air is 344 m/s at 20°C, which gives a go-return

Fig. 7.13 Echo sounding
hardware

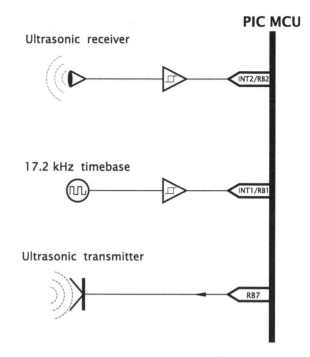

ping time for one cm of 58 μs. Using a 17.2 kHz oscillator as a time base gives one interrupt per 58 μs; that is, a Jiffy per centimeter.

Based on this hardware, the software must implement the following task list:

- Background routine
  1. Zero Jiffy count and New flag.
  2. Pulse the sounder for 1 ms.
  3. Wait until Receive flag is non-zero.
  4. Call subroutine DISPLAY.
  5. Repeat forever.
- Foreground routine.
  1. IF oscillator THEN increment Jiffy count.
  2. IF receiver THEN set Receive flag to non-zero to tell background program that the Jiffy count is the final value.
  3. Repeat until neither interrupt is active.
  4. Return from ISR.

Code the foreground ISR tasked as above, using a GPR labeled NEW as a flag, to tell the background program that the echo has returned and to read the GPR COUNT as the required Jiffy tally in centimeters. The background Main program should call a subroutine to handle the display task, which you do not have to code. Assume that both the Compatible default interrupt mode is used.

7.6 It is proposed to increase the range of the digital echo sounder to 10 m and resolution to 1 mm. What change in the hardware and software would be required?

7.7 The system in SAQ 7.6 has been built and tested. However, readings seem to shift slowly with time. Drift is suspected but the oscillator has been proven to be stable. Thinking laterally, one student wonders if the speed of sound varies with atmospheric conditions. After some research the student arrives at the formula for temperature dependence as

$$V_t = V_0\sqrt{1 + \frac{\Delta t}{273}}$$

where V_0 is the propagation velocity at 20°C and V_t is the velocity at a temperature of t. How much change in temperature Δt will there be to cause an error of 1 mm with the sounder measuring at its maximum range?

Chapter 8
Assembly Language Code Building Tools

We have now been writing programs with abandon since Chap. 3. For clarity these listings have been written in a human-readable form. Thus instructions have been represented as a short mnemonic, such as `return` instead of b'0000 0000 0001 0010'; the Files similarly have names, such as `INTCON`; lines have been labeled and comments attached. Such symbolic representations are only for human consumption. The MCU knows nothing beyond the binary codes making up operation codes and data, such as shown on p. 46.

With the help of the device's instruction set (see p. 66), it is possible to translate by hand from the human-readable symbolic form to machine-readable binary. This is not particularly difficult for a device such as a PIC MCU that has a reduced set of instructions (RISC) and few address modes. However, it is slow and tedious, especially where programs of a significant length are being coded. Furthermore, it is error prone and difficult to maintain whenever there are changes to be made.

Computers are good at doing boring things quickly and accurately; and translating from symbolic to machine code definitely falls into this category. Here we will briefly look at some software packages that aid in this machine-level translation process. In the following chapter we will look at a high-level language alternative.

After reading this chapter you will:

- Know what assembly-level language is and how it relates to machine code.
- Appreciate the advantages of a symbolic representation over machine-readable code.
- Be aware of the function of the assembler.
- Understand the difference between absolute and relocatable assembly; including the role of a linker.
- Appreciate the process involved in translating and locating an assembly-level language program to absolute machine code.
- Understand the structure of a machine-code file and the role of the loader program.
- Be aware of the role of a simulator.
- Appreciate the use of the integrated development environment to automate the interaction of the various software tools needed to convert source code into a programmed MCU device.

S. Katzen, *The Essential PIC18® Microcontroller,*
Computer Communications and Networks,
DOI 10.1007/978-1-84996-229-2_8, © Springer-Verlag London Limited 2010

```
incf    COUNT,f                Translate              0010101000100000
movf    COUNT,w                                       0101000000100000
daw                                                   0000000000000111
movwf   COUNT                                         0110111000100000
return                                                0000000000010010
```

Fig. 8.1 Conversion from assembly-level source code to machine code

The essence of the assembly-level conversion process is shown in Fig. 8.1. Here the program is prepared by the tame human in symbolic form, digested by the computer and output in machine-readable form. Of course, this simple statement belies a rather more complex process, and we want to examine this in just enough detail to help you in writing your programs.

In general, the various translator and utility computer packages are written and sold by many software companies, and thus the actual details and procedures differ somewhat between the various commercial products. In the specific case of PIC MCU devices, Microchip Technology, Inc. as a matter of policy, has always provided their assembly-level software tools free of charge; a large factor in their popularity. For this reason, commercial low-level software for the PIC MCU is relatively rare and what there is usually conforms to the Microchip syntax. For this reason we will illustrate this chapter with the Microchip suite of computer-aided code building tools.

Using the computer to aid in translating code from more user-friendly forms (known as **source code**) to machine-friendly binary code (known as **object code** or **machine code**) and loading this into memory began in the late 1940s for mainframe computers. At the very least it permitted the use of higher-order number bases, such as hexadecimal.[1] In this base the code fragment of Fig. 8.1 becomes:

```
    2A20
    5020
    0007
    6E20
    0012
```

A **hexadecimal loader** will translate this into binary and put the code in designated memory locations. This loader might be part of the software in your PIC-MCU programmer. Hexadecimal coding has little to commend it, except that the number of keystrokes is reduced; but there are more keys and it is slightly easier to spot certain types of errors.

As a minimum, a symbolic translator, or **assembler**,[2] is required for serious programming. This allows the programmer to use mnemonics for the instructions and internal registers; with names for constants, variables and addresses. The symbolic

---

[1] Actually base-8 (octal) was the popular choice for several decades.

[2] The name is very old; it refers to the task of translating and *assembling* together the various modules making up a program.

language used in the source code is known as **assembly language**. Unlike high-level languages, such as **C** or PASCAL, assembly language has a *one-to-one relationship* with the generated machine code; i.e., *one line* of source code produces *one instruction*. As an example, Program 8.1 shows a slightly modified version of Program 6.11 on p. 192. This subroutine computes the square root of a 16-bit variable called NUM, which has been allocated two bytes in the Data store, and returns the 8-bit integer square root in the Working register.

Giving names to addresses and constants is especially valuable for longer programs, which typically comprise several thousand lines of code. Together with the use of comments, this makes code easier to debug, develop and maintain. For instance, in most of our programs up to now we have had statements such as:

```
STATUS   equ   h'FD8'  ; Status register is File h'FD8'
C        equ   0       ; in which the Carry flag is bit 0
```

The pseudo instruction **equ** is a simple example of an **assembler directive**. A directive does not generate code, like a processor instruction; rather, it is a command giving information from the programmer to the assembler concerning its operation. In this case, stating that whenever the name STATUS is encountered in an instruction operand field, it is replaced by the number h'FD8' and that the name C is likewise is to be replaced by the number 0.

The equ directive is best suited to listing names of the SFRs and bits within. As these are fixed for a given member of the PIC MCU family, and therefore are not unique to any particular program, Microchip provide .inc files for each device. These can be *included* in user programs as a Header file.[3] For instance, Table 8.1 shows part of the file p18f1220.inc.

In Program 8.1 the directive **#include**[4] has been used to 'inject' the names of the SFR register set into the program. In addition to saving the programmer having to type in a set of equ directives for each SFR used in a program, any subsequent change in the processor, say from a PIC18F1220 to a PIC18F4520, can be simply realized by changing the Header file; to p18f4520.inc in our instance. We will use this technique from now on. Although we have used #include to insert a Header file, it may be used to insert any relevant type of file, such as a subroutine; for example, see Program 12.8 on p. 407.

The header file of Table 8.1 also includes a few useful definitions. For instance, the name FAST is equated to 1, and thus the instruction call FAST can be used as a more readable equivalent to call 1.

In Chap. 4 we explicitly specified a variable as residing either in Access or Banked RAM. In the former case the a-bit is 0 and in the latter 1—see Fig. 3.5 on p. 50. For instance, in Program 5.4 on p. 116 movf AUGEND_H,w,0 as File h'021' is in Access storage. We can now use the substitution movf AUGEND_H,w,ACCESS which makes more sense. During translation the assembler

---

[3]Of course you can make your own version with additional information.

[4]Plain include also works but is not recommended by Microchip.

Program 8.1 Absolute assembly-level code for our square-root module

```
; Global declarations
            #include  "p18f1220.inc"  ; Header file
            cblock h'060'  ; Begin block of variables @ File h'060'
            NUM:2          ; Hi-byte is in NUM+1. Lo byte is in NUM
            endc           ; End of block

; Dummy Main loop --------------------------------------------------
MAIN        movlw h'10'    ; Set up integer h'2710'
            movwf NUM      ; decimal 10,000
            movlw h'27'    ; as a test
            movwf NUM+1
            call  SQR_ROOT ; Call the subroutine SQR_ROOT
            sleep          ; Stop computing
; End of MAIN ------------------------------------------------------

;
; *****************************************************************
; * FUNCTION  : Calculates the square root of a 16-bit integer  *
; * EXAMPLE   : Number = h'FFFF' (65,535), Root = h'FF' (d'255') *
; * ENTRY     : Number in File h'026:27'                         *
; * EXIT      : Root in W and in COUNT.                          *
; * ENVIR'MENT: Files h'35--037' and Status register altered     *
; *****************************************************************
; Local declarations
            cblock         ; Block of variables
            I:2, COUNT:1   ; 2-byte magic number, 1-byte loop Count
            endc

            org   h'00200'  ; Code begins @ h'00200' in Program store

; Task 1: Zero loop count ------------------------------------------
SQR_ROOT clrf    COUNT

; Task 2: Set magic number I:2 to one ------------------------------
            clrf   I+1
            clrf   I
            incf   I,f

; Task 3: DO -------------------------------------------------------
; Task 3(a): Number - I --------------------------------------------
SQR_LOOP movf    I,w       ; Get low byte magic number
            subwf  NUM,f    ; Subtract from low byte Number
            movf   I+1,w    ; Get hi byte magic number &
            subwfb NUM+1,f  ; subtract with borrow from hi byte

; Task 3(b): IF underflow THEN exit --------------------------------
            bnc    SQR_END  ; No Carry is Borrow. IF true terminate

; Task 3(c): ELSE increment loop Count -----------------------------
            incf   COUNT,f

; Task 3(d): Add two to the magic number I:2 -----------------------
            movlw  2        ; Add two to low byte I
            addwf  I,f
            clrf   WREG     ; Zero Working reg
            addwfc I+1,f    ; and add Carry bit to upper byte I
            bra    SQR_LOOP ; and do another subtract and test

; Task 4: Return loop count as the square root ---------------------
SQR_END  movf    COUNT,w   ; Copy into WREG
            return          ; and return to caller
            end
```

Table 8.1 Part of Microchip's file `p18f1220.inc`

```
        LIST
; Standard Header File, Version 1.0    Microchip Technology, Inc.
        NOLIST

; This header file defines configurations, registers, and other
; useful bits of information for the PIC18F1220 microcontroller.
; These names match the data sheets as closely as possible.


;================================================================
;        18Fxxx Family      EQUates
;================================================================


FSR0              EQU  0
FSR1              EQU  1
FSR2              EQU  2

FAST              EQU  1

W                 EQU  0
A                 EQU  0
ACCESS            EQU  0
BANKED            EQU  1


;================================================================
;
;        Register Definitions
;
;================================================================


;----- Register Files -------------------------------------------
TOSU              EQU  H'0FFF'
TOSH              EQU  H'0FFE'
TOSL              EQU  H'0FFD'
STKPTR            EQU  H'0FFC'
PCLATU            EQU  H'0FFB'
PCLATH            EQU  H'0FFA'
PCL               EQU  H'0FF9'
TBLPTRU           EQU  H'0FF8'
TBLPTRH           EQU  H'0FF7'
TBLPTRL           EQU  H'0FF6'
TABLAT            EQU  H'0FF5'
PRODH             EQU  H'0FF4'
PRODL             EQU  H'0FF3'
INTCON            EQU  H'0FF2'
INTCON2           EQU  H'0FF1'
INTCON3           EQU  H'0FF0'
```

(continued on the next page)

Table 8.1 (*Continued*)

```
;----- INTCON Bits ---------------------------------------------
GIE             EQU    H'0007'
GIEH            EQU    H'0007'
PEIE            EQU    H'0006'
GIEL            EQU    H'0006'
TMR0IE          EQU    H'0005'
INT0IE          EQU    H'0004'
RBIE            EQU    H'0003'
TMR0IF          EQU    H'0002'
INT0IF          EQU    H'0001'
RBIF            EQU    H'0000'

;----- STATUS Bits ---------------------------------------------
N               EQU    H'0004'
OV              EQU    H'0003'
Z               EQU    H'0002'
DC              EQU    H'0001'
C               EQU    H'0000'
```

will silently replace ACCESS by , 0.[5] Actually the assembler is perfectly capable
of working out which mode to use, so since Chap. 5 we have left out this explicit
notation; simply writing in this instance movf AUGEND_H, w we have left out this
explicit notation. However, if desired we could override this automatic selection.
Thus movf AUGEND_H,w,BANKED would use BSR to generate AUGEND_H's
address, and if BSR = 00 then this would be equivalent.

  GPR variables specific to the program, like NUM in Program 8.1, still have to be
explicitly named. Thus in Program 6.11 on p. 192 we have:

```
NUM_L          equ    h'026'   ; Number Low byte
NUM_H          equ    h'027'   ; Number High byte
```

Such names and locations are of course unique to the program rather then any specific device. Program 8.1 uses the alternative directive pair **cblock-endc**, which
lets the assembler take over the job of allocating variables to specific Files, within
given constraints. Sandwiched inside these directives are listed the names of the
variables and how many bytes each occupies. In our example we have:

```
    cblock  h'060'  ; Begin block of variables at File h'060'
    NUM:2           ; Reserve two bytes for NUM
    endc            ; End of block
```

where the following colon-delimited number specifies the number of bytes to be
reserved for that name. Individual bytes within the variable can subsequently be

---

[5]The replacement is dumb, for movf AUGEND_H,w,w would do the same job!

accessed by using the arithmetic + operator; for instance, with a 3-byte variable SUM:3, byte 1 is SUM, byte 2 is SUM+1 and byte 3 is SUM+2.

The first code block in Program 8.1 is directed to begin at File h'060' by the programmer. In any subsequent cblock this specification can be omitted, in which case the new variables simply follow on. Thus I:2 is located at File h'063:62' and COUNT:1 at File h'064'. This approach is much more flexible than the programmer allocating locations by hand, as whenever modules are altered or new elements added, the address allocations are automatically altered. In addition, changing any specific code block location, say from File h'060' to File h'020', will automatically alter the complete program variable set to the new range of locations.

A third way of naming entities is to use the #define directive. For example:

```
#define  h'F81',7  BUZZER
```

enables us to use the string bsf BUZZER instead of bs fh'F81',7 to turn on a Buzzer connected to pin 7 of Port B (File h'F81').

For illustrative purposes the programmer has asked the assembler to place the subroutine beginning at location h'00200' in the Program store. This is done using the **org** directive—see also Program 7.1 on p. 218. Effectively the program label SQR_ROOT has been given the value h'00200'.

The last line of Program 8.1 is the **end** directive. This command tells the assembler to ignore any following text; that is, to cease translation.

―――――――――――――――――

Of course symbolic translators demand more computing power than simple hexadecimal loaders, especially in the area of memory and backup store. Prior to the introduction of personal computers in the late 1970s, either mainframe, minicomputers or special-purpose MPU/MCU development systems were required to implement the assembly process. Such implementations were inevitably expensive and inhibited the use of such computer aids, and hand-assembled coding was relatively common.

Translation software essentially implements two tasks:

- conversion of the various instruction mnemonics and labels to their machine-code equivalents;
- allocation of the instructions and data to the appropriate memory location.

Most programs running on 8-bit PIC MCUs are adequately handled by an **absolute assembler**. To clarify the process, we will take our program through from the creation of the source file to the final absolute machine-code file—as outlined in Fig. 8.2. We will examine relocatable assemblers later on.

Editing

Initially the source file must be created using a **text editor**. A text editor differs from a word processor in that no embedded control codes, giving formatting and other information, are inserted. For instance, there is no line wrapping; if you want

Fig. 8.2 Absolute
assembly-level code
translation

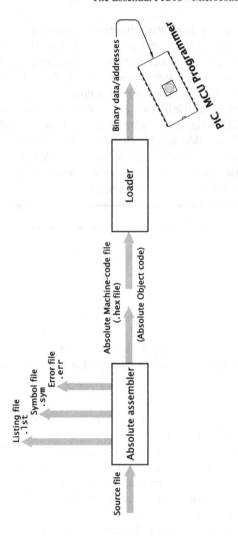

a new line then you hit the [ENT] key. Most operating systems come with a simple
text editor; for example, notepad for Microsoft's Windows. Third-party products
are also available and most word processors have a text mode which can double as
a program editor. Microchip-compatible assembly-level source file names have the
extension .src.

The format of a typical line of source code looks like:

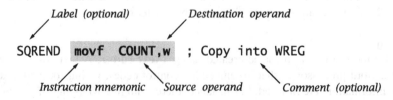

With the exception of comment or label-only lines, all lines must contain an instruction (either executable by the MCU or a directive) and any relevant operand or operands. Any label must begin in column 1, otherwise the first character must be a space or a tab to indicate no label. A label can be up to 32 alphanumeric, underline or question mark characters, with the proviso that the first character be a letter or underscore. Labels are usually case sensitive. A line label names the Program store address of the first following executable instruction. A space, colon or even new line should separate a label from the following instruction or directive.

An optional comment is delineated by a semicolon, and whole-line comments are permitted—see lines 14–23 of Program 8.1. Comments are ignored by the assembler and are there solely for human-readable documentation. Such notes should be copious and should explain what the program is doing, and not simply repeat the instruction. For instance:

```
movf I+1,w  ;Move I+1 into WREG
```

is a waste of energy:

```
movf I+1,w  ;Get high byte of magic number
```

is rather more worthwhile. Not, or doing so only minimally, commenting source code is a frequent failing. A poorly documented program is difficult to debug and subsequently to alter or extend. The latter is sometimes known as program maintenance.

Space should separate the instruction from any operand. Where there are two operands, the source and destination fields are delineated by a comma. In instructions where the destination can be the Working register or the addressed File, the predefined names w or f should appear in the destination fields or numbers 0 or 1 respectively. The assembler will default to destination File if this is omitted and not always warn the programmer!

Assembling
The assembler program will scan the source file, checking for syntax errors. If there are no such errors, the process goes on to translate to absolute object code; which is basically machine code with information concerning the locations in which it is to be placed in program memory. Syntax errors include such things as referring to labels that don't exist or instructions that are not recognized. The output will include an error file giving any such *faux pas*. If there are no syntax errors, a listing file and machine-code file are generated.

Listing
The **listing file** shown in Table 8.2 reproduces the original source code, with the addition of the hexadecimal location of each instruction and its code. The file also provides a symbol table enumerating all symbols/labels defined in the program; for instance, NUM is listed as File h'060'. The memory usage map gives a graphical representation of program memory usage.

This file has only documentation value and is not executable by the processor.

Table 8.2 The listing file `root.lst`

```
MPASM 5.20  PROG8_1_10.ASM 9-8-2008 18:32:26 PAGE  1

LOC  OBJECT CODE     LINE SOURCE TEXT
  VALUE

          01 ; Global declarations
          02         #include "p18f1220.inc"  ; Header file
          01         LIST
          02 ; P18F1220.INC Standard Header File, Microchip Technology, Inc.
          04         LIST
          03         cblock h'060' ; Begin block of variables @ File h'060'
00060     04         NUM:2         ; Hi-byte is in NUM+1. Lo byte is in NUM
          05         endc          ; End of block
          06
          07 ; Dummy Main loop ------------------------------------------------
000 0E10 08 MAIN      movlw h'10'   ; Set up integer h'2710'
002 6E60 09           movwf NUM     ; decimal 10,000
004 0E27 10           movlw h'27'   ; as a test
006 6E61 11           movwf NUM+1
008 EC00
    F001 12           call  SQR_ROOT ; Call the subroutine SQR_ROOT
00C 0003 13           sleep          ; Stop computing
          14 ; End of MAIN ---------------------------------------------------
          15
          16 ; ******************************************************************
          17 ; * FUNCTION  : Calculates the square root of a 16-bit integer  *
          18 ; * EXAMPLE   : Number = h'FFFF' (65,535), Root = h'FF' (d'255') *
          19 ; * ENTRY     : Number in File h'026:27'                         *
          20 ; * EXIT      : Root in W and in COUNT.                          *
          21 ; * ENVIR'MENT: Files h'035--37' and Status register altered    *
          22 ; ******************************************************************
          23 ; Local declarations
          24         cblock        ; Block of variables
0000062   25         I:2, COUNT:1  ; 2-byte magic number, 1-byte loop Count
          26         endc
          27
200       28         org    h'00200' ; Code begins @ h'00200' in Program store
          29
          30 ; Task 1: Zero loop count --------------------------------------
200 6A64 31 SQR_ROOT clrf   COUNT
          32
          33 ; Task 2: Set magic number I:2 to one --------------------------
202 6A63 34          clrf   I+1
204 6A62 35          clrf   I
206 2A62 36          incf   I,f
          37
          38 ; Task 3: DO ---------------------------------------------------
          39 ; Task 3(a): Number - I ----------------------------------------
208 5062 40 SQR_LOOP movf   I,w    ; Get low byte magic number
20A 5E60 41          subwf  NUM,f  ; Subtract from low byte Number
20C 5063 42          movf   I+1,w  ; Get hi byte magic number &
20E 5A61 43          subwfb NUM+1,f ; subtract with borrow from hi byte
          44
          45 ; Task 3(b): IF underflow THEN exit ----------------------------
210 E306 46          bnc    SQR_END ; No Carry is Borrow. IF true terminate
          47
          48 ; Task 3(c): ELSE increment loop Count -------------------------
212 2A64 49          incf   COUNT,f
          50
          51 ; Task 3(d): Add two to the magic number I:2 -------------------
214 0E02 52          movlw  2      ; Add two to low byte I
216 2662 53          addwf  I,f
218 6AE8 54          clrf   WREG   ; Zero Working reg
```

(continued on the next page)

Table 8.2 (*Continued*)

```
21A 2263 55              addwfc I+1,f    ; and add Carry bit to upper byte I
21C D7F5 56              bra    SQR_LOOP ; and do another subtract and test
        57
        58 ; Task 4: Return loop count as the square root -------------------
21E 5064 59 SQR_END movf  COUNT,w  ; Copy into WREG
220 0012 60              return          ; and return to caller
        61
        62              end

SYMBOL TABLE
  LABEL                          VALUE

C                                00000000
COUNT                            00000064
I                                00000062
MAIN                             00000000
NUM                              00000060
SQR_END                          0000021E
SQR_LOOP                         00000208
SQR_ROOT                         00000200
STATUS                           00000FD8
__18F1220                        00000001

MEMORY USAGE MAP ('X' = Used,  '-' = Unused)


0000 : XXXXXXXXXXXXXX-- ---------------- ---------------- ----------------
0200 : XXXXXXXXXXXXXXXX XXXXXXXXXXXXXXXX XX-------------- ----------------

All other memory blocks unused.

Program Memory Bytes Used:    48
Program Memory Bytes Free:  4048

Errors   :    0
Warnings :    0 reported,     0 suppressed
Messages :    0 reported,     0 suppressed
```

Executable Code

The concluding outcome of any translation process is the **object file**, sometimes known as the **machine-code file**. Once the specified code is *in situ* in the Program store, it may be run as the executable program.

As can be seen in Table 8.3, such files consist essentially of lines of hexadecimal digits representing the binary machine code, each preceded by the address of the first byte location of the line. This file can be used by the PIC MCU programmer to put the code into Program ROM memory at the correct place. Because the location of each code byte is explicitly specified, this type of file is known as **absolute object code**. The software component of the PIC MCU programming hardware (see Fig. 16.4 on p. 580), reading, deciphering and placing this code in Program memory is sometimes called an **absolute loader**.

In the MPU/MCU world there are many different formats in common use. Although most of these *de facto* standards are manufacturer-specific, in the main they can be used for any brand of microcontroller. The format of the machine-code file shown here is known as INHX32 for INtel HeX 32-bit address. There are other sim-

ilar Intel formats limited to 16-bit address fields. Figure 8.3 shows two of these lines
from root.hex in more detail.

The loader recognizes that a record follows when the character : is received.
This colon is followed by a 2-digit hexadecimal number representing the number
of machine-code bytes in the record; h'10' = d'16' in the case of the line shown
in Fig. 8.3(b). The next four hexadecimal digits represent the lower 16 bits of the
Program store address in which the first byte of the following data is to be located—
h'0200' in our example. The following 2-digit number is h'00' for a normal Code
record and h'01' for the end-of-file record—see the last line of Table 8.3.

The core of the record is the machine code, with each instruction taking two
2-digit hexadecimal bytes ordered low:high byte. The loader reads this lower byte
first (e.g. h'64') and then 'tacks on' the upper byte (e.g. h'6A') giving a 16-bit instruction word; e.g. h'646A' for clrf h'6A'.[6]

The final byte is known as a **checksum**. The checksum is calculated as the 2's
complement of the sum of all preceding bytes in the record; that is, −sum; ignor-

Table 8.3 The absolute 8-bit Intel INHX32 object-code file root.hex

```
:020000040000FA
:0E000000100E606E270E616E00EC01F0030022
:10020000646A636A626A622A6250605E6350615A1D
:1002100006E3642A020E6226E86A6322F5D7645078
:020220000120 0CA
:00000001FF
```

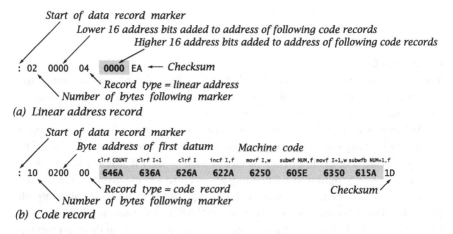

(a) Linear address record

(b) Code record

Fig. 8.3 Structure of INHEX32 records

---

[6]Locating the multibyte code in memory in the Intel way, formatted low:high byte, is known as
little-endian (working up from low to high address, the low byte end comes first) whereas the
high-endian arrangement is favored by, amongst others, Motorola/Freescale.

Table 8.4 The error file

```
Warning[207] ROOT.ASM 56 : Found label after column 1. (br)
Error[122]   ROOT.ASM 56 : Illegal opcode (SQR_LOOP)
```

ing any overflow. As a check-up on transmission accuracy, the loader adds up all received bytes including this checksum for each record. This received count should give zero if no download error has occurred.

Machine-code formats, such as INHX32 (there are several types, such as Motorola), were originally developed for microprocessors with a maximum of 64 kbyte Program store. For processors with larger stores, and therefore needing address fields more than 16 bits, the INHX32 format uses a Linear Address record line. Figure 8.4(a) shows such a line, coded 04. A record of this type carries a complete 32-bit address, with an upper capacity capable of servicing a 4 Gbyte Program store! At the time of writing (autumn 2008) the maximum store size is 128 kbyte; for instance, the PIC18F6722, with an address range spanning h'0 0000–1 FFFF'. All code records following a Linear address record keep this address word as an offset until a new Linear address record is read. For an example with a higher address segment than illustrated here, see p. 317.

Assemblers are very particular that the syntax is correct. If there are **syntax errors**[7] then an **error file** will be generated. For instance, if line 56 was mistakenly entered as:

```
br   SQR_LOOP
```

then the error file of Table 8.4 would be generated.

The assembler does not recognize br as an instruction or directive mnemonic and erroneously assumes that it is a label mistakenly not beginning in column 1. On this basis it assumes that SQR_LOOP is an instruction/directive mnemonic and again does not recognize it.

Most assemblers allow the programmer to define a sequence of processor instructions as a **macro instruction**. Such macro instructions can subsequently be used in a similar manner to native instructions. For example, the following code defines a macro instruction called Delay_1ms[8] that implements a 1 ms delay when executed on a PIC MCU running with a 4 MHz crystal. The directive pair **macro-endm** is used to enclose the sequence of native instructions which will be substituted

---

[7]If the assembler announces that there are no errors then there is a tendency to think that the program will work. Unfortunately a lack of syntax errors in no way guarantees that the program will do anything of the sort!

[8]I have capitalized the first letter of all macro instructions to distinguish them from native instructions.

when the mnemonic `Delay_1ms` is used anywhere in the subsequent program. The mnemonic will be replaced by the assembler with the defined code. Note that this will be in-line code, unlike calling up a subroutine.

```
Delay_1ms    macro
             local    LOOP

             movlw    d'250'    ; Count from 250
LOOP         addlw    -1        ; Decrement
             nop                ; Extra delay
             bnz      LOOP      ; Repeat unless zero

             endm
```

Where labels are used within the body of the macro, they should be declared using the **local** directive. This means that any conflict with labels outside the macro or where a macro instruction is evoked more than once, is avoided.

This example is unusual in that the 'instruction' did not have any operands. Like native instructions, macros can have one or more operands. To see how this is done, consider a macro instruction called `Movlf` for MOVe byte Literal to File. As there is no one native instruction to copy (move) a constant into a File in one go, this will involve more than one native instruction. The definition of `Movlf` is:

```
Movlf    macro        LITERAL,DESTINATION ; Dummy operand names

         push                     ; Save WREG & STATUS
         movwf    TOSL            ; First WREG
         swapf    STATUS,w        ; Then STATUS
         swapf    WREG,f
         movwf    TOSH

         movlw    LITERAL         ; Put constant in WREG
         movwf    DESTINATION ; and copy into File

         movf     TOSL,w          ; Restore WREG
         movff    TOSH,STATUS ; and STATUS
         pop

         endm
```

For instance, if the programmer wished to initialize File h'020' to, say, h'55' then the invocation would be `Movlf h'55',h'020'`.

To make `Movlf` transparent, both WREG and STATUS are saved in the Hardware stack before moving the literal into the former and then to the specified File. These previous values are restored before completion.

Our example involved two comma separated dummy operands, but in general a macro can be of any arbitrary complexity involving many such declared parameters. Macro names should not be the same as for a real instruction; even from a different family. Microchip have available a large number of macros implementing arithmetic operations such as 16-bit × 16-bit and 32-bit × 32-bit multiplication. However, extensive use of macros can make programs difficult to debug, especially when an apparently simple macro instruction hides a number of side effects which alter GPR and SFR File contents and flags. These can be reduced by transparency techniques, such as illustrated in this example, but not always entirely eliminated. For instance, a frequent source of error is to precede a macro instruction with a Skip instruction, intending to go around it on some condition. As the macro instruction is in fact a structure of several native instructions, this skip will actually be into the middle of the macro—with dire consequences.

Macro definitions, whether commercial or/and in-house, may be collected together as a single file and *included* in the user program using the include directive. Thus if your file is called mymacros.mac then the line at the beginning of your program

```
#include    "mymacros.mac"
```

will allow access by the programmer to all macro definitions in the file. Any macros defined in the included file but not used, will have no effect on the final machine code.

The process outlined up to here is known as absolute assembly. Here the source code is in a single file (maybe plus some included files) and the assembler places the resulting machine code in known (i.e., absolute) locations in the Program store. Where many modules are involved, often written by different people or/and coming from outside sources and libraries, some means must be found to *link* the appropriate modules together to give the final single absolute executable machine-code file. For example, you may have to call up one of the modules that Fred is busy writing at the moment. You do not know exactly where in memory this module will reside or where its variables are stored, until the project is nearing its conclusion. What can you do? You should be able to call module FRED and refer to its component objects without knowing exactly what address they will be allocated.

The process used to facilitate this is shown in Fig. 8.4. Central to this modular tie-up is the **object linker** program, which satisfies such external cross-references between the modules. Each module's source-code file needs to have been translated into **relocatable object code** prior to the linkage. 'Relocatable' means that its final location and various addresses of external labels have yet to be determined. This translation is done by a **relocatable assembler**. Unlike absolute assembly, it is the linker that determines where the machine code is to be located in memory, rather than the human programmer; although absolute locations, say of the Ports, can still be specified and the programmer will give guidance.

Treating the linker as a type of task builder, its main functions are:

- To concatenate code and data from the various input module streams.
- To allocate values to symbolic labels which have *not* been given explicit fixed values by the programmer using equ and similar directives.
- To generate the absolute machine-code executable file together with any symbol, listing and link-time error files.

Fig. 8.4 Relocatable
assembly-level code
Translation

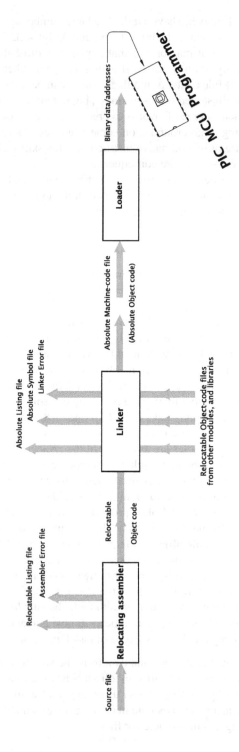

Table 8.5 The `rms.lkr` linker command file

```
// File: rms.lkr
// Simple linker script for the PIC18F1220 Created 19/09/2008

// Architecture of target device
CODEPAGE    NAME=vectors    START=0x0    END=0x29   PROTECTED
CODEPAGE    NAME=page       START=0x2A   END=0xFFF

ACCESSBANK  NAME=scratch    START=0x0    END=0x3F
ACCESSBANK  NAME=accessram  START=0x40   END=0x7f
DATABANK    NAME=gpr0       START=0x80   END=0xFF
ACCESSBANK  NAME=sfr        START=0xF80  END=0xFFF  PROTECTED

SECTION     NAME=STARTUP     ROM=vectors    // Reset & Int vectors
SECTION     NAME=PROGRAM     ROM=page       // ROM program text
SECTION     NAME=AUTO        RAM=scratch    // Used for temporary vars
SECTION     NAME=STATIC_ACS  RAM=accessram  // Used for Global vars
SECTION     NAME=STATIC      RAM=gpr0       // Vars stored in Bank0
```

In relocatable strategy, instructions and data are separated out into **streams**, which are located in Program and Data stores as directed by the programmer. In order to allow the linker to do this job, it must have knowledge of the memory architecture of the target processor; basically where the array of general-purpose register Files start and end, where the vectors reside in program memory and where the code begins and ends. These may be subdivided as directed by the programmer with regard to the resources available to the hardware. For instance, foreground and background code could be placed in separate areas of the Program store. In the case of Microchip's `mplink.exe` linker this information is supplied in the form of a **linker command file**.

A simple example of such a command file for a PIC18F1220 is given in Table 8.5. Four directives are used in the file.[9]

CODEPAGE

The `codepage` directive is used for code in the Program store. Here these directives are used to define two regions, one for the Reset and Interrupt vectors in between h'0000' and h'0029' called `vectors`, and the other called `page` to be used for executable code from h'0002A' through h'0FFF'. Notice the use of the prefix `0x` to denote the hexadecimal base. This is the notation used in the **C** language. The `vector` domain has the attribute PROTECTED, which tells the linker to keep away from this area of memory when allocating space for code unless explicitly commanded to in the program. Microchip recommend that this region be specified oversize to accommodate some limited startup code; for instance, context saving for Low-priority handling when High-priority interrupts are also in use. For this reason this area of the Program store ends at `0x0029` and executable code can begin at `0x002A`.

---

[9]Only a basic range is given here. Microchip's *MPASM*™ *Assembler Users Guide with the MPLINK*™ *Object Linker and MPLIB*™ gives a full list of linker directives.

DATABANK

This is similar to codepage but is used for variable data in RAM where Direct-Banked addressing is to be used. In the script above, Bank 0 is declared to be this type.

ACCESSBANK

This directive is used where the programmer wants the assembler to use Access-Direct addressing for variables in this area of RAM. Here the Access regions of the Data store is split into three streams.

1. Between 0x000 and 0x03F is going to be used as an area to store local temporary variables whose storage is ephemeral; that is can be reused for each module.
2. Between 0x040 and 0x07F completes the lower Access region and will be used for variables which will last for the program run.
3. Between 0xF80 and 0xFFF is where the SFRs are located with the attribute of PROTECTED. This area is specified for completeness, as SFRs are absolute locations and as such will be given fixed addresses in the program; as defined in the header file p18f1220.inc.

SECTION

This linker directive allows the programmer to specify which of the defined memory areas a named stream of source code or data is to be placed. Table 8.5 names two code and three data logical sections.

1. STARTUP will be used by the programmer to store the three vector goto or bra instructions—see p. 257. The source code assembler directive **code** with the appropriate label tells the linker into which stream any following code is to be placed; for example, see Program 8.2.
2. PROGRAM is where the core instruction codes will be streamed.
3. AUTO is where temporary variables will be directed. This resource will be reusable with each module able to overlay previously RAM memory. The assembler source code directive access_ovr (ACCESS uninitialized data OVeRlaid) steers data into this region of RAM.
4. STATIC_ACS is for storage in Access RAM that will persist for the entire program. The assembler directive udata_acs (Uninitialized DATA ACceSs) allows space to be reserved for labels in the Access general-purpose register array.
5. STATIC_0 is a stream for storage in Bank 0 used for permanent data storage. The assembler directive udata (Uninitialized DATA) allows space to be reserved for labels in non-Access (i.e. Banked) RAM areas.

As many code sections from any codepage and data sections for any databank can be created as appropriate. For instance, all subroutines may be placed together in program memory by modifying the linker script file thus:

```
SECTION  NAME=PROGRAM      ROM=page  // ROM code space
SECTION  NAME=SUBROUTINES ROM=page  // ROM subroutine stream
```

In the particular case of the STARTUP stream, the three vectors are not consecutive. To cope with this disjointed code, an entry within a named stream can be explicitly located. For instance:

```
; The three vectors -----------------------------
STARTUP   code              ; Reset vector at beginning of stream
          bra   MAIN        ; Go to the background software

STARTUP   code  0x0008      ; High-priority vector at h'00008'
          bra   HIGH_ISR ; Go and execute the High-priority ISR

STARTUP   code  0x0018      ; Low-priority vector at h'00018'
          bra   LOW_ISR  ; Go and execute the Low-priority ISR
```

This should be compared to the similar absolute assembly equivalent on p. 223.

To illustrate the principle of linking we will implement the mathematical function $\sqrt{NUM\_1^2 + NUM\_2^2}$, known as root mean square. There are three teams working on this problem.[10] Tasks have been allocated by the project manager (a fourth person?) as follows:

1. The main function which sequences the steps:
   (a) **Square signed byte** NUM_1.
   (b) **Square signed byte** NUM_2.
   (c) **Add** $NUM\_1^2 + NUM\_2^2$.
   (d) **Square root item** (c).
2. Design of a subroutine to square a signed byte number in the Working register to give a double-byte outcome in two GPRs.
3. Design of a subroutine to evaluate the square root of a double-byte sum and return it in **WREG**.

The process based on this decomposition of the task is shown diagrammatically in Fig. 8.5.

The main function is shown in Program 8.2. The program commences with the Reset bra MAIN instruction and is located in the STARTUP code stream. From the MAIN label onwards, code is located in the PROGRAM code stream using the directive PROGRAM code. We see from the map file output by the linker in Table 8.6 that MAIN is located at 0x002A (the first location in the PROGRAM stream).

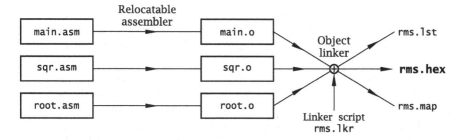

Fig. 8.5 Linking three source files to implement a root mean square program

---

[10]Obviously this is a ridiculously simple problem for teamwork, but it illustrates the principle in a manageable space.

Program 8.2 The main relocatable source file `main.asm`

```
        include   "p18f1220.inc"
        extern    SQR_ROOT, SQR, SQUARE
; ----------------------------------------------------------------
STATIC  udata                          ; Permanent data in Bank 0 RAM
NUM_1   res       1                    ; The first number
NUM_2   res       1                    ; The second number
SUM     res       2                    ; Two bytes HI:LO for the sum
RMS     res       1                    ; One byte for the outcome
; ----------------------------------------------------------------
STARTUP code
        bra       MAIN                 ; The Reset vector

PROGRAM code

MAIN    movf      NUM_1,w              ; Get Number 1
        call      SQR                  ; Square it
        movff     SQUARE,SUM           ; Get lower byte to Sum_L
        movff     SQUARE+1,SUM+1       ; Get high byte to Sum_H
        movwf     SUM+1                ; Is the high byte of Sum

        movf      NUM_2,w              ; Now get Number 2
        call      SQR                  ; Square it
        movf      SQUARE,w             ; Get lower byte
        addwf     SUM,f                ; Add to the low byte of sum
        movf      SQUARE+1,w           ; Get upper byte and add
        addwfc    SUM+1,f              ; with carry to the hi byte of sum

        call      SQR_ROOT             ; Work out the square root
        movwf     RMS                  ; Is the root mean square

        sleep
        global    SUM
        end
```

The main routine uses four variables located in data stream STATIC. These are placed in uninitialized Bank 0 RAM with the directives **udata** and **res** (RE-Serve). A single GPR File is reserved for each of the two input variables NUM_1 and NUM_2, respectively. Two bytes are reserved for SUM, which is used to hold the sum NUM_1 + NUM_2. As this is to be the input for the subroutine SQR_ROOT, it is declared **global** at the end of the file. This means that the location is public; that is, additional files that are linked together can use the label SUM by declaring it **extern** (i.e., external to the file). Variables not declared thus are 'hidden' from the outside world, i.e., are private (or local) variables. In this manner the directive extern at the head of Program 8.2 allows the main routine to call the subroutines SQR_ROOT and SQR without knowing in advance where they are. In the same way the variable SQUARE is used by subroutine SQR to return the square of the byte sent to it in WREG. Space for this is reserved in a GPR File in subroutine SQR and its exact location in the Data store is not known by main.asm but will be allocated later by the linker. From the map file of Table 8.6 it is finally located in File 0x083:82 (high:low byte).

The main body of the code follows the task list enumerated above. The value NUM_1^2 is placed in Files SUM+1 : SUM to which the computed NUM_2^2 is added. The outcome is then used as input to subroutine SQR_ROOT to return the root-mean

Program 8.3 The relocatable source file `sqr.asm`

```
            include  "p18f1220.inc"
; ***************************************************************
; * FUNCTION: Squares one signed byte to give a 2-byte result*
; * EXAMPLE : X = h'10' (16), SQUARE = h'0100' (256)          *
; * ENTRY   : X in WREG                                        *
; * EXIT    : Global SQUARE:2                                  *
; ***************************************************************

STATIC_ACS udata_acs                ; Global data in Access RAM
SQUARE     res    2                  ; High:Low byte of square
; -------------------------------------------------------------
AUTO       access_ovr                ; Auto (temporary) data
X_NUM      res    1                  ; One place for X_NUM
; -------------------------------------------------------------

PROGRAM    code

; Task 1: Make byte in WREG positive and copy into X_NUM -------
SQR        btfsc  WREG,7             ; Skip IF sign bit is positive
           negf   WREG               ; ELSE Negate to make it positive
           movwf  X_NUM              ; and copy as second operand

; Task 2: Multiply to give WREG^2 in PROD_H:PROD_L-------------
           mulwf  X_NUM              ; Multiply

; Task 3: Return 2-byte product in SQUARE:2
           movff  PRODL,SQUARE       ; Low byte of WREG^2
           movff  PRODH,SQUARE+1     ; High byte of WREG^2

           return                    ; Finished

           global SQUARE, SQR
           end
```

square byte in WREG. Finally this is copied to the File named RMS, for which a single byte has been reserved in the STATIC Data stream.

The subroutine `sqr.asm` of Program 8.3 returns a 2-byte square of the signed datum passed in WREG. As by definition the outcome will always be positive and the `mulwf` instruction only deals with unsigned operands, the subroutine first checks the sign bit of WREG and if negative, 2's complements (`negf`) it. Copying it into memory at X_NUM and then multiplying gives X_NUM^2. X_NUM is put into the data stream AUTO with the directive **access_ovr** to indicate to the linker that these Files in Access RAM can be reused by other modules. In the map file of Table 8.6 we see that X_NUM has been allocated File 0x000 as has I; a variable in subroutine SQR_ROOT—see Program 8.3. This will make more efficient use of available data memory. Variables that are only alive within the subroutine that they are declared in are known in the **C** language as **automatic**, as their space is automatically reallocated as needed. The situation where variable space is preserved is known as **static**. Global variables, such as SQUARE, are always static. In this case the variable SQUARE is created by reserving two bytes using the udata_acs directive in the STATIC_ACS stream. It is also published using the global directive, as is the name of the subroutine. The 2-byte outcome in the SFRs PRODH:PRODL are copied into global SQUARE:2 from which they can be seen by the caller routine

Program 8.4 The relocatable source file `root.asm`

```
        include  "p18f1220.inc"
        extern   SUM   ; The 2-byte number Hi:Lo

; **************************************************************
; * FUNCTION    : Calculated the square root of a 16-bit integer  *
; * EXAMPLE     : Number = h'FFFF' (65,535), Root = h'FF' (d'255')*
; * ENTRY       : Number in Global SUM:2                       *
; * EXIT        : Root in WREG                                 *
; * ENVIR'MENT: Overlay I:2, COUNT:1 and STATUS altered       *
; **************************************************************

AUTO      access_ovr         ; Auto (temporary) variables in Acc RAM
I         res      2         ; Magic number hi:lo (two bytes)
COUNT     res      1         ; Loop count (one byte)
; --------------------------------------------------------------

PROGRAM  code

SQR_ROOT clrf     COUNT     ; Task 1: Zero loop count

         clrf     I         ; Task 2: Set magic number I to one
         incf     I
         clrf     I+1

SQR_LOOP movf     I,w       ; Task 3(a): Number - I
         subwf    SUM,f     ; Subtract lo byte I from lo byte Num
         movf     I+1,w     ; Get high byte magic number
         subwfb   SUM+1,f   ; Subtract with Borrow high bytes

         btfss    STATUS,C  ; IF No Borrow THEN continue
          bra     SQR_END   ; ELSE the process is complete

         incf     COUNT,f   ; Task 3(c): ELSE inc loop count

         movf     I,w       ; Task 3(d): Add 2 to the magic number
         addlw    2
         btfsc    STATUS,C  ; IF no carry THEN done
          incf    I+1,f     ; ELSE add carry to upper byte I
         movwf    I
         goto     SQR_LOOP

SQR_END  movf     COUNT,w   ; Task 4: Return loop count as the root
         return

         global   SQR_ROOT
```

MAIN. Another possibility would have been to leave the datum in these SFRs, which as absolute locations are by definition global. However, this makes the subroutine a little less flexible for other applications.

The final source file of the trio is the subroutine coded in Program 8.4. This is virtually identical to the absolute equivalent described in Program 8.1. Com-

Table 8.6 Part of the output linker map file `rms.map`

```
MPLINK 4.20, Linker
Linker Map File - Created Sat Sep 27 19:59:54 2008

                         Section Info
          Section      Type     Address    Location Size(Bytes)
        ---------   ---------  ---------  ---------  ---------
          STARTUP       code   0x000000    program   0x000002
          PROGRAM       code   0x00002a    program   0x00005e
           .cinit    romdata   0x000088    program   0x000002
             AUTO      udata   0x000000       data   0x000003
       STATIC_ACS      udata   0x000040       data   0x000002
           STATIC      udata   0x000080       data   0x000005


                    Program Memory Usage
                      Start          End
                  ---------    ---------
                  0x000000     0x000001
                  0x00002a     0x000089
98 out of 4096 program addresses used, memory utilization is 2%

                    Symbols - Sorted by Name
           Name     Address   Location    Storage File
        ---------  ---------  ---------   --------- ---------
            MAIN   0x00002a    program     static main.asm
             SQR   0x000076    program     extern sqr.asm
         SQR_END   0x000072    program     static root.asm
        SQR_LOOP   0x000056    program     static root.asm
        SQR_ROOT   0x00004e    program     extern root.asm
           COUNT   0x000002       data     static root.asm
               I   0x000000       data     static root.asm
           NUM_1   0x000080       data     static main.asm
           NUM_2   0x000081       data     static main.asm
             RMS   0x000084       data     static main.asm
          SQUARE   0x000040       data     extern sqr.asm
             SUM   0x000082       data     extern main.asm
           X_NUM   0x000000       data     static sqr.asm
```

paring the two, the `org` directive has been replaced by `PROGRAM code` and
`cblock` by `AUTO access_ovr` for the automatic local data in Access RAM.
The data is passed to the subroutine `SQR_ROOT` via the external 2-byte global variable `SUM`, space for which has been allocated in `main.asm`. The subroutine name
`SQR_ROOT` is published as global to make it visible to `main.asm`.

Like all source files, `root.asm` makes use of SPRs such as `STATUS`. For this
reason the file `p18f1220.inc` has been included at the head of each of the source
files. Because this file comprises a set of `equ` directives (see Table 8.1), the names
thus published are absolute and are not allocated or changed in any way by the
linker. Thus the linker map of Table 8.6 does not list such fixed symbols. They are,
however, enumerated in the listing file produced by the linker.

Table 8.7 The resulting absolute object file `rms.hex` in INHX32 format

```
:020000040000FA
:0200000014D01A
:06002A0080513BEC00F0E8
:100030004050826F4150836F81513BEC00F0405043
:100040008227415083232 7EC00F0846F0300026A6B
:10005000006A002A016A0050825F0150835BD8A0C9
:1000600008D0022A0050020FD8B0012A006E2BEFF0
:100070000F002501200E8BEE86C006E0002F3CF00
:0800800040F0F4CF41F0120042
:02008800000076
:00000001FF
```

In order to link the three source files together, the linker program must be given a command line listing the names of the input object files output by the relocatable assembler, the linker command file and the names of the output map and machine code file. In the case of our example this was:

```
mplink "rms.lkr" "main.o" "root.o" "sqr.o"
/m"rms.map" /o"rms.hex"
```

which names the output map file `rms.map` and the absolute machine-code file `rms.hex`.

For documentation purposes the linker generates a composite listing file, similar (but more comprehensive) to that of Table 8.2 and an optional map file. The map file of Table 8.6 shows two lists. The first displays information for each section. This includes its name, type, start address, whether the section resides in program (code) or data memory (data) and its size in bytes. The Program Memory Usage table shows that 98 bytes of program memory are used, including the two bytes of the Reset vector `bra` instruction—or around 2.4% of the possible total.

The second table shows information about the symbols in the composite program. Each symbol's location in either the Program or Data store is given together with the source file where it is defined. Global symbols are noted as `extern`. Local variables are all labeled `static` (not to be confused with the Data stream `STATIC`), including automatic reusable variables such as X_NUM and I, both at File 0x000.

The full map file also includes a table of symbols sorted by address.

The final outcome, shown in Table 8.7, is a normal executable machine code file. The format of this file is described for Table 8.3 and can be loaded into absolute program memory and run in the normal way.

Developing, testing and debugging software requires a large number of software tools, many of which we have discussed earlier, such as an editor, assembler and linker. In practice there are many other packages such as high-level language compilers (see Chap. 9), simulators and in-circuit debuggers; shown diagrammatically in Fig. 8.6. Setting up these tools and interacting on an individual basis can be quite

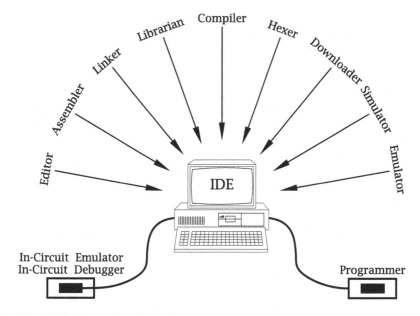

Fig. 8.6   Code building and testing tools

complex, especially where products from differing manufacturers are involved. In this latter case, ensuring compatibility between the various intermediate file formats can be a nightmare.

Many software houses designing code development tools provide a graphical environment which integrates and sequences the process in a logical and easy to use manner. Of relevance to the PIC MCU family, Microchip Technology provides a Microsoft Windows-based **Integrated Development Environment** (**IDE**) called MPLAB®, which brings all compatible code development tools under one roof. Like all Microchip software tools (except **C** compilers) the MPLAB IDE is supplied free of charge.

MPLAB integrates Microchip-compatible tools to form a complete software development environment. Among its features are:

- A project manager which groups the specific files related to a project; for instance, source, object, simulator, listing and hex files.
- An editor to create source files and linker script files.
- An assembler, linker and librarian to translate source code and create libraries of code, which can be used with the linker.
- A simulator to model the instruction execution and I/O on the code development computer—see Fig. 8.8.
- A downloader to work in conjunction with device programmers via the PC's serial or USB ports—see Fig. 16.4 on p. 580.

- Software to emulate PIC MCUs in real time in the target hardware. This is accomplished by driving an In-Circuit Emulator (ICE) or Debugger (ICD)[11] via the PC's serial or USB port, controlling or replacing the target processor.

The Microchip manual *MPLAB® IDE User's Guide* gives a MPLAB IDE tutorial and reference details, which are beyond the scope of this book. However, for illustrative purposes two screen shots taken during the development of our previous example linking main.asm, sqr.asm and root.asm are reproduced in Figs. 8.7 and 8.8.

Figure 8.7 shows the window displaying the Project file rms.mcw after the setup wizard. The three source files, which have already been created using the editor, are specified. Also the name of the linker script file rms.lkr, which has also been previously created and saved. The resulting machine code file will be called rms.hex.

Once the project specified in this manner, the sequence of operations, namely:

1. Assemble main.asm to give main.o.
2. Assemble sqr.asm to give sqr.o.
3. Assemble root.asm to give root.o.
4. Use rms.lkr to link together object files 1, 2 and 3.
5. If no syntax errors, create the absolute executable file of Table 8.7.

can be initiated by choosing from the Project menu (top fourth left in Fig. 8.8) Make Project. If there are syntax errors an Error window will appear listing errors. Double clicking on any specific error will bring up the relevant Source-code window with the cursor set to the line in question. The Output window charts the progress of the Make process, and will display any Linker errors.

Once the program has been successfully made, it may be simulated. Here the PC models the PIC MCU's instruction set and peripheral ports and allows the user to

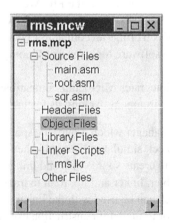

Fig. 8.7 MPLAB® Project window, showing files selected to assemble, link and simulate Program 8.4

---

[11]This is a hardware 'pod' that replaces the PIC MCU in the target circuit and allows the PC to take over the running of the system—see Fig. 16.6 on p. 582.

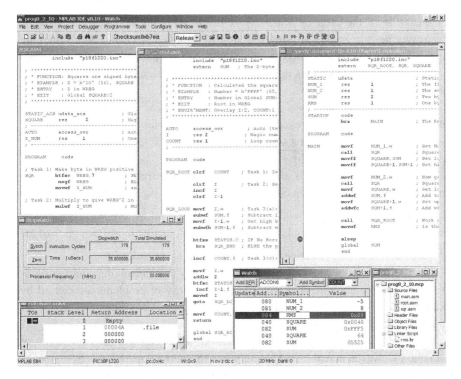

Fig. 8.8 MPLAB screen shot showing the programs selected in Fig. 8.7 being simulated

reset the (simulated) PIC MCU, set break points, single step or run continuously and even inject fake signals into I/O pins. During this process, user-selected File registers or the whole of data memory can be monitored, as can execution time. Of course, simulated execution time by the PC will be several orders of magnitude slower than a real PIC MCU.

A simulation can be actioned via the Debugger menu. This brings up the Debugger toolbar, top right of Fig. 8.8. This allows the operator to:

- Reset the virtual processor by clicking on the 🔳 icon.
- Run ▷ and pause ❚❚ the simulation at top speed.
- Continuously execute; i.e. animate ▷▷, at a rate of several steps per second.
- Single step in three modes; one instruction being simulated on a single click.
  - Step-in 🔁 steps through all code, including subroutines.
  - Stop-over 🔁 runs through subroutines at top speed.
  - Step-out 🔁 runs down to the end of the stepped subroutine at full speed.

Figure 8.8 shows the end result of a simulation of our root-mean square example. As well as windows showing the three source files, a Watch window has been opened using the View menu. This allows the operator to add any named GPRs, such as NUM_1, the value of which can be displayed in binary, signed or unsigned decimal or hexadecimal in bit, single, double or triple byte format. These values are updated as the simulation proceeds in a Single-step or Animate mode. If top speed is used, the Watch window is updated when the simulation is paused or stops at a breakpoint.

Also shown in the diagram is the Stop-watch window. This indicates that the program took 179 cycles to execute with initial values for NUM_1 and NUM_2 of −5 and 8, respectively. With a simulated 20 MHz crystal, execution time is shown as 35.8 µs. As expected $\sqrt{-5^2 + 8^2} = 9$.

As the simulation proceeds, the currently executed instruction is marked by a $\Rightarrow$ in the left pane of the relevant Source-code window. In Fig. 8.8, this is pointing to the final sleep instruction and overlays the breakpoint **B** symbol. Breakpoints can be set or cleared by right-clicking on the relevant instructions. Each click of the **D** icon will run the simulation at top speed to the next breakpoint.

Simulation will not catch all problems, especially those involving complex hardware/software interaction. However, the vast majority of problems are caused by purely software design faults, and simulation is a fast and convenient technique for testing and debugging such code.

Debugging should always, at a first iteration, try largest and smallest values of variables. However, in real-world length programs, correct operation is by no means guaranteed by this test for all possible combinations and sequences of input.

Finally, we review some general information specific to Microchip-compactible assemblers as an aid to reading programs in the rest of the book:

- Number representation.
  - Hexadecimal: Denoted by a leading h with the number delineated by quotes; e.g. h'41' or a following h; e.g. 41h, or a 0x prefix; e.g. 0x41. The assembler normally defaults to this base so some programs show no hexadecimal indicators. However, it is better not to rely on the default behavior.
  - Binary: Denoted by a leading b with a quote delimited number; e.g. b'01000001'.
  - Decimal: Denoted by a leading d with a quote delineated number; e.g. d'65' or a leading period prefix; .65 in our example.
  - 2's complement Sign is indicated by appending a negative sign; e.g. -d'60' or -h'3C'; which is the same as h'C4'.
  - ASCII: Denoted by a quote delimited character; e.g. '?'.
- Label arithmetic.
  - Current position: $; e.g. goto $+2.
  - Addition: +; e.g. goto LOOP+6.
  - Subtraction: -; e.g. goto LOOP-8.
  - Multiplication: *; e.g. subwf LAST*2.
  - Division: /; e.g. subwf LAST/2.

- Directives.
  - org: Places the following code in program memory starting from the specified address; e.g. `org h'00100'`. If no `org` is used, the default reset point is h'00000'. Can only be used for absolute assembly.
  - code: Counterpart to `org` for relocatable assembly. The actual address of the code stream is defined in the linker's command file. More than one code stream may be defined in the linker script file and in this case its name appears in the label field; for instance, `SUBROUTINES code`. An absolute address can be appended in the same manner as `org`.
  - equ: Associates a value with a symbol; e.g. `PORTB equ h'F81`. The `#define` directive may be used instead; `#define PORTB h'F81'`.
  - cblock-endc: Used in absolute assembly to allocate program variables in data memory; e.g.:

```
      cblock  h'020'
      FRED          ; One byte at h'020' for FRED
      JIM:2         ; Two bytes at h'021:22' for JIM
      ARRAY:10      ; Ten bytes for ARRAY at h'023 - 02C'
      endc
```

  The address is optional after the first `cblock` use.
  - udata: Counterpart to `cblock` for relocatable assembler where the datastream specifies Banked RAM. The start address for data memory streams are in the linker's script file. There may be more than one Data stream defined in this script file in which case its name is published in the label field; e.g.:

```
SCRATCHPAD  udata              ; Uninitialized data stream
       FRED    res 1   ; Reserve one byte for FRED
       JIM     res 2   ; Reserve two bytes for JIM
       ARRAY   res 10  ; Reserve ten bytes for ARRAY
```

  - udata_acs: is the equivalent to `udata` where the data stream is targeted to Access RAM.
  - udata_ovr: OVeRlay Uninitialized DATA is similar to `udata` but the linker tries to reuse Files for the specified named variables. Used for data streams in Banked RAM.
  - access_ovr: ACCESS OVeRlay data is similar to `udata_ovr` but used where the uninitialized data is in Access RAM.
  - res: Used with a `udata` type directive to REServe one or more bytes for a variable in a Data stream.
  - extern: Allows the named variables which are defined outside the current file, to be used in the current file and subsequently resolved by the linker.
  - global: Publishes the named variables that have been defined (i.e., space reserved) in the file and that are to be made visible to other files through the linker.
  - macro-endm: Used to allow the specified enclosed sequence of instructions to be replaced by a new macro instruction; e.g.:

```
Addlf   macro   N,datum
        movf    datum,w
        addlw   N
        movwf   datum
        endm
```

adds the literal N to the specified File datum; e.g. to add five to File h'020' the
programmer can use the invocation Addlf 5,h'020'.

- #include: Used to include the specified file at this point; for instance,
  #include "myfile.asm". Plain include is identical.
- end: Normally the last line of an assembly-level source file. Tells the assembler to ignore anything following.

Examples

Example 8.1 The following routine effectively exchanges the byte contents of W
and a File F without needing an additional intermediate File.

```
    xorwf   F,f   ; [File] <- WREG^F
    xorwf   F,w   ; WREG <- WREG^(WREG^F) = 0^F = F
    xorwf   F,f   ; [File] <- F^WREG^F = 0^WREG = WREG
```

where ^ denotes eXclusive-OR.

Wrap the given code within a macro to generate a new instruction Exgwf F
where F is the designated File; e.g. Exgwf h'020'.

Solution Wrapping the code inside a macro gives:

```
Exgwf   macro   FILE

        xorwf   FILE,f
        xorwf   FILE,w
        xorwf   FILE,f

        endm
```

Note that this macro instruction will not affect the **C** flag and will activate the **Z** flag
according to the datum that was in the Working register at entry.

Example 8.2 The mulwf instruction multiplies the unsigned byte in WREG with
that in the target File, putting the unsigned outcome in PRODH:PRODL—see p. 122.
Based on this instruction, design a macro called Mulwfs which will implement a
multiplication of two RAM-based signed operands.

Solution Program 5.6 on p. 124 gives a routine to implement this task. Wrapping
this into a macro gives:

```
Mulffs macro    FILE1,FILE2

        movf    FILE1,w ; Get Number 1 from memory into WREG
        mulwf   FILE2   ; Multiply with Number 2

        btfsc   FILE2,7 ; Test Number 2's sign bit, skip if +ve
        subwf   PRODH,f ; ELSE take away Number 1 from PRODH

        movf    FILE2,w ; Now get Number 2 into WREG
        btfsc   FILE1,7 ; Test Number 1's sign bit, skip if +ve
        subwf   PRODH,f ; ELSE take away Number 2 from PRODH

        endm
```

A typical usage of this macro would be Mulffs h'020',h'060' to multiply the signed bytes in File h'020' and File h'060'. The 16-bit signed Product will be in PRODH:PRODL.

Example 8.3 Write a macro that will create a delay of nominally 100 µs independent of the clock speed, over the range 1 to 40 MHz. The user program must declare the frequency used in the target hardware by defining the constant CLOCK. For instance, #define CLOCK d'20' where the hardware is clocked at 20 MHz.

Solution The macro code below uses a down count of the Working register with each loop taking four instruction cycles. At the top frequency of 40 MHz, each cycle takes 0.1 µs. This gives a total requirement of 1000 cycles, or 250 loops. At the bottom $\frac{250}{40}$ gives 6.25 loop passes. We can only do integer label arithmetic so the constant $(6 \times CLOCK) + \frac{CLOCK}{4}$ gives the closest approximation for our loop count. The $\frac{CLOCK}{4}$ factor gives an approximation for the 0.25 fraction of a loop pass. For instance, at 20 MHz the constant loaded into WREG will be 125, and this will give 500 cycles at 0.2 µs in total as it decrements to zero.

```
Delay_100us macro
            local DLOOP
; 4K~ delay in total = 6.25 * CLOCK us
            movlw (CLOCK*6) + CLOCK/4); Delay parameter K 1~
DLOOP       addlw -1                 ; Decrement          K~
            nop                      ;                    K~
            bnz   DLOOP              ; until zero         2K-1~

            endm
```

The macro label is qualified with the local directive to ensure that each time a macro is used it does not inject DLOOP into the assembler's symbol table. If not so qualified, then an Address label duplicated assembler syntax error will occur.

Example 8.4 Macros may be nested; that is, a macro may use other macros in its definition. For example, consider a macro to create a clock-independent nominal

1 ms delay. Assuming that the macro Delay_100us of Example 8.3 has already been defined, write a suitable macro definition.

Solution One possible solution is:

```
Delay_1ms macro
          local   DLOOP

          push              ; Open up location for 100us count
          movlw   d'10'     ; Initialize counter to ten
          movwf   TOSL
DLOOP     Delay_100us       ; 100us delay
          decfsz TOSL,f     ; One more time
           bra    DLOOP     ; until zero (ten times)
          pop               ; Restore Hardware stack

          endm
```

Our macro definition simply runs the Delay_100us macro ten times inside a loop. As this macro already uses the Working register, the loop counter needs to be located in memory. This can be a temporary GPR File but to reduce side effects a better way is to use a Hardware stack cell. In this case the Stack Pointer is moved to the next cell and TOSL initialized to ten and then decremented to zero after each 100 μs delay. The Stack Pointer is then moved back.

Side effects are a hazard in using macro instructions, especially if the macro has been designed by someone else and hidden in an Include file. At the very least assume that W and STATUS are altered, unless known otherwise. Altering banks in a macro is also potentially hazardous. A better solution would have been to make macro Delay_100us transparent—for instance, see p. 252. In this case WREG could have been used in Delay_1ms.

The real delay will be slightly longer than 1000 μs, due to the overhead of the loop instructions. An actual run at 10 MHz gave 1005.2 μs.

Example 8.5 When using a relocatable assembler, the programmer will not know in advance whether variables declared extern, that is defined in another external module, have been assigned to Access or Banked RAM. At link time (see Fig. 8.5) the generated code will use Access- or Banked-Direct addressing as appropriate. However, if the latter, it will not automatically set up the Bank Select Register accordingly. Thus if a variable VAR_2 has been assigned to a stream in Bank 5; the programmer needs to set up BSR to h'05'; i.e. movlb 5.

In the case of our exemplar PIC18F1220 there is no problem, as only Bank 0 is implemented and BSR is cleared to h'00' or Power-on Reset, and thus never needs modified. However, using a processor such as the PIC18F4520 with five banks, presents a seemingly insuperable problems in dealing with relocatable variables. Not only will the programmer of a module not know the bank assignment of a particular external variable, but also this can change as a project develops!

To get around this problem, Microchip-compatible assemblers provide the **banksel** (BANK SELect) directive. This automatically keeps track of the location of the named variable and issues code to target the appropriate bank. Show how this directive should be used when storing the decimal literals 1, 10, 100 in three GPRs called var_0, var_1 and var_2, respectively. To use this mechanism, the banksel directive should be placed in the source code whenever an external variable is referenced.

Solution A possible sequence of instructions is shown below. The directive issues the appropriate movlb instructions as appropriate before the following instruction.

```
        movlw    1        ; The first literal
        banksel  VAR_0    ; Change to the appropriate bank
        movwf    VAR_0    ; Do it

        movlw    d'10'    ; Literal ten
        banksel  VAR_1    ; Change to the appropriate bank
        movwf    VAR_1    ; Do it

        movlw    d'100'   ; Literal hundred
        banksel  VAR_2    ; Change to the appropriate bank
        movwf    VAR_2    ; Do it
```

Assuming that VAR_0 is in Access RAM, VAR_1 is in Bank 2 and VAR_2 is in Bank 5, the resulting listing file for this fragment of code is:

```
002A 0E01 MOVLW 0x1            7:  MAIN movlw    1
002C 0100 MOVLB 0             8:       banksel VAR_0
002E 6E40 MOVWF 0x40, ACCESS  9:       movwf   VAR_0
                             10:
0030 0E0A MOVLW 0xa          11:       movlw   d'10'
0032 0102 MOVLB 0x2          12:       banksel VAR_1
0034 6F00 MOVWF 0, BANKED    13:       movwf   VAR_1
                             14:
0036 0E64 MOVLW 0x64         15:       movlw   d'100
0038 0105 MOVLB 0x5          16:       banksel VAR_2
003A 6F00 MOVWF 0, BANKED    17:       movwf   VAR_2
```

Notice that using banksel to identify a variable that is in Access RAM generates a movlb 0 instruction, even though the following movwf instruction correctly uses Access-Direct addressing!

Self-Assessment Questions

8.1 Both macros and subroutines inject a series of instructions into the source code. Compare and contrast the two techniques in assembly-level program coding.

8.2 Design a macro of the form `Cpfl_gt file,literal,destn` to jump to a `destn` label if the datum in `file` is greater than `literal`. For instance, `Cpfl_gt h'20',d'80',NEXT` will cause execution to transfer to the instruction labeled NEXT if the contents of File h'020' is greater than 80.

8.3 Code a macro of the form `Lsldp file_h,file_l` that will do a double-precision logic shift left on `file_h:file_l`. A zero is to be shifted in from the right and the leftmost bit will end up in the Carry flag. For instance, `Lsldp h'021',h'020'` will shift the 16 bits in File h'021:20' once left, with b_0 zero.

8.4 Example 8.3 developed a clock-speed independent 100 μs delay macro. Re-engineer the coding to give a fully transparent version.

8.5 A certain electromechanical counter requires a 100 ms pulse (counting rate of 10 per second) to advance. Based on the transparent clock-independent `Delay_100us` macro above, design a transparent 100 ms macro called `Delay_100ms`.

8.6 The following routine based on the macro instruction `Movlf` of p. 252 does not work as intended. COUNT is altered seemingly at random and not consistently with the desired literal 32. Why is this?

```
movf    COUNT,f      ; Test COUNT for zero
btfsc   STATUS,Z     ; IF not Zero THEN skip
Movlf   d'32',COUNT  ; ELSE re-initialize it to 32
```

8.7 The `banksel` approach to selecting a bank is inefficient in that an extra `movlb` instruction is issued even if the PIC MCU is already in the correct bank or in Access RAM. Consider how in a time- or space-critical subroutine this inefficiency can be avoided.

8.8 The PIC24 and dsPIC30/33 16-bit family cores have an array of 16×16 Working registers; as shown in Fig. 8.9. For instance the instruction to copy a 16-bit datum word located in File h'0020' to W8 would be `mov h'0020',W8`. Design a macro called `Movfwi FILE,WI` (MOVe File to Working register i) instruction. You can assume that the simulated array is located in Access mem-

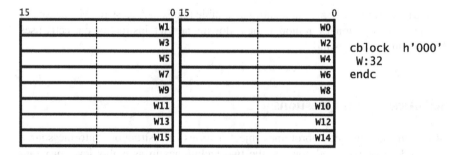

Fig. 8.9 Working register array

ory, with W0 at File h'001:000'. The parameter `FILE` indicates the lower byte of the datum address.

8.9 Figure 8.9 shows the `cblock` directive reserving 32 bytes of Access RAM for the array of Working registers. Based on Table 8.5, add a new stream for the pseudo Working register array into the same area of Access RAM, suitable for a relocatable assembly process with the `udata_acs` directive.

8.10 A programmer with expertise in the Freescale (Motorola) 68HC05 MCU has been converted to the PIC18 MCU family and wishes to design macros to simulate, amongst others, the following 68HC05 instructions. Note that the Accumulator register in the 68HC05 family is the equivalent to the Working register of the PIC MCU.

lda memory
LoaD Accumulator with data from memory.

lda #data
LoaD Accumulator with literal data.

sta memory
STore Accumulator data into memory.

tst memory
TeST memory for zero

tsta
TeST Accumulator for zero

Code suitable macros. Why do you think this approach might not be such a good idea?

Chapter 9
High-Level Language

All the programs we have written in the last six chapters have been in symbolic assembly language. Whilst assembly-level software is a quantum step up from pure machine-level code (see p. 240) nevertheless there is still a one-to-one relationship between machine and assembly-level instructions. This means that the programmer is forced to think in terms of the MCU's internal structure—that is, of registers and memory—rather than in terms of the problem algorithm. Although most assemblers have a macro facility, whereby several machine-level instructions can be grouped to form pseudo high-level instructions, this is only tinkering with the difficulty. What is this difficulty with machine-oriented language? In order to improve the effectiveness, quality and reusability of a program, the coding language should be mostly independent of the underlying processor's architecture and should have a syntax more oriented to problem solving.

We are not going to attempt to teach a high-level language in a single short chapter. However, after completing this chapter you will:

- Understand the need for a high-level language.
- Appreciate the advantages of using a high-level language.
- Understand the problems of using a high-level language for embedded microcontroller applications.
- Be able to write a short program in **C** targeted to a PIC18 MCU.

The difficulty in coding large programs in a computer's native language was clearly appreciated within a few years of the introduction of commercial systems. Apart from anything else, computers became obsolete with monotonous regularity, and programs needed to be rewritten for each model introduction. Large applications programs, even at that time, required many thousands of lines of code. Programmers were as rare as hen's teeth and worth their weight in gold. It was quickly deduced that for computers to be a commercial success, a means had to be found to preserve the investment in scarce programmers' time. In developing a universal language, independent of the host hardware, the opportunity would be taken to allow the programmer to express the code in a more natural syntax related to problem-solving rather than in terms of memory, registers and flags.

S. Katzen, *The Essential PIC18® Microcontroller,*
Computer Communications and Networks,
DOI 10.1007/978-1-84996-229-2_9, © Springer-Verlag London Limited 2010

Of course, there are many different classes of problem tasks which have to be coded, so a large number of languages have been developed since.[1] Amongst the first were Fortran (FORmula TRANslation) and COBOL (COmmon Business Oriented Language) in the early 1950s. The former has a syntax that is oriented to scientific problems and the latter to business applications. Despite being around for over 50 years, the inertia of the many millions of lines of code written has made sure that many applications which were written in these antique languages are still in use. Other popular languages include Algol (ALGOrithmic Language), BASIC, Pascal, Modula, Ada, **C**, **C++**, **C#** and Java—the latter four forming a related family.

Although writing programs in a high-level language may be easier and more productive for the programmer; the process of translation from the high-level source code to the target machine code is rather more complex than the assembly process described in Chap. 8. The translation package for this purpose is called a **compiler** and the process, **compilation**; as shown in Fig. 9.2.

The complexity and cost of a compiler was acceptable on the relatively powerful and extremely expensive mainframe computers of that time. However, until the mid-1980s the use of high-level languages as source code was virtually unknown for microprocessor-controlled circuitry. In the last two decades the easy availability of relatively powerful and cheap personal computers and workstations, capable of running compilers, together with the growing power of MPU/MCU hardware and financial importance of this market, is such that the great majority of software written for such targets is now in a high-level language.

If you are going to code a task in a high-level language to run in a system with an embedded MCU; for instance, a washing-machine controller, then the process is roughly as follows.

1. Take the problem specification and break it up into a series of modules, each with a well-defined task and set of input and output data.
2. Devise a coding to implement the task for each module.
3. Create a source file using an editor in the appropriate high-level syntax.
4. Compile the source file to its assembly-level equivalent.
5. Assemble and link to the machine-code file.
6. Download the machine code to the target's program memory.
7. Execute, test and debug.

This is virtually identical to the process outlined in Fig. 8.4 on p. 254, but with the extra step of compilation. Some compilers go directly from the source file to the machine-code file; however, the extra flexibility of going through the assembly-level phase, as shown in Fig. 9.1, is nearly universal when embedded MPU/MCU circuitry is targeted.

The choice of a high-level language for embedded targets is crucial. Of major importance is the size of the machine code generated by a high-level language implementation, as compared with the equivalent assembly-level solution. Most em-

---

[1] A popular definition of a computer scientist is one who, when presented with a problem to solve, invents a new language instead!

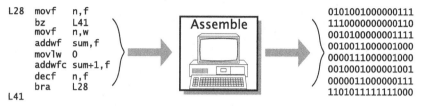

```
while(n>0)
{
sum = sum + n;
--n;
}
```

Compile

```
L28   movf   n,f
      bz     L41
      movf   n,w
      addwf  sum,f
      movlw  0
      addwfc sum+1,f
      decf   n,f
      bra    L28
L41
```

(a) First, compile to assembly-level code.

```
L28   movf   n,f
      bz     L41
      movf   n,w
      addwf  sum,f
      movlw  0
      addwfc sum+1,f
      decf   n,f
      bra    L28
L41
```

Assemble

```
0101001000000111
1110000000000110
0010100000001111
0010011000001000
0000111000001000
0010001000001001
0000011000000111
1101011111111000
```

(b) Second, assemble-link to machine code.

Fig. 9.1   Conversion from high-level source code to machine code

bedded MCU circuitry is 'lean and mean', such as the remote controller for your television. Lean translates to physically limited and mean maps to low processing power and memory capacity—and cost! Most low-cost MCUs have a low-capability processor with a few hundred bytes of RAM and a few tens of kilobytes of ROM Program store at best. Thus to be of any use, the high-level language and the compiler must generate code that, if not as efficient as assembly-level (low-level), at least is in the same ball park.[2]

By far the most common high-level language used to source code for embedded MPU/MCU circuitry is **C**. Historically **C** was developed as a language for writing operating systems. At its simplest level, an operating system (OS) is a program which makes the detailed hardware operation of the computer's terminals, such as keyboard and disk organization, invisible to the operator. As such, the writer of an OS must be able to poke about the various registers and memory of the computer's peripherals and easily integrate with assembly-level driver routines. As conventional high-level languages and their compilers were profligate with resources, depending on a rich and fast environment, assembly language was mandatory up to the early 1970s, giving intimate machine contact and tight fast code. However, the sheer size of such a project means that it is likely to be a team effort, with all the difficulties in integrating the code and foibles of several people. A great deal of self-discipline and skill is demanded of such personnel, as is attention to documentation. Even with all this, the final result cannot be easily transplanted to machines with other processors, needing a nearly complete rewrite.

---

[2]In the author's experience a code size increase factor of $\times 1.25 \cdots \times 2.5$ is typical.

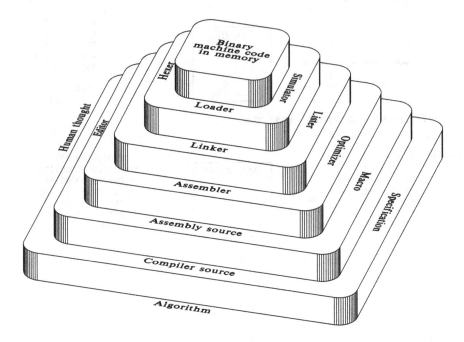

Fig. 9.2 Pyramid view of the steps leading to an executable program

In 1969 Ken Thompson—an employee at Bell Laboratories—developed the first version of the UNIX[3] operating system. This was written in assembler language for a DEC PDP7 minicomputer. In an attempt to promote the use of this operating system (OS) within the company, some work was done in rewriting UNIX in a high-level language. The language CPL (Combined Programming Language) had been developed jointly by Cambridge and London universities in the mid-1960s, and had some useful attributes for this area of work. BCPL (Basic CPL) was a somewhat less complex but more efficient variant designed as a compiler-writing tool in the late 1960s. The language B (after the first letter in BCPL) was developed for the task of rewriting UNIX for the DEC PDP11 and was essentially BCPL with a different syntax.

Both BCPL and B only used one type of object, the natural size machine word—16 bits for the PDP-11. This typeless structure led to difficulties in dealing with individual bytes and floating-point computation. **C** (the second letter of BCPL) was developed in 1972 to address this problem, by creating a range of objects of both integer and floating-point types. This enhanced its portability and flexibility. UNIX was reworked in **C** during the summer of 1973. It comprised around 10,000 lines

---

[3]Originally called UNICS, for UNIplexed operating and Computing System, based on MULTICS (Multiplexed operating and Computing System) an operating system developed in the early 1960s at MIT.

of high-level code and 1000 lines at assembly level, and occupied some 30% more storage than the original version.

Although **C** has been closely associated with UNIX, over the intervening years it has escaped to appear in compilers running under virtually every known OS, from mainframe CPUs down to single-chip MCUs. Furthermore, although originally a systems programming language, it is now used to write applications programs ranging from Computer Aided Design (CAD) packages down to the intelligence behind smart egg-timers!

For over 10 years the official definition was the first edition of *The C Programming Language*, written by the language's originators Brian W. Kernighan and Dennis M. Ritchie. It is a tribute to the power and simplicity of the language that over the years it has survived virtually intact, resisting the tendency to split into dialects and new versions. In 1983 the American National Standards Institute (ANSI) established the X3J11 committee to provide a modern and comprehensive definition of **C** to reflect the enhanced role of this language. The resulting definition, known as Standard or ANSI **C**, was finally approved during 1990 by the International Standards Organisation (ISO).

Apart from its use as the language of choice for embedded MPU/MCU circuits, **C** (together with its **C++** and Java object-oriented offspring) is without doubt the most popular general-purpose programming language at the time of writing. It has been called by its detractors a high-level assembler. However, this closeness of **C** to assembly-level code, together with the ability to mix code based on both levels in the one program, is of particular benefit for embedded targets.

The main advantages of the use of high-level language as source code for embedded targets are:

- It is more productive, in the sense that it takes around the same time to write, test and debug a line of code irrespective of language. By definition, a line of high-level code is equivalent to several lines of assembly code.
- Syntax is more oriented to human problem-solving. This improves productivity and accuracy, and makes the code easier to document, debug, maintain and adapt to changing circumstances.
- Programs are easier to port to different hardware platforms, although they are rarely 100% portable. Thus they are likely to have a longer productive life, being relatively immune to hardware developments.
- As such code is more or less hardware-independent, the customer base is considerably larger. This gives an economic impetus to produce extensive support libraries of standard functions, such as mathematical and communication modules, which can be reused in many projects.

Of course there are disadvantages as well, specifically when code is being produced to run in poorly resourced MCU-based circuitry.

- The code produced is less space-efficient and often runs more slowly than native assembly code.
- A compiler is much more expensive than an assembler. A professional product can cost several hundred pounds/dollars.

Program 9.1 A simple function coded in **C**

```
1:    unsigned int sqr_root(unsigned long number)
2:        {
3:        unsigned int count = 0;
4:        unsigned long i = 1;
5:        while(number >= i)
6:            {
7:            number = number - i;
8:            i = i + 2;
9:            count++;
10:           }
11:       return count;
12:       }
```

- Debugging can be difficult, as the actual code executed by the target processor is the generated assembler code. The processor does not execute high-level code directly.

Program 9.1 is an example of a **C** function (a function is the counterpart to a subroutine) that evaluates the square root of a 16-bit number using the same process outlined in Fig 6.17 on p. 191. Essentially, the task list to be implemented is:

1. Subtract a series of integers i from the number, beginning at 1, and incrementing in steps of 2 while keeping a count.
2. When the residue drops below i then the count of subtractions is the nearest square root of the number.

In the implementation number is the 16-bit integer passed to the function, which computes and returns the 8-bit integer count, as defined in the task list. Let us dissect it line by line. Each line is labeled with its number. This is for clarity in our discussion and is not part of the program.

Line 1: This line names the function (subroutine) sqr_root. It declares that it returns an unsigned integer; which in the compiler used to illustrate this book, is a single byte. It expects an unsigned long integer (a 16-bit unsigned object) to be passed to it, to be used in the body of the function. This is to be named number.

Line 2: A left brace { means begin. All begins must be matched by an end, which is designated by a right brace }. It is good practice to indent each begin from the immediately preceding line(s). This makes it easier to ensure each begin is paired with an end. However, the compiler is oblivious of the style the programmer uses. In this case line 12 is the corresponding end brace. Between lines 2 and 12 defines the body of the function sqr_root().

Line 3: There are two variables that are local to our function. The first is named and typed (defined) in this line. Thus count is of type unsigned int. In **C** all objects have to be defined before they are used. This tells the compiler what properties the named variable has; for example, its size (8 bits), to allocate storage and its arithmetic properties (unsigned). At the same time count is given an *initial* value of zero. The complete statement is terminated by a semicolon, as are all statements in **C**.

Line 4: In this statement, variable i is defined as being an unsigned long int. This tells the compiler that this is to be treated as a 16-bit integer with unsigned attributes. The initial value of this variable is to be 1; that is 0x0001. Giving a local variable an initial value is optional. If not done, its value is not specified.

Line 5: In evaluating count we need to *repeat* the same subtract and increment process as long as the residue of number is greater or equal to the ever growing i. This is the purpose of the while construction introduced in this line. The general form of this loop construct is:

```
while(true)
    {
    do this;
    do that;
    do the other;
    }
```

The body of the loop, i.e., the set of statements that appears between the following left and right braces of lines 6 and 10, is continually executed as long as the expression in the brackets evaluates as non-zero. Anything non-zero is considered *true* by **C**. This test is done before each pass through the body. In our case the expression number >= i, i.e. (number ≥ i), is evaluated. If true, then number is updated by subtracting i, which is then augmented by 2 and count incremented. Eventually number >= i computes to *false* when number drops below i and then execution breaks out of the loop to the statement following the closing brace at line 11.

Line 6: The opening brace defining the while body. Notice that for style it is indented.

Line 7: The expression to the right of the = operator is evaluated to number − i and the outcome assigned to the left variable number; that is number is updated from its original value less i. Both variables are double-byte, and so the arithmetic is done accordingly—Table 9.1 instructions at locations h'001A– 0021'. If one of the right-side variables was single-byte, then it would automatically be extended to 16-bits.

Line 8: The value of i is augmented by adding two and assigning it back to i at the left side of the expression. An alternative statement uses the expression i += 2;.

Line 9: The value of count is incremented using the ++ operator. This is equivalent to the statement count = count + 1;[4]

Line 10: The end brace for the while body. Again note how the opening (line 6) and closing braces line up. The compiler does not give a hoot about style; this is solely for human readability and to reduce the possibility of errors.

---

[4]The ++ operator has given the name **C++** to the next development of the **C** language.

Line 11: The `return` instruction passes *one* parameter back to the caller; in this
case the final value of `count`. The compiler will check that the size of this
parameter matches the prefix of the function header in line 1, that is `unsigned
int`. This returned parameter is the value of the function, i.e., the function can
be used as a variable in the same way as any other. If the function is called by
another function, then its returned value can be assigned by the caller to a normal
variable. For instance:

```
root = sqr_root(100);
```

would give the returned value 10 to the caller's variable `root`—see Fig. 9.3.

Line 12: The closing brace for function `sqr_root()`.

We see from Fig. 9.1 that the output from the compiler is assembly-level code,
which can then be assembled and linked with other modules[5] in the normal way. To
illustrate this process, Table 9.1(a) shows the assembly-level code generated when
the **C** code of Program 9.1 is passed through the Custom Computer Services (CCS),
Inc cross-C compiler Version 4.[6] This is a low-cost **C** compiler ($\approx$\$200) that can be
integrated with MPLAB; see Fig. 9.3.[7] This listing file shows each line of **C** source
code as a comment together with the resulting assembly-level code.

It is instructive to look at how the compiler has translated this program.

unsigned int count = 0

The CCS compiler reserves a single byte for an `int` object. Table 9.1(b) shows
the variable `sqr_root.count` (`count` in function `sqr_root()`) stored in
File h'009'. To zero this GPR, the compiler has generated a `clrf` instruction:

```
clrf    h'009'     ; Clear count
```

unsigned long i = 1;

Two bytes at File h'00B:0A are reserved for this 2-byte `long` object `sqr_root_i`.
To set these GPRs to h'0001', the literal h'01' is first copied into the low byte `i` in
File h'00A' and the high byte in `i+1` is cleared:

```
movlw   01      ; Constant one
movwf   h'00A'  ; Low byte of i is 01
clrf    h'00B'  ; High byte of i is 00
```

while(number >= i){

The while loop is implemented by comparing the double-byte variable `number`
to the like-sized variable `i`. If the former is not greater or equal to the latter then
execution will break out of the loop to the exit `return` statement.

---

[5]Some of which can be functions hand-coded in native assembly-level language for efficiency, and
from libraries supplied with the compiler or bought in.

[6]See http://www.ccsinfo.com/picc.shtm.

[7]MPLAB V. 8 comes with a free CCS compiler for the base-family range of devices.

Table 9.1 Resulting assembly-level CCS compiler output after linking

---

```
CCS PCH C Compiler, Version 4.066, 42523              11-Oct-08 13:40

                  Filename: sqrt.lst

                  ROM used: 98 bytes (2%)
                            Largest free fragment is 3998
                  RAM used: 6 (2%) at main() level
                            11 (4%) worst case
                  Stack:    1 locations

*
0000:  GOTO   MAIN
..................... #include <18f1220.h>
..................... // Standard Header file for the PIC18F1220 device
..................... #device PIC18F1220
..................... #list
.....................
.....................
..................... unsigned int sqr_root(unsigned long number)
.....................   {
.....................   unsigned int count = 0;
0004:  CLRF   count
.....................   unsigned long i = 1;
0006:  MOVLW  01
0008:  MOVWF  i
000A:  CLRF   i+1
.....................   while(number>=i)
.....................     {
000C:  MOVF   i+1,W
000E:  SUBWF  number+1,W
0010:  BNC    002E
0012:  BNZ    001A
0014:  MOVF   i,W
0016:  SUBWF  number,W
0018:  BNC    002E
.....................     number = number - i;
001A:  MOVF   i,W
001C:  SUBWF  number,F
001E:  MOVF   i+1,W
0020:  SUBWFB number+1,F
.....................     i = i + 2;
0022:  MOVLW  02
0024:  ADDWF  i,F
0026:  MOVLW  00
0028:  ADDWFC i+1,F
.....................     count++;
002A:  INCF   count,F
.....................     }
002C:  BRA    000C
.....................   return count;
002E:  MOVFF  count,01
.....................   }
0032:  RETLW  00
```
(a): Assembly-level code listing file generated by the CCS compiler.

(*continued on the next page*)

The compiler has passed the caller's parameter, listed in Table 9.1(b) as sqr_root.number, in File h'008:07'. The 2-byte comparison is coded in two stages. Initially the high bytes are subtracted (number+1) – (i+1). If there is

Table 9.1 *(Continued)*

```
000        @SCRATCH
001        @SCRATCH
001        _RETURN_
002      · @SCRATCH
003        @SCRATCH
004        @SCRATCH
006        MAIN.root
007-008 sqr_root.number
009        sqr_root.count
00A-00B sqr_root.i

ROM Allocation:
0004    sqr_root
0034    MAIN
0034    @cinit

Compiler Settings:
    Processor:        PIC18F1220
    Pointer Size:     16
    ADC Range:        0-255
    Opt Level:        9
    Short,Int,Long:  UNSIGNED: 1,8,16
    Float,Double:     32,32
```

(b): Partial symbol file.

```
:020000040000FA
:100000001AEF00F0096A010E0A6E0B6A0B50085CC9
:100010000EE303E10A50075C0AE30A50075E0B5047
:10002000085A020E0A26000E0B22092AEFD709C031
:1000300001F0000CF86AD09EEA6AE96AC29CC29E8E
:10004000C150800B7F09C16E086A650E076EDADF4A
:1000500001C006F0086A660E076ED4DF01C006F024
:0200600003009B
:020000040030CA
:0E00000000CF0E1E0080810003C003E003400D
:00000001FF
;PIC18F1220
```

(c): Executable Intel machine code file.

a borrow-out (**C** is 0) then number < i and execution breaks out of the loop at
the instruction at h'0002E'; i.e. bnc 002E. If these high bytes are then not equal
then number > i and the following statement is entered. Where both tests are
false, then the high bytes are the same and the low bytes subtracted NUMBER − i.
A borrow-out in this situation signals that number < i and the while loop is exited to the return statement at h'0002E'. Otherwise (number >= i) is true and
the first statement in the loop at h'0001A' is then executed.

```
        movf   h'00B'   ; Get the high byte of i
        subwf  h'008'   ; Subtract from the high byte of number
        bnc    BREAK    ; IF borrow THEN number < i and BREAK
        bnz    NEXT     ; IF not equal THEN CONTINUE
        movf   h'00A'   ; Equal so try the low bytes to give
        subwf  h'007'   ; number - i
        bnc    BREAK    ; IF borrow THEN number < i and BREAK
NEXT
```

number = number - i;
This is implemented as a double-byte subtraction thus:

```
   movf    h'00A',w  ; Get low byte i
   subwf   h'007',f  ; Subtract from number gives new number
   movwf   h'00B',w  ; Get high byte i+1
   subwfb  h'008',f  ; Subtract from number+1 gives new number+1
```

Many **C** programmers use the alternative statement number $-=$ i; which states number *augmented by* i.

i = i + 2;
Two is added onto the low byte of i and any carry-out to the high byte:

```
   movlw  h'02'     ; Two into WREG
   addwf  h'00A',f  ; Added onto the low byte of i
   movlw  0         ; Clear WREG and
   addwfc h'00B',f  ; is added with any carry-out to high byte
```

Alternatively i $+=$ 2;.

count++;
Now increment the single byte in File h'009'.

```
   incf    h'009',f  ; Increment count
```

In more complicated expressions the placement of the ++ Increment operator (and the analogous – – operator) before or after the object can affect the outcome. Where it appears before, such as in:

```
   number = ++n - i;
```

then the value of n is *first* incremented *before* i is subtracted from it. In the following case:

```
number = n++ - i;
```

i is subtracted from n and only then is n incremented.

In our example the logic of the program is unaffected if the operator is pre- or post-incremented. However, the compiler in the latter case adds an extra instruction to bring n down into the Working register before it is incremented *in situ* as it thinks that some computation involving the original value of n is to be performed.

}

The while loop is repeated by going back to the loop test, which is located starting at h'0000C'.

```
    bra    h'0000C'          ; Repeat While loop
```

return count;

At the end of a function returning an int object the CCS compiler places the byte in the fixed GPR File h'0001'. This is listed in Table 9.1(b) as the system generated symbol _RETURN_. Thus this code fragment simply copies the byte in File h'009', i.e. sqr_root.count, into the return location.

```
BREAK
  movff  h'009',h'001'  ; Copy count to _RETURN_ and come back
  retlw  0              ; with an error code of 00 (success)
```

The final machine code file is shown in Table 9.1(b) for a total length of only 24 bytes. This compares with 17 bytes in Program 6.11 on p. 192.

Every **C** program *must* at the very least have a main() function. main() is a little unique in that it alone sets up a known state in which the core 'useful' code will run from Reset. A **C** program typically comprises many functions, but only main() will set up the starting environment.

As an example, consider the dummy main() function listed in Program 9.2. We are assuming that this is part of Program 9.1 and calls sqr_root() twice. In both cases the 1-byte variable root is assigned the returned integer value. In both instances this will be ten.

If we look at the assembly-level code generated in Table 9.2 we see that before the main code proper several SPRs are initialized.

Program 9.2 A main() function calling sqr_root() twice
_____
```
void main(void)
{
unsigned int root;
root = sqr_root(101);
root = sqr_root(102);
}
```
_____

Table 9.2 Resulting assembly-level CCS compiler output after linking

```
. . . . . . . . . . . . . . . . . . . .  void main(void)
. . . . . . . . . . . . . . . . . . . .  {
0034:   CLRF    TBLPTRU
0036:   BCF     RCON.IPEN
0038:   CLRF    FSR0H
003A:   CLRF    FSR0L
003C:   BCF     ADCON0.VCFG0
003E:   BCF     ADCON0.VCFG1
0040:   MOVF    ADCON1,W
0042:   ANDLW   80
0044:   IORLW   7F
0046:   MOVWF   ADCON1
. . . . . . . . . . . . . . . . . . . .  int root;
. . . . . . . . . . . . . . . . . . . .  root = sqr_root(101);
0048:   CLRF    number+1
004A:   MOVLW   65
004C:   MOVWF   number
004E:   RCALL   0004
0050:   MOVFF   01,root
. . . . . . . . . . . . . . . . . . . .  root = sqr_root(102);
0054:   CLRF    number+1
0056:   MOVLW   66
0058:   MOVWF   number
005A:   RCALL   0004
005C:   MOVFF   01,root
. . . . . . . . . . . . . . . . . . . .  }
0060:   SLEEP
```

clrf TBLPTRU

This zeros the upper byte of the 3-byte register TBLPTRU used to point into the program store when reading and writing data during program execution. This is normally clear on reset, but it is possible that it may have been altered by another program previously run and the processor not reset. The use of the TABeL PoinTeR is described in Fig. 15.4 on p. 550.

bcf RCON,IPEN

Clearing this bit sets the interrupt mode to Compatible; as shown in Fig. 7.2 on p. 209. Again this is the default reset mode but may have been altered since the last reset. As we see in Program 9.3, the CCS directive #device HIGH_INTS=TRUE is used to set IPEN to 1 on startup.

As indicated in this program, individual functions can be designated High priority.

clrf FSR0H:clrf FSR0L

Initializes FSR0 to zero.

bcf ADCON0,VFG0:bcf ADCON0,VFG1

This initializes the Analog to Digital Convertor (ADC) module to use internal power-supply voltages for reference—see Fig. 14.12 on p. 510.

```
movf ADCON1,w:andlw b'10000000':iorlw b'011111111':
```

```
movwf ADCON1
```

Effectively this sets to 1 all bits in ADCON1 and thus by default all Port A and Port B pins are set to be digital—see Fig. 14.12 on p. 510.

Once this environment is set up, the 'useful' code simply initializes the variable number, which we already know is located in File h'008:07', to the appropriate constant and calls (rcall) the subroutine/function sqr_root(). On return the content of the return location h'001' is simply copied to the location of the variable main.root at h'006'—see Table 9.1(b). Specifically the main() function is terminated by the sleep instruction—see p. 318. Normally a function is terminated by a return to the caller function. If a function is only called once or is very short, the CCS compiler will usually implement such an orphan function using in-line code. This is why Program 9.2 called sqr_root() twice; to illustrate function calling. The directives #INLINE and #SEPARATE can be used to request that a following function be implemented as an in-line or separate subroutine.

If a function does not return a value, then the return type is indicated as void. No parameters are sent to main() and thus the passed parameter field is also annotated as void. Thus the function header in Program 9.2 is void main(void). If it is

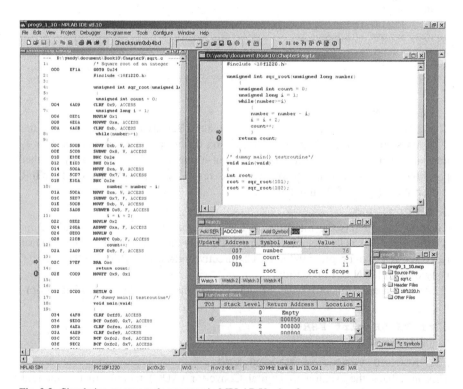

Fig. 9.3 Simulating our example program in MPLAB Version 8

written as plain `main()`, the CCS compiler will complain, but will nevertheless generate the correct code.

C-level programs can be compiled and simulated in the IDE environment of Microchip's MPLAB—see p. 264. The screen shot of Fig. 9.3 shows windows into both the **C**-level source code (top right) and the resulting assembly-level code (left). The program can be optionally stepped in either window, depending on which is active. The execution arrow ⇒ appears in both windows in tandem in the appropriate place. The Watch window shows the state of the three **C** objects visible in the function `sqr_root()`; namely `number`, `count` and `i`. `root` is defined inside the `main()` function and is therefore invisible outside that function. In the snapshot in the Watch window, as execution is in `sqr_root()`, this variable is designated as `Out of Scope`. When execution returns to `main()` the value of its local variable `root` will be displayed, but that of the others will then be out of scope. Variables defined outside any function will be global and accessible from anywhere.

In the screenshot variables are shown in decimal. Any base as appropriate can be chosen by right clicking the variable's entry and choosing `[Properties]`. The screen shot shows the situation where `number` has reached 76, with `i` now 11; that is $101 - 1 - 3 - 5 - 7 - 9 = 76$. At the end of the simulation `count` reached 10, with `i` at 21.

Actually we can generate the square root by ignoring `count` and simply shifting `i` right once and ignoring the carry-out (effectively -1). In **C** this shift-right operation is implemented using the `>>` n operator, where `n` is the number of places. Thus in our example, we could terminate the `sqr_root()` function by `return (i >> 1);`.

Using **C** to implement source code gives the programmer access to structures, operators and library functions appropriate to a modern high-level language. Nevertheless, to be of use in a microcontroller environment it is necessary to permit the programmer to easily access specified locations in the Data store and individual bits within. In this manner Special-Purpose Registers, such as the parallel ports, may be initialized, monitored and controlled in order to allow the processor to interact with its peripheral modules and the outside world. It is possible to do this using standard **C** operators. However, many compilers targeted to microprocessors and microcontrollers have non standard extensions to facilitate this 'bit twiddling'. As we are using the CCS product as the exemplar for this text, we will use the syntax appropriate to this compiler.

As an example, consider a routine that is to continually pulse Port A pin 0 (that is pin RA0 or named `pin_A0` in the CCS compiler's terminology) as long as pin 7 of Port B (RB7 or `pin_B7`) is high—see p. 137. To do this we must tell the compiler that Port A and Port B is really an descriptor for the contents of File h'F80' and File h'F81' respectively. That is, we must create a *pointer to* the addresses 0xF80 and 0xF81; using the **C** language's hexadecimal prefix 0x.[8] In standard **C** we can do this by using a **cast** thus:

---

[8]Decimal is the default base in **C** but beware, because a leading zero is interpreted as octal; e.g., 026 is octal 26 (which is decimal $2 \times 8 + 6 = 22$).

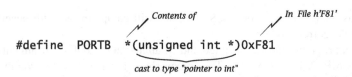

A cast operator gives an object of one type the properties of another kind. For instance, to promote an 8-bit object var to a 16-bit equivalent we have (long)var. In this case the cast (* unsigned int) gives the constant 0xF80 the characteristics of a pointer to the byte[9] at File h'F80'; that is an address.

The leftmost * operator means "contents of". Therefore the complete definition reads "The name PORTA means the contents of File h'F80'". The object PORTA can henceforth be used as a normal global int unsigned variable.

This is how our example might be coded in standard **C**.

```
#define   PORTA *(unsigned int *)0xF80
#define   PORTB *(unsigned int *)0xF81

while(PORTB & 0x80)         /* Isolate bit7; is it non-zero?*/
  {
  PORTA = PORTA | 0x01;   /* IOR with 00000001; RA0 -> hi */
  PORTA = PORTA & 0xF7;   /* AND with 11111110; RA0 -> lo */
  }
```

The routine above ANDs (&) Port B with b'10000000' to determine if bit 7 is set, which if so will give a non-zero (true) outcome—see p. 126. If this is the case, the body of the while loop will be executed. This body uses Inclusive-OR (|) to set a bit in Port A (see p. 127) and AND to clear a bit in Port A. As we see from the following assembly-level code generated by the CCS Version 4 compiler, this has been interpreted as a *single-bit* set or cleared respectively, and correctly uses the btfss, bcf and bsf instructions.

If several bits had been tested set or cleared then the appropriate ior and and instructions would have been used.

```
WHILE_LOOP
      btfss  h'F81',7        ; Test bit7 of Port B
      bra    NEXT            ; IF 0 THEN break out of loop
      bsf    h'F80',0        ; Pin RA0 high
      bcf    h'F80',0        ; Pin RA0 low
      bra    WHILE_LOOP      ; Go again
NEXT  .....    .....
```

This executable code is the same as a hand-coded assembly version.[10]

---

[9]Some compilers either use an unsigned short int or unsigned char to hold an 8-bit datum.

[10]Many compilers are unable to distinguish between single bits which use efficient 'bit twiddling' instructions and use the less efficient logic instructions; for instance, Version 2 of this compiler.

In the specific case of the CCS compiler, the non-standard directive **#byte** can be used to name the contents of a fixed Data store address; e.g. #byte INTCON = 0xFF2 assigns the contents of File h'FF2' the name INTCON. In a similar manner, an individual bit can be named in the CCS compiler using the **#bit** directive. For instance, #bit INT0IF = 0xFF2.1 names bit 1 of File h'FF2'. Alternatively, if INTCON has already been named as above, #bit INT0IF = INTCON.1 does the same thing. Such defined objects can only have the value 0 and 1.[11] Thus the statement INT0IF = 0; will clear bit 1 of the INTCON File.

Using this CCS syntax gives us the equivalent code:

```
#byte PORTA = 0xF80 /* Port A is File h'F80'              */
#byte PORTB = 0xF81 /* Port B is File h'F81'              */
#bit  RA0 = PORTA.0 /* Bit 0 of File h'F80' now named RA0 */
#bit  RB7 = PORTB.7 /* Bit 7 of File h'F81' now named RB7 */

while(RB7)
    {
    RA0 = 1;          /* Pin RA0 high                      */
    RA0 = 0;          /* Pin RA0 low                       */
    }
```

which generates exactly the same executable code as our rather more long-winded standard **C** version. Moreover, where a compiler has a special notation like this it gives a stronger message that efficient 'bit twiddling' instructions should be generated. This is at the expense of portability. We will use this notation from now on.

Functions in the **C** language fulfill the same purpose as a subroutine at assembly level.[12] In the situation where interrupt-driven code is to be written in **C** the relevant function needs to be implemented as an Interrupt Service Routine (ISR). In addition, the global and local interrupt masks and flags, priority and multiple source service polling issues need to be addressed.

In order to illustrate the main principles involved in coding a real-time process, we will repeat Example 7.3 on p. 229 but using a CCS **C** implementation. As in the assembly-level solution there are two functions in Program 9.3. The main() function represents the background routine. Its initial job is to set up the INT-CON SRF using the CCS built-in function enable_interrupts() function. enable_interrupts(int_ext) effectively sets the INT0IE mask bit and thus enables the device to respond to interrupts from the INT0 pin. Similarly, there are equivalents for all the various sources of interrupts the target device can handle; for instance, enable_interrupts(int_timer1) to qualify requests from Timer 1. All these are listed in the appropriate header file for the device.

---

[11] A 2-valued object of this kind is sometimes called a Boolean. In the CCS compiler a short int is a Boolean, but this is unusual.

[12] The CCS compiler will usually implement a function called from one point only as in-line code. This is the reason why the function sqr_root() was called from two points from main() in Program 9.2. The directives #INLINE and #SEPARATE can be used to encourage the compiler to implement the following function either as in-line code or as a subroutine.

Although it would be possible to map the INTCON register with its various masks and flags using the byte and bit directives, it is recommended that the built-in functions be used where available to set up and service peripheral services. This makes the code more portable between devices and even families.

enable_interrupts(GLOBAL) sets the global mask bit GIE/GIEH. This activates the overall interrupt system and allows the processor to respond to any unmasked service requests—in our case an external (hardware) interrupt from pin INT0/RB0. Figures 7.2 and 7.4 on pp. 209 and 214 detail the arrangements of the interrupt logic for the default Compatible and Priority modes respectively. As we have already noted on p. 287, the main() function defaults the interrupt mode to the former.

Using this approach, the function set_tris_a(0xF7) in Program 9.3 puts the code b'11111110' into TRISA, thereby making pin RA0 an output. Function setup_adc_ports(NO_ANALOGS) makes all pins digital as opposed to the default analog—see Program 7.1 on p. 218 and Fig. 14.12 on p. 510. However, as we have seen on p. 287, the CCS compiler already configures these parallel-port pins as digital I/O in the environmental set up of the main() function. Therefore, strictly this function call is not needed.

Finally, pin RA0 is initialized as low and global variable Batch zeroed. The first alpha character of global variables have been capitalized for emphasis, with local names (there are none in this program) using lower case. The unsigned qualifier is not needed a this is the CCS compiler's default for integer objects.

The core of main() is a DO forever loop where the state of Batch is continually monitored for non-zero (truth) and if this is the case it is reset and RA0 is pulsed high once for 1 ms. The CCS built-in function delay_ms() provides an easy means of generating precise delays of up to 65,535 ms in this implementation of the language. To facilitate this the #use delay directive must be used to tell the compiler what the actual clock frequency is—8 MHz in our case. The delay_us() and delay_cycles() functions can be used for shorter delays. For compilers without similar non-standard functions, assembly-level delay subroutines can be used with care! The directives #asm and #endasm can be used to sandwich such code into the C source file at the appropriate place.

The function can_count() is signaled to the compiler as an interrupt service routine using the prefix directive #int_ext, and there are similar directives for each of the various sources of interrupt request. For instance, #int_timer0 to service a request from TMR0—see Program 13.2 on p. 462. The compiler handles setting up an appropriate interrupt response to handle interrupts from multiple sources and context saving and retrieval.

As can_count() is an ISR, values are not passed in the normal way, as indicated by the void keyword. Instead, any variable monitored or changed is global. In our program, both Batch and Event are defined *outside* a function and are therefore known to all functions, both foreground and background. Any functions *following* the main() function must have a prototype in the program preamble. This allows the compiler to check if the function definition uses and returns the correct types of object. In Program 9.3 these types are shown as void. There was no

Program 9.3 Program in **C** for the pea-canning packer

```
#include <18f1220.h>          /* Device specific information   */
#use delay (clock=8000000)    /* Target to use 8 MHz clock     */
#bit  RA0 = 0xF80.0           /* RA0 mapped to bit0 of PortA   */
void can_count(void);         /* Tells compiler function details*/
int Event, Batch;             /* Global vars to pass parameters */

void main(void)               /* Startup function: background   */
{
enable_interrupts(INT_EXT);   /* Enable interrupts from INT0    */
enable_interrupts(GLOBAL);    /* Enable interrupt system        */
set_tris_a(0xFE);             /* Set up pin RA0 to Output       */
setup_adc_ports(NO_ANALOGS);  /* All pins digital               */

RA0 = 0;                      /* Pin RA0 starts low             */
Batch=0;                      /* Batch initialized to zero      */
while(1)                      /* DO forever (1 is always TRUE)  */
    {
    if(Batch)        /* DO nothing as long as Batch 0 (FALSE) */
        {
        Batch = 0;   /* Re-zero it                             */
        RA0 = 1;     /* and pulse RA0 high for 1ms             */
        delay_ms(1);
        RA0 = 0;
        }            /* Repeat the while loop                  */
    }
}

/****************************************************************
 * This is the foreground interrupt service routine           *
 ****************************************************************/

#int_ext                      /* Make the function an INT0 ISR  */
void can_count(void)
    {
    if(++Event == 24)         /* Increment Event count          */
        {                     /* and IF 24 THEN DO              */
        Event=0;              /* Reset Event count and signal   */
        Batch++;              /* background with non-zero Batch */
        }
    }                         /* Finish and return              */
```

prototype used in Programs 9.1 and 9.2 as sqrt() was defined *before* main(). However, traditionally main() is placed as the first function in the list and thus *all* other functions subsequently defined should be listed in the preamble in prototype form.

The kernel of can_count() *first* increments Event—the ++ appears *before* the variable. If this produces a value of 24, then it is zeroed and Batch is incremented to tell the background function that a batch of 24 cans has been recorded.

Compared to the seven instructions of the assembly-level coding ISR of Program 7.3 on p. 229, this high-level implementation generates 41 executable instructions. This is because the **C** implementation needs to be able to cope with all types of ISR function. Thus as well as saving the Status and Working registers, all three double-byte File Select Registers and the PRODH:PRODL pair are also preserved. In this case there was only one source of enabled interrupt, but the compiler always sets up a polling routine to check which source requested service. This dispatch code adds to the overhead.

A table of all standard **C** operators is given in Appendix C for reference.

Examples

Example 9.1 Write a **C** function to compute the relationship:

$$sum = \sum_{k=1}^{n} k$$

For example, if $n = 5$ then we have $sum = 5 + 4 + 3 + 2 + 1$.

Solution The function summation() implementing our relationship $\sum_{k=1}^{n} k$ is listed in Program 9.4. A single 8-bit variable (unsigned int) is passed to the function by the caller to represent the upper limit of the series index. Locally it will be known as n. When the 16-bit (unsigned long int) series summation has been evaluated, it is returned to the caller as the effective value of the function. Thus a typical call might be something like arg = summation(26);. The script unsigned long int sqr_root(unsigned int) will need to appear in the preamble before main() to declare that function summation() will return an unsigned long int value and one unsigned int object will be passed to it.

The body of summation() after defining and zeroing the unsigned long int variable sum, consists of a while loop adding a decrementing n to sum as long as it remains above zero; i.e. (n>0) is true. In adding an 8-bit to a 16-bit

---

Program 9.4 Calculating the sum of all integers to n

```
unsigned long summation(unsigned int n)
    {
    unsigned long sum = 0;
    while(n>0)
        {
        sum = sum + n;
        --n;
        }
    return sum;
    }
```

object, **C** will automatically extend the former to 16 bits before the addition. This 16-bit outcome to the right of the assignment = is given to the left variable sum. In general in mixed type arithmetic, the lesser types are promoted to the greatest.

The Decrement operator -- can be pre- (as shown) or post- (i.e. n--;) in this case as order is irrelevant. Actually, most **C** programmers would incorporate this action into the while test expression thus while(--n.0). In this case it is important to pre-decrement *before* the test; whose outcome will be either true or false.

Example 9.2 On p. 250 we implemented a root mean square program to evaluate the mathematical relationship $\sqrt{\text{NUM\_1}^2 + \text{NUM\_2}^2}$. Write a **C** function to implement this relationship, where the two signed 8-bit objects num_1, num_2 are passed to the function which returns the 8-bit value rms.

Solution The solution shown in Program 9.5 uses the internal unsigned long 16-bit variable sum to hold the addition of the two squared signed 8-bit variables. The squaring operation is simply implemented using the **C** multiplication operator * rather than coding a squaring function of the manner of Program 8.3 on p. 259. However, the programmer needs to force the compiler to do its arithmetic in 16-bit precision to match the 16-bit sum. This is done by casting one of each of the multiplication operands thus (signed long). The function developed in Program 9.1 is used to generate the square root of the 16-bit sum object and is called from line 6 of the function variance() with the return value being assigned to the variable rms as part of the call. In compiling the source code using the CCS **C** compiler, 110 machine-level instructions are needed to implement this problem. This compares to 48 instructions for the assembly-code version of Chap. 8. This gives an code factor of 2.29.

Program 9.5 Generating the root-mean-square value of two variables

```
unsigned int variance(signed int num_1, signed int num_2)
  {
  unsigned long sum;
  unsigned int rms;
  sum = (signed long)num_1*num_1 + (signed long)num_2*num_2;
  rms = sqr_root(sum);
  return rms;
  }
```

Example 9.3 A K-type thermocouple is characterized by the equation:

$$t = 7.550162 + 0.0738326 \times v + 2.8121386 \times 10^{-7}v^2$$

where t is the temperature difference across the thermocouple in degrees Celsius and v is the generated emf spanning the range 0–52,398 μV. This is represented by a 14-bit unsigned binary number, for a temperature range of 0–1300°C. Write a **C** function which will take as its input parameter a 14-bit output from an analog to digital converter and return the rounded-up integer temperature in Celsius measured by the thermocouple.

Solution Our function, named `thermocouple()` in line 1 of Program 9.6, takes one unsigned long integer (16-bit) parameter, named `emf`, and returns a similar 16-bit value. The internal variable `temperature` is defined in line 3 to be a floating-point object[13] to cope with the complex fractional mathematics of line 6. Because we are told that only the 14 lower bits of `emf` have any meaning, line 5 ANDs the 16-bit object with h'3FFF' (0x3FFF) to clear the upper two bits. Finally, an `unsigned long` (cast) version of the `float` object `temperature + 0.5` is made and returned in line 8. The 0.5 offset gives rounding to the nearest integer up or downwards.

Program 9.6 Linearizing a K-type thermocouple

```
unsigned long thermocouple(unsigned long emf)
  {
  float temperature;
  unsigned long outcome;
  emf = emf & 0x3FFF;            /* Clear upper two bits */
  temperature = 7.550162+0.073832605*emf+2.8121386e-7*emf*emf;
  outcome = (unsigned long)(temperature + 0.5);
  return outcome;
  }
```

The resulting executable code running on a enhanced-range core takes 324 program words; that is, around $\frac{1}{3}$ of the Program store's capacity of a PIC18F1220 device! Because of the size penalty of using floating-point objects, fixed-point (integer) arithmetic is used wherever possible in embedded microcontroller implementations.

Example 9.4 Write a function that will shift each bit in a specified RAM address passed to the function, out of pin RA0 in turn, working right to left. As each bit is presented in turn to RA0, pin RA1 is pulsed ⎍ to tell the outside world that a new bit is ready.

Solution The code given in Program 9.7 specifies a parameter of type **pointer-to** an `unsigned int`. The pointer-to operator * can also be read as "contents-of"; that is the contents of `address` is an `unsigned int`. In either case, the variable `address` is a File address. Subsequently, the contents of this File can be accessed by prefixing the variable address with a * contents-of operator; as in line 16. As

---

[13] Having a mantissa and exponent of the form $m \times 10^e$. The CCS compiler uses a 32-bit representation (signed 24-bit mantissa with an offset 127 8-bit mantissa) giving a potential range of approximately -1.5×10^{-45} to $+3.3 \times 10^{38}$.

Program 9.7 A simple serial data transmitter

```
#byte  PORTA   = 0xF80        /* Port A is File h'F80'        */
#bit   SER_OUT = PORTA.0      /* in which pin0 is named       */
#bit   CLOCK   = PORTA.1      /* and pin1 is named            */

/* Returns nothing and accepts a File address (pointer)      */
void put_char(unsigned int *address)
{
int i;                        /* Loop count                  */
for(i=0; i<8; i++)            /* DO eight times              */
    {
    if(*address & 0x01)
        {SER_OUT = 1;}        /* IF Datum bit0 is 1, RA0 <- hi*/
    else
        {SER_OUT = 0;}        /* ELSE make RA0 low           */
    CLOCK = 1;                /* Pulse pin RA1 high          */
    CLOCK = 0;                /* and then low                */
    *address= *address >> 1;  /* Shift Datum right one place */
    }
}
```

an example, if the caller wanted to transmit the datum byte in File h'020' then the function call put_char(0x020); would give the variable address the value h'020'.

The core of the function uses a For loop to shift the contents of address (i.e. *address) right one place eight times using the >> Shift Right **C** operator. Before each shift, pin RA0 (named SER_OUT) is either set or cleared, depending on the state of bit 0 of the datum in address using a if-else decision structure. This bit is isolated by ANDing with b'00000001'; that is, *address & 0x01. In either case pin RA1 (named CLOCK) is pulsed. The resulting code shown in Program 9.7 implements a simple synchronous serial communications link—see Chap. 12.

Example 9.5 Arrays of identically sized objects can be defined in **C** using the notation fred[n], where fred is the name of the array (actually the address of the first element) and n is the nth element. For instance, the object declared as unsigned int fred[16]; will be allocated 16 Files in the Data store by the compiler.

It is possible to give each element of an array an initial value; e.g. for svn_seg[10]:

```
unsigned int svn_seg[10] = {0x3f, 0x06, 0x5b, 0x4f, 0x66,
                            0x6d, 0x7d, 0x07, 0x7f, 0x6f};
```

defines an array of ten bytes initialized with the 7-segment patterns described in Fig. 6.8 on p. 173.

These ten values for svn_seg[0] through svn_seg[9] will be placed in ten consecutive Files. Most PIC MCUs have a severely limited Data store capacity and, as in this example, where the values will not be subsequently changed, it makes more sense to place these ten constants in Program ROM as a table of data—as will

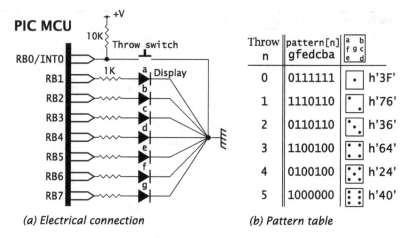

(a) Electrical connection   (b) Pattern table

Fig. 9.4 The active-low die patterns

be described in Chap. 15. The `tblrd` instruction can then be used to read out the
pointed-to byte. In this compiler, this can be done by qualifying the array with the
keyword `const`; giving the definition:

```
unsigned int const svn_seg[10] = {0x3f, 0x06, 0x5b, 0x4f, 0x66,
                                   0x6d, 0x7d, 0x07, 0x7f, 0x6f};
```

Using the techniques outlined here, write a program to implement an electronic
die, with seven LEDs connected to the top seven pins of Port B—as shown in
Fig. 9.4(a). The main routine is simply going to increment a global integer as rapidly
as possible. The throw switch is connected to the `INT0`/`RB0` pin and when it pulses,
the program is to transfer to an interrupt service function—see Program 9.3. This
ISR is to display one of six die patterns and after 10 s blank out the display to save
battery life. Running the PIC MCU using a Watch crystal frequency of 32,768 Hz
also reduces the energy requirements—as illustrated in Fig. 10.3 on p. 309.

Solution The LED patterns are listed as a global array `const array[6]` of six
patterns in Program 9.8, following the tabulation of Fig. 9.4(b). The patterns listed
in the array are shifted left one place compared to the truth table, to align with the
top seven bits in Port B.

The main routine simply increments the byte variable `throw` and continually
resets the count to zero when it goes beyond five to give a modulo-6 count; that is
$0, 1, 2, 3, 4, 5, 0, \ldots$.

The CCS version 4 compiler generates six instructions for the endless loop in
`main()` and this gives an incrementation rate of about 1000 per second at the stated
clock frequency. This effectively gives a random selection when the outside human
throws the switch connected to the `INT0` pin and thus causes an interrupt. In any
case such a switch will mechanically bounce giving multiple interrupts and adding
to the apparent randomness.

Program 9.8 The electronic die

```
#include <16f84.h>
#use delay (clock=32768)
#byte PORTB = 0xFE1

void die(void);
unsigned int const array[6] = {0x7e, 0xec, 0x6c, 0xc8, 0x48, 0x80};
unsigned int throw;

main()
{
set_tris_b{0x01};               /* All PortB pins ex RB0 Output*/
setup_adc_ports(NO_ANALOGS);    /* All pins digital           */
enable_interrupts(INT_EXT);     /* INT1IE set to 1            */
enable_interrupts(GLOBAL);      /* GIE set to 1               */

while(1)                        /* Forever DO                 */
    {
    PORTB = 0;                  /* LEDs off                   */
    if(++throw > 5) {throw=0;}  /* Increment modulo-6         */
    }
}

#int_ext                        /* Hardware ISR               */
void die(void)
    {
    PORTB = array[throw];       /* Display nth element pattern */
    delay_ms(10000);            /* for 10,000 ms              */
    }
```

The ISR function die() copies the *n*th element of our array of constants to Port B and then delays 10 s before returning to the background function. As Port B is cleared in main(), the display will then be blanked. This helps keep the battery drain down to a minimum.

Example 9.6 In Figure 7.2 on p. 224 we saw how the integrity of the teeth on a drive belt could be monitored. Repeat the coding, this time using **C**, with the same approach as this assembly-level implementation.

Solution As in the assembly-level solution, the coding in Program 9.9 is structured as three functions, with two interrupt service and a background routine. Essentially on each 10 kHz High-priority interrupt at INT2 the 2-byte variable Timer is incremented. A tooth pulse interrupting at INT1 copies this as the inter-tooth period and updates the moving average. The background routine monitors these variables and activates the alarm if the period significantly exceeds this average.

jiffy()
This handles the 10 kHz pulse train from the timing oscillator. The associated directive int_ext2 HIGH designates the following function as a High-priority INT2 ISR. As we saw on p. 287, the interrupt logic defaults to Compatible mode. The directive in the program preamble #device high_ints=true sets up the interrupt mode to Priority; i.e. IPEN = 1.

Program 9.9 Belt-drive monitor software

```
#include <18f1220.h>
#device HIGH_INTS=TRUE          /* Set Priority int mode    */
#use delay (clock=4M)
#bit  ALARM = 0xF80.0

/* Function declarations                                    */
void jiffy(void);
void tooth(void);

/* Global variables                                         */
unsigned long int Timer=0, Period = 0;
unsigned long long int EMA256 = 0;

void main(void)
{
unsigned long int temp;
enable_interrupts(INT_EXT1);
enable_interrupts(INT_EXT2);
enable_interrupts(GLOBAL);
set_tris_a(0xFE);
setup_adc_ports(NO_ANALOGS);

ALARM = 0;

while(1)
    {
    temp = (Period/4)*3;        /* temp = Period*3/4        */
    if(temp > (EMA256>>8))      /* Period too long?         */
        {
        ALARM=1;                /* IF yes THEN alarm        */
        delay_ms(256);          /* sounded for 0.25s        */
        }
    else
        {ALARM=0;}              /* IF not THEN no alarm      */
    }
}

#int_ext1                       /* Low priority INT1 ISR    */
void tooth(void)
{
Period = Timer;                 /* Count is the new period  */
Timer = 0;                      /* Reset the time count     */
EMA256 = (EMA256 - (EMA256>>8)) + Period;
}

#int_ext2 HIGH                  /* High priority INT2 ISR   */
void jiffy(void)
{Timer++;}                      /* Record a new 100us tick  */
```

The core of the ISR simply increments the long int global variable Timer, recording another 100 µs clock tick.

tooth()
This Low-priority INT1 ISR does three things.

1. Updates the 16-bit global variable Period with the contents of Timer giving the inter-tooth period in 100 µs increments.
2. The global variable Timer is zeroed to initialize the next inter-tooth count.
3. The Exponential Moving Average period is updated. In the assembly-level implementation the relationship:

$$EMA_n = \frac{255}{256} EMA_{n-1} + \frac{1}{256} Period_n$$

was implemented. To cope with fractional period quantities the 2-byte EMA_H: EMA_L variable was extended with an additional byte EMA_DEC to hold any fractions. In our **C** coding, rather than using a floating-point object we use the equivalent relationship:

$$256 \times EMA_n = 255 \times EMA_{n-1} + Period_n.$$

This allows us to use a 32-bit integer (fixed-point) representation holding $256 \times EMA$, which we call EMA256 in the program. The CCS compiler has a non-standard type specifier int32 or long long int to designate a 32-bit variable. In our statement implementing this relationship

```
EMA256 = (EMA256 - (EMA256>>8)) - Period;
```

we have shifted right eight times to implement the $\frac{1}{256}$ fraction. Subtraction gives the required fraction $\frac{255}{256}$. This shift-right operation is in parenthesis (-(EMA256>>8)) to ensure that this operation is carried out *first* before the subtraction; as we see from Appendix C that subtraction has a higher precedence than shifting. The statement EMA256 - EMA256>>8; actually gives zero!

main()
The background routine core is an endless while loop generating a reduced duration of $\frac{3}{4}$ Period. Here the division ÷4 is carried out first to prevent the possibility of overflow of the 16-bit arithmetic.

If our reduced period is greater than the running average (which is $\frac{EMA256}{256}$) then the alarm is sounded for 256 ms as specified.

Self-Assessment Questions

9.1 Driving the die of Example 9.5 requires seven parallel port lines and a particular electronic game needs to drive two die displays. By inspection of the patterns of Fig. 9.4, how could you reduce the requirement to four bits per die?

9.2 As part of an electronic game, a function is to be written to return the next pseudo random number in the 127 sequence defined by the generator configuration of Fig. 6.22 on p. 203. The current number is to be passed to the function and the next number in the sequence returned. Assume that this passed datum is non-zero.

How could you modify the function to send the entire sequence of random numbers out of Port B beginning with the passed number?

9.3 A PIC MCU-based digital thermometer is to display temperatures between 0°C and 100°C. To be able to market the device to USA the thermometer is to have the option to display the temperature in degrees Fahrenheit. Write a function for a PIC-MCU based thermometer that is to convert Celsius integers to the equivalent Fahrenheit integer. The input is to be an unsigned int byte representing Celsius and the return Fahrenheit is also to be an unsigned int datum. The relationship is:

$$\texttt{fahrenheit} = (\texttt{celsius} \times 9)/5 + 32$$

and 16-bit arithmetic should be forced to avoid overrange errors.

9.4 A cold-weather indicator in an automobile dashboard display comprises three LEDs, which are connected to the lower three bits of Port A. Bit 2 of this location is connected to the red LED, which is to light if the Fahrenheit temperature is less than 34. Bit 1 is the yellow LED for temperatures below 40°F, and bit 0 is the green LED. You may assume that the appropriate port pins have already been set as outputs and that a LED is illuminated when the driving pin is low. Write a function, whose input is °F, that activates the appropriate LED.

Part III
The Outside World

Up to this point we have confined our deliberations to the internal structure and software of the PIC18 microcontroller family core, with some limited allusion to parallel ports. This final part looks at how the MCU core reacts with the various peripheral functions and the environment, physically beyond the confines of its pins.

The original base-range PIC MCU family has little in the way of peripheral functions beyond parallel ports and an 8-bit timer. These were inherited in the first mid-range PIC16 introductions, to which was added interrupt handling and an additional module; typically a Data EEPROM (PIC18C84) or analog-to-digital converter (PIC16C71). As new family members were introduced, the range and number of peripheral devices became much more extensive, with additional timers, serial ports, comparators and more specialist functions, such as USB.

The PIC18 enhanced-range family makes use of these same peripheral modules; in some cases with additional functionality. An example of this is the 18-pin PIC18F1220 of Fig. 4.1 on p. 71. In Part III we will also use the 40-pin PIC18F4520 as an exemplar, which has a slightly larger inventory of peripheral interface capability. On the way you will:

- Look at support issues such as the power supply, clock, power management and device configuration.
- Consider parallel and serial digital data input and output.
- Examine the Timer and Watch-dog subsystems.
- See what is involved in dealing with analog signals.
- Design an embedded MCU-based viva timer.
- Consider how a system may be tested and debugged.

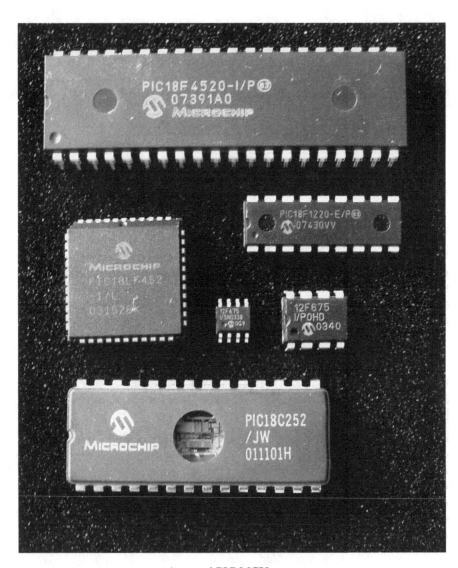

Assorted PIC MCU parts

Chapter 10
The Real World

Up to this point we have mainly concentrated on how the software has interacted with the processor's internal registers and data memory. Now, as a prelude to how the MCU relates to its internal peripheral devices and hence monitors and controls its external environment, i.e. the *real world* outside its pins, we need to look at external support issues, such as power requirements, clocking and resetting.

After reading this chapter you will:

- Be familiar with the permitted range of power supply, brown-out and input/output voltages.
- Distinguish between quiescent and dynamic power dissipation and recognize that the latter is directly proportional to both frequency and to the square of the supply voltage.
- Understand the nuances of the various reset processes.
- Understand the basics of the integral clock oscillators.
- Be aware of how the Run, Sleep and Idle modes are configured, invoked and exited, and their effect on the processor.
- Know how the PIC MCU's option configuration can be set-up.

As a prelude to our discussion on real-world issues, Fig. 10.1 shows the architecture of the PIC18F4420 and 18F4520 MCUs, which we are going to use as one of our exemplars for most of the rest of the book. Apart from the latter's larger Program and Data stores, the two devices are identical and so we will concentrate on the latter. The PIC18F2420 and 2520 MCUs are corresponding 28-pin variants and therefore support a somewhat truncated inventory of peripherals. We will refer to these four devices as the PIC18FXX20 series.

Except for issues relating to memory capacity, the core of these processors are very similar in all enhanced-range devices. Their instruction set described in Chap. 5 are identical. In comparing the PIC18F1220 of Fig. 4.1 on p. 71 to Fig. 10.1 we see that the main difference is in the latter's more extensive set of peripheral modules; which we will be describing in the following chapters. Of course, even 40 pins is not enough to go round and give each peripheral its own separate I/O connection to the outside world. Thus the majority of pins are a shared resource. For instance,

S. Katzen, *The Essential PIC18® Microcontroller*,
Computer Communications and Networks,
DOI 10.1007/978-1-84996-229-2_10, © Springer-Verlag London Limited 2010

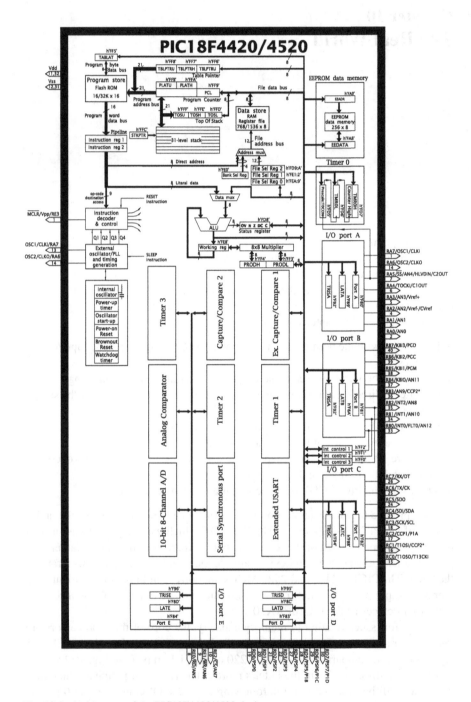

Fig. 10.1 Architecture of the PIC18F4420/4520 devices

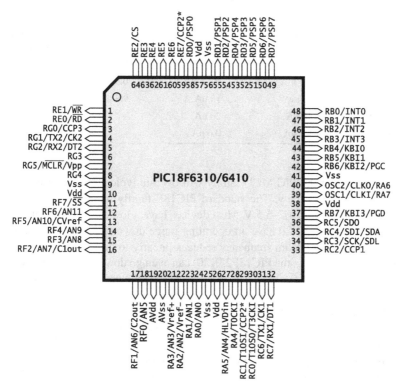

Fig. 10.2 Pinout of the 64-pin PIC18F6410 MCU

pin RA3 is bit 3 of Port A but can also be used as ANalog channel 3 AN3 or even as an external positive reference voltage input V_{ref+} for the analog-to-digital converter module. PIC MCUs with smaller form factors, such as the 18-pin PIC18F1220[1] can still support a rich set of peripheral modules, but in such cases pin sharing is more extensive, with a consequently greater restriction on what can be used in any given application. Using a fully internal clock oscillator and foregoing the external Master CLear ($\overline{MCLR}$) will save precious pin resources.

The mid-range family have members with form factors down to six pins. Obviously low-cost options like this come with severe penalties! Conversely, the PIC18 family supports devices with 64-, 80- and 100-pin packages; e.g. see Fig. 10.2. Generally these support additional functions requiring an outlet to the outside world. For instance, the PIC18F8410 80-pin version of the PIC18F6410 has a 20-wire Address/Data bus to support an external Program memory of up to $2^{20} = 1$ Mbyte capacity. Other resources, such as LCD-segment drivers (e.g. the 80-pin PIC18F8490), USB (e.g. the 80-pin PIC18F86J50) and Ethernet (e.g. 100-pin PIC18F96J60) are usually to be found in devices with larger form factors.

---

[1]The PIC18F1320 is identical but with double the Program store size at 2 kwords.

Table 10.1 Power supply operating current for the PIC18FXX20

Oscillator	Run @ 5 V	Run @ 2 V	Idle @ 5 V	Idle @ 2 V
40 MHz	23 mA	–	9.1 mA	–
4 MHz	2.6 mA	0.74 mA	0.9 mA	0.25 mA
32 kHz	90 μa	15 μA	9.8 μA	3.1 μA
Sleep @ 25°C	40 nA	20 nA	–	–
Sleep @ 85°C	1.7 μA	0.6 μA	–	–

All members of the PIC MCU family will operate typically with a supply voltage V_{DD} of nominally 5 V. The standard PIC18F family members can run up to 40 MHz over the range 4.2–5.5 V. Most devices have a low-voltage variant. For instance, the PIC18LFXX20 have an operating range down to 2 V. However, below $V_{DD} = 4.2$ V the maximum frequency reduces linearly to only 4 MHz at 2 V. The PIC18FXXK range (e.g. the PIC18F25K20) can even go down to $V_{DD} = 1.8$ V. With a V_{DD} of 3.6 V in this family, a top speed of 64 MHz is possible. However, below 3 V the maximum frequency is reduced to 20 MHz.

To the outside world, the electrical characteristics of a PIC MCU are similar to that of any other electronic digital circuit. In terms of voltages, a pin configured to be an output which has been set to the Low state by the PIC MCU will normally be no more than $V_{OL} = 0.6$ V if sinking (accepting) a current up to 8.5 mA, over the temperature range −40°C to +85°C. A pin set to the High state by the PIC MCU can source (supply) up to 3 mA and not drop more than 0.7 V below the supply; e.g., a V_{OH} of 4.3 V with a 5 V supply.

A port pin configured to be an input will generally (with a few exceptions) recognise a voltage less than 15% (20% for Schmitt trigger buffered inputs) of the supply voltage as being a Low-state input; for instance, $V_{IL} = 0.75$ V for a 5 V supply. An input pin will normally[2] recognise a voltage more than 25% plus 0.8 V of the supply (80% for Schmitt trigger inputs) as being in the High state; for instance, $V_{IH} = 2$ V for a 5 V supply—see Fig. 11.5 on p. 344.

Table 10.1 shows the supply current for the PIC18F/LFXX20 series over a range of clocking and supply voltages. Unless otherwise indicated, values are typical at an operating temperature of 25°C. Normal industrial devices (indicated by an -I suffix to the part number) have a working temperature range of −40°C to +85°C. Extended devices (-E) can operate up to +125°C.

Many microcontroller applications are battery powered and in such situations power consumption is critical. In general, power efficiency considerations are important and as these bare figures from the data sheet show a variation range of more than a million, it is important that the factors influencing current consumption be understood. Most of the newer PIC16 and PIC18 family members are described by Microchip as part of the nanoWatt™ range.

---

[2]The main exceptions are oscillator I/O and RC[4:3] pins.

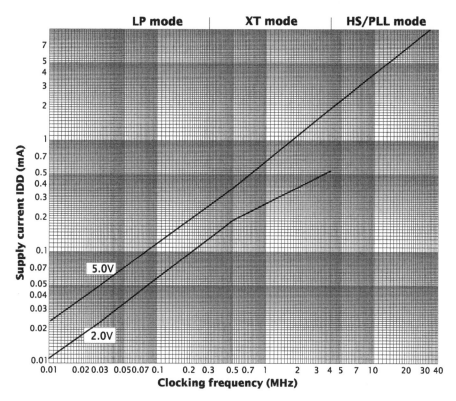

Fig. 10.3 Typical supply current versus clocking frequency

From Table 10.1 we see that supply current is radically related to the clocking frequency as well as operating supply voltage. These relationships are more clearly seen in the graphs of Fig. 10.3. Clearly power dissipation $V_{DD} \times I_{DD}$ is directly proportional to operating frequency. For instance, 100 times more current is required at 10 MHz as compared to 100 kHz.

To see why this is so, consider a switch charging and discharging a capacitive load C, as in Fig. 10.4. The switch is implemented by a transistor and the load is due to the stray capacitance of the connection to the next transistor and its input gate. R_S represents the on-resistance of the switching transistor.

When the switch opens, the capacitance charges up exponentially to V volts with a time constant $\tau = CR_L$. In steady state $\frac{1}{2}CV^2$ joules of energy is stored. Energy is dissipated in the load by this charging current as follows:

Initial charging current ($V_c = 0$): $i_o = V/R_L$

Instantaneous current: $i_c = i_o e^{-\frac{t}{\tau}}$

Instantaneous power in R_L: $i_c^2 R_L = i_o^2 R_L e^{-2\frac{t}{\tau}} = (V^2/R_L)e^{-2\frac{t}{\tau}}$

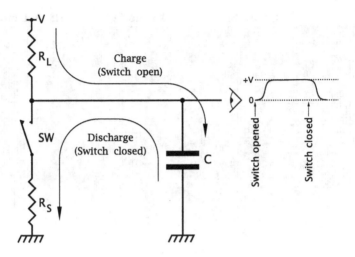

Fig. 10.4 Equivalent output circuit, where C represents both intrinsic and external load capacitance

$$\text{Total energy dissipated in } R_L: E = V^2/R_L \int_0^\infty e^{-2\frac{t}{\tau}} \, dt$$

$$= V^2/R_L \left| -\frac{\tau}{2} e^{-2\frac{t}{\tau}} \right|_0^\infty$$

$$= V^2/R_L \left(\frac{\tau}{2}\right) = \frac{1}{2} CV^2$$

Thus in going high, $\frac{1}{2}CV^2$ joules are dissipated in the load resistance (irrespective of its value R_L!) and $\frac{1}{2}CV^2$ joules are stored in the capacitor's electric field. On discharge, this stored energy is dissipated in $R_S//R_L$ (once again irrespective of value). The energy dissipated in one switching cycle thus CV^2 joules. The total power is this figure multiplied by the number of cycles per second $(CV^2 f)$, plus any quiescent dissipation due to leakage through the switches.

The preceding relationship $CV^2 f$ shows that dissipated power is proportional to frequency for any given supply voltage. Furthermore, it is proportional to the square of the supply voltage, so halving V_{DD} from 5 V to 2.5 V should quarter the power dissipation $V_{DD} \times I_{DD}$.[3]

The dynamic power dissipation derived above should be added to that due to the quiescent current the device consumes when the clocking rate has dropped to zero. From the lower two rows of Table 10.1 this base or Power-Down current, listed in data sheets as I_{PD}, is typically less than 1 μA. These figures assume that peripheral modules that (sometimes optionally) have their own private clock oscillator, such as the Watchdog timer, and any Brown-out reset circuitry are disabled.

---

[3] This is why most current microprocessors used as the PC's CPU, such as the Intel Pentium IV, are powered at under 3 V rather than the standard 5 V of older devices.

Table 10.2 Oscillator operation modes

LP crystal	Low Power for frequencies up to 200 kHz	OSC1:OSC2
XT crystal	Crystal (XTAL) between 200 kHz and 4 MHz	OSC1:OSC2
HS crystal	High Speed for frequencies above 4 MHz	OSC1:OSC2
HSPLL	Uses a Phase Locked Loop to give ×4 frequencies	OSC1:OSC2
RC	Resistor-Capacitor at OSC1. OSC2 is CLocK Out	OSC1:OSC2
RCIO	As above but OSC2 is RA6 for I/O	OSC1
INTIO1	Internal clock with OSC2 as CLKO. OSC1 is RA7	OSC2
INTIO2	As above with OSC1:OSC2 as RA7:RA6	
EC	External Clock into OSC1 with OSC2 as CLKO	OSC1:OSC2
ECIO	As above with OSC2 as RA6	OSC1

Of course, not clocking the processor is rather unproductive, in that nothing happens! However, many embedded systems only need a processing capability on a sporadic basis, and it would be advantageous to be able to put the processor in a standby mode when no action is required. For instance, a MCU-based radio telemetry transducer at the bottom of a lake may need to measure the temperature only once an hour and have a battery life of a year. Microchip gives an example of a PIC18LF8722 powered by a 2450 Lithium cell with a V_{DD} of 3 V and a clock rate of 1 MHz. With its real-time calendar/clock (RTC) module running continuously and active 600 ms in every minute (99% powered-down) the battery life is 9+ years. With 10% active running the battery life drops to 1 year and running continuously at 1 MHz yields a one month life.

Microcontrollers generally facilitate the provision of a master clock to sequence the fetch and execute process and the various peripheral modules, by providing internal oscillator electronics. When optionally augmented with external timing elements, this provides the Primary clock source.

The first PIC MCU family allowed quartz crystals or ceramic resonators to be used across the OSC1 and OSC2 pins to generate oscillation frequencies up to 20 MHz, giving up to five million instructions per second (MPIS) execution—see Fig. 4.5 on p. 76. For low-cost implementations, an external resistor and capacitor could be used at the OSC1 pin or even a completely external oscillator into this pin. Newer PIC16 devices increased the number of options; for instance, in low pin-count members, removing the need for external components entirely.

Table 10.2 shows the ten clock options for our PIC18 exemplars. These have been presented as four groups.

External crystal/resonator modes

These four types use either a crystal or ceramic resonator across the OSC1 and OSC2 pins; as shown in Fig. 10.5(a). We see from Fig. 10.11 that the internal oscillator electronics comprises an inverting amplifier across which user-supplied timing elements give the desired clock rate. The key differences between the three crystal ranges is the value of the inverting amplifier's gain and drive. In the LP mode

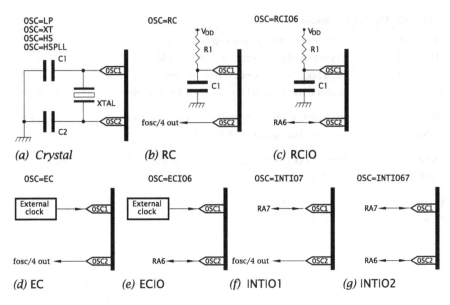

Fig. 10.5 Clock modes

the gain is lowest and, as we have seen, power consumption is minimized. The HS mode is used for high frequencies and has the largest current requirement. The maximum crystal frequency for the standard enhanced-range family is 20 MHz. A phase-locked loop is available and configured to give a ×4 frequency multiplication. Used with a 10 MHz crystal, a bounding top clocking range of 40 MHz is available.[4]

A typical 10 MHz system uses a 10 MHz AT-cut crystal with a C_1 of 22 pF and a C_2 of 33 pF. A 32 kHz crystal needs a C_1 of 68 pF and a C_2 of 100 pF in the LP mode. Although both capacitors may have the same value, making C_2 larger improves the oscillator start-up characteristics. Some crystals in the HS mode may require a series resistor at the OSC2 pin. Details are given in Microchip's application note AN588 *PIC16/17 Oscillator Design*. Ceramic resonators are less expensive than crystals but have an inferior frequency accuracy of the order of 0.5%, and temperature stability is poorer. Ceramic resonators may come with integral capacitors to reduce the part count. Microchip's application note AN588 gives a comparison between ceramic resonators and crystals used in this application.

Resistor-Capacitor modes
Using an external resistor and capacitor as timing elements, as shown in Fig. 10.5(b) & (c), is a low-cost alternative. In the standard RC mode, the OSC2/CLKO pin provides a buffered output clock signal $f_{osc}/4$ which can be used to synchronize external digital circuits. If this is not required, the RCIO mode releases this pin for use as pin 6 of Port A; RA6.

---

[4]The PIC18FXXK subfamily has a maximum clocking frequency of 64 MHz.

These modes are useful for applications where the clocking rate and stability are not of importance. The actual rate is dependent on the external resistor R_1 and C_1, as well as temperature and supply voltage V_{DD}, in a complex manner. Generally, the chosen device's data sheet will give tables and graphs showing typical frequencies against these variables. For example, the PIC18F452 device will have an average clocking rate of 1.6 MHz $\pm$ 10% for a V_{DD} of 5 V, R_1 of 3.3 kΩ, and C_1 of 100 pF at 25°C. The frequency peaks at 1.9 MHz for a V_{DD} of 2.7 V under the same conditions. Of course the tolerance and temperature variation of the timing components and V_{DD} must be considered.

External Clock

It is possible to clock a PIC MCU from an external oscillator up to 40 MHz. This can be useful if several devices are to be synchronized to the one clock. The oscillator should have a Low level V_{IL} below $0.2V_{DD}$ and a High level above $0.8V_{DD}$.[5] As shown in Fig. 10.5(d) & (e), the OSC2/CLKO pin can optionally be assigned to output the instruction cycle clock rate $f_{osc}/4$ or as RA6.

Internal Clocks

Both our exemplars have a completely internal oscillator block. As can be seen from Fig. 10.5(f) & (g), using this as the clock source can release both OSC1 and OSC2 pins for use as Port A I/O pins RA7 and RA6 respectively.

The main internal clock oscillator INTOSC runs at 8 MHz. This is calibrated by the factory with typically $\pm$1% (worst case $\pm$2%) accuracy. This frequency is temperature sensitive and over the range -40°C to $+85$°C can vary by up to $\pm$10%. It is possible to tune the frequency slightly, up to $\pm$12%, by writing to the lower five bits of the OSCTUNE register TUN4:0; as shown in Fig. 10.6.

This 8 MHz clock can be divided to give an alternative range of frequencies; from 4 MHz down to 31.25 kHz. This is directed by the three IRCF2:0 bits in the OSCCON register; as shown in Fig. 10.7. Program 11.1 on p. 338 gives an example where the clocking frequency is set to 4 MHz.

The internal oscillator block also contains a second oscillator, known as INTRC (INTernal RC oscillator). This runs at a nominal 31 kHz rate (minimum 26.5 kHz, maximum 36 kHz). This is primarily used as a clock source for the Watchdog and Power-on timers and as a system clock for Two-Speed Startup or as an emergency backup the Primary oscillator fails. In the case of the PIC18FXX20 series, INTOSC and INTRC are separate and therefore tuning the former will not affect the latter. The INTRC clock can be used as the system clock if the 31.25 kHz IRCF2 : 0 = 000 ratio is chosen and the INTSRC bit in OSTUNE[7] is 0 (the reset default)—see Fig. 10.11. In the case of the PIC18F1220/1320 (PIC18F1X20) this division ratio always selects the INTRC clock rather than INTOSC/256. Also, in this device there is a linkage between the two oscillators.

---

[5]If using a TTL-compatible oscillator, then a pull-up resistor may be needed to ensure a high enough V_{IH}.

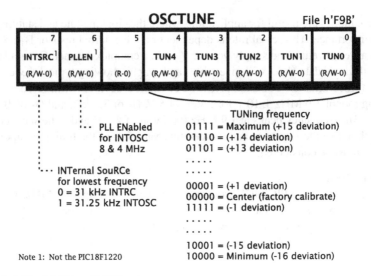

Fig. 10.6 The OSCillator TUNE register

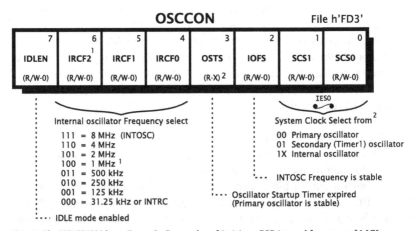

Fig. 10.7 The OScillator CONtrol register

The various clock modes are just one of an extensive range of options and settings that can be selected at the same time as the software is blasted into the Program store—see Appendix B. For instance, the XINST fuse in CONFIG4L[6] enables the Extended instruction set outlined on p. 188. The actual electrical process involved in this code blasting process is not an issue unless you are designing your own device programmer. Normally you will use a commercial programmer, such as the Microchip PICSTART® Plus of Fig. 16.4 on p. 580.

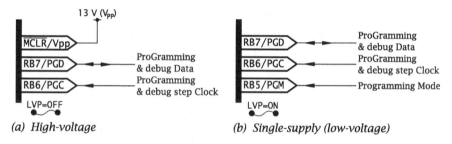

(a) *High-voltage*   (b) *Single-supply (low-voltage)*

Fig. 10.8 Programming the PIC18 family

For background information, Fig. 10.8(a) shows the High-Voltage Programming process. This special Program/Verify state is initiated by raising the Master CLeaR ($\overline{\text{MCLR}}$) pin to +13 V whilst holding both RB7/**PGD** and RB6/**PGC** pins Low. Subsequently the Programming data may be read in from the former as synchronized by the incoming clock signal on the latter pin. This data may be command instructions or machine code. Conversely, the contents of unprotected Program store may be read out and compared with the original code for correctness. When normal voltages are used at the $\overline{\text{MCLR}}$ pin, RB6 and RB7 can be used as normal Port B I/O pins.

One of the options available to the designer is to **code protect** portions of the Program store. Protection prevents existing code from being read out as a security precaution where industrial espionage is a problem. CONFIG5L in Appendix B gives more details.

To obviate the need for a separate power supply, an alternative **Low-voltage programming** (LVP) technique is available. This single-supply mode is especially useful for In-Circuit Serial Programming (ICSP™) where the Program store can be reprogrammed *in situ* on the circuit board. As shown in Fig. 10.8(b), pin RB5 actions entry to this state. Initially held Low during reset, when RB5 is brought High, programming via RB7 and RB6 can begin. The problem with this mode is that pin RB5 cannot subsequently be used as a normal port pin; that is, it is permanently out of action.

In either case, pins RB6 : 7 can also be used for data and clock in a special Debug mode, where execution can be monitored and controlled under control of software running on a remote terminal. Typically, this will be a PC hosting Microchip's MPLAB IDE.

With the PIC MCU in one of these special Programming modes, the Device programmer has access to the Program store and can burn in the application code. The Device programmer also has access to certain private Program store locations from h'30000' to h'3FFFF'; known as **Configuration memory**. This area of memory is beyond the normal executable code space; for instance, h'00000–07FFF' for the PIC18F4520 device—see Fig. 15.4 on p. 550. This zone chiefly comprises an array of Configuration bytes holding the various options, or **fuses**. These are listed in full

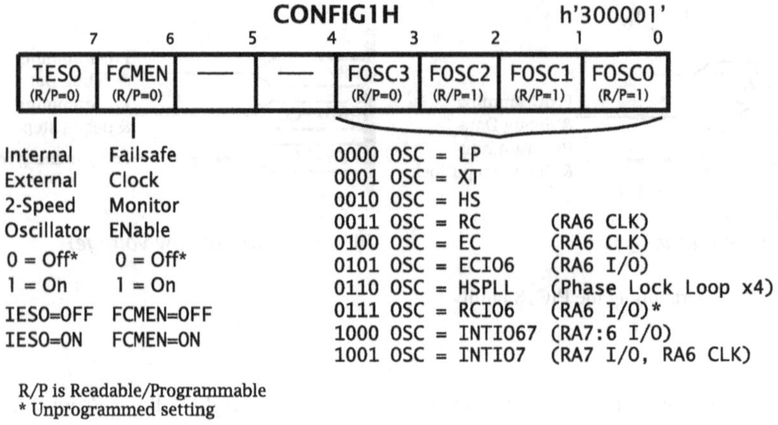

CONFIG1H   h'300001'

	7	6	5	4	3	2	1	0
	IESO (R/P=0)	FCMEN (R/P=0)	—	—	FOSC3 (R/P=0)	FOSC2 (R/P=1)	FOSC1 (R/P=1)	FOSC0 (R/P=1)

Internal	Failsafe	0000 OSC = LP	
External	Clock	0001 OSC = XT	
2-Speed	Monitor	0010 OSC = HS	
Oscillator	ENable	0011 OSC = RC	(RA6 CLK)
		0100 OSC = EC	(RA6 CLK)
0 = Off*	0 = Off*	0101 OSC = ECIO6	(RA6 I/O)
1 = On	1 = On	0110 OSC = HSPLL	(Phase Lock Loop x4)
IESO=OFF	FCMEN=OFF	0111 OSC = RCIO6	(RA6 I/O)*
IESO=ON	FCMEN=ON	1000 OSC = INTIO67	(RA7:6 I/O)
		1001 OSC = INTIO7	(RA7 I/O, RA6 CLK)

R/P is Readable/Programmable
* Unprogrammed setting

Fig. 10.9 The CONFIG1H fuses defining the PIC18FXX20 series clock options

in Appendix B for the PIC18FXX20 series.[6] These are not visible to the software
when the PIC MCU is running normally.

Usually the Configuration fuses are set when the device is programmed. However, it is possible to alter most of these option bits from within the executable code
using the tblrd and tblwt instructions; as described on p. 553, and thus dynami-
cally alter various options on the fly. These facilities can be optionally disabled—see
CONFIG6H in Appendix B.

Figure 10.9 shows a close up of the CONFIG1H register as an example of a Configuration byte. This configuration register holds option fuses relating to the system
clock. These are:

Primary oscillator mode
Bits CONFIG1H[3:0] set up one of the ten oscillator modes. For instance, to use the
Internal oscillator with the OSC1 and OSC2 released for use as Port A I/O pins, we
need to set the **FOSC3:0** fuses to b'1000'.

Fail Safe
The FCMEN fuse enables the Fail Safe clock feature when 1. If the main clock
source fails, the processor is switched to the internal oscillator block and the OSCFIF
(OSCillator Fail Interrupt Flag) of PIR2[7] in Fig. 7.2 on p. 209 is set to optionally
generate an interrupt request. The Watchdog timer is also restarted. This gives a
'soft landing' in the event of a catastrophic failure. The default state of this feature
is off.

Internal/External oscillator Switch Over
When power is applied or the processor comes out of a Sleep mode, a crystal-based

---

[6]Locations h'3FFFE:F' are read-only Device registers, holding the part number and hardware version details. Eight locations h'20000–20007' are designated ID, where the user can store code
identification data; such as company and version information.

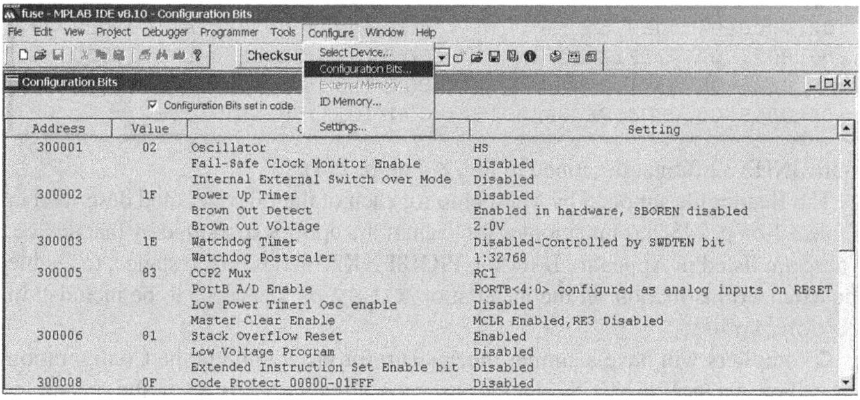

Fig. 10.10 Configuration menu in the MPLAB IDE

oscillator can take a considerable time to stabilize. Normally when in an appropriate mode, an OScillator Timer (OST) is used to count 1024 clock cycles before allowing the oscillator to clock the processor—as shown in Fig. 10.15.

In order to minimize the startup delay, a Two-Speed Startup may be optionally enabled by setting the IESO fuse. In this situation the Internal oscillator block oscillator is used to clock the processor until such time as the Primary oscillator stabilizes; as indicated by the OSTS bit in OSCCON[3]. The default state of this feature is off.

Most device programmers' software will allow the operator to set the required fuses 'manually' before beginning the actual Program store burn process; e.g. see Fig. 10.10. However, it is better to embed this desired fuse state in the program code to automatically action this every time the device is programmed. As an example, consider a PIC18F4520 device which is to have the following configuration:

Oscillator in HS mode
CONFIG1H[3:0] = 0010

Watchdog timer off
CONFIG2H[0] = 0

Low-Voltage Programming off
CONFIG4L[2] = 0

Master CLeaR enabled
CONFIG3H[7] = 1

Then the directive

```
config WDT=OFF, OSC=HS, LVP=OFF, MCLRE=ON
```

in the assembly-level source file will result in the lines machine code loading the Configuration registers thus:

```
:02 0000 04 0030 CA            ; Address record base 0030 0000
:03 0001 00 021F1E BD          ; CONFIG1H:2L:2H @ h'00300001'+
:02 0005 00 8381 F5            ; CONFIG3H:4L @ h'003000005' +
:06 0008 00 0FC00FE00F40 E5 ; CONFIG5L:5H:6L:7L:7H
```

to the INHX32 format described in Fig. 8.4 on p. 254).

The Header file supplied by Microchip for each of their devices, and described in Table 8.1 on p. 243, has mnemonics for each of the options supported by that device. These are listed in Appendix B for the PIC18FXX20 series. For instance, to enable the extended instruction set the incantation XINST = ON needs to be included in the config list.[7]

C compilers will have a similar mechanism for programming the Configuration fuses. For instance, the CCS compiler uses the directive #fuses in the preamble. For our example this is:

```
#fuses    WDTOFF, HS, NOPROTECT, NOLVP, MCLR
```

One of the factors to be considered in designing electronic system is energy efficiency. We have already seen that the power dissipated by a digital system is directly proportional to its switching rate. Thus, a simple strategy is to clock the circuitry at the lowest possible frequency. For instance, reducing the rate from 40 MHz to 32 kHz at 25°C with a 5 V supply lowers the power from typically 60 mW to 0.04 mW. Unfortunately, the processing rate falls in tandem, and where this is not acceptable, the designer must use a more sophisticated approach.

A basic approach is to simply turn off the clock when not required. With no dynamic (switching) power dissipation, only the quiescent (leakage) current will be left. From the lower row of Table 10.1, this Power-Down current, listed in data sheets as I_{PD}, is typically less than 1 μA. These figures assume that peripheral modules that (sometimes optionally) have their own private clock oscillator, such as the Watchdog timer, are disabled.

Of course, not clocking the processor is rather unproductive, in that nothing happens! However, many embedded systems only need a processing capability on a sporadic basis, and it would be advantageous to be able to put the processor in a standby mode when no action is required. For instance, a MCU-based radio telemetry transducer at the bottom of a lake may need to measure the temperature only once an hour and have a battery life of a year.

To expedite situations like this, all PIC MCU families feature a **Sleep mode**, which effectively turns off the clock oscillators.[8] This switch is actioned in software

---

[7]Older versions of the assembler required the programmer to specify each Configuration register when listing the various options. For instance, config config41, XINST=ON.

[8]If the Watchdog timer is enabled, the INTRC oscillator continues to operate, as does the Secondary oscillator if Timer 2 requires it.

using the sleep instruction. Once asleep, the contents of the Data store are retained provided that the supply voltage remains above 1.5 V (V_{DR} in the data sheet). The PIC MCU can be awakened either by resetting the device (see p. 323), by an enabled interrupt request or a Watchdog timer overflow.

When the processor executes a sleep instruction it will clear the $\overline{PD}$ (**Power Down**) bit in the RCON (Reset CONtrol) register (see Fig. 10.14) and the Primary clock oscillator is turned off. If the Watchdog timer (see p. 455) is enabled at that time then it will be restarted and will continue to run, as it uses the INTRC internal oscillator which remains operational in this situation. At this time the $\overline{TO}$ (**Time Out**) flag will be set (i.e., no Time Out). All Files, including the various port settings, remain unchanged.

In the case of an interrupt-actioned awaking, the relevant interrupt flag needs to be cleared and the corresponding interrupt mask bit set to enable requests from that source. If the Global Interrupt Enable mask (GIE/GIEH; see Fig. 7.2 on p. 209) is set to enable the entire interrupt system, then *after* the instruction following sleep is executed, the processor will go to the interrupt service routine as a normal interrupt response. However, if GIE/GIEH is clear, hence disabling the interrupt response, then the processor will not vector to the ISR, but will simply execute the instruction following sleep and continue on as normal. In either case, the programmer should clear the relevant local interrupt flag following the sleep instruction.

If the interrupt logic is disabled and where an unmasked interrupt occurs and sets its associated interrupt flag *before* a sleep instruction is executed, then it is executed as a nop (No Operation). In this situation the $\overline{PD}$ bit will not be cleared, so if necessary the program can determine after a sleep instruction if the PIC MCU really did go through a dormant period. The software can also find out if the processor was awakened by a Watchdog time-out, by checking to see if the $\overline{TO}$ bit in RCON[3] has been cleared. Normally in Watchdog-enabled applications, the sleep instruction is followed by a clrwdt (CLeaR WatchDog Timer) instruction to restart it. Checking the appropriate interrupt flag in the appropriate INTCON or PIR register will determine if the source of the awakening was an interrupt. If this local flag is not cleared, then any subsequent sleep instruction will be implemented as a nop indefinitely!

Whatever the source of the awakening, if the system is running in one of the crystal modes, there will be a delay of 1024 clock cycles f_{osc} before processing the instruction following the sleep breakpoint. This is to ensure that the crystal clock oscillator has started up and stabilized. This oscillator startup delay, illustrated in Fig. 10.15, is not implemented if the PIC MCU is using any of the non-crystal modes, or as we shall see, the secondary or internal oscillator. If the Two-Speed Startup option is enabled, then on awakening, the internal oscillator will drive the processor until the crystal oscillator stabilizes. In all cases there will be a short delay T_{CSD} of not more than 10 μs irrespective of the clock type.

To the base Power-down current I_{PD} given in the data sheet, has to be added the current drain of any enabled peripheral that can operate without the system clock. For instance, if the Watchdog timer (which is clocked by the INTRC oscillator) is running, then for the PIC18FXX20 series at 25°C the typical incremental current

ΔI_{WDT} is typically 1.4 µA at 2 V rising to 5.5 µA at 5 V. Corresponding maximum values are 8 µA and 15 µA respectively at the same temperature. Brown-out Reset, High/Low Voltage Detect, Timer 1 and the ADC module can each add tens of µA to the Power-down budget. These values assume the I/O ports are set to input, with pins tied to either V_{DD} or V_{SS}; usually ground.

Completely turning off the MCU is only appropriate in limited situations. An alternative strategy is to switch the clock to a lower frequency whenever the reduced software throughput is appropriate. For instance, it may be adequate to scan various peripheral modules looking for external activity at 32.78 kHz, and only switch to 20 MHz to handle newly arrived data.

Clock switching was introduced in later PIC16 family members and is available in all PIC18 family members. To see how this works, consider the structure of the PIC18FXX20 clock system shown in Fig. 10.11. Here we can identify three clock sources; any one of which can be chosen to drive the core and peripheral modules. These are:

Primary oscillator

This is the main oscillator using external timing elements at the OSC1 and OSC2 pins. The high-speed (HS) mode can optionally make use of a Phase Locked Loop to frequency multiply by four, with a maximum crystal frequency of 10 MHz giving an equivalent clocking rate of 40 MHz (16 MHz crystal giving a 64 MHz clock for the PIC18FXXK line).

The appropriate mode is chosen using the FOSC3:0 fuses, as outlined in Fig. 10.9. This oscillator is disabled when not selected as the processor clock source.

Secondary oscillator

Timer 1 has its own private crystal oscillator at pins T1OSCI and T1OSCO—see Fig 13.4 on p. 467. Typically this will be a 32.78 (2^{15}) kHz watch crystal, but other LP range values can be used.

Internal Oscillator block

A self-contained internal 8 MHz oscillator INTOSC can be frequency divided down to give a range of eight clocking frequencies, as shown in Figs. 10.7 and 10.6. A separate[9] nominally 31 kHz INTRC oscillator is used for the Watchdog timer, Two-Speed Startup and PoWer-on Reset Timer and in the PIC18FXX20, even as the system clock (if INTSRC (INTernal SourCe) in OSCTUNE[7] is 0; the default setting).

The PIC18FXX20 series optionally allows the use of the Phase Locked Loop sourced with the 4 and 8 MHz INTOSC frequencies if PLLEN in OSCTUNE[6] is set to 1. This gives the option for an internal 16 and 32 MHz internal clock frequency. The default for this option is off.

The source of the system clock is selected using the SCS1:0 (Select Clock Source) bits in OSCCON[1:0]; as shown in Fig. 10.7. The setting of these bits defines three **Run modes**.

---

[9]In some family members, such as the PIC18F1220, INTRC and INTOSC are linked.

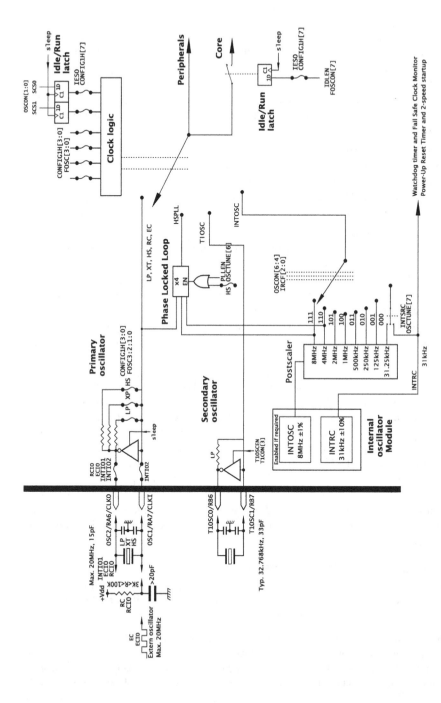

Fig. 10.11 The clock infrastructure for the PIC18FXX20 series

Primary Run mode; SCS1:0 = 00

This is the normal default full-power execution state with the Primary oscillator used as the system clock.

Where the Primary oscillator is crystal timed, there will be a start-up delay to allow for frequency stabilization. This will occur when switching from another Run or Sleep mode or on a Power-on Reset. In such cases the Primary oscillator will only begin clocking after this delay; as signaled with the OSTS (Oscillator Start-up Time-out Status) bit in OSCCON[3] going to 1. If the IESO fuse is set to 1, enabling the Two-Speed Startup, then the Internal oscillator allows processing to start with a minimal delay.

Secondary Run mode; SCS1:0 = 01

In this situation the Timer 1 oscillator is used to clock the processor. The Primary os-cillator is switched off and the OSTS bit cleared to reflect this situation. If switching back to the PRI_RUN mode, the system continues to be clocked by this Secondary oscillator until the Primary oscillator stabilizes; as signaled by OSTS going to 1.

RC Run mode; SCS1:0 = 10

In the RC_RUN mode,[10] the Internal oscillator supplies the system clock. The fre-quency of this clock can be altered on the fly in the normal way. In devices that have separate INTOSC and INTRC oscillators (such as the PIC18FXX20) switching to the latter gives the lowest power consumption at the expense of the poorest frequency accuracy. In this case INTOSC is switched off.

The Primary oscillator will be switched off, and if switching back to a crys-tal Primary oscillator, then the same switch-over transition sequence will occur as detailed for the SEC_RUN mode.

Newer processors have an additional series of **Idle** power management modes. These are based on an extension to the behavior of the sleep instruction.

The normal response to this instruction is to disconnect the complete system from the clocking source, and disable the Primary oscillator. In the case of our exemplar devices, this legacy behavior is actioned with the IDLEN (IDLe ENable) bit at its default 0 setting. If IDLEN is set to 1 then when a sleep instruction is executed the processor core, or CPU, is not clocked. However, the Peripheral modules con-tinue to be clocked; as selected by the SCS1:0 bits. This gives three Idle modes[11] which parallel the Run modes listed above, but are only entered as a result of a sleep instruction and are exited on an awakening. This behavior is illustrated in Fig. 10.12.

Primary Idle mode; IDLEN = 1, SCS1:0 = 00

PRI_IDLE is entered from the PRI_RUN mode via a sleep instruction. The Pe-ripheral modules continue to be clocked from the Primary oscillator selected by the

---

[10]The SCS0 bit is don't care for our exemplar devices, but it is recommended that it should be 0 for compatibility with future device enhancements.

[11]The PIC24 family extend this approach with Doze modes. These allow peripheral processing at full speed and core execution at a reduced rate. In addition, each Peripheral module can be opted out of the Idle modes.

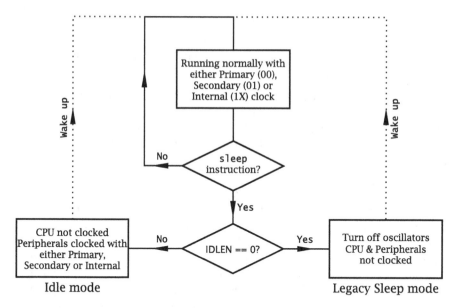

Fig. 10.12   Idle and Sleep modes triggered by the `sleep` instruction

FOSC3:0 fuses. When a wake-up event occurs, the system returns to the PRI_RUN mode; i.e. the CPU begins to operate, after a short (10 µs maximum) delay T_{CSD}. As the Primary oscillator never stops, crystal-mode stabilization delays are avoided.

Secondary Idle mode; IDLEN = 1, SCS1:0 = 01
The CPU is disabled, but the peripherals continue to be clocked with the Timer 1 oscillator. When a wake-up event occurs, the system returns to the SEC_RUN mode after a delay of T_{SCD}.

RC Idle mode; IDLEN = 1; SCS1:0 = 10
From the RC_RUN mode, the RC_IDLE mode is entered by setting the IDLEN bit and executing a `sleep` instruction.

All sequential digital systems must come out of their non-powered or other mal-functioning state in an orderly manner; as it were, up and running. The PIC18 family can be powered up or restarted in several different ways.

- Manually by using an external switch connected to the $\overline{\text{MCLR}}$ pin; as shown in Fig. 10.13(a).
- On application of power; as shown in Fig. 10.15.
- Where the Watchdog timer of Fig. 13.1 on p. 455 times out; due to a software bug or perhaps a glitch in the power supply.
- When a `reset` instruction is executed.

- On a stack overflow if the STVREN fuse in CONFIG4L[0] is on (default); as described on p. 180.
- Where the power supply of a normally running PIC MCU dips below a threshold; as shown in Fig. 10.17.

Looking first at the external or manual mechanism. All family members have the option of re-assigning a parallel port pin to be an active-low $\overline{\text{MCLR}}$ input. For instance, the PIC18FXX20 uses RE3 and the PIC18F1220 pin RA5. In all cases $\overline{\text{MCLR}}$ defaults to enabled, and the MCLRE fuse in CONFIG3H[7] must be cleared to 0 (MCLRE=OFF) to disable this function. In this latter case, $\overline{\text{MCLR}}$ can be used as an *input-only* parallel port pin.

Bringing $\overline{\text{MCLR}}$ below $0.2V_{DD}$ for at least 2 μs will be recognized as a legitimate reset request, and when the voltage rises above $0.8\,V_{DD}$ the processor will begin executing from the instruction at the Reset vector h'00000'. Using a switch to ground this pin, in the manner shown in Fig. 10.13(a), realizes a manual reset. The value $<40\,\text{k}\Omega$ is the maximum recommended pull-up resistor to ensure that leakage current into $\overline{\text{MCLR}}$ of $\pm 5\,\mu A$ when the switch is open, will not drop the pin voltage below $0.8V_{DD}$. The $\geq 1000\,\Omega$ series resistor gives a measure of protection, by limiting current if a negative-going noise spike breaks down the input protection diodes.

Apart from restarting code execution from the bottom of the Program store, a $\overline{\text{MCLR}}$ reset will initialize the various SFRs. Whilst many of these will be unchanged from their pre-reset state, there are important effects. For instance, if $\overline{\text{MCLR}}$ is used to awaken the processor from its Sleep or Idle states then $\overline{\text{PD}}$ will be 0; processor was powered down. If the reset was caused by the Watchdog timing out, then $\overline{\text{TO}}$ will be 0 (Watchdog Time Out); otherwise these bits will be unchanged. These status flags are located in the RCON (Reset CONtrol) SFR, as shown in Fig. 10.14.

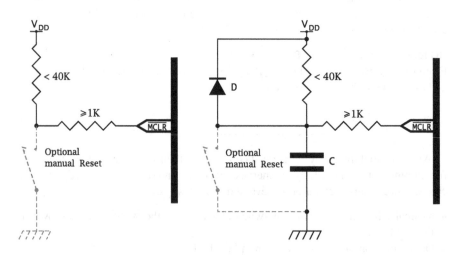

(a) *Power within specification*   (b) *Slow-rise time power supply*

Fig. 10.13 Externally resetting the PIC MCU

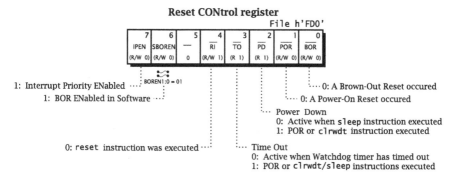

Fig. 10.14   The RCON register for the PIC18 family

Apart from the IPEN bit controlling the Priority interrupt mode (see p. 213) the RCON register holds information concerning the type and circumstances of the Reset event.

The `reset` instruction is a software instruction that mimics a $\overline{\text{MCLR}}$ event. Although by definition it cannot awaken a `sleep` induced state; all the SFR are set in an identical way. For instance, all the parallel ports are set to be inputs; that is, the TRIS registers are set to 1s. The $\overline{\text{RI}}$ **Reset Instruction** flag will be cleared in this event.

Starting a digital engine up from cold is a somewhat more troublesome task than the warm restart discussed above. Not only does the power supply and maybe oscillator need to stabilize, but initialization and synchronization of the various CPU elements needs to be established.

One solution to our problem is to hold the $\overline{\text{MCLR}}$ pin low for a sufficient time to allow the PIC MCU to settle down. This approach is shown in Fig. 10.13(b), where the capacitor holds $\overline{\text{MCLR}}$ low as it charges up while V_{DD} approaches steady state. The value of capacitor should be chosen so that the time constant CR is several times greater than that taken by the power supply to stabilize. With the resistance given, a 2.2 μF capacitor will give a time constant of approximately 100 ms. More details are given in Microchip's application notes AN522: *Power-up Considerations* and AN607: *Power-up Trouble Shooting*. The diode ensures that the capacitor rapidly discharges if the supply voltage falls significantly.

In addition to the External reset initiated at $\overline{\text{MCLR}}$, all PIC MCUs have a **Power-on Reset** (**POR**). This internal resetting mechanism automatically detects when the power supply rises beyond 0.7 V; as shown in Fig. 10.15. In this somewhat idealized situation, V_{DD} rises exponentially to its final values. Once this internal reset signal is generated, the following sequence of events is triggered.

1. A *fixed* delay T_{PWRT} (PoWer-on Reset Timer) of nominally 66 ms is generated by clocking an 11-bit counter with the internal INTRC oscillator. This delay can be by-passed if the $\overline{\text{PWRTEN}}$ fuse in CONFIG2L[0] is set from its off default value to 1 (PWRT=ON).

2. If one of the crystal modes is used; on completion of T_{PWRT} a further delay of 1024 Primary clock pulses is launched. This Oscillator Start-up timer comprises

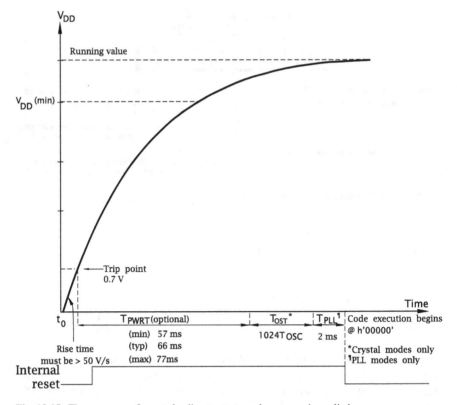

Fig. 10.15 The sequence of events leading to startup when power is applied

a 10-bit counter clocked from the internal crystal oscillator circuit. It ensures that the main oscillator has started up and is functioning correctly before processing begins. T_{OST} is dependent on the crystal frequency; for instance, a 32 kHz crystal will give a base 32 ms delay whilst a 10 MHz configuration gives a base 102 µs delay.[12] This T_{OST} is also activated when changing back to a crystal Primary run mode from a Sleep or Idle state; again to ensure that the crystal oscillator restarts and is running normally before processing commences. Using either a non-crystal mode Primary oscillator or Two-Speed Startup will side-step this delay.

3. If a HSPLL oscillator mode is used, then a further fixed T_{PLL} of nominally 2 ms is added to the T_{OSC} period to allow the Phase Locked Loop frequency multiplier time to stabilize.

4. Whenever a Power-on Reset occurs, the **POR** bit in RCON[1] is set, so that the software can determine that this is the origin of the reset action; see Table 10.3.

---

[12]32 kHz crystal oscillators have a typical start-up time of 1–2 seconds. Crystal oscillators $\geq$100 kHz have a typical start-up time of less than 10–20 ms and ceramic resonators are typically less than 1 ms. Times are voltage dependent.

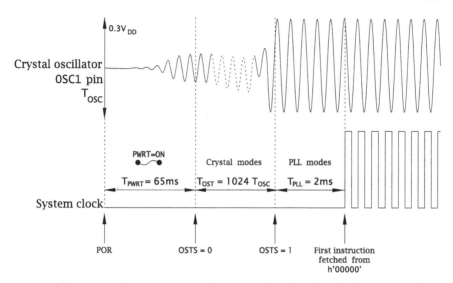

Fig. 10.16 Starting up a HSPLL mode Primary clocked processor

A POR will also set the $\overline{BOR}$ flag in RCON[0]. Both these flags should be reset to 1 in software at the end of the startup routine, so that further resets from these sources can be distinguished. In addition, the $\overline{TO}$ and $\overline{PD}$ flags are deactivated to 1 to cancel any past Sleep-induced mode or Watchdog time-out indication.

5. At the end of this flurry of activity, illustrated in Fig. 10.16, code execution commences from the Reset vector at h'00000' in the Program store.

A normally running MCU can malfunction if its power supply falls below its rated value. This could be due to a momentary blip on V_{DD} when switching in a large current load or due to battery exhaustion. In either case, the PIC MCU may operate erratically due to this **brown-out**.[13] This may have serious consequences; for instance, a dishwasher's heating element may be turned on with no water in the reservoir!

From Fig. 10.17 we see that if the **Brown-out Reset** module (**BOR**) is enabled, an internal reset will be generated if V_{DD} falls below the threshold voltage V_{BOR}.[14] In our exemplar devices, there are four user-selectable thresholds, as set by the **BORV1:0** fuses and listed in the diagram.

The diagram shows the supply subsequently rising back a little above the threshold trip voltage. Provided that the shaded time is more than 200 μs, the PoWer-on Reset Timer, if enabled, kicks in for the nominal 66 ms, before the processor comes

---

[13]The term is from the same phenomena in the mains supply that causes the lights to dim and give a brownish hue to the surroundings!

[14]If the supply falls below 0.7 V, then a Power-on Reset will occur.

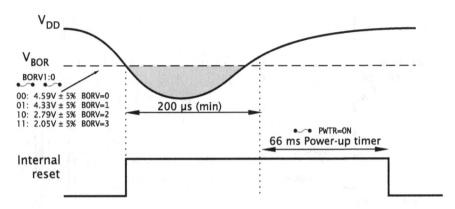

Fig. 10.17 A Brown-out reset due to a blip on the power supply

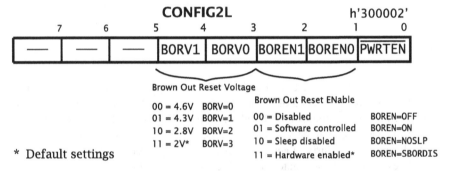

Fig. 10.18 The CONFIG2L configuration register used for Brown-out options

out of reset. Enabling the PoWer-on Reset Timer (the PWRTEN fuse) reduces the
possibility that a slowly rising V_{DD} may give rise to multiple triggers due to noise
on the supply line.

The Brown-out Reset module can be configured to act in four ways; as determined by the BOREN1:0 fuses and shown in Fig. 10.18.

BOREN1:0 = 00 config BOREN=OFF
Brown-out reset is disabled in all circumstances.

BOREN1:0 = 01 config BOREN=ON
Brown-out reset is enabled and controlled by the **SBOREN** switch bit in RCON[6].
This mode allows the executing program to determine when and if the Brown-out
Reset module can operate on the fly.

BOREN1:0 = 10 config BOREN=NOSLP
Brown-out reset is enabled, but does not operate when the process is in its Sleep
state.

The BOR module requires current to bias its Fixed Voltage Reference (FVR)
source (a 1.2 V bandgap diode), from which its V_{BOR} threshold is derived. This

Table 10.3 Reset conditions

Reset	From sleep	Execution starts at	$\overline{\text{TO}}$	$\overline{\text{PD}}$	$\overline{\text{RI}}$	$\overline{\text{POR}}$	$\overline{\text{BOR}}$
External	No	h'00000'	1	U	U	U	U
External	Yes	h'00000'	0	U	U	U	U
reset	–	h'00000'	U	U	0	U	U
Power-on	–	h'00000'	1	1	1	0	0
Brown-out	–	h'00000'	1	1	1	U	0
Stack	—	h'00000'	U	U	U	U	U
Watchdog	No	h'00000'	0	U	U	U	U
Watchdog	Yes	PC+2	0	0	U	U	U
Interrupt[a]	Yes	PC+2	U	0	U	U	U

? Not known: U Unchanged

[a]When Globally disabled; otherwise go to relevant interrupt vector

quiescent current ΔI_{BOR} of $\approx 40\ \mu A$ has to be added to the Sleep current budget, and is large compared to other components; see Table 10.1. It therefore makes sense to disable it in such circumstances.

BOREN1:0 = 11 config BOREN=SBORDIS

This default state enables the BOR module in all circumstances; that is, disables the software control of this facility.

Whenever an enabled Brown-out Reset module detects an undervoltage, the $\overline{\textbf{BOR}}$ flag in RCON[0] is cleared. However, as a POR event also activates this flag, it is recommended that both the $\overline{\text{POR}}$ and $\overline{\text{BOR}}$ flags are checked to reliably monitor a BOR event.

For both a Watchdog timer reset and a globally disabled interrupt event; when the processor is in its Sleep state, execution simply continues on with the instruction following the `sleep` instruction, rather than at the Reset vector. If the interrupt system is globally enabled, then execution will be from the appropriate interrupt vector.

Examples

Example 10.1   There are some instances where the device's internal programmable BOR trip point levels may be unsuitable for the application. Figure 10.19 shows an example of external circuitry which resets the device when V_{DD} drops below a value primarily determined by the zener diode. With the values shown, determine this trip voltage and the approximate quiescent current taken by the circuit.

Solution   The BC477 is a low-current general purpose PNP bipolar transistor with a current gain better than 100. As shown in the diagram, with a 2.7 V drop across

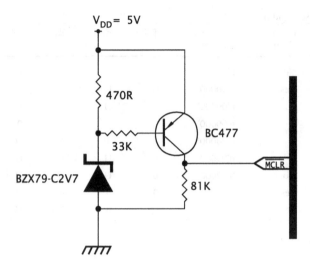

Fig. 10.19 External Brown-out protection circuit

the zener diode, the transistor is forward biased (base negative with respect to the emitter) and with the stated current gain will saturate and act as a switch. With the transistor switch on (conducting), $\overline{MCLR}$ will be $V_{DD} - V_{CE(sat)}$. The saturated emitter-collector voltage (drop across the switch) is 0.25 V at 10 mA, and thus the $\overline{MCLR}$ pin will be logic high at 4.75 V.

The current through the 81 KΩ collector resistor is $\frac{4.75}{81 \times 10^3} \approx 60\ \mu A$. Taking the base-emitter forward voltage as 0.7 V, the current through the 33 kΩ base resistor is $\frac{2.7-0.7}{33 \times 10^3} \approx 60\ \mu A$; which is very much beyond the necessary current to saturate the transistor. This leaves a considerable margin to allow for a falling value of V_{DD}. Ignoring this small base current, the current into the zener diode will be $\frac{5-2.7}{470} \approx$ 5 mA; which is the stated test current for the BZX79 zener diode.

If V_{DD} should fall to below the zener voltage of 2.7 V plus the 0.7 V required to forward bias the transistor, that is 3.4 V, then the transistor will turn off and $\overline{MCLR}$ will drop low through the 81 kΩ pull-down collector resistor, and the MCU will reset.

Example 10.2 A useful enhancement of the circuit shown in Fig. 10.19 would be the ability to switch it off whenever the MCU enters its Sleep mode. This would remove the approximately 5 mA standing current from the energy budget.

Show how the circuit of Fig. 10.19 could be modified with a minimum of additional circuit components.

Solution One possibility would be to use a spare parallel port pin to sink the zener diode and transistor current in place of a direct connection to ground. In Fig. 10.20, pin RA0 acts as the current sink. If this pin is configured as an output, and is brought low (e.g. bsf PORTA, 0) then it will take the place of the ground connection.

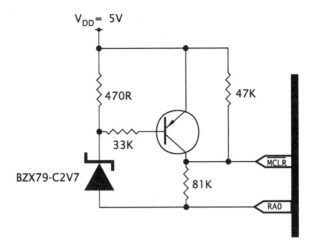

Fig. 10.20 Switchable external Brown-out protection circuit

A single port pin can sink up to 8.5 mA without the logic level rising above 0.6 V, then the 60 µA transistor collector current is well within the specification limit.

If RA0 is set high, then the transistor is switched off, as the collector is brought above the base voltage, and $\overline{\text{MCLR}}$ will be pulled high through the 47 kΩ resistor. The zener diode will not conduct and the current will be no more than the 5 µA $\overline{\text{MCLR}}$ leakage value.

When a PIC MCU is reset, all parallel port pins are automatically configured as input. In this situation the 47 kΩ pull-up resistor will turn off the Brown-out circuitry.

This technique can be used to allow the MCU to powerup and control low-current support circuitry, such as a radio transmitter, on demand. Higher power devices can be switched on and off using a buffer transistor. If several parallel port pins are used concurrently, in order to boost the controlled load, then the absolute maximum current must not exceed 200 mA or 25 mA into any single pin—see Fig. 11.7 on p. 346.

Self-Assessment Questions

10.1 In an attempt to reduce the current consumption of the circuit when in reset, a student has used a 1 MΩ resistor as a pull-up resistor in the Manual reset circuit of Fig. 10.13. Why does the PIC MCU not come out of reset?
10.2 The current consumption of a PIC MCU operating in the RC_RUN mode at 4 MHz with a V_{DD} of 5 V, is measured as 1200 µA with no loading at the port pins. What will be the current consumption if the device were to be clocked at 125 kHz and powered by a 4 V supply?

10.3 A certain 5 V-based design using the INTOSC oscillator at 1 MHz boosts the
     execution rate to 8 MHz by altering the IRCF2:0 bits in OSCCON[6:4] (see
     Fig. 10.7) from the default b'100' to b'111' during a time-critical routine. As
     an upgrade to the design, the power-supply has been reduced to 2 V. However,
     the processor seems to run correctly for a period and then lock up. Apart from
     the change in supply voltage, there have been no other changes. What might
     be the problem?

Chapter 11
One Byte at a Time

The ability of the software to activate or monitor the state of pins connected to circuitry in the outside world is the most fundamental of the various input and output capabilities provided by a microprocessor or microcontroller. These input/output pins are generally gathered in groups of up to the size of the internal Data bus. In the PIC18 MCU family these **parallel ports** allow up to eight bits of external data to be directly read into or sent out of the processor core *one byte at a time*. The total number of such parallel lines available on any specific family member depends on the package footprint and on how much shared resources are used. This parallel-pin budget varies from up to 16 for the 18 pin PIC18F1220 through 36 for the 40-pin PIC18F4520, to a maximum of 70 for the 80-pin PIC18F8410.

When you have completed this chapter you will:

- Appreciate the function of a parallel input/output (I/O) port.
- Be able to configure an I/O port line.
- Understand the structure of a parallel I/O port and differentiate between a LATch and PORT register.
- Comprehend how read–modify–write instructions interact with parallel I/O ports.
- Appreciate the electrical and power characteristics of an I/O port.
- Know how to enable internal port pull-up resistors.
- Understand how the function of the Change in Port B interrupt operates.

Conceptually a parallel I/O port can be considered as a File with its contents visible to and accessible by the outside world. This somewhat simplified view is represented in Fig. 11.1, which is based on a magnified section of the Data store shown in Fig. 4.10 on p. 83. Port A bits 7:6 are shown dotted, as their respective pins RA7:6 are only available for certain fuse selectable oscillator options; see Fig. 10.9 on p. 316. In the specific case of the 18-pin PIC18F1220, pin RA5 defaults to the $\overline{\text{MCLR}}$ function, but can be fuse enabled to act as an *input* to PORTA[5]. Other devices use different port bits to share with $\overline{\text{MCLR}}$; for instance, RE3 in the PIC18FXX20 series. However, in all cases the shared pin can only ever act as a parallel port input. As we see from Fig. 4.1 on p. 71 and Fig. 10.1 on p. 306, in practice all device pins are shared between other functions. For instance, RB0

S. Katzen, *The Essential PIC18® Microcontroller*,
Computer Communications and Networks,
DOI 10.1007/978-1-84996-229-2_11, © Springer-Verlag London Limited 2010

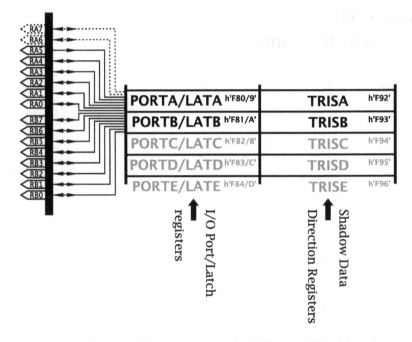

Fig. 11.1 A boiled-down view of the enhanced-range parallel Ports A and B

also functions as the `INT0` interrupt input. All PIC18 enhanced-range family members have at the very least the depicted 16-port lines arranged as two 8-bit parallel ports.

The 28-pin devices, such as the PIC18F2420/2520 have an additional Port C; with 40-pin variants, such as the PIC18F4420/4520 pair, adding an 8-bit Port D and 4-bit Port E (shown grayed out in the diagram). Larger footprint members have further parallel ports, which are also shared with other facilities. Some of these are listed in Table 11.1.

Despite the depiction of Fig. 11.1, an I/O port does not behave quite like any other internal File. For instance, each parallel port bit can be configured individually to either reflect the electrical state of its associated pin or else drive that pin high or low to follow the pin's logic state. That is, each port bit has to be set either to *read* the state of its associated pin or be able to *write* to that pin. Furthermore, we need to determine how this configuration interferes with the action of software that tries to alter or read the state of the port.

In Fig. 11.1 each of the parallel ports is shown paired with a **TRIS** register. Fig. 4.10 on p. 83 this pairing is seen to be a characteristic of all such ports. Most microcontrollers label these as Data Direction registers, but Microchip enigmatically use the term TRIS as short for TRI-State, for reasons we will see later in the chapter. Each bit *n* in a parallel port has a shadow bit *n* in its TRIS register, whose

Table 11.1 Overview of several 28+ pin PIC18 MCU's parallel I/O provision

Port	Size	Characteristics
A	5–8 I/O	RA4 is often an open-drain output and common with Timer 0's input
		Shared with analog modules
B	8 I/O	RB0 : 2 is shared with Hardware interrupts. Weak pull-up resisters
		RB7 : 4 can generate a Changed interrupt
C	8 I/O	28 pin+ PIC MCUs shared with Serial ports
D	8 I/O	40 pin+ PIC MCUs shared with Parallel Slave port
E	4–8 I/O	40 pin+ PIC MCUs shared with analog modules
F	8 I/O	68 pin PIC18F6310, shared with mixed functions
G	5 I/O	68 pin PIC18F6310, shared with mixed functions
H	8 I/O	80-pin PIC18F8310, shared with external memory address/data
J	8 I/O	80-pin PIC18F8310, shared with external memory control

function is to configure the associated pin either as an input (TRIS[n] is 1) or an output (TRIS[n] is 0).[1]

As an example, consider a situation where pin RA0 and pins RB[7 : 0] are to be outputs and the rest of Port A are to be inputs. The following code fragment normally appears at the beginning of the main routine; see Program 11.1.

```
movlw b'1111110'    ; Pin RA0 = Output
movwf TRISA         ; Rest of Port A pins are Inputs
clrf  TRISB         ; All Port B pins are Outputs
```

In **C** we could ape the assembly-level code above; for instance, in our exemplar CCS **C** language:

```
#byte TRISA  = 0xF92 /* The TRISA Data Direction register */
#byte TRISB  = 0xF93 /* The TRISB Data Direction register */
main()
{
TRISA = 0xFE;        /* Pin RA0 = Output, rest are Inputs */
TRISB = 0;           /* All PortB are Outputs            */
```

However, some compilers may come with built-in functions to support port set-up and usage. In the case of the CCS compiler we have the function set_tris_x() for Port X, giving:

---

[1] *Aide-mémoire*: 0 for Output and 1 for Input.

```
main()
{
/* Define any variables first                              */
set_tris_a(0xFE);    /* Pin RA0 = Output, rest are Inputs */
set_tris_b(0);       /* All PortB are Outputs             */
```

Any reset will set all TRIS bits to 1; that is, after a reset all pins will come on stream as inputs. The choice of a starting configuration as input is deliberate, as if a pin were set to output before the software has had a chance to set the port pin to its initial state, then such a pin will come out of reset with an unpredictable voltage state. This could activate any driven circuitry in an undesirable manner. For instance, a latch actuating a switch turning on the heating element of a washing machine may be triggered before any water is in the tank. Where this kind of catastrophe could occur, the state of the appropriate port bits should be set to their appropriate initial value *before* configuring the TRIS registers.

Once the directional properties of a port's pins have been set-up, then the software can either read from or write to a port in a comparable manner to a normal File and hence interact with the outside world. Specifically:

- To *monitor* the state of any pin set as an input, use the btfss or btfsc instruction. For instance, btfss PORTA,1 skips if pin RA1 is High (that is, if PORTA[1] is set to logic 1). Several pins at a time can be read by copying the complete File into W; e.g., movf PORTA,w. If required, the byte could be copied into a GPR File for further processing; e.g. movff PORTA,h'030'.

- To *change* the state of any pin set as an output, use the bcf, bsf or btg instruction. For instance, bcf PORTA,0 will force pin RA0 to its Low state (that is, PORTA[0] is cleared to logic 0). Several pins at a time can be changed by copying the contents of a File to the port. For instance, if all Port B's pins are set as outputs, then to bring pins RB[7:6] to its High state and RB[5:0] Low we have movlw b'11000000' : movwf PORTB.

We will look at the electrical characteristics of ports later in the chapter; for instance, what happens if you read a pin that is set to be an output? First we will illustrate the usage of parallel ports with two examples.

When the temperature of an oven reaches the set point, a LED is to continually blink. The LED is connected to pin RA0 and the thermostat switch to pin RB7; as shown in Fig. 11.2. The thermostat opens when the set temperature is reached or exceeded and the blink duration in total is to be nominally 200 ms. The processor is to be a PIC18F1220 running at 4 MHz and using a 5 V supply.

Figure 11.2 shows the hardware implementing the oven display. The LED is connected to pin RA0 via a current limiting resistor. When RA0 is high the diode is forward biased and the resulting current generated light. When Ra0 is low no current flows and the LED is off. Using the typical forward biased junction voltage drop for a LED of 2 V and an operating current of 20 mA gives a value for the resistor of $\frac{5-2}{20\times10^{-3}} = 150\,\Omega$.

The thermostat is a bi-metallic switch connected to pin RB7 and with one end grounded. When below the set temperature the switch is closed then RB7 is low.

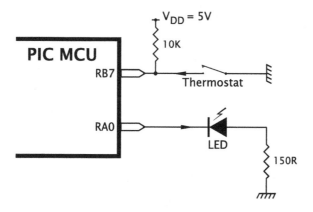

Fig. 11.2 Pulsing an LED when temperature setting reached

With an over-temperature situation, the open switch results in RB7 being pulled high through the pull-up resistor. The value of this resistor is not critical. If it is too low then the resulting current through a closed thermostat will be unnecessarily high and wasteful. Conversely if it is too high, leakage current from RB7 will cause the pin voltage to drop, even with the thermostat open. Also, very high resistance values are prone to pickup noise voltages from electro-magnetic interference; for instance, from the heating elements. A value between nominally 10 and 100 kΩ is suitable.

Program 11.1 shows an assembly language implementation of the software. This listing can be broken-down into three phases.

Initialization
This phase selects appropriate configuration options, sets up the parallel port to act a digital I/O and sets the two pins as input or output.

- As we don't require precision timing, the internal oscillator is used as the Primary clock source, as configured with the OSC = INTIO67 FOSC3:0 fuse setting. The IRCF2:0 bits in the OSCCON register are set to b'110' to give the required 4 MHz frequency—see Fig. 10.7 on p. 314.
- The Watchdog timer is disabled.
- Several Ports A and B pins share with the ADC module and on a Power-On Reset default to analog inputs. To change their function to digital, the ADCON1 register has to be set to b'11111111'—see Fig. 14.12 on p. 510.
- Pin RA0 is configured as an output to drive the LED, by clearing TRISA[0]. All the other pins default to inputs on a POR.

Switch monitoring
The state of the RB7 pin is monitored by testing bit 7 of PORTB and skipping out if that bit goes to 1; reflecting pin RB7 going high.

LED pulsing
The LED is controlled by bringing bit 0 of PORTA to logic 0 (pin goes low and the

Program 11.1 Set temperature indication

```
        #include "p18f1220.inc"
; Fuses for INTernal OSCillator with RA6:7 free
        config  WDT=OFF,OSC=INTIO67 ; and no Watchdog timer
        org     0              ; Code starts @ Reset vector

STAT    equ     7              ; Thermostat is bit 7 of PORTB
LED     equ     0              ; LED is bit 0 of PORTA

; First initialize ports --------------------------------------
MAIN    movlw   b'01100000' ; 1st set up the internal oscillator
        movwf   OSCCON         ; to 4MHz
        setf    ADCON1         ; Make ports digital (default analog)
        bcf     PORTA,LED      ; Start with LED off
        bcf     TRISA,0        ; Make pin RA0 (LED) an Output

; Now continually monitor thermostat & activate LED if needed -
M_LOOP  btfss   PORTB,STAT  ; Skip IF open (high)
        bra     M_LOOP         ; ELSE try again

; This is the blinking routine -------------------------------
        bsf     PORTA,LED      ; Turn on LED
        rcall   DELAY_100MS ; for 100ms
        bcf     PORTA,LED      ; and off
        rcall   DELAY_100MS ; for 100ms
        bra     M_LOOP         ; & go again while thermostat open
```

LED is off) and logic 1 (pin goes high and the LED is on). The 100 ms subroutine DELAY_100MS of Program 6.3 on p. 170 is used to create the 100:100 ms cycle for the flash. After each 200 ms pulse, the thermostat is again checked.

As an alternative, Program 11.2 gives a CCS **C** coded equivalent. The program algorithm follows that of the assembly-level equivalent.

Initialization

- The #fuses directive (see p. 318) used with the INTRC_IO parameter specifies the Primary clock as the INTOSC, with pins RA6 and RA7 free for digital I/O— see p. 313.
- The set-up_oscillator() function used with the parameter OSC_4MHZ, configures **OSCCON** to give the specified 4 MHz clock rate.
- The #fuse parameter NOWDT disables the Watchdog timer (actually allowing software control—see p. 454).
- The set-up_adc_ports() with the parameter NO_ANALOGS makes all shared analog/digital pins all digital—see Fig. 14.12 on p. 510.
- The set_tris_a(0xFE); function call sets **TRISA** to b'11111110', effectively making pin RA0 an output to drive the LED.

Program 11.2 Oven monitoring in **C**

```
#include <18f1220.h>
#use delay (clock=4000000)
#byte PORTA = 0xF80          /* Port A address                   */
#byte PORTB = 0xF81          /* Port B address                   */
#bit  LED   = PORTA.0        /* LED is connected to RA0          */
#bit  STAT  = PORTB.7        /* Thermostat is connected to RB7   */

#fuses NOWDT, INTRC_IO

void main(void)
{
set-up_oscillator(OSC_4MHZ); /* Set internal oscillator to 4MHz */
set_tris_a(0xFE);            /* Set RA0 to be an output          */
set-up_adc_ports(NO_ANALOGS);/* All parallel pins are digital    */

while(1)                     /* DO forever                       */
    {
    if(STAT == 0)
    {LED=0;}                 /* Keep LED off as long as RB7 is 0 */
    else
        {
        LED = 1;             /* 'stat has opened (RB7 was high)  */
        delay_ms(100);       /* LED on for 100ms                 */
        LED = 0;             /* LED off                          */
        delay_ms(100);
        }
    }
}
```

Switch monitoring

If the thermostat is closed, the if operand STAT == 0 will be true and the LED will stay off indefinitely (LED = 0). STAT has been defined as PORTB.7, reflecting the state of pin RB7 and LED as PORTA.0.

Else the LED is turned on (LED = 1) for 100 ms, using the delay_ms() CCS **C** function and then off for the same period. After this 200 ms blink, control is returned to the beginning of the endless loop (while(1)) and the thermostat checked again.

As a slightly more sophisticated example; consider the situation shown in Fig. 11.3, where any external peripheral device (maybe a printer) wishes to read the byte contents of File h'020' via Port B on request, every time it brings pin RA1 to its Low state. This signal from the peripheral is labeled $\overline{RFD}$ (Ready For Data). When the PIC MCU responds some time later, it copies the datum to Port B and then it pulses ⎍ pin RA0, which is labeled $\overline{DAV}$ (Data AVailable), to inform the peripheral that the datum is now available. On a Power-on Reset, Port B pins are to be in their Low state and RA0 is to be High.

Signaling in this manner using semaphores, such as $\overline{RFD}$ and $\overline{DAV}$, is known as **handshaking**. The term comes from the protocol when beginning and ending a conversion. Handshaking allows separate non-synchronized devices to converse with each other without missing data.

Program 11.3 shows how the handshake is implemented in assembly code. As in Program 11.1, the two parallel ports are set to be all digital, rather than analog.

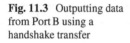

Fig. 11.3 Outputting data
from Port B using a
handshake transfer

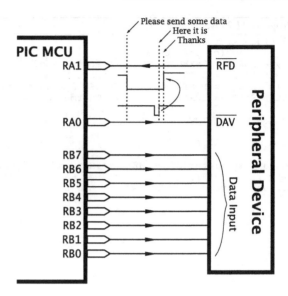

Preceding the port direction configuration code, the initial state of Port A and B are set-up by writing the appropriate pattern to each port. Once this is done, the port pins set to output will take on the initial state of RA0 High (corresponding to bit 0 of PORTA being 1) and pins RB[7:0] Low (corresponding to all PORTB bits being 0).

After this initialization code, the state of pin RA1 is checked for a Low-state voltage, which reflects as a 0 in bit 1 of PORTA. When this occurs, the datum in File h'020' is copied out to PORTB via the Working register and pin RA0 dropped Low. A single nop intervenes before it is brought High again, to give an extra cycle's delay. The specification did not give a duration of the $\overline{\text{DAV}}$ pulse, so in practice nop would be replaced by a call to a delay subroutine; say, to give a 10 μs delay.

Finally, the state of pin RA1 is again monitored (as imaged in bit PORTA[1]) until the peripheral brings its $\overline{\text{RFD}}$ line High to indicate that the transaction is over. Of course, this is a potential hazard; as if the peripheral fails to respond, the PIC MCU will hang. It would be safer to have a time-out; perhaps if there is no response after 65,536 tries then go to some error handling subroutine.

Program 11.4 gives an equivalent routine using CCS **C**. This follows a similar structure, but notice in particular how an input pin is tested using constructions like while(RFD == 0) {;} which does nothing ({;} is a null statement) as long as it is true that the pin named RFD is 1. When RFD does go to 0, that is, pin RA1 goes Low, then the loop exits to the next statement.

The built-in function delay_cycles() gives an additional 1-cycle delay, and it can be replaced by an appropriate delay function if this is not satisfactory; e.g. delay_us(10) for 10 μs.

These programs may seem rather useless, as the datum in File h'020' is never set-up or changed. However, in a real situation the value could be changed if an interrupt occurs, maybe on a regular basis as dictated by an internal or external timer. This

Program 11.3 Implementing a parallel port handshake data transfer using assembly-level code (15 instructions)

```
       include   "p18f1220.inc"
       config  WDT = OFF    ; No Watchdog timer
       org     0

       cblock  h'020'
        DATUM:1              ; File h'020'
       endc

; Initialize ports and set up the pins ------------------------
MAIN    setf    ADCON1       ; Make ports all digital
        clrf    PORTB        ; Starting value of Port B is all 0s
        clrf    TRISB        ; Port B is all Outputs
        bsf     PORTA,0      ; Initial state of DAV is 1
        movlw   b'1111110'   ; Pin RA0 = Output
        movwf   TRISA        ; Rest of pins are Inputs

; Monitor pin RA1 looking for a low voltage -------------------
RFD_YES btfsc   PORTA,1      ; Bit 1 of Port A: Is it 0?
        bra     RFD_YES      ; IF not THEN try again

; Copy the requested Datum to Port B --------------------------
        movff   DATUM,PORTB  ; Copy File h'20' contents to Port B

; Now pulse the DAV pin RA0 low to signal "Here it is" --------
        bcf     PORTA,0      ; DAV (pin RA0) low
        nop                  ; for a short time
        bsf     PORTA,0      ; and then high

; Now hang around until the RFD signal goes high --------------
RFD_NO  btfss   PORTA,1      ; Skip if RA1 is high
        bra     RFD_NO       ; IF not THEN keep trying

        bra     RFD_YES      ; Repeat forever
```

could trigger an analog-to-digital converter module, which dumps its outcome in this holding File. We will be looking at Timer and ADC modules in subsequent chapters.

In order to fully understand the characteristics of parallel I/O ports we need to look at its hardware implementation. A somewhat simplified version of a single I/O port bit n together with its associated Data Direction bit is shown in Fig. 11.4. The two key elements in this circuit are the Data D flip flop and Data tristate (3-state) buffer.

- *Writing* to this port will trigger the Data D flip flop, causing the data on the internal Data store line to be clocked in and held as long as the MCU is powered—see Fig. 2.16(c) and (d) on p. 31. For instance:

Program 11.4 Implementing a parallel port handshake data transfer using CCS **C** code (32 instructions)

```
#include <18f1220.h>

#byte DATUM = 0x020          /* File 0x20 holds the datum byte  */
#byte PORTA = 0xF80          /* Port A address                  */
#byte PORTB = 0xF81          /* Port B address                  */
#bit  DAV   = PORTA.0        /* Pin RA0 is the \DAV line        */
#bit  RFD   = PORTA.1        /* Pin RA1 is the \RFD line        */

#fuses NOWDT

void main(void)
{
set-up_adc_ports(NO_ANALOGS);/* All parallel pins are digital   */
DAV   = 1;                   /* Start with \DAV line not active */
PORTB = 0;                   /* Start value of Port B is all 0  */
set_tris_a(0xFE);            /* Pin RA0 (\DAV) is Output        */
set_tris_b(0);               /* All Port B pins are Outputs     */

while(TRUE)                  /* DO forever loop                 */
    {
    if(RFD == 1) {;}         /* Wait until \RFD goes FALSE (low)*/
    else
        {
        PORTB = DATUM;       /* Copy Datum out to Port B        */
        DAV = 0;             /* \DAV (pin RA0) low              */
        delay_cycles(1);     /* For a short time                */
        DAV = 1;             /* and then high                   */
        while(RFD == 0)      /* Hang around until \RFD goes low  */
            {;}              /* Null statement means do nothing */
        }
    }
}
```

```
movlw  b'11110000'      ; Pattern in Working register
movwf  h'F81'           ; Send to Port B (File h'F81')
```

will set the top four Data flip flops in Port B to logic 1 and the bottom four to logic 0.

Setting the port bits will occur irrespective of whether its associated I/O pins are configured as input or output. However, to pass the flip flop's state through to the I/O pin, the TRIS (TRIState) buffer must be enabled. In this situation, as shown in Fig. 11.6(b), the Data flip flop is directly connected to the outside world.

- *Reading* from this port enables the Data buffer, causing the state of the staticizer latch[2] to be gated through to the internal Data store line. When the port is idling, i.e., not being read, the D latch is transparent and its output follows the state

---

[2]There is no staticizer latch in the low-range 12-bit family.

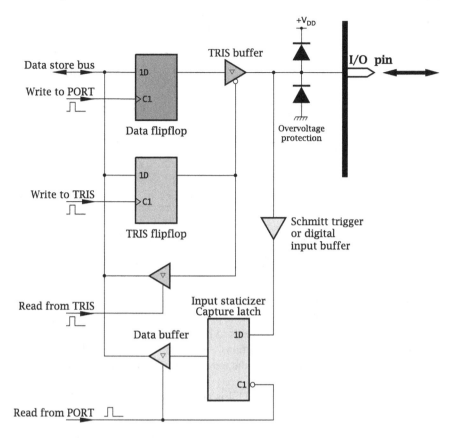

Fig. 11.4 A rudimentary generic I/O port line

of the pin—see Fig. 2.16(a) and (b) on p. 31. When the port is being read, the D latch clock enable goes High and the data into the 3-state Data buffer is frozen, effectively holding its state constant while being read; that is, staticizing it. For instance, to read the state of Port B we have:

```
movf   h'F81',w    ; Read all eight input PortB lines into W
```

This reading action, shown in Fig. 11.6(a), will occur independently of whether the associated I/O pin is configured as an input or output.

From Fig. 11.4 we see that a TRIS bit can be read from as well as written to. Although this may be rather useless, consider a programmer wishing to alter pin RB7 to an output; for instance, see p. 406.

```
bcf   h'F93',7      ; Clear bit 7 of TRISB
```

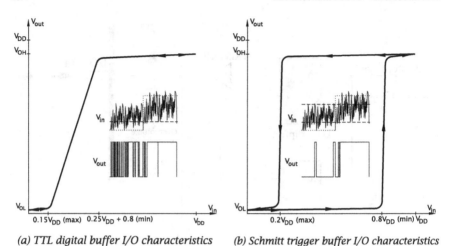

(a) TTL digital buffer I/O characteristics   (b) Schmitt trigger buffer I/O characteristics

Fig. 11.5 Input buffer voltage transfer characteristics; showing the response to a noisy input signal

bcf (Bit Clear File) is an example of a **read-modify-write** instruction (see p. 121) whereby the state of TRISB is *read* into the processor, modified and then *written* out to TRISB. To do this, the processor needs to be able to both read from and write to the File. In fact nearly all write instructions, such as movwf, do a read cycle before the key write cycle. The exception is the movff instruction which dispenses with the destination read.

The voltage at the port pin in Fig. 11.4 is buffered from the Capture latch input. Two types of buffer are used.

(a) The standard TTL digital buffer shown in Fig. 11.5(a) is basically a high-gain analog amplifier which rapidly saturates if the input voltage rises above V_{IL} (maximum of $0.15V_{DD}$ or 0.8 V if $V_{DD} > 4.5$ V). The minimum V_{OH} is $0.25V_{DD} + 0.8$ for $V_{DD} < 4.5$ V and 2 V otherwise.

This type of buffer is satisfactory for well behaved logic signals; that is, with fast rise times and low noise. Most[3] PIC18 devices use this type of buffer for Ports A & B.

(b) All ports above Port B use Schmitt trigger buffers. Such buffers have hysteresis coupled with a snap action. As shown in Fig. 11.5(b), when the input voltage is rising, the output will not respond until it reaches the upper threshold, which is around 80% of V_{DD} (4 V for a 5 V supply). Should the input subsequently fall the output will not respond until it drops below 20% of V_{DD}, or 1 V for a 5 V supply. This is due to positive feedback around the analog buffer amplifier, and this also gives a magnified gain, or slew rate, when the response does come.

This snap action lowers the risk of oscillation where the input voltage has a slow rate of change. The hysteresis gives a better noise immunity. For this reason RA4, which is shared with the Timer 0 input, always uses a Schmitt buffer to

---

[3] Exceptionally, the PIC18F1220 only uses these buffers on Port B.

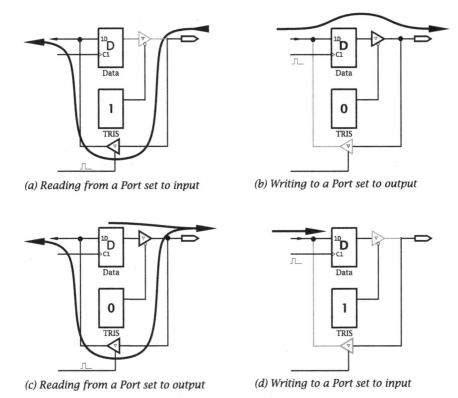

(a) Reading from a Port set to input   (b) Writing to a Port set to output

(c) Reading from a Port set to output   (d) Writing to a Port set to input

Fig. 11.6 Reading from and writing to a port bit with linked I/O pin set to input or output

help avoid Timer 0 counting multiple instances for the one logic cycle. Indeed, all shared digital functions also use this type of buffer; for instance, the interrupt inputs INT0, INT1 and INT2 which share with the RB0, RB1 and RB2 pins respectively.

Because a parallel port may be configured as an input, or output, or a mixture of both, it is important to know what restrictions are introduced when reading or altering the state of such special Files. For instance, what would happen if the software read from a port bit which has been configured as an output? The four possibilities enumerated in Fig. 11.6 are:

(a) **Reading from a port pin set as input (TRIS = 1)**

Here the TRIS buffer is disabled and the state of the Data flip flop remains unchanged. For instance, movf h'F81',w reads the state of Port B's *input pins* into the Working register.

(b) **Writing to a port pin configured as an output (TRIS = 0)**

Here the TRIS buffer is enabled and the Data flip flop altered by the processor writing to the port. The state of this flip flop is imaged at the output pins. For instance, if all Port B pins RB[7:0] are set as output, movlw b'10101010' followed by `movwf h'F81'` sets the Port B pins to HLHLHLHL (H = High, L = Low).

(c) **Reading from a port pin configured as an output (TRIS = 0)**

In this situation the TRIS buffer is enabled and so the applicable I/O pins are driven from their associated Data flip flop. Normally, reading port pins set to output will effectively copy their flip flop and associated pin states into the CPU.

(d) **Writing to a port pin configured as input (TRIS = 1)**

In this situation the state of the Data flip flop will be altered in the appropriate manner. However, as the TRIS buffer is disabled, any change will not be reflected at a linked I/O pin until the direction of the port pin is subsequently changed to output. This ability to set-up the state of a port in a manner invisible to the outside world was used in Program 11.3 to initialize the parallel ports after reset and before any pins are set to output. Remembering that on reset, all ports are set to input; in other words, all TRIS registers are set to b'11111111'.

The basic parallel port structure of Fig. 11.4 is actually that of the PIC16 family. The structure used in the PIC18 family, shown in Fig. 11.8, has been slightly enhanced to avoid one serious failing of this circuit. When the TRIS buffer is enabled (that is the pin is set to be an output), the voltage actually read is that of the pin. In a well designed circuit, the pin voltage corresponds to that of the TRIS buffer. However, if the current sourced or sunk by a pin is too high, then an erroneous logic level can be sensed.

As an example, consider the situation of Fig. 11.7(a) where a bipolar transistor's base is directly connected to a port pin. This will take sufficient current from the TRIS buffer to drag the pin voltage to ≈0.7 V; the forward conducting voltage of a typical transistor base-emitter.[4] The situation in Fig. 11.7(b) is similar, with current flowing through the light-emitting diode (LED) into the port pin,[5] and the TRIS buffer will be pulled up to ≈3 V, assuming a conducting LED offset of 2 V.

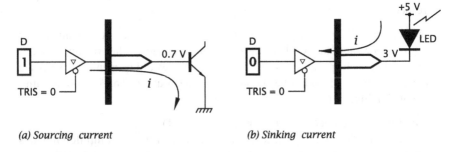

(a) Sourcing current   (b) Sinking current

Fig. 11.7 Sinking and sourcing current

---

[4]Typically somewhere between 25 mA and 45 mA (see Example 11.1). Note the 25 mA maximum rating for any one pin limit!

[5]Typically around 45–80 mA; see Fig. 11.22.

In these situations the outcome of reading a port pin set as output is often not the state of that port bit's Data flip flop, due to the improper voltage levels. For instance, btfsc PORTB, 7 in purporting to skip if bit 7 of Port B is zero may fail to function as expected if the linked pin RB7 is sinking or sourcing too much current.

In some cases the effect of overloaded pins can be rather bizarre. Consider the situation where pin RB6 is to be set high; e.g. bsf PORTB, 6. Unfortunately, any instruction[6] that writes to a File will first read the data—see p. 121. In our instance, *all* eight bits of Port B will be read, bit 6 will be set to 1 and the modified byte sent out again. However, overloaded pin RB7 well result in bit 7 being read as 0, and this is the value sent out. Thus, twiddling bit 6 (maybe to light a LED) also turns off the transistor!

Unintended interactions can also occur where several changes are made to the state of port pins in quick succession. In some cases, resistor-capacitor transients delays may cause misreading of pin voltages. Example 11.5 gives an instance of this difficult to diagnose phenomena.

To avoid the unintended consequences of badly behaved pin voltages, the PIC18 family port introduced an additional 3-state buffer to enable the processor to read the state of the port Data bits directly, rather than the pin voltages. This alternative vision of Port*n* is named LAT*n*. For instance, LATB at h'F8A' is the counterpart to PORTB at h'F81'. Writing to LAT*n* is identical to writing to PORT*n*. In most cases reading will give exactly the same results, but using LAT*n* is safer in more extreme cases.

The majority of ports have TRIS buffers implemented as shown in Fig. 11.9(a) which uses a series N-channel/P-channel field effect transistor totem-pole structure.

1. When the TRIS flip flop is logic 1 the lower AND gate has a logic 0 output and the upper OR gate has a logic 1 output. In this situation, both transistors TRN and TRP are non-conducting and the state of the Data flip flop is isolated from the I/O pin. The port pin is thus isolated from the linked Data flip flop and is configured as an input.
2. When the TRIS flip flop is logic 0 then the complement state of the Data flip flop is gated through to both totem-pole transistors. With D Low, TRN conducts and TRP is off, giving a Low pin voltage.
3. When the TRIS flip flow is logic 0 and D high, TRP conducts and TRN is off giving a High pin voltage.

In the latter two states, the pin (normally) follows the state of the Data flip flop, with current being sourced or sunk through the relatively low resistance active conducting transistors.

These *three states*; namely off, low and high, give the buffer its **tristate** adjective—see Fig. 2.4 on p. 20. Microchip name their Data Direction registers TRIS registers from these interface buffers.

As an example, consider the situation where an electromagnetic relay is to be activated, requiring a 200 mA activation current at 12 V. For currents and voltages

---

[6]Except movff.

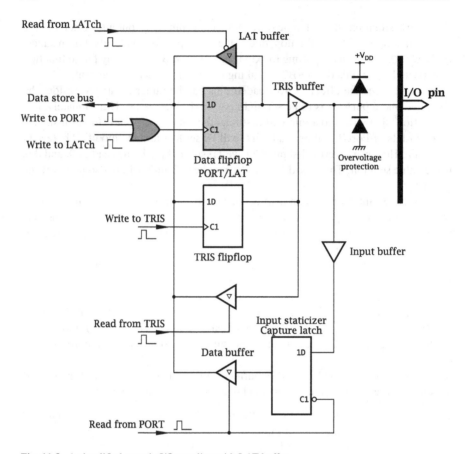

Fig. 11.8 A simplified generic I/O port line with LAT buffer

of this magnitude we need external buffering. In Fig. 11.9(b) a bipolar transistor acts as an external switch. If the minimum gain of this transistor is 100, then a 1.8 kΩ resistor will give a base current of 2 mA, assuming a base-emitter conduction voltage of 0.7 V and a PIC MCU V_{OH} of at least 4.3 V.

Most PIC16 and PIC18 devices[7] implement the TRIS buffer driving pin RA4 shown in Fig. 11.9(c) in a somewhat different manner, in that only the bottom totem-pole transistor is implemented. As opposed to the 3-state structure of Fig. 11.9(a), this structure has only two states; that is, active logic 0 and open-circuit. This type of output is known as **open drain** (or open-collector)—see Fig. 2.3 on p. 19.

1. When the TRIS flip flop is logic 1 (its reset state) then the AND output is Low and TRN is off with the output pin high resistance. The buffer is effectively off and RA4 can be read without any interference from the associated Data flip flop.

---

[7]But not the PIC18F1220/1330 device.

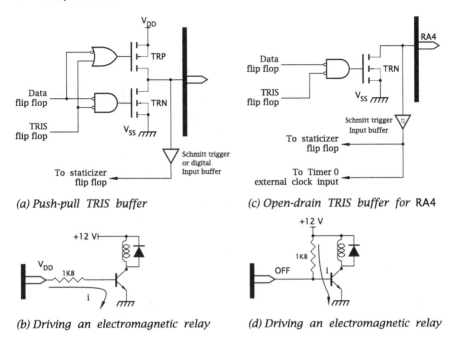

(a) Push-pull TRIS buffer

(c) Open-drain TRIS buffer for RA4

(b) Driving an electromagnetic relay

(d) Driving an electromagnetic relay

Fig. 11.9 Output driver structures

2. When TRIS flip flop is logic 0, the output transistor conducts when the Data flip flop is logic 0, giving an active-Low output. When the Data is logic 1, TRN is off and the output pin floats.

An open-drain output cannot source current; either the load itself must be connected from the output pin to a positive voltage or an external pull-up resistor used as a load for the on-chip transistor. This is the case in Fig. 11.9(d), where the base current for the external transistor is derived from the 1.8 kΩ pull-up resistor when RA4 is off.

If RA4 is to be used as the Timer 0 input, it is usually configured as an input. If configured as an output, then in this situation RA4 must be set to logic 1, which will disable the open-drain transistor and prevent interaction between it and the external clock input to the Timer. Notice that the input buffer for pin RA4 is *always* a Schmitt trigger, to reduce the chance of noise and oscillations being counted by the Timer.

Any pin set-up to be an output has to be able to carry an appropriate current to activate the driven load. In most situations a port pin configured as an output will only be required to source or sink a few milliamps of load current. Nevertheless, it is important to be aware of the limitations of the drive capabilities of port output pins.

Fig. 11.10 Power dissipation
model

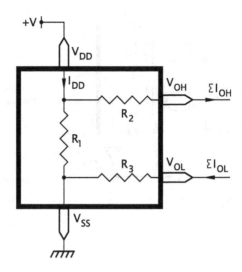

Generally two situations are tabulated in a device's data sheet.

1. Sink current I_{OL} into a pin when an output is in a Low state should not exceed +8.5 mA if the Low voltage V_{OL} is not to rise above 0.6 V.
2. Source current I_{OH} out of a pin which is High should not exceed −3 mA if the High-state voltage is not to drop more than 0.7 V below V_{DD}. The negative current denotes source; i.e., out of the device.

Larger currents may be sourced or sunk, as in Fig. 11.7, if degradation of logic levels are acceptable, subject to an absolute limitation that it must be within the range ±25 mA for any single I/O pin to avoid damage. Where more than one I/O pin is involved in driving current, an overall global limit must be observed. The maximum sunk or sourced from all the ports together should not exceed 200 mA.

Each output pin in sourcing or sinking current will dissipate energy, which appears in the package as heat. From the simplified model of Fig. 11.10 we see that there are three components to this dissipation that are modeled as resistors.

1. From the V_{DD} power pin we have a current I_{DD}. However, the current through the body resistor R_1 is less by the source currents from the port pins, giving a $V \times I$ dissipated power of $V_{DD} \times (I_{DD} - \sum I_{OH})$.
2. The voltage drop across the equivalent resistance R_2 between output pins and the power pin is $\Delta V = V_{DD} - V_{OH}$. Thus the power dissipated is $\Delta V \times \sum I_{OH}$.
3. Current sunk into the output pins through R_3 to ground via the V_{SS} pin dissipates $V_{OL} \times \sum I_{OL}$.

Adding these components gives the formula quoted in the data sheet:

$$P_{DIS} = V_{DD} \times \left(I_{DD} - \sum I_{OH} \right) + \sum \left((V_{DD} - V_{OH}) \times I_{OH} \right) + \sum (V_{OL} \times I_{OL})$$

but note that the output voltages at each pin will differ with different currents. The figure given in the data sheets for P_{DIS} is 1 W. In all cases the maximum current into the V_{DD} power pin should not exceed 250 mA and 300 mA out of the V_{SS} pin.

In practice, the equivalent resistance R_2 is not linear and varies in a rather complex way; as illustrated in Fig. 11.16. That is, V_{OH} does not drop with current in a straight line. Data sheets show graphs of this current–voltage relationship; for instance, see Figs. 11.16 and 11.22. However, a worst-case scenario with large currents would be to assume that V_{OH} had dropped to zero and V_{OL} had been pulled up to the supply V_{DD}. In this situation the excess of I_{DD} over $\sum I_{OH}$ supplying the CPU and other peripheral modules would be minimal and could be ignored. The total power dissipated would then be:

$$P_{DIS} = V_{DD} \times \left(\sum I_{OH} + \sum I_{OL} \right).$$

Many applications involve reading the state of arrays of switches. Rather than use the relatively more expensive single-pole double-throw (SPDT) switch arrangement of Fig. 11.11(a) to give the two logic states, most switches—for instance, those in the keypad of Fig. 11.13—are single-pole single-throw (SPST) types. In these situations an external pull-up resistor is needed to convert the open state to a High-state voltage; as shown in Fig. 11.11(b). A similar situation arises when open-drain/collector electronic devices, such as phototransistors, are to be read by a port. The value of such pull-up resistors should not be too low, as a large current will flow through the switch when closed, nor too high, or else noise will be induced by electromagnetic means from external sources. A good compromise is in the range 10–100 kΩ.

In order to simplify the interface of such devices, Port B *inputs* have optional internal pull-up resistors. These internal resistors are called **weak pull-ups**, as their typical equivalent values of around 20 kΩ is high enough not to interfere with devices being read which have 'normal' logic Low and High outputs.

We see from Fig. 11.12 that the internal pull-up resistors (actually a P-channel FET) are switched in only if $\overline{RBPU}$ ($\overline{\text{Register B Pull Up}}$, bit 7) of the Interrupt Control 2 register is 0. Although all eight pull-ups are qualified by $\overline{RBPU}$, those pins configured as outputs ($TRIS[n] = 0$) will have the resistor switched off. $\overline{RBPU}$ resets to 1, and so the pull-up resistors default to being off.

As a typical application of weak pull-ups, consider the problem of reading a keypad, such as that illustrated in Fig. 11.13(a). In this particular instance there are

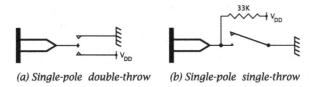

(a) *Single-pole double-throw*   (b) *Single-pole single-throw*

Fig. 11.11 Interfacing switches to a port pin

INTerrupt CONtrol 2 (INTCON2)   File h'FF1'

7	6	5	4	3	2	1	0
RPBU	INTEDG0	INTEDG1	INTEDG2	0	TMR0IP	0	RBIP
(R/W 1)	(R/W 1)	(R/W 1)	(R/W 1)		(R/W 1)		(R/W 1)

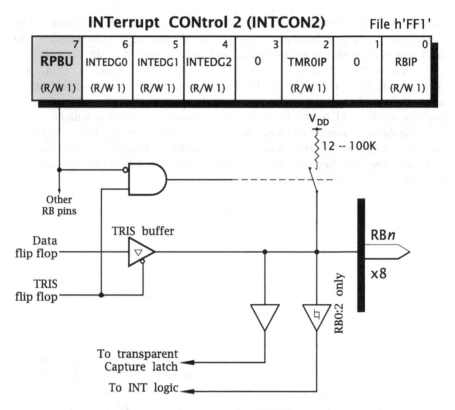

Fig. 11.12 Port B's weak pull-up resistors controlled by $\overline{\text{RBPU}}$ in INTCON2[7]

12 switches, and rather than use up all these scarce I/O pins it is hardware efficient to connect these switches in the form of a 4 × 3 matrix, as illustrated in Fig. 11.13(b). This 2-dimensional array reduces the I/O pin count to 7. Larger keypads show an even greater efficiency gain, with a 64-contact 8 × 8 keyboard needing only 16 I/O pins.

Although there are variations on this theme, the topology shown here is typical. The three columns are read in via RB[7:5], with internal pull-up resistors enabled. The four rows connected to RB[3:0] can be individually selected in turn by driving the appropriate pin Low, thus scanning through the matrix. The sequence is shown in Fig. 11.13(c). The switch contacts are normally open and, because of the pull-up resistors, read as logic 1. Should a switch connected to a Low row line be closed then the appropriate column line will be driven Low. This means that once the closed key column has been detected the column:row intersection is known. The 330 Ω protection resistors limit the current through the switch, should one of the RB[7:5] pins accidentally give a High-state output due to erroneous software.

In order to tie these concepts together, consider a subroutine to interrogate the keypad and return either with the key pressed (or at least the first key found if more

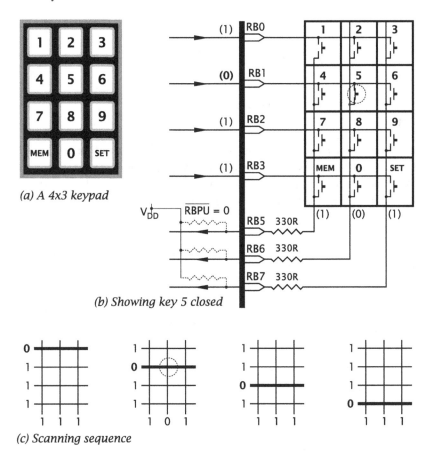

(a) A 4x3 keypad

(b) Showing key 5 closed

(c) Scanning sequence

Fig. 11.13 Interfacing to a keypad

than one) or if no key then −1 (i.e., h'FF'). Before looking at the coding, we can assume that somewhere in the main software Port B has been configured appropriately with the correct input and outputs assigned and that bit $\overline{RBPU}$ in the INTCON2 register has been cleared. Something like:

```
        include "p18f1220.inc"
MAIN setf  ADCON1      ; Make all port pins digital
     movlw b'11110000' ; Make RB[7:4] inputs
     movwf TRISB       ; RB[3:0] outputs
     bcf   INTCON2     ;Activate internal pull-ups
```

The listing of Program 11.5 is based on the task list:

1. Set KEY_COUNT to one.
2. For i = 0 to 3.

   • Activate row i.

- For j = 0 to 2.
  - Check column j.
  - IF zero THEN BREAK to step 4.
  - ELSE increment KEY_COUNT.

3. Set KEY_COUNT to −1 if no key found.
4. Return KEY_COUNT.

Basically the sequence of operations is to begin with a count of one; i.e., key[1], and bring row[0] Low. As each column is checked for a zero, the count kept in the Working register is incremented. If no closure (that is, a 0) is found, the next row is tried by shifting the initial test pattern one position.

Program 11.5 Scanning the keypad

```
; ***********************************************************
; * FUNCTION: Scans 4x3 keypad & returns with a key identifier*
; * ENTRY    : None                                         *
; * EXIT     : Key in W [MEM]=10, [0]=11, [SET]=12          *
; * EXIT     : Return -1 (h'FF') if no key detected         *
; * ENVIRON  : KEY, PATTERN byte vars                       *
; ***********************************************************
          cblock                    ; Two global variables
            KEY_COUNT:1, PATTERN:1
          endc

SCAN_IT clrf   KEY_COUNT            ; Key 1 is the first key
        incf   KEY_COUNT,f
        movlw  b'11111110'          ; The initial scan pattern
        movwf  PATTERN

SLOOP   movff  PATTERN,PORTB        ; Scan pattern sets row low
        movf   KEY_COUNT,w          ; Get Key count
; Now check each column for a zero
        btfss  PORTB,5              ; Check column 1
         bra   GOT_IT               ; IF zero THEN found the key!
        incf   KEY_COUNT,f          ; ELSE inc Key
        btfss  PORTB,6              ; Check column 2
         bra   GOT_IT               ; IF zero THEN found the key!
        incf   KEY_COUNT,f          ; ELSE inc Key again
        btfss  PORTB,7              ; Check column 3
         bra   GOT_IT               ; IF zero THEN found the key!
        incf   KEY_COUNT,f          ; ELSE inc Key again

; Reach here if no closed key
        rlncf  PATTERN,f            ; Shift scan pattern once <-
        btfsc  PATTERN,4            ; Check; has the 0 arrived at RB4?
         bra   SLOOP                ; IF not DO another column

; ELSE no key found anywhere
        movlw  -1                   ; Return -1
        goto   S_EXIT

GOT_IT  movf   KEY_COUNT,w          ; Copy Key count into W
S_EXIT  return                      ; and return
```

There are two ways out of the loop.

- If a 0 is found during the scan, the count in W is the desired value and the subroutine immediately returns after copying the Key count from memory into W.
- If the row pattern shift results in the sample 0 arriving at bit 4, then the subroutine returns h'FF' to tell the caller that no key has been found.

In the real world a subroutine like this would often read in rubbish, due to switch bounce and possibly noise induced in the connections between keypad and the electronics. One way of filtering out this unpredictability is shown in the subroutine of Program 11.6. Here the state of the keypad is interrogated using the SCAN_IT subroutine of Program 11.5. By keeping the state of the previous reading in Data memory, any change can be detected. Only if no change occurs over 256 readings will subroutine GET_IT return with the keypad state. Depending on the quality of the keypad, ambient noise and processor speed, the outcome can be improved at the expense of response time by including a short delay in the loop, or by using a 2-byte stability count to increase the number of readings.

Program 11.7 shows the equivalent coding in CCS **C** to that in Programs 11.5 & 11.6. This assumes that Port B has already been initialized as follows:

Program 11.6 Noise filtered keypad scanning

```
; ************************************************************
; * FUNCTION: Scans 4x3 keypad and returns with a debounced *
; * FUNCTION: key identifier                                *
; * ENTRY   : None                                          *
; * EXIT    : Key in W [MEM]=10, [0]=11, [SET]=12           *
; * EXIT    : Return -1 (h'FF') if no key detected          *
; * ENVIRON : COUNT, NEW_KEY, OLD_KEY                       *
; * RESOURCE: Subroutine SCAN_IT                            *
; ************************************************************

           cblock                    ; Three global variables
            COUNT:1, NEW_KEY:1, OLD_KEY:1
           endc

GET_IT     clrf   COUNT       ; The no-change count zeroed
GLOOP      rcall  SCAN_IT     ; Raw value returned in W
           movwf  NEW_KEY     ; Is new value
           cpfseq OLD_KEY     ; New and old the same?
            bra   NOT_EQUAL   ; IF same go to Not Equal

; IF readings are the same THEN
           incfsz COUNT,f     ; Increment count; IF not
            bra   GLOOP       ; rolled around to 00 repeat
           movf   OLD_KEY,w   ; ELSE thats it!
           return             ; and return with Key value

; Readings are different, so:
NOT_EQUAL  movff  NEW_KEY,OLD_KEY ; Make old key = new key
            bra   GET_IT          ; and start all over again
```

Program 11.7 Scanning the keypad in **C**

```
/*****************************************************************
 * FUNCTION: Scans 4x3 keypad & returns with a debounced        *
 * FUNCTION: key identifier                                      *
 * ENTRY    : None                                              *
 * EXIT     : key, [MEM] = 10, [0] = 11, [SET] = 12. -1 if none*
 * RESOURCE: Function scan_it()                                 *
 *****************************************************************/
unsigned int get_it(void)
{
unsigned int count, old_key, new_key;
count = 0;
while(count<255)
    {
    new_key = scan_it();
    if(new_key == old_key)
        { count++;}
    else
        {
        old_key = new_key;
        count = 0;
        }
    }
return (old_key);
}


/*****************************************************************
 * FUNCTION: Scans 4x3 keypad & returns with a key identifier*
 * ENTRY    : None                                             *
 * EXIT     : key, [MEM] = 10, [0] = 11, [SET] = 12. -1 if none*
 *****************************************************************/
unsigned int scan_it(void)
{
unsigned int key, pattern;
key=1; pattern = 0xFE;          /* Initial pattern b'01111111' */
while(key<13)
    {
    PORT_B = pattern;
    if(!COL1) {break;}
    key++;
    if(!COL2) {break;}
    key++;
    if(!COL3) {break;}
    key++;
    pattern = pattern << 1;  /* Shift pattern left once       */
    }
if(key==13) {key = 0xFF;}
return key;
}
```

```
#include  <18f1220.h>
#byte PORT_B = 0xF81
#bit  COL1  = PORT_B.5      /* Column 1 is RB5              */
#bit  COL2  = PORT_B.6      /* Column 2 is RB6              */
#bit  COL3  = PORT_B.7      /* Column 3 is RB7              */

unsigned int scan_it(void);
unsigned int get_it(void);
int main()
{
int reading;
set-up_adc_ports(NO_ANALOGS);/* All parallel pins digital  */
set_tris_b(0xF0);
port_b_pullups(TRUE);
```

Notice the use of the function port_b_pullups(TRUE) as an alternative to setting the $\overline{\text{RBPU}}$ bit in the INTCON2 register.

The logic of the program is very similar to our assembler coding, with a shifting pattern zeroing each row in turn. The only difference is that the loop is executed a fixed number of times using a count, rather than testing bit 4 of the test pattern. This makes the process more transparent, although the latter is more efficient.

To facilitate interfacing switches and keypads, PIC MCU families possess the ability to detect whenever a variation happens at some Port B inputs. The logic of this **Port B Change** feature is shown in Fig. 11.14.

The top four Port B I/O pins have a second D latch in parallel with the main Capture latch. This main latch is updated every instruction cycle on the first phase Q_1—see Fig. 4.5 on p. 76. The new Capture latch is only updated whenever Port B (not LATch B) is read, and this occurs on the third phase Q_3. When the reading action is over, the Change latch freezes and captures the pin state as it is at the time of reading. The outputs of both the Capture and Change latch are Exclusive-ORed together. As we have seen on p. 14, an XOR gate detects *differences* between its two inputs. As the Capture latch is now following the state of the associated pin, any subsequent variation at this pin input will cause the output of the associated XOR gate to go to logic 1. Each of the four Port B cells RB[7:4] has a Change feature and the four XOR gates are ORed to give a composite signal which sets the RBIF (Register B Interrupt Flag) in the INTCON of Fig. 7.2 on p. 209. If the RBIE (Register B Interrupt Enable) bit is set to 1, then this is a convenient way of awaking a PIC MCU slumbering in its Sleep state. If the GIE (General Interrupt Enable) bit is set as well, an alteration in the top nybble of Port B will cause an interrupt. Each XOR gate is ANDed with the appropriate TRIS line, so that only bits that are programmed as an input can contribute to the Change signal.

In the specific case of our keypad of Fig. 11.13, if *all* the row lines are set to the Low state, then when any switch is pressed a column line will change state. If RBIE has been set to enable the Port B Change facility, then when the RBIF flag sets an

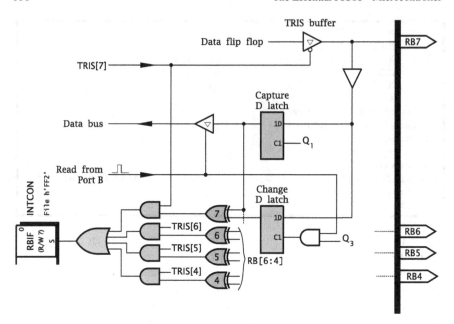

Fig. 11.14 The PIC18 Port B Change feature, with detail for RB7

interrupt will occur. The keypad may then be scanned in the ISR to determine which key has been closed.

Care must be taken in using this facility. For instance, using the lower (non-Change) part of Port B (e.g., bcf PORTB,0) can affect the Change facility by forcing all the latches to resample. Remember that instructions that write to memory, except movff, also do a preliminary read.

Once the PIC MCU has responded to the Change interrupt, the Change signal setting RBIF should be removed by reading Port B again, which equates the state of the two D latch arrays. Only then should RBIF be cleared. Failure to do this initial read will result in this interrupt flag being immediately set again.

As an instance, using the keypad to awaken the PIC MCU with the assumption that GIE is zero (no interrupt) should be implemented as:

```
movf    PORTB,w         ; Read Port B to cancel any difference
bcf     INTCON,RBIF     ; Clear the Change interrupt flag
bsf     INTCON,RBIE     ; Enable the Change interrupt enable
sleep                   ; Go to sleep
; zzzzz
call    DELAY           ; On wakening let things settle
movwf   PORTB,w         ; before canceling any difference
bcf     INTCON,RBIF     ; Clear the Change interrupt flag
bcf     INTCON,RBIE     ; Disable Change interrupt facility
```

As an example, consider the coding of a Port B Change interrupt service routine for a dishwasher with four front-panel momentary push switches. These switches are labeled AUTO, ECONO, QUICK and RINSE and are connected as shown in Fig. 11.15. Whenever the operator wishes to select a wash program, the appropriate switch is depressed. Once the PIC18F1220 senses a change to the switch settings, the software is to transfer its focus to the designated subroutine AUTOMATIC, ECONOMY, QUICK_RINSE or PRE_RINSE. A buzzer is to be sounded for 0.5 s, and an LED is to be lit above the appropriate switch and remain illuminated for the duration of the wash program.

Our background software needs to initialize the processor to:

- Configure all parallel port pins digital.
- Make pins RB[7:4] and RA0 outputs; the rest remaining inputs.
- Switch in the Port B integral pull-up resistors.
- Clear any possible Port B mismatch and clear the associated interrupt flag.
- Enable both the RB and General interrupt functions.

```
        org    0                    ; Reset vector
        goto   MAIN                 ; Jump to background software

; Background software ------------------------
MAIN    setf   ADCON1               ; Make all pins digital
        movlw  b'11110000'
        movwf  TRISB                ; RB7:4 outputs, rest inputs
        bcf    TRISA,0              ; Make RA0 an Output to Buzzer
        bcf    INTCON2,NOT_RBPU     ; Activate internal pull-ups
        bsf    INTCON,RBIE          ; Enable Change PortB int
        movf   PORTB,w              ; Reset the Change mechanism
        bcf    INTCON,RBIF          ; Clear the interrupt flag and
        bsf    INTCON,GIE           ; the General interrupt system
        ....   .....                ; Rest of background routine
```

The code fragment above closely follows the task list above. The only point of note is the setting up of the Port B Change interrupt system. This is implemented by setting the RBIE (INTCON[3]) RB-enable mask bit. Resetting the Change logic, which may be triggered at reset, is accomplished by clearing Port B. clrf is an example of a read-modify-write instruction (despite the target value being irrelevant) and the read cycle resets any mismatch condition. It also turns off all LEDs. After (not before) the RBIF (INTCON[0]) flag can be cleared.

The foreground ISR listed in Program 11.8 is based at the default (Compatible) interrupt vector at h'00008'. This is entered whenever there are any changes in the state of the four switches (this is the only kind of interrupt enabled). The core of this process tests each switch in turn. If a closed switch (logic 0 on the tested Port B pin) is detected, the associated front panel LED is lit and the buzzer sounded for 500 ms before calling the appropriate wash program subroutine. On return, the ISR is exited by reading Port B to reset the change logic and turn off the LEDs; resetting the RBIF flag and returning to the background routine. Using the Fast version of the retfie instruction automatically returns the context.

Program 11.8 The dishwasher RB-interrupt handler

```
        org   h'00008'  ; The Interrupt vector
; Foreground software --------------------------------------
; **********************************************************
; * FUNCTION: ISR to handle Change Port B interrupts       *
; * FUNCTION: Calls subroutine appropriate to closed switch *
; * FUNCTION: Activates buzzer & fitting LED for chosen program*
; * ENTRY   : Whenever any switch changes RB[7:4]          *
; * EXIT    : Change function & LEDs cleared after subroutine *
; * RESOURCE: Subroutines BUZ, AUTOMATIC, ECONOMY, QUICK, RINSE*
; * ENV'MENT: None                                         *
; **********************************************************
ISR_DISHWASHER
        btfsc PORTB,4     ; Check, was it the AUTO switch?
        bra   TRY_ECONO   ; IF not THEN try the ECONO switch
        bsf   PORTB,0     ; ELSE turn on the A_LED
        call  BUZ         ; Buzz for 500ms
        call  AUTOMATIC   ; Go for it
        bra   DISH_EXIT   ; and exit

TRY_ECONO
        btfsc PORTB,5     ; Check, was it the ECONO switch?
        bra   TRY_QUICK   ; IF not THEN try the QUICK switch
        bsf   PORTB,1     ; ELSE turn on the E_LED
        call  BUZ         ; Buzz for 500ms
        call  ECONOMY     ; Go for it
        bra   DISH_EXIT   ; and exit

TRY_QUICK
        btfsc PORTB,6     ; Check, was it the QUICK switch?
        bra   TRY_RINSE   ; IF not THEN try the RINSE switch
        bsf   PORTB,2     ; ELSE turn on the Q_LED
        call  BUZ         ; Buzz for 500ms
        call  QUICK       ; Go for it
        bra   DISH_EXIT   ; and exit

TRY_RINSE
        btfsc PORTB,7     ; Check, was it the RINSE switch?
        bra   DISH_EXIT   ; IF not THEN no switch was set!
        bsf   PORTB,3     ; ELSE turn on the R_LED
        call  BUZ         ; Buzz for 500ms
        call  RINSE       ; Go for it

; Exit point -----------------------------------------------
DISH_EXIT
        clrf  PORTB       ; Turns off all LEDS and resets Change B
        bcf   INTCON,RBIF; Cancel Change PortB flag
        retfie FAST       ; Return with saved context

; **********************************************************
; * FUNCTION: Sounds buzzer (RA0) for 500ms               *
; * ENTRY   : None                                        *
; * EXIT    : Buzzer off                                  *
; * RESOURCE: DELAY_500MS subroutine                      *
; * ENV'MENT: None                                        *
; **********************************************************
BUZ     bsf   PORTA,0     ; Turn on buzzer
        call  DELAY_500MS ; For 0.5s
        bcf   PORTA,0     ; Turn off buzzer
        return
```

Fig. 11.15 Hardware for the
dishwasher control panel

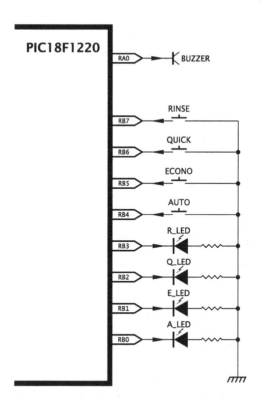

As the buzzer can be sounded at four possible points in the ISR, it is handled
with a separate subroutine. There is no problem calling a subroutine from an ISR;
which in this respect is no different than nesting subroutines. The 500 ms subroutine
listing is not shown. A modified version of Program 6.10 on p. 190 can be used to
code this task.

Coding the dishwasher control software in **C** again requires the environment to
be set-up in the background routine, according to our tasklist.

```
void main(void)
{
set-up_adc_ports(NO_ANALOGS);/* All parallel pins are digital*/
set_tris_a(0xFE);          /* RA0 = Output, RA[7:1] Inputs  */
set_tris_b(0xF0);          /* RB[7:4] Inputs, rest Outputs  */
port_b_pullups(TRUE);      /* Pullups for switches          */
PORT_B = 0;                /* Reset Change mechanism        */
clear_interrupt(int_RB);   /* Clear RBIF flag               */
enable_interrupts(int_RB); /* Enable Change_B interrupts    */
enable_interrupts(GLOBAL); /* Interrupt system enabled      */
/* Rest of background software                              */
```

The code fragment above is similar to the assembly-level equivalent.

Program 11.9 Coding the RB-interrupt handler in **C**

```
#int_RB
void isr_dishwasher(void)
{

if(!AUTOMAT)    {A_LED = 1; buz(); automatic();}
else if(!ECONO) {E_LED=1; buz(); economy();}
else if(!QUIK)  {Q_LED=1; buz(); quick();}
else if(!RINS)  {R_LED=1; buz(); rinse();}
PORT_B = 0;                  /* Reset Change mechanism & LEDs   */
}

void buz(void)
{
BUZZER = 1;
delay_ms(500);
BUZZER = 0;
}
```

The foreground function listed in Program 11.9 is designated a Port B change ISR using the #int_RB qualifier. Each LED and switch has been named (not listed here) corresponding to a specific bit in Port B. Some of the switch names have been altered slightly to avoid using the same name as a wash-program function; e.g. QUICK and quick().

The core of the function is a 4-way if-else tree. Each branch tests one switch and if the test is true (which because of the ! NOT operator will be the case if the switch is closed) the associated LED lit, the buzzer function will be called followed by the appropriate wash-program.

When this string of events is complete, execution will fall through to the end of the if-else tree. There the Port B change logic and LEDs are cleared by accessing PORT_B. When the ISR terminates, the RBIF flag is automatically cleared.

Examples

Example 11.1 A 2N3055 NPN bipolar transistor is to be used to activate the field coils of a small stepper motor. Taking into account the minimum gain of the transistor over the range $+85 \rightarrow -40°C$, it has been calculated that the base current must be at least 10 mA. The transistor is to be controlled from a port pin and the processor is to be powered with a V_{DD} of 5 V. The 2N3055's base-emitter voltage can be assumed to be no more than 0.7 V. What is the maximum value of the base resistor R_B, and, given this value, what will be the worst-case maximum base current?

Solution For currents of this magnitude we can assume that the pin voltage will be less than 5 V. The data sheet specifies a minimum voltage of 4.3 V (a drop of 0.7 V) for a I_{OH} of -3 mA, but for currents greater than this we must resort to graphical techniques.

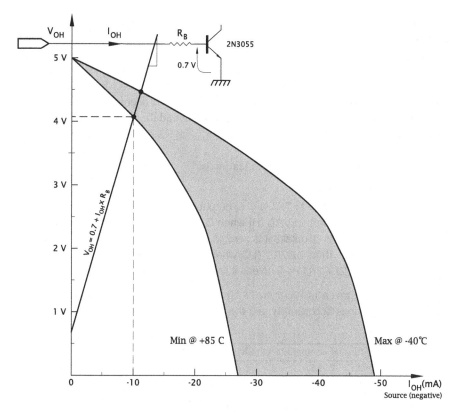

Fig. 11.16 Source current against voltage

Figure 11.16 shows the graphical relationship of output source current I_{OH} for a High-output voltage state V_{OH}. The gray area is bounded by the minimum situation, which is at +85°C and maximum condition at −40°C.

V_{OH} is also a function of the transistor input base resistor circuit according to the equation $V_{OH} = 0.7 + I_{OH} \times R_B$. This straight line relationship (called a **load line**) is shown on the graph from (0,0.7) drawn to intersect the minimum locus at a current of −10 mA. This crossover is the only point that satisfies both current–voltage relationships. The slope of the load line $\frac{\Delta V}{\Delta I}$ is the resistance in kΩ (as current is in mA) and measures 280 Ω. Notice that the High output state has fallen to 4 V (−10, 4.0).

Extending the load line onwards gives the maximum current as the X co-ordinate of the intersection with the maximum locus, which is approximately 11.5 mA; not much different. If the current requirement had been larger, then the minimum/maximum currents diverge showing a significant temperature sensitivity. For instance, a 20 mA minimum base current requires a base resistor of ≈120 Ω (assuming a base voltage of 0.8 V) and the maximum base current would be 28 mA.

Example 11.2 A PIC18F1220 MCU is to be used as a digital comparator where a parallel-input 8-bit word P is to be compared to a byte datum located in a File named TRIP. Outputs are to indicate Lower-Than, Equivalent, and Higher-Than. The comparator is to have an hysteresis of ± 1 bit. That is, if previous comparisons showed $P <$ TRIP then the trigger level is increased to TRIP $+ 1$ for equality. Similarly, on a downward trajectory the equality level is to be decreased to TRIP $- 1$.

Datum P is to be input via Port B set-up as input and the lower three Port A pins give the active-High comparator outputs $<, ==, >$ at RA2, RA1, RA0, respectively.

Solution The task list for such a specification is:

1. Subtract P from LEVEL.
2. IF Equal (EQ when Z = 1) THEN $==$ output active.
3. ELSE IF P Higher than LEVEL (HI when C = 0, Borrow) THEN $>$ output active AND LEVEL=TRIP-1 unless it is zero, in which case LEVEL=TRIP.
4. ELSE IF P Lower than LEVEL (LO when C = 1, No Borrow) THEN $<$ output active AND LEVEL=TRIP+1 unless it is h'FF', in which case LEVEL=TRIP.

The subroutine given in Program 11.10 assumes that the main program has set-up the port directions accordingly and the fixed value is in TRIP. Initially LEVEL

Program 11.10 A digital comparator with hysteresis

```
COMP        movf    PORTB,w      ; Get input N
            subwf   LEVEL,w      ; LEVEL - N
            btfss   STATUS,Z     ; Skip if equality
            bra     CONTINUE     ; ELSE IF not THEN try alternative

; This code for equality ---------------------------------------
            movlw   b'11111010'  ; Make == output logic 1
            movwf   PORTA        ; Other outputs logic 0
            bra     COMP_END     ; and exit

CONTINUE    btfsc   STATUS,C     ; Skip if borrow (N higher than)
            bra     LO           ; ELSE N < LEVEL

; This code if N > LEVEL ---------------------------------------
HI          movlw   b'11111001'  ; Set > output RA0 to logic 1
            movwf   PORTA        ; Rest to 0
            decf    TRIP,w       ; Copy TRIP-1 to w
            btfsc   STATUS,C     ; Skip if underflows
            movwf   LEVEL        ; ELSE is new comparator level
            bra     COMP_END     ; and exit

; This code when N < LEVEL -------------------------------------
LO          movlw   b'11111100'  ; Set < output RA2 to low
            movwf   PORTA        ; Rest to 0
            incfsz  TRIP,w       ; Copy TRIP+1 to w.  IF not 00
            movwf   LEVEL        ; gives the new comparator level

COMP_END    return
```

Program 11.11 Coding the digital comparator in **C**

```
void compare(int trip)
{
EQ = HI = LO = 0;
if      (PORTB == LEVEL) {EQ = 1;}
else if (PORTB > LEVEL)
      {
      HI = 1;
      if(LEVEL > 0) {LEVEL = trip - 1;}
      }
else
      {
      LO = 1;
      if(LEVEL < 0xFF) {LEVEL = trip + 1;}
      }
}
```

would have been set to the same value as TRIP but would subsequently vary by ± 1 *as per* the specification—the hysteresis band—unless LEVEL would under- or over-flow.

Software solutions to traditional hardware functions, such as comparison, have the advantage of greater flexibility, albeit at the price of a lower data throughput. Using low-cost 'computing engines', such as the PIC MCU, means that relatively simple functions traditionally implemented by dedicated hardware can be replaced by embedded processors.

In this instance, flexibility could be replacing the fixed trip level by a variable datum input via, say, Port C—see SAQ 11.4. Example 12.1 on p. 440 shows how an external datum can be serially acquired. Alternatively, an analog signal could represent one or both of the levels in devices with integral A/D converters—see Chap. 14. In all these situations the hysteresis may advantageously be made a fraction of the trip voltage, e.g., $\pm\frac{1}{32}$, rather than a fixed ± 1 bit.

In the case of the **C** function equivalent coding listed of Program 11.11, the names EQ, HI and LO have been defined externally as the appropriate bits in Port A. Here we are assuming that trip is a variable, acquired elsewhere, and passed to the function. The body of the function does the comparison and actuates the appropriate pin. If required, the global variable LEVEL is altered in order to shift the trigger level. If trip is fixed, then it need not be passed to the function and could be a literal.

Example 11.3 The principle of a stepper motor is shown in Fig. 11.17. In essence there are four coils, labeled A, B, C, D, which may be selectively energized either singly or in pairs, to generate a magnetic field in one of eight directions in divi-

Fig. 11.17 The stepper
motor; showing a
north-easterly field

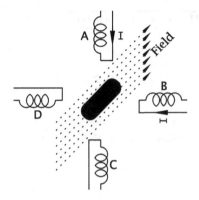

sions of 45°C.[8] Thus Coil A alone gives a northerly field, A and B together give a
north-easterly direction, B alone is east, etc. The rotor follows the field as it changes
direction, provided that inertial considerations allow it to keep up during accelera-
tion and de-acceleration.

Write a subroutine originating at h'00050' in the Program store to advance the
rotor by a passed value of one to 256 steps. Assume that the Port A pins RA[3:0]
are connected respectively to coils A, B, C, D. The rate is to be nominally 100 steps
per second, based on a 10 ms delay, which is to be written to be largely independent
of the crystal frequency. The latter is to be indicated by the programmer by the
constant FREQ, which is the multiple of 1 MHz; e.g., d'4' for a 4 MHz crystal.

Solution Our first step is to devise a table showing energization patterns for the
eight possible field directions, as shown in Table 11.2.

The coding shown in Program 11.12 comprises three subroutines.

Table 11.2 Energization
pattern for the eight field
directions

Position	A	B	C	D	Bearing
0	1	0	0	0	↑
1	1	1	0	0	↗
2	0	1	0	0	→
3	0	1	1	0	↘
4	0	0	1	0	↓
5	0	0	1	1	↙
6	0	0	0	1	←
7	1	0	0	1	↖

[8] A real stepper motor repeats the coil set several times around the peripheral motor stator giving a
finer mechanical step resolution. Thus, if there are four sets of stator coils the 45°C electrical step
translates to 11.25°C mechanical.

Program 11.12 Driving a stepper motor

```
          #define FREQ   d'40'; Programmer gives value in 1M steps
                          ; 40MHz in our example
; ****************************************************************
; * FUNCTION: Advances stepper motor 1 -- 256 steps            *
; * ENTRY   : Step number in STEP                              *
; * ENTRY   : Current field position in POSITION               *
; * EXIT    : POSITION updated, STEP = -1, W destroyed         *
; * RESOURCE: Subroutine PATTERN, DELAY_10MS                   *
; ****************************************************************
          org    h'00050'    ; Begins at h'00050'
MOTOR     incf   POSITION,w  ; Advance field direction
          andlw  b'0111'     ; Module-8
          movwf  POSITION    ; updated
          call   PATTERN     ; Get the energization pattern
          movwf  PORTA       ; Send to stepper motor
          call   DELAY_10MS  ; Hold off 10ms
          decfsz STEP,f      ; Decrement step count
          bra    MOTOR       ; until zero
          return


; ****************************************************************
; * FUNCTION: Maps an integer 0 --7 to field pattern           *
; * ENTRY   : Modulo-8 integer in W                            *
; * EXIT    : Stepper energization pattern in W                *
; ****************************************************************
PATTERN addwf  PCL,f        ; Increment Program Counter
        retlw  b'1000'      ; North
        retlw  b'1100'      ; North east
        retlw  b'0100'      ; East
        retlw  b'0110'      ; South east
        retlw  b'0010'      ; South
        retlw  b'0011'      ; South west
        retlw  b'0001'      ; West
        retlw  b'1001'      ; North west


; ****************************************************************
; * FUNCTION: Delays 10 ms delay independent of clock freq     *
; * ENTRY   : Clock frequency in steps of 1MHz in TEMP         *
; * EXIT    : 10ms delay; DELAY zero, W destroyed              *
; ****************************************************************
DELAY_10MS
        movlw  FREQ*5       ; The programmers statement x 5
        movwf  TEMP         ; Gives the PIC frequency
; Delay loop 2ms @ f = 1MHz xtal  (1 cycle = 4us; x5 gives 10ms)
DLOOP1 movlw  d'167'        ; Loop count
        movwf  DELAY
DLOOP2 decfsz DELAY,f       ; (168) * (FREQ*5) * 4us
        bra    DLOOP2       ;  2*(166) * (FREQ*5) * 4us

        decfsz TEMP,f       ; Decrement frequency parameter
        bra    DLOOP1       ; and repeat until zero
        return
```

MOTOR

The main subroutine simply modulo-8 increments the position vector by post-ANDing with b'00000111' to give a wrap around from 7 to 0. This vector is then converted to the appropriate energizing pattern and sent out to the motor after a nominal 10 ms delay. The process is repeated until the decrementing STEP datum reaches zero; if initially zero then 256 steps will be actioned.

PATTERN

Returns one of eight energization patterns corresponding to the field vector as listed in Table 11.2. The mechanism of this look-up table coding has been described in Program 6.6 on p. 175. As this suite of subroutines originates at h'00050' in the Program store, the 8-bit addition to the Program Counter will not result in roll-over across page boundaries.

DELAY_10MS

This subroutine gives a nominal 10 ms delay independent of the processor crystal frequency. This is defined by the programmer in the program header as the constant FREQ which denotes the number of multiples of 1 MHz. For instance, for a 8 MHz crystal FREQ is set to d'8' using the #define directive, before the program is assembled.

The core of the subroutine is a inner loop giving a 2 ms delay (4 μs instruction cycle) at 1 MHz; that is, 500×4 μs. An outer loop multiplies this by the constant $5 \times$ FREQ. The factor $\times 5$ effectively magnifies the inner loop to 10 ms at 1 MHz. The factor $\times$ FREQ compensates for the actual cycle period, which is inversely proportional to the actual clock frequency.

Example 11.4 A reaction meter is to be designed to act as a crude blood-alcohol level indicator. The principle of the device is that a buzzer is sounded for 100 ms when the unseen tester closes his switch. The subject is to respond to the sound by immediately pressing his/her switch. An 8-LED barograph display is to indicate the passage of time by progressively lighting a new LED every 50 ms. The number of lit LEDs at the conclusion of the test is the reaction time in 50 ms steps.

Show how a PIC18F1220 could be configured and design software coded in **C** to read the switch and activate the LEDs and buzzer.

Solution The software listed in Program 11.13 is based on the LED array connected to Port B, the subject's switch to RA0 and the buzzer to RA2. On reset, as long as RA0 is high the barograph variable display is updated every 50 ms. When this display reaches b'00000011'; that is after 100 ms the buzzer is turned off.

Example 11.5 A PIC MCU is to switch in a 24 V 100 mA solenoid switch, as shown in Fig. 11.18. As this is to operate in a noisy environment, a 220 pF capacitor is to be placed across the port pin to act as a low-pass filter.

The software is to activate a LED for a short time everytime the solenoid is activated. A test run on the prototype with the MCU running at 4 MHz executes

Program 11.13 Measuring reaction time in 50 ms steps

```
#include   <18f1220.h>
#fuses NOWDT, HS
#use    delay(clock=20000000)    /* 20MHz clock                    */
#byte   PORT_A = 0xF80
#byte   PORT_B = 0xF81
#bit    BUZZER  = PORT_A.2
#bit    SUBJECT = PORT_A.0
main()
{
unsigned int display=0;          /* Initial LED display pattern   */
setup_adc_ports(NO_ANALOGS);     /* All port bits are digital     */
set_tris_a(0x1B);                /* Make RA2 output to buzzer      */
set_tris_b(0);                   /* Port B is output to LEDs       */

BUZZER = 1;                      /* Turn on buzzer                 */
while(SUBJECT)                   /* DO as long as switch is == 1 */
     {
     PORT_B = display;           /* Activate LEDs                 */
     delay_ms(50);               /* Call 50ms delay               */
     display = display << 1;     /* Create next barograph display*/
     display = display | 0x01;   /* Feeding in ones               */
     if(display == 0x03)         /* After 2 shifts                */
        {BUZZER = 0;}            /* Turn off the Buzzer           */
     }
}
```

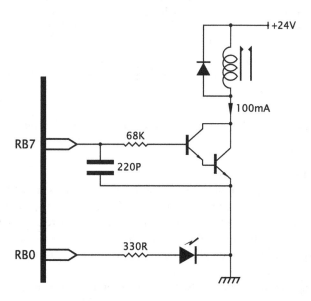

Fig. 11.18 Activating a high-voltage high-current solenoid

correctly. However, the production version gives erratic operation, with the LED sometimes being illuminated with no solenoid activation. A storage oscilloscope monitoring pin RB7 indicates a short-lived pulse whenever this malfunction occurs.

The only difference between the prototype and the final version, is the use of a 40 MHz clock in the latter. What is going on?

Solution Although the software is not at fault, it is instructive to examine the fragment operating the solenoid.

```
        bsf     PORTB,7     ; Turn on the solenoid
        bsf     PORTB,0     ; Turn on the LED
        call    DELAY       ; for a short time
        bcf     PORTB,0     ; and turn it off again
        ....    ......      ; continue on
```

The logic of the process is flawless.

Examination of the data sheet gives a transition time at a port pin of 25 ns (10% to 90%) with a load capacitance of 50 pF. The bsf instruction turning on the solenoid will order pin RB7 to go high on the final quadrature of the machine cycle, as shown in Fig. 4.5 on p. 76. The next bsf PORTB, 0 turning on the LED, as a read-modify-write instruction, reads the whole of Port B's contents in phase 1 of the following instruction cycle. In the following phase 4 this datum byte is sent out again, but this time with bit 0 logic 0.

At a clock frequency of 4 MHz, each phase lasts 250 ns. However, with a 40 MHz clock this reduces to 25 ns. In this latter situation the voltage at pin RB7 may not yet have risen enough to be recognized as a logic 1. Thus the dummy read will sense a 0 at bit 7 and this is the value written out back to pin RB7. This bsf PORTB, 0 will actually change pin RB7 as well as RB0.

The solution is to use LATB instead of PORTB when turning on the LED. An alternative would be to implement a short delay before activating the LED, or better still to turn on the LED before the solenoid.

Example 11.6 Despite the increasing use of liquid-crystal alphanumeric readouts, discrete 7-segment LED displays are commonly used to show multiple numerical digits. Such readouts are particularly effective in low ambient light situations and where large displays are needed.

Assuming each display requires eight lines (seven segments plus decimal point) then a budget of $8 \times n$ parallel lines are required for an n-digit display. The straightforward solution to this problem is shown in Fig. 11.19, where a 3-digit display is driven from three parallel-in parallel-out registers on a local bus—see Fig. 2.18 on p. 32. The principle can be extended as required by using the appropriate number of registers.

The displays shown in the diagram are common cathodes and the appropriate LED is illuminated when the register output is High, with the source current limited by the series resistor. In practice some logic circuitry can sink more current into a Low-state output as compared to sourcing current from a High state, and because of this, common anode displays are often used with the LEDs, activated on a Low state. In some larger displays, e.g., 5 cm (2″), several LEDs may be paralleled or

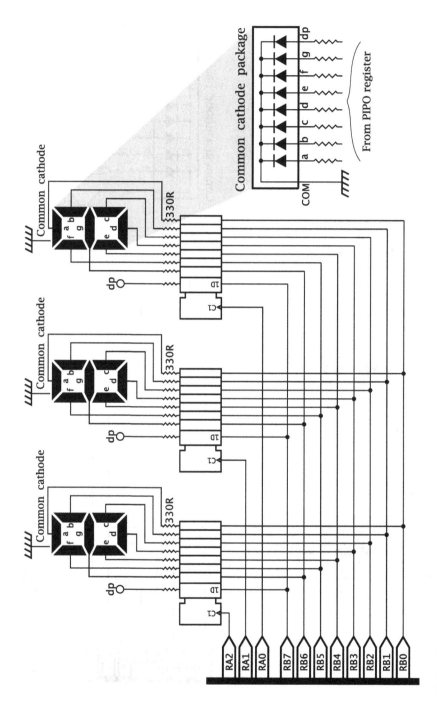

Fig. 11.19 Using port expansion to drive three 7-segment displays

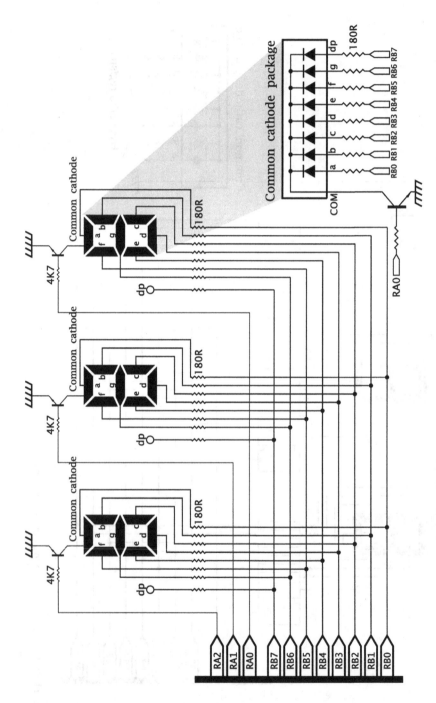

Fig. 11.20 Scanning a multiplexed 3-digit 7-segment array

in series in each segment. In this situation larger voltages and/or currents may be needed, and suitable drivers used to boost the register outputs.

An alternative approach, shown in Fig. 11.20, is frequently used with LED-based displays. Instead of using a register for each digit, all readouts are connected in *parallel* to the *one* MCU port. Each readout is enabled in turn for a short time with the appropriate data from the output port. Provided that the scan rate is greater than 50 per second (preferably greater than 100) the brain's persistence of vision will trick the onlooker into visualizing the display as flicker free.[9] Of course the current flowing through the segment must be increased to compensate for the mark:space ratio, but LEDs are more efficient when pulsed in this manner and the reduction of series resistance need not be *pro rata*.

Discuss the *pros and cons* of these arrangements, with reference to the tradeoff of software and hardware. Illustrate your answer by displaying the decimal equivalent of the binary byte in File h'020'. For instance; if the contents of BINARY were h'FF' then the display should be ⎕⎕⎕.

Solution From the software perspective, two main functions can be identified. First, the binary code in a File called BINARY has to be decomposed into three BCD digits; HUNDREDS, TENS and UNITS. Program 11.14 is a straightforward extension of Program 5.11 on p. 149 to hundreds.

Once this is done then each BCD digit ranging from 0 to 9 must be converted to 7-segment code to illuminate the relevant segments to form the appropriate characters. We already have a subroutine to implement this mapping in Program 6.6 on p. 175.

Based on these subroutines *in situ*, we have as a task list for software to interact with the hardware of Fig. 11.19:

1. Convert the binary byte into BCD.
2. DO
   (a) Copy contents of HUNDREDS into W and convert to 7-seg.
   (b) Copy 7-segment code to Port B.
   (c) Pulse ⌐‾⌐ RA2.
3. DO
   (a) Copy contents of TENS into W and convert to 7-seg.
   (b) Copy 7-segment code to Port B.
   (c) Pulse ⌐‾⌐ RA1.
4. DO
   (a) Copy contents of UNITS into W and convert to 7-seg.
   (b) Copy 7-segment code to Port B.
   (c) Pulse ⌐‾⌐ RA0.

The coding implementing this task list is shown in Program 11.15.

---

[9]Of course this is how the brain interprets a series of 24 still frames per minute in a movie as a moving image. Each frame is shown twice using a 2-bladed shutter, giving a flicker rate of 48 per second.

Program 11.14 8-bit binary to 3-digit BCD conversion

```
; ****************************************************************
; * FUNCTION: Converts a binary byte in W to three BCD digits*
; * EXAMPLE : Binary = h'FF' (d'255'), HUNDREDS = h'02'      *
; * EXAMPLE : TENS = h'02', UNITS = h'05'                    *
; * ENTRY   : Binary  in W                                   *
; * EXIT    : HUNDREDS = Hundreds digit, TENS = Tens digit   *
; * EXIT    : UNITS = Units digit. W holds units             *
; ****************************************************************
; First divide by a hundred --------------------------------
BIN_2_BCD  clrf   HUNDREDS   ; Zero the Hundreds loop count

LOOP100    incf   HUNDREDS,f ; Record one hundred subtracted
           addlw -d'100'     ; Subtract decimal hundred
           bc     LOOP100    ; IF no borrow (C==0) THEN DO again
           decf   HUNDREDS,f ; ELSE compensate for one inc too many
           addlw d'100'      ; by adding a hundred to residue

; Next divide by ten ---------------------------------------
           clrf   TENS       ; Zero the Tens loop count

LOOP10     incf   TENS,f     ; Record one ten subtracted
           addlw -d'10'      ; Subtract decimal ten
           bc     LOOP10     ; IF no borrow (C==0) THEN DO again

; Retrieve last remainder for units ------------------------
           decf   TENS,f     ; Compensate for one inc too many
           addlw d'10'       ; by adding ten to residue
           movwf UNITS       ; which gives the remainder
           return            ; and return to caller
```

The interaction of the software to the hardware of Fig. 11.20 is not so straight-forward as there are no registers to dump the data and run! Instead, data has to be continuously sent out in sequence with the appropriate display being enabled. If we use a scan rate of 100 updates each second, then this data should be held for 10 ms before moving on. Thus we have as our new task list:

1. Convert the binary byte into BCD.
2. DO forever:
   (a)
   - Copy contents of HUNDREDS into W and convert to 7-segment code.
   - Copy 7-segment code to Port B.
   - Bring RA2 Low ⌐_.
   - Delay 10 ms.
   - Bring RA2 High _/ .

   (b)
   - Copy contents of TENS into W and convert to 7-segment code.
   - Copy 7-segment code to Port B.
   - Bring RA1 Low ⌐_.
   - Delay 10 ms.
   - Bring RA1 High _/ .

   (c)
   - Copy contents of UNITS into W and convert to 7-segment code.

- Copy 7-segment code to Port B.
- Bring RA0 Low ‾⌐.
- Delay 10 ms.
- Bring RA0 High ⌐‾.

The coding in Program 11.16 makes use of the 10 ms delay subroutine illustrated in Program 11.12 to regulate the scanning rate. Apart from the length of the enabling pulse, the core of the program is identical to our previous situation. However, the code must run continually to give the impression of a constant display. This illustrates the trade-off between hardware and software. Reducing the hardware has led to greater loading on the software. Indeed, as illustrated here, the entire existence of the PIC MCU will be to service the display! However, in practice the situation can be redeemed somewhat by interrupting the PIC MCU at 10 ms intervals to avoid the need for time-wasting delay routines. The listing on p. 472 shows how this can be done, but of course the Timer cannot be used for anything else. Alternatively an external 100 Hz oscillator can be used in its place, but some of the hardware advantages are then lost. With a 10 ms digit rate, up to ten digits may be handled with no additional interface hardware and still have a scan rate no worse than 100 per second.

Another issue that can occur with scanning, is noise introduced by pulsing relatively large currents on a continual basis. This can be a particular problem where

Program 11.15 Displaying the decimal equivalent of a binary byte

```
; Task 1 -----------------------------------------------------
DISPLAY movf   BINARY,w      ; Get binary byte
        call   BIN_2_BCD     ; Convert to 3-digit BCD

; Task 2 -----------------------------------------------------
        movf   HUNDREDS,w    ; Get Hundreds nybble
        call   SVN_SEG       ; Convert to 7-segment code
        movwf  PORTB         ; Send out to PortB
        bsf    PORTA,2       ; Clock into register
        bcf    PORTA,2

; Task 3 -----------------------------------------------------
        movf   TENS,w        ; Get Tens nybble
        call   SVN_SEG       ; Convert to 7-segment code
        movwf  PORTB         ; Send out to PortB
        bsf    PORTA,1       ; Clock into register
        bcf    PORTA,1

; Task 4 -----------------------------------------------------
        movf   UNITS,w       ; Get Units nybble
        call   SVN_SEG       ; Convert to 7-segment code
        movwf  PORTB         ; Send out to PortB
        bsf    PORTA,0       ; Clock into register
        bcf    PORTA,0
```

Program 11.16 Displaying a 3-digit decimal number on a scanning readout

```
; Task 1 --------------------------------------------------------
DISPLAY movf   BINARY,w      ; Get binary byte
        call   BIN_2_BCD     ; Convert to 3-digit BCD

; Task 2(a) -----------------------------------------------------
LOOP    movf   HUNDREDS,w    ; Get Hundreds nybble
        call   SVN_SEG       ; Convert to 7-segment code
        movwf  PORTB         ; Send out to PortB
        bcf    PORTA,2       ; Enable Hundreds display
        call   DELAY_10MS    ; for 10ms
        bsf    PORTA,2       ; and turn off

; Task 2(b) -----------------------------------------------------
        movf   TENS,w        ; Get Tens nybble
        call   SVN_SEG       ; Convert to 7-segment code
        movwf  PORTB         ; Send out to PortB
        bcf    PORTA,1       ; Enable Tens display
        call   DELAY_10MS    ; for 10ms
        bsf    PORTA,1       ; and turn off

; Task 2(c) -----------------------------------------------------
        movf   UNITS,w       ; Get Units nybble
        call   SVN_SEG       ; Convert to 7-segment code
        movwf  PORTB         ; Send out to PortB
        bcf    PORTA,0       ; Enable Units display
        call   DELAY_10MS    ; for 10ms
        bsf    PORTA,0       ; and turn off


        bra    LOOP          ; DO forever
```

analog circuitry is adjacent. Good power-supply decoupling can reduce this problem to some extent.

Self-Assessment Questions

11.1 Many situations call for more parallel I/O port lines than are available from one device; especially when other peripheral modules are using the shared I/O pin budget. One solution to the problem is to use MSI logic; such as decoders and registers, to expand a single port.[10]

An alternative approach is to use additional MCUs solely to expand the bus. Figure 11.21 shows a mooted expander, based on a 40-pin device. This gives three I/O ports, at the cost to the master of 12 lines. Eight of these implement

---

[10]For an example, see my *The Quintessential PIC® Microcontroller* Fig. 11.12.

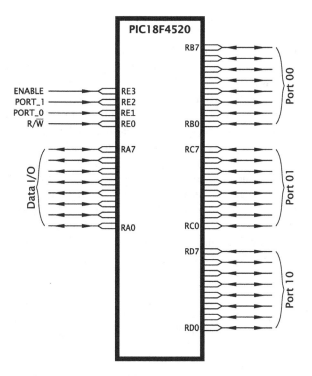

Fig. 11.21 A proposed bus expander

the data pathway and four target the expanded port, data direction and enable the expander.

Design code that will integrate with this hardware to allow the master device to set-up the data pathway when the expander is enabled; that is when RE3 = 1. The actual port is a function of RE2:1.

11.2 Pins RC[1:0] are to be configured as outputs with an initial value of 0 on Power-on Reset. The following code is designed to clear both flip flops before changing the port bits to output. On testing, the end result for RC0 is the opposite to the desired outcome. Why is this so and can you modify the code to rectify the situation?

```
bcf    PORTC,0     ; Clear flip flop 0
bcf    PORTC,1     ; Clear flip flop 1
movlw b'11111100'  ; Make RC[1:0] outputs
movwf TRISC
```

11.3 A certain system needs to be able to both activate eight LEDs and to read the state of up to eight normally-open (N.O.) push switches. It has been proposed that a single Port B might be able to combine these functions—the former when set to output, the latter when set to input. Can you devise a suitable circuit?

11.4 Extend the digital comparator of Example 11.2 to compare two *external* digital bytes presented to a 28-pin footprint PIC MCU, with byte P being input at Port B and Q at Port C.

11.5 In a low-power wireless data logging system, placing the PIC MCU in its Sleep mode will not affect the current consumption of the radio transmitter. It is proposed to use a port pin to supply current to the transmitter and in this way the auxiliary circuitry can be switched on and off as necessary. Discuss.

11.6 The variation of logic 0 output voltage V_{OL} against sink current I_{OL} for the two extremes of the commercial temperature range is shown in Fig. 11.22. Using this graphical relationship, determine the maximum value of a series resistor to ensure that a current of no less than 20 mA will flow through an LED connected to +5 V, as shown in the diagram, for any temperature. With this value what will be the current at $-40°$C? Assume that the conducting voltage across the LED is a constant 2 V.

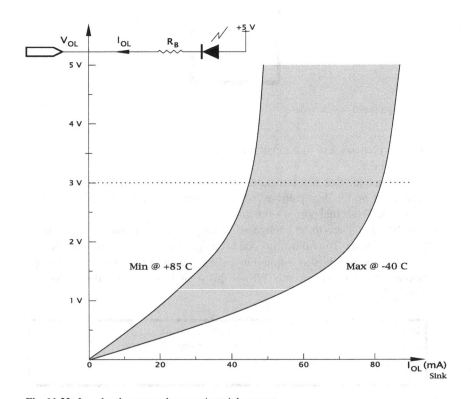

Fig. 11.22 Low-level output voltage against sink current

Chapter 12
One Bit at a Time

Parallel data transmission is fast, with a minimum of software overhead. Nevertheless, there are many circumstances where its use is inappropriate, either because of the additional hardware cost of multiple conductors, or more commonly where the receivers are geographically distant, with the concomitant cost or non-availability of multiple communication channels and their necessary interface hardware. In such situations data can be sent *one bit at a time* and assembled by the remote device into the original data bytes. In this manner a comparison can be made with the Parallel port on a PC, commonly used for local peripherals, such as a printer, and the Serial or USB ports frequently used with a modem to link into the Internet via a single telephone line.

In this chapter we will examine a range of techniques used to serially transmit data, both using bespoke shift register circuits and modules using standard communication protocols. After reading this chapter you will:

- Understand the need for serial transmission.
- Be able to design serial ports and associated software routines to communicate with standard parallel peripheral devices.
- Be capable of interfacing serial peripheral devices using both the SPI™ and I²C protocols.
- Appreciate the need for asynchronous serial communication and be able to write software drivers conforming to this protocol.
- Be able to use the integral Enhanced Universal Synchronous Asynchronous Receiver-Transmitter module (EUSART) for asynchronous protocols.
- Understand the necessity for buffering long-distance communication circuits.

As an example of serial data communications, consider the smart cards in your wallet. Each card has an embedded microcontroller, typically 8-bit, giving it its intelligence. Cost constraints are severe to give a manufacturing price of under $1, and a large component of this is accounted for by the non-corrosive gold-plated contacts, via which the microcontroller is powered and clocked when in contact with the card reader. In order to keep the mechanical precision of the reader low and hence reliability high, the number of contacts must be minimized and the pad size maximized.

S. Katzen, *The Essential PIC18® Microcontroller,*
Computer Communications and Networks,
DOI 10.1007/978-1-84996-229-2_12, © Springer-Verlag London Limited 2010

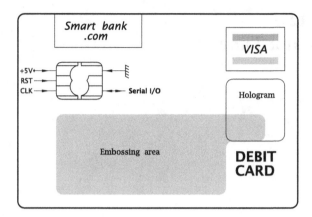

Fig. 12.1 The smart card

The standard arrangement shown in Fig. 12.1 uses contacts to provide the two power nodes, Reset and Clock, and *one* line to allow data to be shifted in or out *one* bit at a time. Although this is relatively slow, in comparison to the human–mechanical constraints speed is not an issue. Furthermore, contact between the reader/automatic teller and the central computer, perhaps several thousands of miles/kilometers away, will typically be via a single channel telephone line.

Check the parallel 3-digit 7-segment display interface of Fig. 11.19 on p. 371, which uses both Parallel Ports A and B. Although this is a working circuit, most of the parallel port budget of an 18-pin footprint device has been used up. Speed is certainly not a factor here, so a slower mode of data transmission is acceptable.

Consider the serial equivalent shown in Fig. 12.2. Here only two port pins are used. One labeled **SDO** (**Serial Data Output**) outputs the data bit by bit, with the most significant bit first. The other, labeled **SCK** (**Serial ClocK**), is used to clock the three shift registers at the same time, and hence shift the data left one bit at a time, in the manner of Fig. 3.8 on p. 59.

Each display has an associated 74HCT164 8-bit shift register[1]—see Fig. 2.22 on p. 36. The 74HCT164 has a positive-edge triggered shift input clock C1 and two serial data inputs ANDed together at 1D. One of these data inputs can be used to gate the other input, but in our example they are both connected together to give a single serial input. There is also an active-Low Reset input to clear the register contents, which are held High in the diagram. If desired, another port line can be used to drive R.

To change the display, a total of 24 bits will have to be shifted into the register array. To see how this can be done we will repeat the 7-segment driver routine of Program 11.15 on p. 375, which converts a binary byte to an array of BCD digits in HUNDREDS, TENS and UNITS. These are mapped to 7-segment code and then sent out to each digit 8-bits at a time.

---

[1] All data outputs are simultaneously available and thus the 74HCT164 is best described as a serial-in parallel-out (SIPO) register as well as a SISO shift register.

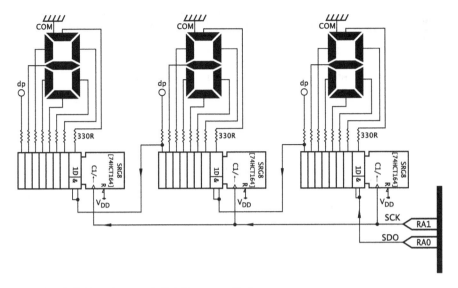

Fig. 12.2 Serial interface to a 3-digit 7-segment display

To serialize this process we need to design a subroutine (which we will call DATA_OUT) to copy each bit of a File to the RA0/SDO pin; beginning with the leftmost bit. At the same time the RA1/SCK pin is pulsed to clock the data. A task list for such a subroutine is:

1. Bring SCK to its Low state.
2. COUNT = 8.
3. WHILE COUNT > 0 DO:
   (a) Copy most-significant bit of DATA_OUT to SDO.
   (b) Shift DATA_OUT left one place.
   (c) Pulse SCK ⎍.
   (d) Decrement COUNT.

Program 12.1 shows two subroutines. The first called DISPLAY is closely akin to Program 11.15, in that it calls the subroutines BIN_2_BCD and then sends the 7-segment coded bytes out to the interface registers. In this instance the Hundreds byte is sent first, as this will eventually be shifted to the far end of the chain, followed by the Tens and finally the Units byte.

The actual serial transmission is handled by the subroutine SPI_WRITE, which implements our task list. The state of bit 7 of the datum pre-placed by the caller in File DATA_OUT is tested and its state used to make the Serial Data Out pin RA0 High or Low. The Serial ClocK pin RA1 is then toggled once ⎍ to shift the data into the shift register chain. The data byte is then shifted left and the process repeated in total eight times, to complete the transaction. This takes a maximum of 87 cycles to complete, depending slightly on the data pattern. A complete update of the 3-digit display will take around 120 μs with a processor clock of 8 MHz; excluding the time spent in doing the data conversion.

Program 12.1 Displaying the decimal equivalent of a binary byte using a serial data stream

```
DISPLAY bcf     PORTA,SCK    ; Initialize the clock line
        movf    BINARY,w     ; Get binary byte
        call    BIN_2_BCD    ; Convert to 3-digit BCD
        movf    HUNDREDS,w   ; Get HUNDREDS nybble
        call    SVN_SEG      ; Convert to 7-segment code
        movwf   DATA_OUT     ; Copy into the serial register
        call    SPI_WRITE    ; Shift it out

        movf    TENS,w       ; Get Tens nybble
        call    SVN_SEG      ; Convert to 7-segment code
        movwf   DATA_OUT     ; Copy into the serial register
        call    SPI_WRITE    ; Shift it out

        movf    UNITS,w      ; Get Units nybble
        call    SVN_SEG      ; Convert to 7-segment code
        movwf   DATA_OUT     ; Copy into the serial register
        call    SPI_WRITE    ; Shift it out
        return

; ****************************************************************
; * FUNCTION: Clocks out a byte in series, MSB first           *
; * ENTRY   : Datum in DATA_OUT                                *
; * EXIT    : DATA_OUT unchanged                               *
; ****************************************************************
; Task 1 --------------------------------------------------------
SPI_WRITE
        bcf     PORTA,1      ; Make sure clock starts at Low

; Task 2 --------------------------------------------------------
        movlw   8            ; Initialize loop counter to 8
        movwf   COUNT

; Tasks 3(a)&(b) ------------------------------------------------
LOOP    bcf     PORTA,0      ; Zero data bit
        btfsc   DATA_OUT,7   ; Skip if MSB is 0
         bsf    PORTA,0      ; ELSE make data bit 1
        rlncf   DATA_OUT,f   ; Shift datum right one place

; Task 3(c) -----------------------------------------------------
        bsf     PORTA,1      ; Pulse clock
        bcf     PORTA,1

; Task 3(d) -----------------------------------------------------
        decfsz  COUNT,f      ; Decrement count
         bra    LOOP         ; and repeat until zero
        return
```

Program 12.2 shows a possible **C** implementation of our output subroutine. Function spi_wrt()[2] accepts a data byte and in a loop of eight copies bit 7 out to SDO,

---

[2]I have used this name as we shall see, there is a built-in function called spi_write().

Program 12.2 A **C** implementation of the `SPI_WRITE` subroutine

```
void spi_wrt(datum)
{
int k;
for(k=0;k<8;k++)                   /* DO eight times            */
    {
    if((datum & 0x80)) {SDO = 1;}
    else {SDO = 0;}
    SCK = 1;                       /* Clock the external receiver */
    SCK = 0;
    datum = datum << 1;            /* Shift datum left one place  */
    }
}
```

whilst shifting left. The two SPI pins have been previously defined as the appropriate port pin.

Where a long chain of shift registers is being serviced, speed may be improved a little if each register has its own data feed but all clocked with the same SCK pin or sharing the same lines but each with a separate Enable. This latter technique is the method used in Fig. 12.7.

One problem with our shift register technique is that for the period where shifting is in process, the data appearing at the port outputs are not valid—for 23 clock pulses in our example. Of course in this situation the response of the eye to microseconds-long changes in illumination makes this observation spurious. However, this may not always be the case and in such instances the shift register may be buffered from the parallel outputs using an array of D flip flops or latches. These can be loaded after the shifting process has been completed to give a single update.

Rather than employing a separate buffer register, many devices optimised for serial data transmission have integral PIPO registers. For example, the 74HCT595 shown in Fig. 12.3, is a latched shift register with integral 8-bit parallel-in parallel-out (PIPO) register between the shift register and the outside world. A rising edge _⌐ on the RCK (Register ClocK) pin transfers the serialized data to the parallel outputs. The last stage output of the shift register is made available to allow cascading to any length. In this situation, all RCK pins can be pulsed together to allow the entire chain to update simultaneously.

One example where rippling of data may be undesirable is where a digital datum is to be converted to its analog equivalent. In Fig. 12.3 the conversion is carried out using a National Semiconductor DAC0800. Essentially the analog voltage is a linear function of the 8-bit digital input varying from −9.96 V for an input of b'00000000' through +9.96 V for b'11111111'—see Fig. 14.18 on p. 525.

Using a 74HCT595 registered shift register, the digital input does not change until the new datum is in place and the PIC MCU pulses the RCK Register Clock. This gives clean changes in the data presented to the DAC and corresponding analog output.

Data can be input serially in a similar manner using parallel-in serial-out (PISO) shift registers. The example shown in Fig. 12.4 is a serialized intruder alarm us-

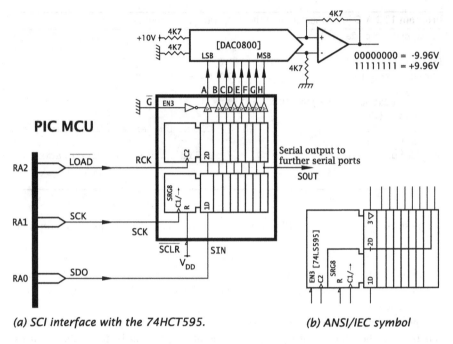

(a) SCI interface with the 74HCT595.   *(b) ANSI/IEC symbol*

Fig. 12.3 Serially interfacing to a DAC digital-to-analog converter using a 74HCT595 octal shift register with output register

ing only three lines to connect to eight zones of eight sensors each; a considerable economy compared to a parallel equivalent.[3]

Each sensor group is attached to a 74HCT165 8-bit PISO shift register, with the serial output of the further register feeding the serial input of the next left register. Once the data has been loaded in, it may be shifted into the **SDI (Serial Data In)** Parallel Port A input pin RA1 and assembled in software bit by bit. In the specific case of the multi-zone intruder alarm, after each eight shifts the assembled byte can be tested for non-zero and the appropriate action taken.

Also shown in Fig. 12.4 is the single output port used to display any active zone. As both input SDI and output SDO serial channels share the same shift clock SCK, then shifting data in will simultaneously clock this serial output port. Conversely, sending data to the output port will shift data in from the Zone ports. In this example there is no problem, as microsecond fluctuations in the Zone lamps are of no consequence, and the sequence of operations ends with the output port being loaded with the earmarked data. Where this interaction is undesirable, then either a latched register, such as the 74HCT595, should be used to staticize the display data or separate serial clock lines could be utilized.

The core serial interface software SPI_READ is the input counterpart of subroutine SPI_WRITE in Program 12.1, and implements the following task list:

---

[3]See Fig. 11.12 in my *The Quintessential PIC® Microcontroller*.

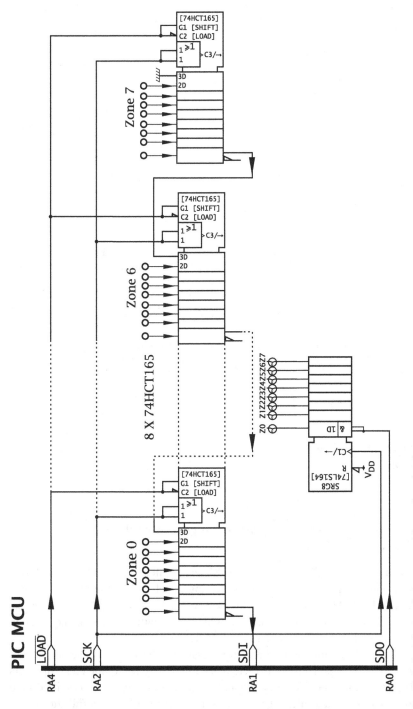

Fig. 12.4 Serially interfacing to a multi-zone intruder alarm

Program 12.3 Input serial byte subroutine

```
; *****************************************************************
; * FUNCTION: Clocks in a byte in series, MSB first             *
; * ENTRY    : None                                             *
; * EXIT     : Datum in DATA_IN; COUNT = 0                       *
; *****************************************************************
; Task 1: Bring SCK Low -----------------------------------------
SPI_READ     bcf      PORTA,SCK    ; Make sure clock starts at Low

; Task 2: COUNT=8 -----------------------------------------------
             movlw    8            ; Initialize loop counter to 8
             movwf    COUNT

; Task 3: WHILE COUNT>0 DO: -------------------------------------
; Task 3 (a): Pulse SCK -----------------------------------------
SER_IN_LOOP bsf      PORTA,SCK
             bcf      PORTA,SCK

; Task 3(b): Shift datum left -----------------------------------
             bcf      STATUS,C     ; Zero the Carry flag
             rlcf     DATA_IN,f    ; Shift it in and datum once left

; Task 3(c): IF SDI is 1 THEN set bit 0 (rightmost bit) -------
             btfsc    PORTA,SDI    ; Skip if SDI == 0
             bsf      DATA_IN,0    ; ELSE set bit0 to 1

; Task 3(d): Decrement COUNT and repeat Task3 WHILE>0 ---------
             decfsz   COUNT,f      ; Decrement count
             bra      SER_IN_LOOP  ; and repeat until zero

             return
```

1. Bring SCK to its Low state.
2. COUNT = 8.
3. WHILE COUNT > 0 DO:
   (a) Pulse SCK _/‾_.
   (b) Shift DATA_IN left once place.
   (c) Copy input state of pin SDI into least-significant bit of DATA_IN.
   (d) Decrement COUNT.

This task list is similar to that on p. 381, except that File DATA_IN is shifted left once and the state of the SDI pin following the clock pulse at pin SCK copied as the new bit 0. After eight clock-shift-test loops the datum in DATA_IN is the parallelized byte assembled from the serial input port, with the first bit ending up in the leftmost significant placeholder in DATA_IN.

The SPI_READ subroutine coded in Program 12.3 is similar to the output subroutine SPI_WRITE of Program 12.1. Indeed they may be combined so that data is shifted out of the specified output data File at the same time as it is shifted in to the specified input data File. This type of scheme is referred to as **full duplex**, as

Program 12.4 A **C** implementation of a SPI input read

```
unsigned int spi_read()
{
unsigned int k;
for(k=0;k<8;k++)                /* DO eight times              */
    {
    SCK = 1;                    /* Clock Slave TX bit to SDI */
    SCK = 0;
    DATA_IN = DATA_IN << 1;     /* Shift left one place        */
    if(SDI)
        {DATA_IN = DATA_IN | 0x01;} /* Set bit 0 IF SDI is 1 */
    else
        {DATA_IN = DATA_IN & 0xFE;} /* ELSE make it a 0        */
    }
return data_in;                 /* Return complete byte        */
}
```

opposed to **half duplex** where only one direction at a time is possible. A serial link where data flow can only be in one fixed direction is known as **simplex**.

The **C** coding of Program 12.4 follows the same coding strategy as the assembly counterpart. Note how Inclusive-OR'ing with b'0000001' using the **C** | operator is used to set bit 0 of the variable DATA_IN. Similarly AND'ing with b'11111110' clears bit 0. Specifically in CCS **C** the non-standard integral functions bset(DATA_IN,0) and bclr(DATA_IN,0) can be used to set or clear any bit in a variable, and when single bits are involved, is often more efficient than using logic operators.

The serial protocol similar to that described in this example is commonly known as **serial peripheral interface (SPI[TM])**.[4] Microwire[TM] is a similar, but not identical, serial protocol.[5] SPI is a sufficiently standardized protocol used by most microcontrollers to allow manufacturers to produce a range of peripheral devices specifically designed to directly interface to this bus without the necessity to add external shift registers. As an example of this genre, the MAX549A of Fig. 12.5 is a dual digital-to-analog converter (DAC) which is powered with a V_{DD} of +2.5 V to +5.5 V. Its operating current is typically 150 µA per DAC at 5 V and either or both DACs can be shut down to reduce the current drain to less than 1 µs in its Standby mode. Data can be clocked in at a rate of up to 12.5 MHz. All this functionality is available in an 8-pin package and should be contrasted with the 20-pin MAX506 of Fig. 14.17 on p. 524, designed for direct parallel port connection.

The simplified functional model of the MAX549A shown in the diagram shows an integral 16-stage shift register clocked from SCLK and fed data via DIN using the normal SPI protocol. The additional eight locations are used to store four control bits, with the following functionality:

----

[4]SPI[TM] is a trademark of Motorola/Freescale, Inc.

[5]Microwire[TM] is a trademark of National Semiconductor Corporation.

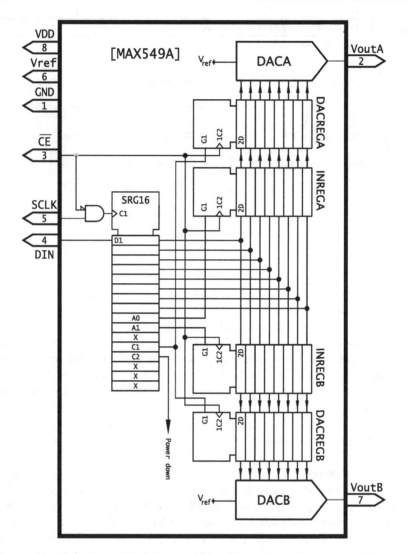

Fig. 12.5 The MAXIM MAX549A SPI dual 8-bit DAC

A0
Enables the input PIPO register for channel A and which is clocked on a rising edge at the $\overline{CE}$ pin.

A1
Enables the input PIPO register for channel B and which is clocked on a rising edge at the $\overline{CE}$ pin.

C1
Gates both DAC registers; allowing them to be updated simultaneously by a ⌐⌐ on $\overline{CE}$.

C2

When 1 will power down any DAC selected with A0 or/and A1. This disconnects the reference voltage V_{ref} from the DAC's resistor network (see Fig. 14.16 on p. 523) and leaves only a residual current of less than 1 µA to activate the internal registers, whose contents remain unchanged.

Both DACs have a 2-layer register pipeline isolating them from the shift registers. The first layer is the In registers, which are gated when A0 or A1 as appropriate is 1. The data sitting in the first byte of the shift register can then be clocked in by pulsing $\overline{CE}$ (pin 3) Low. This change will be stored but will not appear at the input of the DAC until the next layer of PIPO registers are clocked. These registers are enabled when C1 is 1 and $\overline{CE}$ is pulsed. This means that one data byte can be sent to, say, DACA and then another to DACB. The DAC registers can then be updated together, resulting in both outputs V_{outA} and V_{outB} changing simultaneously—see Program 12.5. This can even be done when the MAX549A is asleep, as the registers are not affected by this power-down state. From this discussion we see that each transition from the PIC MCU takes two 8-bit transfers | Control | Data | followed by a $\underline{\diagup}$ on the $\overline{CE}$ pin.

For our example we will send the contents of File h'020' to Channel A and then the contents of File h'021' to Channel B; at that point updating both DAC registers and hence outputting the analog equivalent of File h'020' to pin V_{outA} and File h'021' to pin V_{outB}.

Our implementation will involve the transmission of four bytes of information:

1. Control byte 1: b'XXX00X01'
   No power down, update Channel A, no output change.
2. Data byte 1:
   Contents of File h'020'.
3. Pulse $\overline{\diagdown\diagup}$ $\overline{CE}$.
4. Control byte 2: b'XXX01X10'
   No power down, update Channel B, both outputs change.
5. Data byte 2:
   Contents of File h'21'.
6. Pulse $\overline{\diagdown\diagup}$ $\overline{CE}$.

The hardware–software interaction is shown in Program 12.5. Four bytes are transmitted using subroutine SPI_WRITE, with the MAX549A's $\overline{CE}$ being pulsed $\diagup\overline{\diagdown}$ after each | Control | Data | byte pair. The final process sets C1 High, which transfers both data bytes to the DAC registers. At the same time the Channel B In register is updated.

Looking at the three pins on the MAX549A would give a waveform similar to that of Fig. 12.6 for the transmission of the first | Control | Data | byte pair. During the transmission $\overline{CE}$ remains Low, with the data shifting into the MAX549A's integral shift register. After the second byte, i.e., the 16th clock pulse, bringing $\overline{CE}$ High activates the selected internal registers, executing the instruction.

Program 12.5 Interacting with the MAX549A dual-channel SPI DAC

```
CE       equ 2

; *****************************************************************
; * FUNCTION: Sends out Channel A & B data in SPI protocol to    *
; * FUNCTION: MAX549A simultaneously updating outputs            *
; * RESOURCE: Subroutine SPI_WRITE                               *
; * ENTRY    : Channel A in File h'020', Channel B in File h'021'*
; * EXIT     : Both analog outputs updated                       *
; *****************************************************************
MAX549A movlw  b'00000001'      ; Control byte 1
        movwf  DATA_OUT         ; Put in designated location
        call   SPI_WRITE        ; and send out to MAX549A and get
        movff  h'020',DATA_OUT  ; ChannelA data to named location
        call   SPI_WRITE        ; and send out to MAX549A
        bsf    PORTA,CE         ; Pulse CE
        bcf    PORTA,CE

        movlw  b'00001010'      ; Control byte 2
        movwf  DATA_OUT         ; Put in designated location
        call   SPI_WRITE        ; and send out to MAX549A
        movff  h'021',DATA_OUT  ; Get ChannelB data to named place
        call   SPI_WRITE        ; and send out to MAX549A
        bsf    PORTA,CE         ; Pulse CE
        bcf    PORTA,CE
        return
```

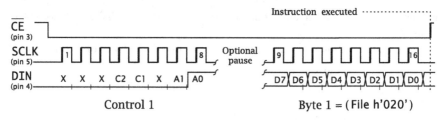

Fig. 12.6 SPI waveforms for the MAX549A

The diagram shows transitions on the DIN line from the PIC MCU's SDO pin, occurring sometime before the active rising edge on SCK. Sometime is a vague term; obviously it must occur no later than a minimum time before _/‾ and be held for a short time after. The MAX549A data sheet gives the minimum set-up time t_{DS} of 30 ns and hold time t_{DH} of 10 ns. Even at a PIC MCU clock rate of 40 MHz an instruction cycle takes 100 ns, so timing will not be violated.

By judicious use of the MAX549A's $\overline{CE}$ input, several DACs may be connected to the SCK/SDO lines, with a serial transmission only being shifted into the device which has its $\overline{CE}$ Low. Figure 12.7 shows two MAX549As sharing the one SPI link, giving four analog output channels in total. Using a 2- to 4-line decoder in conjunction with RA3:2 would enable up to four MAX549As, with a total budget of only four port lines.

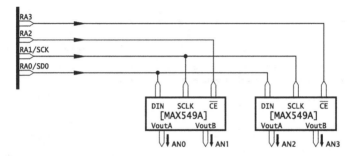

Fig. 12.7 Multiple MAX549As on the one SPI circuit

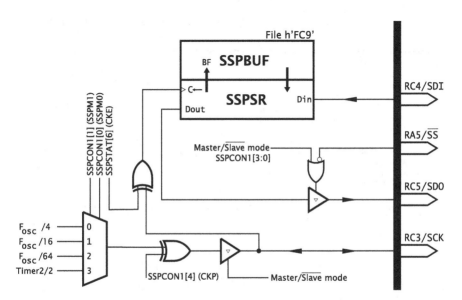

Fig. 12.8 The basic SPI Serial Synchronous Port set to implement SPI. Pinning is shared with Parallel Port C for 28-pin+ devices

Most mid-range and all extended-range PIC MCUs feature an integral synchronous serial port (SSP) which implements, amongst others, the SPI protocol. The PIC18 family, in common with later members of the PIC16 range, implement this protocol using the **Master Synchronous Serial Port** (**MSSP**) module.

A somewhat simplified representation of the MSSP module set-up for the SPI protocol is shown in Fig. 12.8. The heart of the Master Synchronous Serial Port is the SFR **SSPBUF** (**SSP BUFfer**). A datum byte written into this SFR will automatically be transferred into the **SSP Shift Register** (**SSPSR**) and shifted out of the PIC MCU's dedicated SDO pin. At the *same* time, eight bits of data will be shifted in from the SDI pin. When this frantic burst of activity is completed, the new byte is automatically transferred to SSPBUF, from where it can be read. This transfer is

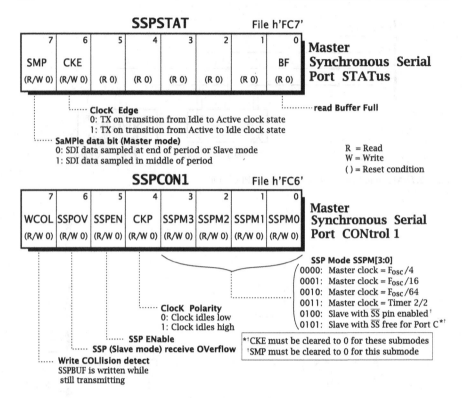

Fig. 12.9 The MSSP module's CONtrol and STATus registers as appropriate for the SPI mode

signaled by setting the **BF** (**Buffer Full**) flag in the **SSPSTAT** (**SSP STATus**) register, shown in Fig. 12.9. Once SSPBUF is read, BF is automatically cleared.

Apart from parallel ports, in general interface modules are configured and monitored with a set of associated Control and Status registers. In addition, interrupt mask bits and flags are located either in one or two Peripheral Enable and Interrupt registers, such as shown in Fig. 7.2 on p. 209. Peripheral control, status and interrupt registers are normally set-up as part of the startup initialization routine, where parallel ports are configured as input/output. As such modules invariably multiplex their pins with the parallel ports; even if these latter are not used, such shared pins often need to be considered in this initialization phase. Pin I/O settings may be automatically overridden if the peripheral module is enabled or may need to be 'manually' set-up by the software. Unfortunately the settings are not always obvious and the data sheet should be consulted for specific information.

Returning to the specific case of the MSSP, Fig. 12.9 shows the **SSPCON1** (**SSP CONtrol 1**) register and **SSPSTAT** (**SSP STATus**) registers set-up for the SPI protocol and I/O pinning. In connecting to the outside world, four pins need to be considered.

RC5/SDO
Bit TRISC[5] must be cleared to 0 to allow this pin to output data.

RC4/SDI
This pin is overridden by the MSSP irrespective of the state of its associated TRISC[4] bit.

RC3/SCK
When in one of the Master modes, bit TRISC[3] must be cleared to allow this pin to output the clock signal. Conversely, when in one of the Slave modes, TRISC[3] must be set to allow this pin to input the clock signal from an external Master.

RA5/$\overline{SS}$
In Slave mode b'0100' this pin should be configured as an input; i.e., TRISA[5] should be set to 1, in order to allow the external Master to select this device.

On any kind of reset both SSPCON1 and SSPSTAT registers are cleared and the internal data bit counter is zeroed. In this situation the module is disabled and if the programmer wishes to use the MSSP then the various control bits[6] must be set-up.

SSPEN
Setting SSPCON1[5] to 1 enables the Synchronous Serial port. If disabled the associated pins can be used as normal parallel port lines.

SSPM[3:0]
The four SSP Mode switch bits located in SSPCON1[3:0] are used to set the communication protocol and various Master/Slave options. The diagram shows the six combinations relevant to the SPI protocol.

The four Master submodes only differ in selecting the one of four internal clocking frequencies. Three of these frequencies are derived from the main PIC MCU oscillator. For example, with a 20 MHz crystal the SCK shift rate can be selected as 5, 1.25 MHz and 312.5 kHz (200, 800 ns and 3.2 μs). The final selection gives the shift rate as half the frequency generated by Timer 2 overflowing—see Fig. 13.6 on p. 471. This option is used where slow or variable shift rates are required.

The Slave options use a clock coming from an outside Master driving the SCK pin. Optionally the $\overline{SS}$ pin can be used by this external Master device to select one of several Slaves—see Fig. 12.12.

SSPOV
In a Slave mode this status bit indicates that a new byte has been received before the previous byte has been read; that is, a byte or bytes have been lost. To zero this bit, SSPBUF must first be read and then SSPOV cleared in software. The SSP OVerflow status bit does not operate in a Master mode.

WCOL
When software attempts to write a byte into the SSPBUF before the last byte in

---

[6]Some of which are in the Status register due to lack of space!

the SSPSR has been completely shifted out, the action is aborted and the Write COLlision bit is set. This bit can be tested, and if a collision is confirmed, should be cleared in software and the process subsequently tried again.

CKP, CKE, SMP

These three bits work in tandem to ensure that the correct clock edges are used to shift data into and out of the remote receivers and transmitters and that incoming data is sampled only after stabilization.

In order to illustrate the various possibilities, consider the situation when the MSSP is set-up as a Master. As a Master device the MSSP has complete control of the clocking signal at SCK, which is used to clock *both* the remote Slave transmitter and receiver shift registers. The Slave receiver shift registers require an active edge on this clock when the Master data at pin SDO is stable. In addition, Slave transmitter shift registers need to be clocked so that their data bit is stable when the MSSP reads it at its SDI pin. An example of this situation is shown in Fig. 12.12, where the PIC MCU-based Master SPI device can select one of two Slaves. Each enabled Slave can both transmit and receive data simultaneously. The Slaves can be other PIC microcontrollers (as shown) or any SPI circuit.

Each byte transmission is broken up into eight clock phases; as shown in Fig. 12.11. In all situations, the next data bit D_n will be presented at the SDO pin shortly (in Industrial devices, not more than 50 ns) after the beginning of each clock phase; see top of diagram. The remote Slave receiver should then be in a position to clock it mid-phase. Similarly, the remote Slave transmitter should present its data bit d_n to pin SDI in time to be sampled by the Master.

Figure 12.10 is split up into two broad situations. The top two SCK waveforms are used when the remote transmitters and receivers have opposite active shift clock edges. As the transmitter is clocked at the beginning of each phase it should be sampled mid-phase by setting SMP = 0.

CKE:CKP = 0:0

When the remote transmitter is clocked by a rising edge ⎍ then its data at SDI should be sampled mid-phase. Such data should be present at least 100 ns before this point and be held for at least 100 ns afterwards. The remote Slave receiver clocks in its data from SDO on a falling edge ⎍ on SCK, also mid-phase. In standard SPI terminology, this is described as mode 0,1.

CKE:CKP = 0:1

SPI mode 1,1 is similar, but the remote transmitter is clocked by a ⎍ edge and the remote receiver by a ⎍ edge.

Where all remote transmitters and receivers have the same active shift clock edge, then the bottom waveforms are applicable. As the transmitter is clocked mid-phase, its data at SDI should be sampled at the end of the clock phase by making SMP = 1.

CKE:CKP = 1:0

SPI mode 0,0 is used where both Slave transmitters and receivers are clocked

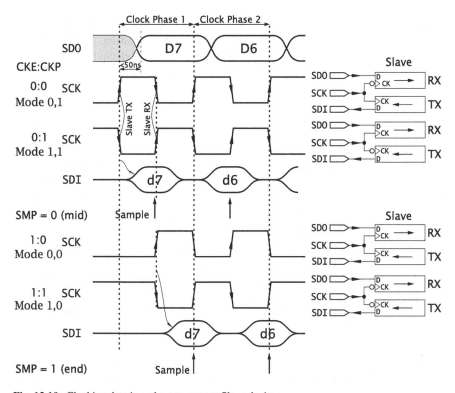

Fig. 12.10 Clocking data in and out to remote Slave devices

together mid-phase on a ⟋ edge. By that time the Master data bit D_n will be stable to be clocked into the remote Slave. The remote Slave's data should be ready for sampling by the end of the clock phase.

CKE:CKP = 1:1

SPI mode 1,0 generates a ⟍ edge mid-phase to trigger both Slave transmitters and receivers.

When the MSSP is being used in one of the Slave modes, the clock comes from a remote device. As before, any data previously loaded into the SSPBUF will be clocked out from the SDO pin at the beginning of each clock phase. The CKE and CKP bits still need to be set according to whether the remote sender's data is output on a ⟋ or ⟍ edge of its clock and on the active edge of a remote receiver. Also relevant, is if the remote Master has its first D_n bit presented before or after the first clock pulse. In all such cases the Slave MSSP-configured module should sample such data as it presented at its SDI pin at the end of each clock phase; that is, SMP = 0. The reader should refer to the device data sheet for specific waveforms.

When a PIC MCU is set-up as a Slave SPI device, the $\overline{SS}$ ($\overline{Slave\ Select}$) pin can be used by the remote Master to select it for an 8-bit transfer. When $\overline{SS}$ goes High, even in the middle of a transmission, the internal bit counter is reset to zero. Also the SDO pin goes open-circuit, so that another device can take over the line.

BF, SSPIF

When a complete frame of eight bits has been shifted in and been dumped into the SSPBUF Buffer register, BF goes to 1 to indicate that a new datum is ready for collection. This transfer also sets the SSPIF flag in PIR1[3], (see Fig. 7.2 on p. 209) and this can be used to initiate an interrupt if the companion SSPIE mask bit in PIE1[3] has been set. If the MSSP has been configured as a Slave and the PIC MCU is sleeping this can be used to waken the device. This is possible, as the SCK pin is clocked by the external Master device and thus the PIC MCU need not be active; that is, the system oscillator can be off.

Reading the newly arrived datum from SSPBUF automatically clears this Buffer Full bit. If not read on time, the datum will be lost and the SSPOV flag will be set to record this. The SSPIF interrupt flag needs to be manually cleared in any polled or interrupt service routine.

Using Figs. 12.8 and 12.9 as a programmer's model we can now deduce the hardware–software interaction task list in order to action a transmission of a byte and/or receive a new byte:

1. Configure SSP module.

   - Set-up RC3/SCK, RC5/SDO as outputs and RC4/SDI, and if appropriate RA5/$\overline{SS}$, as input.
   - Set-up Master/Slave mode with appropriate clock source.
   - Choose active clock edges with CKP:CKE:SMP.
   - Enable the SSP by setting SSPEN.

2. Move datum to SSPBUF to initiate transmission.
3. IF WCOL = 1 THEN reset WCOL and go to item 2.
4. Poll BF for 1.
5. Move RX data from SSPBUF, which also resets BF.

To illustrate this process, consider a subroutine SPI_IN_OUT, which combines the function of SPI_READ and SPI_WRITE; that is, it transmits the datum in File DATA_OUT, whilst at the same time returning the consequently received byte to DATA_IN. Assume that the remote shift registers are all ⎽╱⎺ triggered; that is, SPI mode 0,0.

The implementation of this subroutine depends on setting up the MSSP during the initialization phase of the main software after a reset. In the following code fragment we are using the $f_{OSC}/4$ clock rate Master mode:

```
          .include  "p18f4520.inc"
MAIN      movlw b'11010111'     ; RC5/SDO, RC3/SCK outputs
          movwf TRISC           ; RC4/SDI input
          movlw b'11000000'     ; Make CKE and SMP = 1
          movwf SSPSTAT
          ..... .....
          movlw b'00100000'     ; Enable SSP, TX clock idles Low
          movwf SSPCON          ; SPI Master, Fosc/4 rate
```

The coding shown in Program 12.6 follows the task list exactly. Data to be transmitted is moved from the designated File to SSPBUF and status bit WCOL checked to see that it got there. If there is a transmission in progress then the datum is not stored in SSPBUF and WCOL is set. If this subroutine is the only code to access the MSSP then this should rarely be the case and in most instances this check is omitted, but its inclusion makes the system more robust.

Once the transmit datum is *in situ*, the transmit sequence is immediately initiated, as shown in Fig. 12.11, and progresses to its conclusion. When the Buffer Full status flag BF is set, the received datum can be moved out of SSPBUF to its ordained location. This automatically resets BF.

Apart from a slight reduction in the code length, the advantage of using this hardware is the increase in speed. The actual transmit/receive takes eight SCK cycles, which in our case is eight instruction cycles. With an f_{osc} of 40 MHz, the clocking rate is 10 MHz (that is, a bit rate of 10 million bits per second; commonly written as 10 Mbit/s or 10 Mbps), giving a total time of 0.8 μs per byte.

Figure 12.11 shows the SPI mode timing for our subroutine. As we have cleared CKP and set CKE then SCK is idling Low. As soon as SSPBUF is written to, the

---

Program 12.6 Using the MSSP for SPI data input and output

```
; ****************************************************************
; * FUNCTION: Transmits and simultaneously receives one byte *
; * FUNCTION: from the SSP using the SPI protocol            *
; * ENTRY   : Data to be transmitted is in DATA_OUT          *
; * EXIT    : Data received is in DATA_IN                    *
; ****************************************************************
SPI_IN_OUT
      movff  DATA_OUT,SSPBUF    ; Get datum for TX into SSPBUF
      btfss  SSPCON,WCOL        ; Did it make it?
      bra    SPI_IN_OUT_CONT    ; IF so THEN continue
      bcf    SSPCON,WCOL        ; ELSE reset WCOL and try again
      bra    SPI_IN_OUT

SPI_IN_OUT_CONT
      btfss  SSPSTAT,BF         ; Check for Buffer Full
      bra    SPI_IN_OUT_CONT    ; IF not then poll again

      movff  SSPBUF,DATA_IN     ; ELSE get the RX'ed datum put away
      return
```

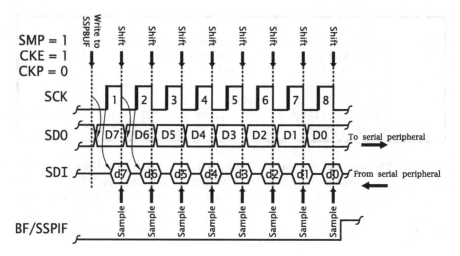

Fig. 12.11 SSP SPI-mode Master waveforms

MSB of the TX datum appears at SDO. In mid-phase the rising edge clocks this data into the remote receiver.

With the remote receiver also clocked at mid-phase there is plenty of time for its data to be presented to the PIC MCU's SDI. This data is then sampled by the PIC MCU at the end of each clock phase; that is SMP = 1.

One use of serial transmission is to connect a number of devices together in one multiprocessor network. For instance, a robot arm may have a MCU controlling each joint, communicating with a master processor. A simple multidrop circuit of one Master and two Slave processors is shown in Fig. 12.12.

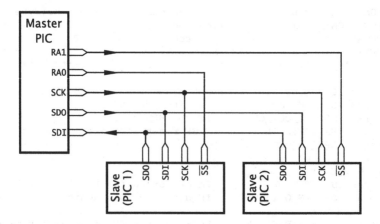

Fig. 12.12 A multidrop SPI communications network

In this configuration the Master PIC MCU externally drives the SCK of both Slaves; thus controlling when and how fast transmission occurs across the network. Both Slaves are configured in Mode 0100, so that the $\overline{\text{Slave Select}}$ inputs are enabled. Thus, if the Master wishes to read a datum from Slave 2, the latter's $\overline{\text{SS}}$ is brought Low and the Master clocks the eight bits from Slave 2's SSPBUF/SSPSR, into its own SSPBUF/SSPSR. At the same time any data transmitted by the Master will be received by the Slave.

SPI transactions may be coded in **C** either by mimicking the assemble-level code and setting/reading the appropriate registers, or by using built-in functions specific to the task. The key CCS compiler internal functions used for the SSPort in its SPI mode are:

setup_spi(spi_master|spi_h_to_l|spi_clk_div_4);
This example function instance configures the SSP as an SPI Master, with clock polarity rising edge and a ÷4 clock frequency. These scripts, and others such as spi_slave, spi_sample_at_end and spi_xmit_1_to_h, are part of the included header file; e.g. 18f4520.h. This function also sets the direction of the appropriate Port A and Port C pins.

spi_write(value);
This is used to write out the value from the SSP. It checks that the BF flag is set before returning.

spi_read();
This is virtually identical to spi_write() except that it returns the value read by the SSP. If a value is passed to this function then it will be clocked out of SDO. For this instance, spi_read(0x0A); will transmit the byte in the same manner as subroutine SPI_IN_OUT of Program 12.6.

spi_data_is_in();
This function returns non-zero if a datum has been received over the SPI connection; that is, if BF is set.

To illustrate this technique, consider our interface to the MAX549A coded in Program 12.5. In order to do this the SSP needs to be configured using code of the form:

```
#include <18f4520.h>
#bit     CE = 0xF80.2    /* Port A, bit 2 to MAX549A's CE    */
void MAX549A(unsigned int channel_A, unsigned int channel_B);
main()
{
set_tris_a(0xFB);        /* CE = RA2 output                  */
setup_adc(NO_ANALOGS);   /* Ports A & E all digital          */
setup_spi(spi_master|spi_1_to_h|spi_clk_div_4);
```

Program 12.7 Interfacing to the MAX549A in **C**

```
void MAX549A(unsigned int channel_A, unsigned int channel_B)
{
spi_write(0x01);        /* Send out Control 1              */
spi_write(channel_A);   /* Send out Data 1                */
CE=0; CE=1;             /* Pulse CE                       */
spi_write(0x0A);        /* Send out Control 2              */
spi_write(channel_B);   /* Send out Data 2                */
CE=0; CE=1;             /* Pulse CE                       */
}
```

---

in which we are assuming that the MAX549A's $\overline{\text{CS}}$ is connected to Port A's RA2 pin; as shown in Fig. 12.7.

The program comprises four `spi_write()` calls, with CE being pulsed between | Control | Data | pairs. This function may be called with an evocation something like `MAX549A(data_x, data_y);`

Although the SPI protocol is relatively fast, it requires a minimum of three data lines plus one select line for each duplex Slave device. Apart from the cost; adding a device to an original design will require some hardware modification. By increasing the intelligence of the Slave device, it is possible to send both control, address and data in the one serial stream. The **inter-integrated circuit (I^2C^{TM})** protocol developed by the Philips/Signetics Corporation (now NXP Semiconductors)[7] in the early 1980s embodies this concept and also reduces the interface to only two lines, by permitting bidirectional transmission—see Fig. 12.13.

SCL

This is the clock line synchronizing data transfer, serving the same function as SCK in the SPI protocol. SCL is bidirectional, to allow more than one Master to take control of the bus at different times.

The original I^2C specification set an upper limit on shift frequency of 100 kHz; that is, 100 kbit/s, but the specification was augmented in 1993 by a Fast mode with

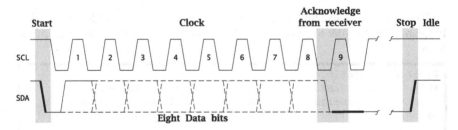

Fig. 12.13 Data transfer on the I^2C bus

---

[7]I^2C^{TM} is a trademark of the Philips/NXP Corporation.

an upper data rate of 400 kbit/s, which is the current *de facto* standard. In 1998 a compatible High-Speed mode was added with an upper bit rate of 3.4 Mbit/s.

SDA

This I^2C data line allows data flow in either direction. This bidirectionality allows communication from Master to Slave (Master-Write) or from Slave to Master (Master-Read). Furthermore it allows the receiver to signal its status back to the transmitter at the end of each byte.

The I^2C protocol is relatively complex and its full specification can be viewed at the NXP Semiconductors' web site.[8] Before looking at the basic protocol, we need to examine the SCL and SDA lines in more detail. When no data is being transmitted, both lines should be High; the **Idle condition**. A device wishing to seize control of an idling bus must bring its SDA output Low. This is known as the **Start condition**. In order for the would-be Master to be able to pull this line Low, all other devices hung on the line must have their SDA pins open circuit and the line as a whole pulled up to the High state through a single external resistor; see Fig. 12.14(a). To implement this, SDA outputs must be open-collector or open-drain—see Fig. 2.2(b)

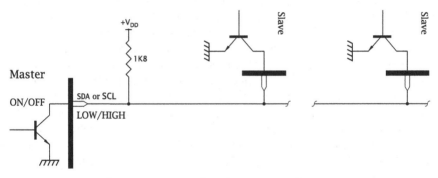

(a) Connection of I^2C devices to the I^2C bus

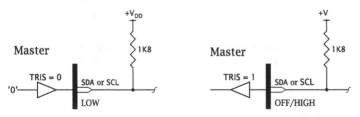

i Output is low   *ii Output is pulled high*

(b) Using the PIC MCU to simulate open collector

Fig. 12.14 Sharing the SCL and SDA bus lines

---

[8]www.nxp.com/acrobat/literature/9398/39340011.pdf.

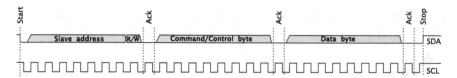

Fig. 12.15 A I²C packet transmission

on p. 19. This means that any device hung on the bus is able to pull its line Low by outputting a logic 0. SCL is also implemented in this way to permit separate Masters to clock the network. This **Multi-Master** I²C bus allows for more than one device to take over as a Master, but of course not simultaneously. Both bus lines will therefore need a pull-up resistor in the normal manner.

In addition to generating the Start event, the Master is responsible for generating the clock signal and also for sending an address code to the other entities on the bus, to establish communications with one or more Slave devices. Along with this address, a single bit tells the Slave if the data flow is to be from the Master to the Slave (Master-Write) or Slave to Master (Master-Read).

Each packet sent between Master and Slave comprises nine bits. Eight of these are data synchronized by the clock. Changes on SDA *must* only occur when the clock line is *Low*. Data is clocked into the receiver on the following SCK rising edge. These bytes may represent address or control information from the Master or data from either Master or Slave. The I²C protocol includes a handshaking mechanism—see Fig. 11.3 on p. 340. During the ninth clock pulse, the transmitting device releases the SDA line and the receiving device on the bus *acknowledges* the data sent by the transmitter. SDA is held Low by the receiver if the datum has been successfully acquired, giving an **ACK** state; as shown in Fig. 12.15. The alternative, where the receiver either signals a problem or that it doesn't want any more data, is called a Not ACKnowledge, or **NACK**. Normally in this latter situation, the transmitter will try again for a number of times before giving up.

Rather less drastic, the Slave device can hold the clock line Low. **Clock stretching** is useful where the Slave device cannot process incoming data fast enough. The Master will attempt to send clock pulses until SCL is released by the Slave.

In any situation, it is the responsibility of the Master to terminate the conversion by bringing the SDA line High when the clock line SCK is High; signaling a **Stop** condition. Another conversion can be started, if desired with a different Slave, by the Master sending another Start signal. It is possible for the Master to send out repeated Starts without first stopping. For instance, a Master may wish to send (Master-Write) an internal address to an I²C memory device (see Example 12.3) and then do a Master-Read of the pointed-to data. This requires a change in the conversation direction, which is done by doing another Start with a new Slave Address:Direction packet being sent—see Fig. 12.29. The difference between using repeated Starts and a Stop condition is that the latter signals other devices that the Master has relinquished the bus and another device can become a Master on a Multi-Master bus system.

In using a PIC MCU to implement the I^2C standard in software, a problem arises in as much as port outputs are not open-drain; that is, the logic 1 output state is not open-circuit as called for in Fig. 12.14(a). However, it is possible to get around this; simulating the high impedance state by switching the port line between output and input. For example, if we wish to use RA2 as the SCL data line, then to pulse SCL Low and then off/High we have:

```
bcf    PORTA,2      ; Sometime during set-up make RA2 = 0
....   .....
bcf    TRISA,2      ; RA2 is output = 0
nop                 ; Short delay
bsf    TRISA,2      ; Float RA2 by making it an input
```

where the High state is a consequence of the external pull-up resistor and the high *input* impedance; as shown in Fig. 12.14(b)*ii*.

A complete transmission between Master and Slave comprises a packet of several byte/Acknowledge transfers sandwiched between Start and Stop conditions. To some extent the form of this packet depends on the requirements of the Slave device; however, all packets conform to the general sequence Slave address:Control/Command:Data shown in Fig. 12.15.

The essence of the I^2C protocol is the requirement that each type of Slave device has an **address**. This address is allocated[9] to the manufacturer of the I^2C peripheral and is factory programmed. To allow more than one device of the same kind to share the same bus, most I^2C-compatible devices allow up to four bits of this address to be set locally by the designer; usually by connecting Slave address pins to the appropriate logic levels. On receipt of a Start bit, all Slaves on the bus will examine the first seven bits for their personal address. If there is no match then the rest of the conversation is ignored until the next Start bit. Bit 8 is a direction bit; $R/\overline{W}$ is Low if the Master is to be the transmitter; that is, to Write to the Slave, and High if the Master wishes to Read from the Slave.

Not all 7-bit addresses are valid. All addresses matching b'0000XXX' or b'1111XXX' are reserved for special situations; leaving 224 valid addresses in total. For instance, the address b'0000000' indicates a General Call broadcast to *all* Slaves on the bus, rather than to one specific device. Along with the introduction of a Fast mode, the I^2C protocol was extended to permit a 10-bit address. This is signaled by the reserved address b'11110$A_9A_8$0'. The following packet is interpreted as the lower byte of the address $A_7 \ldots A_0$. For instance, address h'2A3' would be sent as b'11110<u>10</u>0' followed by b'10100011'.

After the address byte(s), the next byte is usually treated by the addressed Slave as a Command/Control word, passing configuration information. For instance, a I^2C memory may require the internal address where the data is to be written to—see Example 12.3. Bytes following this are usually pure data or a mixture of data and control bytes.

---

[9]By the I^2C-bus committee.

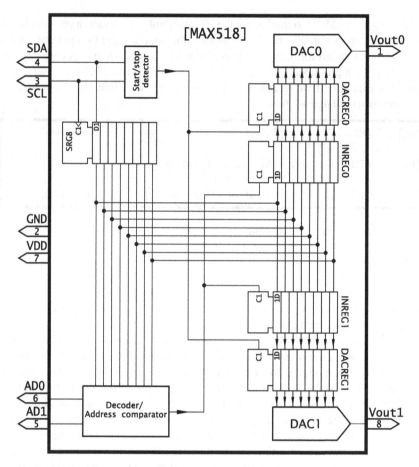

Fig. 12.16 The MAXIM MAX518 I²C dual digital to analog converter

In order to illustrate these concepts, we will use the Maxim MAX518 DAC, shown in Fig. 12.16, as our exemplar. This is the I²C counterpart to the SPI protocol MAX549, with a 2-layer register pipeline, two channels and a power-down feature.

The MAX518 has a 7-bit Slave address of the form **01011**AD1AD0 where AD1 and AD0 should match the logic state of pins 5 and 6, respectively. If we assume that both pins are connected to GND then the Address byte sent out by the Master will be 01011 00 0. R/$\overline{W}$ is 0, as this device can only be written to.

The Command byte has three active control bits, and is of the form 000 RST PD XX A0:

A0

This enables the input PIPO register for Channel 0 if 0 and Channel 1 if 1.

PD

When 1 this control bit will power down both DAC channels, reducing the supply

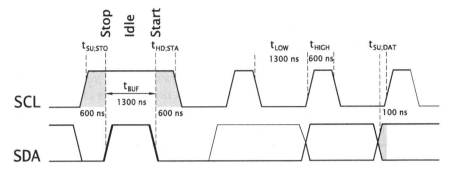

Fig. 12.17 Minimum timing relationships for the Fast I²C mode

current to typically 4 μA. The contents of the internal registers remain unchanged and data may be shifted in and registers updated in this condition. The state information is only executed whenever a Stop bit is sent by the Master, at which point the last transmitted value of PD is acted upon.

RST

All internal registers are cleared irrespective of the following data byte which may be treated as a dummy byte. Analog outputs go to zero after the Stop condition.

In all cases the Stop condition updates the analog outputs according to the commands and data byte. If there have been several Command:Data byte pairs since the last Stop then the most recent command and data are reflected in the state and output of the device.

In order to interface to the MAX518, we will need to design subroutines to send out a Start condition, a Stop condition and a Master-Write byte. To design the device driver we need to look more closely at the time relationship between Clock and Data signals, which generally are more tightly defined than in the SPI protocol.

The MAX518 and most current I²C-compatible devices are designed to the Fast mode specification and the values given in Fig. 12.17 relate to this 400 kHz clocking rate. Of particular note is the requirement that the clock SCL should be held High not less than 0.6 μs ($t_{HD;STA}$) after the active ‾‾_ of SDA to signal a Start condition. Similarly, a Stop condition requires that the clock be set-up High at least 0.6 μs ($t_{SU;STO}$) before the active _/‾ of SDA. A minimum of 1.3 μs is required with the bus free (t_{BUF}) in the Idle state between a Stop and a following Start condition. These requirements allow time for the Slave devices to detect these synchronizing events without ambiguity.

During a data byte transmission the clock should be Low (t_{LOW}) no less than 1.3 μs and High (t_{HIGH}) no less than 0.6 μs within the 2.5 μs overall duration limitation imposed by the 400 kHz clock rate. Data changes only when the clock is Low, and any change should be complete no less than 100 ns ($t_{SU;DAT}$) before the clock goes High.

Not shown in the diagram is the maximum rise and fall times, which should not exceed 300 ns with a maximum bus capacitance of 400 pF. To keep within this transition restriction, the pull-up resistors of Fig. 12.14 should not be more than 1.8 kΩ with this value of capacitance. With short bus runs and few Slave devices this value of resistance can be increased by up to a factor of ten, to reduce energy dissipation.

In implementing the I²C timings, a PIC MCU with a clock frequency above 3.2 MHz, with an execution time of less than 1.25 μs, may need to insert short delays between actions. For example, a 20 MHz crystal-driven PIC MCU implementing the instruction pair:

```
    bcf   TRISA,SCL
; Drag Clock Low by making pin an output to logic 0
    bsf   TRISA,SCL
; Float clock into the High state by making pin an input
```

would give High and Low durations of only 0.2 μs. Short delays are conveniently implemented using nop (No OPeration) instructions; each taking one instruction cycle ($f_{OSC}/4$). For instance, to give a nominally 400 kHz clock at 20 MHz we have:

```
    bcf   PORTA,SCL    ; Clock Low
    nop                ; 0.2us
    nop                ; 0.4us
    nop                ; 0.6us
    nop                ; 0.8us
    nop                ; 1.0us
    nop                ; 1.2us
    bsf   PORTA,SCL    ; Clock High
    nop                ; 1.6us
    nop                ; 1.8us
    nop                ; 2.0us
    nop                ; 2.2us
    nop                ; 2.4us
    nop                ; 2.6us
```

Of course slower clock speeds require less nops, but rather than tailor our subroutines for one particular crystal we will use the assembler macro called Delay_600, coded in Program 12.8, that will expand to the appropriate number of nops to give a nominal 600 ns (0.6 μs) delay, depending on the value of the constant XTAL defined by the programmer at the head of the source file. For instance, to alter the coding of Program 12.9 to suit a 12 MHz crystal system then the one line #define XTAL d'20' should be altered to #define XTAL d'12' and the program reassembled.

The coding of Program 12.8 makes use of the conditional assembler directive if-else-endif. This is similar to the **C** language statement if(true) {do this;} else {do that;} of p. 297. In our example, if(XTAL <= 6) states that if the constant XTAL is less than or equal to 6 then insert one nop instruction whenever the macro Delay_600 is expanded. At 6 MHz this will be

Program 12.8 A crystal frequency-independent short delay macro

```
Delay_600 macro                    ; Delays by nominally 0.6us
             if(XTAL <= 6)
               nop
             else
               fill (nop),2*(3*XTAL/d'20'+1)
             endif
           endm
```

approximately 600 ns. Notice the use of the assembler directive `fill` which in-serts the 2-byte op-code for `nop` (h'0000') in proportion to frequency, with an extra `nop` for each increase of 6 MHz. In practice, extra delays will be introduced by instructions toggling the bus lines and executing housekeeping tasks. Thus some fine-tuning can be undertaken if maximum speed is a criterion.

Based on the macro of Program 12.8 and the following initialization code:

```
        include "p18f4520.inc"
        #define XTAL d'20' ; Eg. 20MHz. Replace with actual data

S_CL equ  0
S_DA equ  1

MAIN setf  ADCON1    ; Make Port A all digital
     bcf   PORTA,SCL ; Preset Clock & Data pins to 0
     bcf   PORTA,SDA ; so that line can be dragged Low
     bsf   TRISA,0   ; Float Clock line High RA0 (TRISA[0])
     bsf   TRISA,1   ; & Idle the Data line RA1 (TRISA[1])
```

which is assuming that we are using Port A bits 0 and 1 of a 20 MHz PIC18F4520 to implement our SCL and SDA lines, we can code the three subroutines outlined in Program 12.9 to allow us to communicate with the I^2C MAX518.

START

This subroutine releases both the SCL and SDA lines which are then pulled High to ensure the bus is in its Idle state for the minimum duration 1.3 μs t_{BUF}. Bringing SDA Low gives the characteristic Start ‾_, which is followed by a 0.6 μs delay to implement $t_{HD;STA}$ (HolD; STArt—see Fig. 12.17) before the subroutine exits with both SCL and SDA Low.

STOP

The Stop condition is implemented by ensuring that both SCL and SDA lines are Low (which should be the case after an Acknowledge condition) and then releasing the SCL line which is then pulled High. After a 0.6 μs delay to implement $t_{SU;STO}$ (Set-Up; STOp), SDA is released to give the characteristic Stop _/‾ . The subrou-tine exits with both lines released idling in preparation for the next Start condition.

I2C_OUT

This subroutine clocks out the eight bits placed in DATA_OUT by the caller, MSB first, and then checks that the Slave has Acknowledged the transaction.

Program 12.9 Low-level I^{2}C subroutines

```
; ****************************************************************
; * FUNCTION: Outputs the Start condition                      *
; * ENTRY   : None                                             *
; * EXIT    : Start condition and SCL, SDA pins low            *
; ****************************************************************
START      bsf    TRISA,S_DA    ; Ensure that we start with the
           bsf    TRISA,_CL     ; Data and Clock lines pulled high
           Delay_600            ; 1.3us delay in Idle state
           Delay_600
           bcf    TRISA,S_DA    ; Low-going edge on Data line
           Delay_600            ; Wait for Slave to detect this
           bcf    TRISA,S_CL    ; Exit with the Clock line low
           return

; ****************************************************************
; * FUNCTION: Outputs the Stop condition                       *
; * ENTRY   : None                                             *
; * EXIT    : Stop condition and SCL, SDA pins high (Idle)     *
; ****************************************************************
STOP       bcf    TRISA,S_CL    ; Make sure that Clock line is low
           bcf    TRISA,S_DA    ; and the Data line is low
           bsf    TRISA,S_CL    ; Bring Clock line high
           Delay_600            ; for a minimum of 0.6us
           bsf    TRISA,S_DA    ; Rising edge on Data signals Stop
           return               ; including the return time

; ****************************************************************
; * FUNCTION: Transmits byte to Slave and monitors Acknowledge*
; * ENTRY   : 8-bit data to be TXed is in DATA_OUT            *
; * RESOURCE: START and STOP subroutines                      *
; * EXIT    : Byte transmitted. ERROR is 01 IF no Ack received*
; * EXIT    : from Slave ELSE 00. SCL low                     *
; ****************************************************************
I2C_OUT    bcf    TRISA,S_CL    ; Make sure that Clock line is low
           clrf   ERR           ; Start with no error
           movlw  8             ; Loop counter = 8
           movwf  COUNT
I2C_OUT_LOOP
           bcf    TRISA,S_DA    ; Start with Data bit low
           rlncf  DATA_OUT,f    ; Shift data left once into Carry
           btfsc  STATUS,C      ; Is C 0 or 1?
            bsf   TRISA,S_DA    ; IF the latter THEN make Data high
           Delay_600            ; Delay plus extra instructions OK
           Delay_600
           bsf    TRISA,S_CL    ; Bring Clock pin high
           Delay_600            ; for at least 0.6us
           bcf    TRISA,S_CL    ; Bring Clock low
           decfsz COUNT,f       ; Decrement loop count
            bra   I2C_OUT_LOOP; and repeat eight times

; Now check Acknowledge from Slave
           bsf    TRISA,S_DA    ; Release Data line
           Delay_600            ; Keep Clock line low
           Delay_600            ; long enough for Slave to respond
           bsf    TRISA,S_CL    ; Bring Clock line high
           btfsc  TRISA,S_DA    ; Check if Data is low from Slave
            incf  ERR,f         ; IF not THEN ERROR1
           bcf    TRISA,S_CL    ; Now finish ACK by bringing CLock low
           return
```

The first part of this process is implemented by repetitively shifting the datum in DATA_OUT and inspecting the Carry flag. SDA is set to mirror **C** and the SCL line toggled to accord with the t_{LOW} and t_{HIGH} parameters illustrated in Fig. 12.17.

Once the loop count reaches zero, the Data line is released with SCL Low for the duration t_{LOW}. SCL is then released High and the state of SDA, which should have been dragged Low by the Slave, checked. If not Low, the No ACKnowledge (NACK) situation is returned with ERR = h'01'; otherwise it will be zero.

Our use of errors here is very rudimentary. For instance, errors can also occur if some other device has locked either line Low; that, is the bus is busy.

We have not coded a Master-Receive I^{2}C counterpart to subroutine I2C_OUT, as the MAX518 only demands a Master-Transmit data interchange. However, Program 12.19 gives the I2C_IN mirror.

As our example we will send the contents of File h'040' to the MAX518 Channel 0 and then the contents of File h'041' to Channel 1; at that point updating both DAC registers and hence simultaneously outputting the analog equivalent of File h'040' to pin V_{out0} and File h'041' to pin V_{out1}. We assume that both AD0 and AD1 pins are connected to Ground.

Our implementation task list will involve the transmission of a group of five packets of information, sandwiched between a Stop and a Start condition.

1. Start.
2. Address byte: b'01011000'
   Slave address b'01011(00)', Write.
3. Command byte 1: b'00000XX0'
   No ReSeT, no Power Down, Channel 0.
4. Data byte 1:
   Contents of File h'040'.
5. Command byte 2: b'00000XX1'
   No ReSeT, no Power Down, Channel 1.
6. Data byte 2:
   Contents of File h'041'.
7. Stop and update both DAC registers.

The listing of Program 12.10 follows our itemization exactly. On return from each call to I2C_OUT the Error datum is tested for zero. If not zero then the process is restarted. Repeated Starts (without Stops) are allowed by the I^{2}C protocol. However, if there was a hardware fault with the bus or Slave then this process would continue indefinitely. Thus, for robustness a time-out mechanism should be implemented to prevent hang-ups.

The MSSP module supports both Slave and Multi-Master I^{2}C protocols with 7- and 10-bit addressing. Avoiding a bus collision situation complicates matters, and the MSSP Master I^{2}C operation is correspondingly complex and beyond the scope of this text. Details are given in the Microchip application note AN7578, *Use of the*

Program 12.10 Interacting with the MAX518 dual-channel I²C DAC

```
ANALOG     call    START      ; Start a transmission packet
; Address byte ------------------------------------------------
           movlw   b'01011000'; Slave address Master-Write
           movwf   DATA_OUT   ; Copied to pass location
           call    I2C_OUT    ; Send it out
           tstfsz  ERR        ; IF no Error THEN continue
           bra     ANALOG     ; ELSE try again
; Command byte 1 ----------------------------------------------
           movlw   b'00000000'; No ReSeT, No Power Down, Channel0
           movwf   DATA_OUT   ; Copied to pass location
           call    I2C_OUT    ; Send it out
           tstfsz  ERR        ; IF no Error THEN continue
           bra     ANALOG     ; ELSE try again
; Data byte 1 -------------------------------------------------
           movf    h'40',w    ; Channel0's datum from memory
           movwf   DATA_OUT   ; Copied to pass location
           call    I2C_OUT    ; Send it out
           tstfsz  ERR        ; IF no Error THEN continue
           bra     ANALOG     ; ELSE try again
; Command byte 2 ----------------------------------------------
           movlw   b'00000001'; No ReSeT, No Power Down, Channel1
           movwf   DATA_OUT   ; Copied to pass location
           call    I2C_OUT    ; Send it out
           movf    ERR,f      ; Check for an error
           tstfsz  ERR        ; IF no Error THEN continue
           bra     ANALOG     ; ELSE try again
; Data byte 2 -------------------------------------------------
           movf    h'41',w    ; Channel1's datum from memory
           movwf   DATA_OUT   ; Copied to pass location
           call    I2C_OUT    ; Send it out
           tstfsz  ERR        ; IF no Error THEN continue
           bra     ANALOG     ; ELSE try again

           call    STOP
```

SSP Module in the I²C Multi-Master Environment. Here we will confine ourselves to the use of the MSSP module as an I²C 7-bit address Slave device.

Figure 12.18 shows a block diagram of a MSSP configured as an I²C Slave. Typically pin RC4 is used as the bidirectional I²C SDA data channel and RC3 implements the SCL clock line. Both pins need to be set as inputs for the I²C Slave protocol.

Internally, data I/O is via the SSPSR shift register, which is used both for Slave transmission or reception.

Transmission

Where the Slave is sending data (Slave-Write) to a remote Master (Master-Read) the datum placed in the SSPBUF buffer register will automatically be transferred into the SSPSR (if empty) whence it is shifted out of SDA in eight clock pulses. If

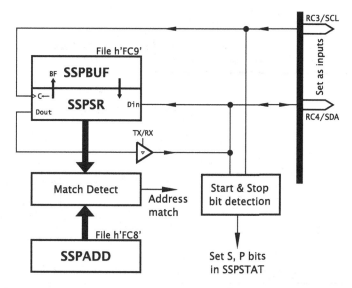

Fig. 12.18 Block diagram of a MSSP module set-up as an I^{2}C Slave device

the SSPSR is not empty, this transfer does not happen and a Write-collision error is set.

Reception
If the Slave expects to read a packet from the remote Master, then the data is shifted in via SDA and when collected is transferred into the SSPBUF register. The MSSP module then automatically ACKnowledges the safe reception of the datum during the ninth clock pulse, unless an overflow error has occurred. This happens where the previously received byte has not been read from the SSPBUF register in time. This can be used by the Slave as a mechanism to tell the Master to stop sending any more data.

Once a Start condition is sensed, all Slaves on the bus shift in the first packet from the Master, looking for a match with the pattern set-up in software in the **SSP ADDress register (SSPADD)**. If the top seven bits match (bit 0 is R/$\overline{W}$) then it is ACKnowledged and the addressed Slave is now ready to converse with the remote Master. Both the BF and SSPIF flags will be set to signal an I^{2}C event and the address byte loaded into SSPBUF. As we have seen in Fig. 12.15, bit 0 of this first address packet tells the Slave to either receive or transmit as directed until the next Start or Stop condition. This will set or clear the R/$\overline{W}$ bit in SSPSTAT[2] as appropriate.

As in the case of the SPI mode, the Control and Status registers need to be set-up to configure and monitor the operation of the MSSP module. Figure 12.19 shows the situation applicable for the four possible I^{2}C Slave modes. These should be compared to those shown for the SPI modes of Fig. 12.9. The same SSPSTAT and SSPCON1 registers are used and indeed, rather confusingly, some of the bit names

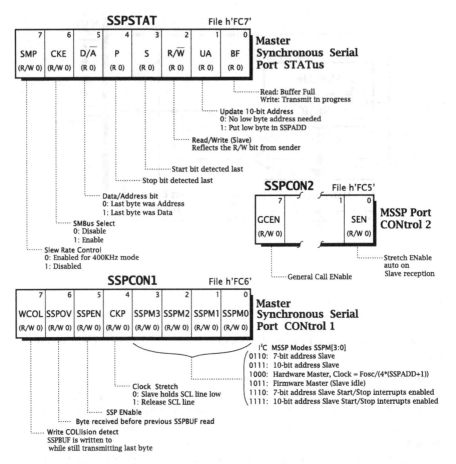

Fig. 12.19 The MSSP module's CONtrol and STATus registers as appropriate for the I²C Slave modes

have been retained, such as CKE, even though their function is very different. The MSSP has a second control register named **SSPCON2**. Apart from bits 7 and 0, SSPCON2 deals solely with the Master I²C environment.

SSPEN

Setting SSPCON1[5] enables the Synchronous Serial port. On any type of reset, the MSSP is disabled and pins RC3 and RC4 may be used for normal Port C I/O duties.

SSPM3:0

Four combinations of these MSSP Mode switches are relevant to this discussion. For simplicity, we will assume a 7-bit Slave address mode. Where a 10-bit address mode is used, the software must first place the top address byte b'11110$A_9A_8$0' in SSPADD for comparison and then follow it with the bottom 8-bit address byte b'$A_7A_6A_5A_4A_3A_2A_1A_0$'. Modes b'0110' and b'1110' differ only in that the SSP interrupt flag SSPIF is optionally set when a Start or Stop situation occurs.

BF, SSPIF

Buffer Full generally indicates that something is still happening with data in the SSPBUF register. SSPIF is the interrupt flag associated with the MSSP module, and is set whenever an I^2C event has occurred. Where an address is sent, BF and SSPIF will only be set if it matches the value of SSPADD.

Slave-Receive

When a complete frame is received from the Master and has been dumped into the SSPBUF register, BF is set to indicate that a new datum is ready for collection and an ACKnowledge is sent back during clock pulse 9. SSPIF in PIR[3] is also set to potentially generate an interrupt. When this byte is read by the software, BF is automatically cleared (it is a read-only flag), but SSPIF has to be manually reset in the normal interrupt flag manner.

Should a complete new byte be shifted in before the previous datum is read; that is, if BF is still 1, then it will not be transferred to SSPBUF and the SSPOV flag will be set to show an OVerflow condition has occurred. In this case an ACKnowledge is Not sent; that is, NACK.

Slave-Transmit

While a byte is being shifted out to the Master, BF is set to show that a transmit is in progress. If a new byte is written into SSPBUF, it will not be transferred into SSPSR and the WCOL flag is set to indicate a Write COLlision has occurred. That is, there is no double-buffering action during a transmit.

SSPOV

In a Slave-Receive situation, failure to read the SSPBUF register before the next byte has arrived is signaled by setting this flag.

A NACK is sent to the Master to indicate an overflow. The NACK state can be used by the Slave deliberately to inform the Master that it should try again later. This condition is cleared by reading the datum from the SSPBUF register to clear BF and manually zeroing SSPOV.

WCOL

An attempt to write to the SSPBUF while a transmission is in progress sets the WCOL flag to indicate the Write COLlision has occurred. This flag has to be manually cleared.

S, P

These flags indicate that a Start or stoP bit, respectively, was detected last. They are normally the inverse of each other except that they are both zero after any Reset or when the module is enabled—SSPEN → 1. A set P indicates that the bus is free; which is useful if a listening device wishes to take over as a Master.

D/$\overline{A}$, R/$\overline{W}$, UA

These flags all relate to the packet(s) following a Start condition, which contains information regarding the Slave address and direction of subsequent data packets.

- D/$\overline{A}$ indicates whether the byte sitting in the SSPBUF is Data or Address.

- R/W̄ informs the software whether the message is going to be Master-Read (R/W̄ = 1) or Master-Write (R/W̄ = 0). Effectively, it is bit 0 of the (first) address packet.
- UA is used only in the 10-bit address modes. In this case the seven MSBs of the first address byte b'11110A$_9$A$_8$0' are matched first. the LSB represents R/W̄ and is 0 to indicate that the next address packet is to be written from the Master. UA is then automatically set to 1, to indicate that the software must now place the lower address byte into SSPADD for match detection. When this is done, UA is automatically cleared.

GCEN

When the General Call Enable bit is 1, the SSPIF interrupt flag will be set when the General Call address b'00000000' is received; irrespective of an address match. This indicates that the Master wishes to make a general broadcast to *all* Slaves.

CKP, SEN

When CKP is cleared, the Slave device holds the SCL pin Low, so that the Master device is unable to send clock pulses. When CKP is set to 1, the SCL pin is released and the Master can clock a new packet. **Clock stretching** is implemented *automatically* during a Slave-Transmit sequence. If BF is clear at the end of a transmit frame, indicating that another byte is not ready to be sent yet, then the CKP bit is automatically cleared, resulting in the SCL pin being held low. When the Slave subsequently copies the next byte into SSPBUF, then it should manually set CKP to release the clock line and allow the next transmit frame to be shifted out.

Clock stretching during a Slave-Receive mode transaction is optional, and is controlled by the **SEN (Stretch ENable)** bit in SSPCON2[0].

SEN = 0

There is no automatic clock stretching when data is being received from a Master.

SEN = 1

When SEN is set, automatic clock stretching is enabled for Slave reception. If the BF bit is set, indicating a full buffer, then the CKP bit is cleared. By thus holding the SCL pin low, the Slave has time to to read the contents of the SSPBUF before the Master can initiate another transmit sequence. This is useful if the Master is sending packets at too fast a rate for the Slave to process. The CKP bit should be set manually then the Slave is ready for the next data frame from the Master.

CKE

This should be set to alter the electrical characteristics of SDA and SCL to conform to the SME bus standard.

Before looking at an example, the MSSP module needs to be initialised in the start-up routine. Typical set-up code for a PIC18F4520 device acting as an interrupt-driven Compatible mode 100 kHz Slave with 7-bit address h'06' would be something like:

```
       include "p18f4520.inc"

SETUP movlw b'00110110' ; Enable MSSP, no clock stretch, CKP=1
      movwf SSPCON1      ; 7-bit Slave mode 0110

      bsf   SSPSTAT,SMP ; Slew rate for 100kHz
      bsf   SSPCON2,SEN ; SEN=1 for auto clock stretch on RX
      movlw h'0C'       ; h'0C' is address 06 shifted left 1
      movwf SSPADD      ; for matching

      bsf   PIE1,SSPIE  ; Enable the SSP interrupt (SSPIE)
      bsf   INTCON,PEIE ; Enable the Peripheral module group
      bsf   INTCON,GIE  ; Globally turn on enabled interrupts
```

where:

1. The three MSSP Status/Control registers are set-up appropriately for our specification.
2. The Slave address h'06' aligned with the top seven bits of the Address packet) is copied into the SSPADD for matching.
3. The SSP interrupt is enabled by setting to 1 the specific SSPIE as well as the general Peripheral group mask bit PEIE and Global mask GIE—see Fig. 7.2 on p. 209.

With our MSSI module set-up as above, we can code the bottom layer of functions that service the reading and writing of single bytes from a bus Master. These will be the counterpart to the subroutines of Program 12.9.

Subroutine I2C_OUT simply keeps checking for Buffer Full (BF == 1?) until any byte in the SSP Shift Register (SSPSR) has been shifted out. When this is the case, any active Write Collide is cleared (WCOL = 0 and the datum in WREG on entry is copied into the SSP BUFfer for onward transmission. Provided WCOL is clear, the process is completed.

The converse subroutine I2C_IN first checks to see if a byte too many has been received since the last read of SSPBUF, causing an OVerflow (SSPOV == 1?). If this is the case then I2C_ERROR is set to −2 to indicate that the returned byte may be suspect. With a zero SSPOV flag, the datum is read from the SSPBUF whenever BF shows a byte has been copied from the SSPSR. This is the value returned from the subroutine.

The layer of functionality above this treats the MSSP Slave as a sequential machine, responding to each of five possible I^{2}C states as they occur. Although the SSPIF flag in the PIR1 register may be polled, our exemplar software will use an interrupt-driven approach with the processor in the default Compatible mode.

Once the MSSP and interrupt system are initialized, the interrupt service routine (ISR) can be written to recognise the various I^{2}C events. Any legitimate event will transfer processing from the background process to the ISR. Even if the Slave is asleep, a I^{2}C happening will set SSPIF and awaken it in the usual way. These events are:

Program 12.11 Bottom layer I²C functions

```
;   ************************************************************
;   * FUNCTION: Slave write to bus Master                     *
;   * ENTRY    : Byte in WREG                                 *
;   * EXIT     : Byte transmitted when ready                  *
;   ************************************************************
I2C_OUT   btfsc   SSPSTAT,BF    ; Check for Buffer Full
          bra     I2C_OUT       ; IF still full THEN try again

I2C_OUT_LOOP
          bcf     SSPCON1,WCOL  ; Clear any Write COLlision
          movwf   SSPBUF        ; Start the transmission
          btfsc   SSPCON1,WCOL  ; Shouldn't be any WCOL here
          bra     I2C_OUT_LOOP  ; but to be sure

          return                ; Done

;   ************************************************************
;   * FUNCTION: Slave read from bus Master                    *
;   * ENTRY    : None                                         *
;   * EXIT     : Byte received is in WREG.  ERROR -2 IF OVerflow*
;   ************************************************************
I2C_IN    btfss   SSPCON1,SSPOV ; Test for an overflow condition
          bra     I2C_IN_LOOP   ; IF none THEN continue
          bcf     SSPCON1,SSPOV ; ELSE clear it
          movf    SSPBUF,w      ; and reset BF by doing a read
          decf    I2C_ERROR,f   ; Indicate an OVerflow error
          decf    I2C_ERROR,f   ; by returning an ERROR code of -2
          return                ; and exit

I2C_IN_LOOP
          btfsc   SSPSTAT,BF    ; Check for Buffer Full yet?
          bra     I2C_IN_LOOP   ; Keep checking

          movf    SSPBUF,w      ; Get byte
          return                ; Exit with datum in WREG
```

1: Master-Write: Packet received was an Address

The bus Master has initiated a Start and has sent an Address packet with R/$\overline{W}$ = 0 to indicate that the Master will begin to send Data packets to the addressed Slave(s). This packet is ignored unless the address matches or it is a General Broadcast.

S = 1   Start condition occurred last.

R/$\overline{A}$ = 0   Master-Write conversation pending.

D/$\overline{W}$ = 0   This packet is an Address.

BF = 1   Buffer is full.

It is important to read the SSPBUF register to clear the BF flag, even though the address byte sent by the Master is going to be discarded. If this is not done, the next byte sent by the Master will cause an SSP OVerflow (SSPOV $\to$ 1) and the MSSP will Not ACKnowledge (NACK) the byte. The SSPBUF will not be updated on a mismatch.

As SEN is 1 in our instance, then CKP will automatically be cleared and the clock stretched. When appropriate, CKP must be set to 1 to allow the Master to proceed. The ISR shown in Program 12.12 always sets this bit to release the clock line when a new event occurs.

2: Master-Write: Packet received was Data

After the Address packet (State 1), the Master will send one or more Data packets. The Slave must read each one to avoid MSSP OVerflow and ensure an ACK at the end of each packet. If any previous received byte has not yet been read the SSPOV bit will be set and the MSSP module will NACK the byte. The Status settings differ from State 1 only in that D/$\overline{\text{A}}$ is 1.

S = 1   Start condition occurred last.

R/$\overline{\text{W}}$ = 0   Master-Write conversation pending.

D/$\overline{\text{A}}$ = 1   This packet is Data.

BF = 1   Buffer is full.

3: Master-Read: Packet received was an Address

The bus Master device begins a new read operation by initiating a Start and sending an Address packet with R/$\overline{\text{W}}$ = 1 to indicate that the Slave is expected to start sending Data packets back. Only when a Slave recognizes its personal address, or a General Broadcast if GCEN is active, will the key SSPSTAT settings will be:

S = 1   Start condition occurred last.

R/$\overline{\text{W}}$ = 1   Master-Read conversation pending.

D/$\overline{\text{A}}$ = 0   This packet is an Address.

BF = 1   Buffer is full.[10]

Otherwise this packet is ignored. As in State 1, the SSPBUF register should be read to clear the BF bit, even though the address byte sent by the Master is going to be discarded. If this is not done, the next byte sent by the Master will cause an SSP OVerflow (SSPOV $\to$ 1) and the MSSP will NACK the byte.

Once its Address has been recognized, the Slave can send the first byte to the Master by copying it into SSPBUF. The CKP bit is cleared in the ISR to release the SCL pin. Clock stretching is always automatically asserted on a Master-Read packet, independent of the setting of the SEN bit.

---

[10]In older PIC18 devices, and also the PIC18F1X20, BF is cleared in this event and it is not necessary to read the SSPBUF to clear it. See App. Note AN734B for further details.

4: Master-Read: Packet received was was Data

This is similar to State 3, with the same treatment of the CKP clock stretching switch. A new data byte should not be copied into SSPBUF whilst BF is 1 otherwise the WCOL bit will set to show a Write COLlision.

The SSPSTAT settings are identical to State 3 except that D/$\overline{\text{A}}$ is 1 to show that a Data packet was sent last:

S = 1   Start condition occurred last.

R/$\overline{\text{W}}$ = 1   Master-Read conversation pending.

D/$\overline{\text{A}}$ = 1   This packet is Data.

BF = 0   Buffer is free for transmission.

5: Master-Read: Master sent a NACK

This occurs typically when the Master does not wish to receive any more bytes from the Slave. The NACK signals the end of the message, and when sensed by the Slave MSSP resets the I²C logic.

CKP = 1   Clock is released.

S = 1   Start condition occurred last.

R/$\overline{\text{W}}$ = 1   R/$\overline{\text{W}}$ bit remains set.[11]

D/$\overline{\text{A}}$ = 1   This packet is Data.

BF = 0   Buffer is free for transmission.

The NACK event is identified because the CKP bit remained set when the status bits indicated that a data byte has been received from the Master. Normally CKP is 0 with SCL being automatically stretched at the end of the byte reception.

For our example, let us assume that the Slave PIC18F4520 set-up to respond to address h'06' is monitoring eight temperature transducers using its Analog-to-Digital Convertor (ADC) module—see Chap. 14. The Master wishes to read any one of these digitized channels via its I²C port by first of all initiating a Master-Write sending the channel number N to the Slave and then launching a Master-Read of the Slave; which then sends the designated datum. We assume the subroutine GET_ANALOG of Program 14.1 on p. 517 is *in situ*.

A suitable task list would be:

1. Master sends a Start pulse followed by a call to Slave 06 requesting that it read the next packet (Master-Write).
2. Master sends one Data packet, specifying a channel number 0–7.

---

[11] In older PIC18 devices R/$\overline{\text{W}}$ was cleared.

3. Master sends a Start pulse (repeat Start) for Slave 06, this time requesting that it write the next packet (Master-Read).
4. Slave stretches clock while its ADC digitizes its specified analog input.
5. Slave sends the requested datum.
6. Master NACKs Slave to signal end of transmission.

For clarity, our foreground software is shown as two separate routines. Program 12.12 gives the ISR framework software. On entry, the SSPIF interrupt flag is checked and if not set the ISR exits. In a real-world situation, interrupts may come from several sources, and this part of the code would be elaborated to check for each of the applicable interrupt flags.

If SSPIF is 1 then a copy of the state of the SSPSTAT Status register is made in I2C_STATUS, with the non-critical bits first cleared. The program variable named I2C_ERROR is also zeroed. This is used to advise the background that an erroneous situation has occurred. We have already seen in Program 12.11, Error -2 indicates an overflow situation when reading a packet.

Once initialization is complete, the actual I^2C interpretation software is called as the subroutine shown in Program 12.13. This software is structured as a series of five separate cases; each one corresponding to one of the listed I^2C states. This is done by using the xorlw instruction to check for no differences (see p. 128) between the copy of the SSPSTAT Status pattern and the listed I^2C state byte. Where a match is found, an action is taken corresponding to our task list, or in some cases just housekeeping to enable the MSSP to continue in the correct state. For instance, nothing needs to be done for I^2C State 5, where the Master NACKs the Slave, as the MSSP will be automatically reset. If no state match is found, I2C_ERROR is decremented to return Error state -1.

Program 12.12 The I^2C temperature acquisition ISR

```
; ****************************************************************
; * FUNCTION    : ISR to send digitized channel N via I2C link*
; * ENTRY       : An I2C event has occurred                    *
; * EXIT        : One of five states implemented               *
; * RESOURCE    : Subroutines GET_ANALOG, I2C_HANDLER          *
; * ENVIRONMENT: Variables I2C_STATUS, I2C_ERROR               *
; ****************************************************************
; Check the SSPIF flag is set? -----------------------------
ISR       btfss   PIR1,SSPIF ; Is this a MSSP interrupt?
          bra     ISR_EXIT   ; IF not THEN exit
          movf    SSPSTAT,w  ; ELSE get the settings from SSPSTAT
          andlw   b'00101101'; Zero all but S, D/A, R/W and BF
          movwf   I2C_STATUS ; and copy into a Temporary File
          clrf    I2C_ERROR  ; Zero Error flag
          call    I2C_HANDLER; Now respond to the I2C event

; Restore the context -------------------------------------
ISR_EXIT bcf      PIR1,SSPIF ; Clear interrupt flag
          bsf     SSPCON1,CKP; Release clock line and
          retfie  FAST       ; return to background with context
```

Program 12.13 The I²C temperature acquisition handler subroutine

```
; *****************************************************************
; * FUNCTION   : Interprets I2C state & responds appropriately*
; * ENTRY      : Copy of SSPCON1 in I2C_STATUS.  SCL released *
; * EXIT       : Appropriate action taken                     *
; * EXIT       : I2C_ERROR = -1 IF state not recognized       *
; * RESOURCE   : I2C_STATUS, TEMP                             *
; *****************************************************************
I2C_HANDLER
; Are we in State 1? (Start: Address packet, Master-Write)
        movf    I2C_STATUS,w; Get copy of SSPSTAT status
        xorlw   b'00001001' ; Check for S=1, D/A=0, R/W=0, BF=1
        bnz     STATE2      ; IF not equal THEN try for State 2
         movf   SSPBUF,w    ; ELSE do a dummy read to clear BF
         return             ; Exit

; Are we in State 2? (Data packet, Master-Write) -------------
STATE2  movf    I2C_STATUS,w; Get copy of SSPSTAT status
        xorlw   b'00101001' ; Check for S=1, D/A=1, R/W=0, BF=1
        bnz     STATE3      ; IF not equal THEN try for State 3
        call    I2C_IN      ; ELSE read in Channel number
        call    GET_ANALOG  ; Digitize Channel N's analog input
        movwf   TEMP        ; and copy outcome into TEMPerature
        return              ; Exit

; Are we in State 3? (Start: Address packet, Master-Read) -----
STATE3  movf    I2C_STATUS,w; Get copy of SSPSTAT status
        bcf     WREG,0      ; Ignore BF which varies with device
        xorlw   b'00001100' ; Check for S=1, D/A=0, R/W=1
        bnz     STATE4      ; IF not equal THEN try for State 4
         movf   SSPBUF,f    ; Dummy read to clear BF
         movf   TEMP,WREG   ; Get TEMPerature
         call   I2C_OUT     ; and send to the Master
         return             ; Exit

; Are we in State 4? (Data packet, Master-Read) --------------
STATE4  movf    I2C_STATUS,w; Get copy of SSPSTAT status
        xorlw   b'00101100' ; Check for S=1, D/A=1, R/W=1, BF=0
        bnz     STATE5      ; IF not equal THEN try for State 5
         return            ; Data has been sent, so do nothing!
; Are we in State 5? (Master NACKs Slave) --------------------
STATE5  btfsc   SSPCON1,CKP ; Master has NACKed?
         bra    S5_ERROR    ; IF yes something is wrong!
        movf    I2C_STATUS,w; ELSE get copy of SSPSTAT status
        bcf     WREG,2      ; Mask out RW to ignore device diffs
        xorlw   b'00101000' ; Check for S=1, D/A=1, R/W=0, BF=0
        btfss   STATUS,Z    ; Equal, so exit?
S5_ERROR decf   I2C_ERROR,f ; IF not THEN signal ERROR -1
        return
```

Program 12.14 Interfacing to the MAX518 in **C** with a 4 MHz PIC18F4520

```
/* Set device as Master with the SSP module and Fast timing. */

#include <18f4520.h>
#use delay(clock=4000000)
#use i2c(master, scl=PIN_C3, sda=PIN_C4, fast, FORCE_HW)

#fuses NOWDT, XT
#byte  DATA_X = 0x20
#byte  DATA_Y = 0x21


void MAX518(unsigned int channel_0, unsigned int channel_1);

main()
{
/* Various code lines here                                  */
MAX518(DATA_X,DATA_Y); /* Send out the two bytes            */
/* More code                                                */
}

void MAX518(unsigned int channel_0, unsigned int channel_1)
{
i2c_start();            /* Start condition                   */
i2c_write(0x58);        /* Send out Slave address; Master-write*/
i2c_write(0);           /* Send out Command 1                */
i2c_write(channel_0);   /* Send out datum to Channel 2       */
i2c_write(0x01);        /* Send out Command 1                */
i2c_write(channel_1);   /* Send out datum to Channel 2       */
i2c_stop();             /* Stop condition                    */
}
```

On exit, the CKP switch bit is set to 1 to lift any clock stretching and the SSPIF is cleared in the normal way.

Another example of the use of the MSSP module is given in the Microchip application note AN734B, *Using the PIC Microcontroller SSP for Slave I^2C Communications.*

As for the SPI protocol, many **C** compilers targeted to the PIC MCU have built-in functions to implement the I^2C protocol and avoid the necessity for user-defined functions.

To illustrate the technique, consider Program 12.14 which replicates the assembly-level coding of Programs 12.9 and 12.10 using the CCS compiler.

i2c_start();
Generates the Master Start condition.

i2c_stop();
Generates the Master Stop condition.

i2c_read();
Reads a byte over the bus. If an optional parameter of 0 is used then will NACK the received data. In Master mode, will generate a clock.

i2c_write(value);

Sends a single byte over the bus. In Master mode will generate a clock. Returns a 0 if the addressed Slave acknowledged, 1 if no acknowledged and 2 if there was a Multi-Master bus collision.

#use i2c(master, scl=PIN_C3, sda=PIN_C4, fast,

FORCE_HW)

This is a directive by which the programmer informs the compiler which pins are used for the I^2C lines, the fast or standard protocols and Master or Slave mode. The compiler will use the MSSI module if the FORCE_HW option is specified, otherwise a bit-banging routine (such as in Program 12.11) is used.

More than one I^2C channel can be hosted by the one device, perhaps one using the MSSP module and the other a software approach, with a #use i2c(stream, ...) naming each stream. For instance:

```
#use i2c(master,scl=PIN_A0,sda=PIN_A1,stream=main_channel)
#use i2c(slave,scl=PIN_C3,sda=PIN_C4,FORCE_HW,stream=s_channel)
/* More code ........................................... */
i2c_slaveaddr(s_channel,0x06);/* Set the Slave's addr to 06*/
/* More code ........................................... */
i2c_write(main_channel,0x55); /* Send byte to this channel */
i2c_write(s_channel,0x99);    /* Send byte to this channel */
```

Notice the use of the **i2c_slaveaddr()** function to set-up the Slave address.

The key characteristic of the various serial protocols discussed up to now is that a clock signal is transmitted by the Master, which allows the various Slaves to receive or transmit data in perfect synchronization. An alternative approach is to send data under the assumption that the transmitter and receivers are running at approximately the same frequency. This **asynchronous** protocol has been in use for data communications systems for over a century to send alphanumeric data over telegraph, telephone, and radio links to implement the Telex system.

One of the features of early computer development in the 1940s/1950s was the extensive use of existing technology. An essential adjunct of any computer-oriented installation is a data terminal. At that time the communications industry made considerable use of the teletypewriter (TTY).[12] Serial data were converted between serial and parallel formats in the terminal itself, as well as providing keyboarding and printing functions.

Until the early 1980s, TTYs were electromechanical machines, driven by a synchronous electric motor. This meant that synchronization between remote terminals

---

[12]Literally a "typewriter from afar"; Greek, tele = far.

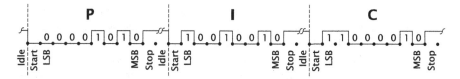

Fig. 12.20 Transmitting the message string "PIC" in the asynchronous serial mode, with an 8-bit code word with a minimum of one stop bit

could only be guaranteed for short periods. To get around this problem, each word transmitted was proceeded by one Start bit and followed by one or more Stop bits. A typical example is shown in Fig. 12.20. While the line is idling, a logic 1 (break level) is transmitted. A logic 0 signals the start of a word. After the word has been sent, a logic 1 terminates the sequence. Electromechanical terminals typically print ten characters per second, and require a minimum of two Stop bits. For 8-bit words, this requires a transmission rate of 110 bits per second, or 110 **baud**.[13]

The first purely electronic terminals required only one Stop bit, and could print at 30 characters per second, giving a rate of 300 baud. Traditionally communication channels use multiples of 300; e.g., 1200, 2400, 4800, 9600, ... PC Serial ports can run up to 115,200 baud. However, this $300 \times n$ series is not mandatory, as long as receiver and transmitter are running at the same nominal rate.

Typically a receiver, on detecting an incoming datum, will try and sample each bit at approximately the mid-point. This means that a frequency drift of ± 0.5 bit time can be tolerated in the space of 10 bits. Thus the receiver and transmitter local sample clocks must be within $\pm 5\%$. The two will be resynchronized at the start of each datum.

Although not the most efficient of techniques, the asynchronous protocol outlined here has the major advantage of being an international standard. There are several variants; for instance, the word can typically be from 5 to 9 bits long. In our example the word length is 8 bits.

The original teleprinter code developed by Emile Baudot in 1875 is only 5 bits long.[14] Here the string "PIC" is coded as 10110 00110 01110. Although limited in capability, its key advantage over Morse code (Samuel Morse, 1840) was its fixed length (compare with · − − · ·· − · − ·) which considerably simplifies the design of the transmitter and receiver. However, Morse code is more efficient, as the number of bits is approximately inversely proportionally to a letter's statistical frequency of use.

---

[13] Strictly speaking the baud rate is a measure of information flow. For a simple baseband system this is equal to the bit rate. However, this equality is not always true. For instance, a telephone modem can use a di-bit modulation scheme where groups of bits two at a time give a carrier tone phase shift of 0°, 90°, 180° and 270° for the patterns 00, 01, 10, 11, respectively. In this case the baud rate is twice the bit rate.

[14] Actually the first documented binary coded alphanumeric code was devised by Francis Bacon in around 1600. It too was a 5-bit code.

The 7-bit ASCII code of Table 1.1 on p. 5, first adopted in 1963, was the first code specifically developed for computer communication systems. In 8-bit systems the extra bit is usually utilized to add a selection of 128 accented, mathematical and graphic symbols. However, the extra bit can be used to provide a limited error-checking capability. This **parity bit** is set so that the number of 1s in the word is always odd or even. This can be checked at the receiver (see SAQ 5.13 on p. 156) to detect a single bit error. Alternatively, parity can be accommodated by using a 9-bit $(8 + 1P)$ word format.

For our example we have adopted a format of one Start, eight data bits with no parity and one Stop bit. Using a bit-banging approach, as we have already done for our SPI and I^{2}C protocols, is straightforward provided that we have an accurate $\frac{1}{2}$-bit delay. For instance, for a 9600 baud link this would be $(1 \div 9600) \times 0.5 = 52$ μs. As the delay is so short we can use an in-line approach using a macro, in the same manner as in Program 12.8 on p. 407, rather than the subroutine approach of Example 6.1 on p. 189.

As the listing below has a total of 4K cycles, the delay will be

$$4K \times \frac{4}{\text{XTAL}} = \frac{16K}{\text{XTAL}} \text{ μs.}$$

Equating this to $0.5 \times \frac{1000000}{\text{BAUD}}$ gives

$$K = \text{XTAL} \times \frac{31250}{\text{BAUD}}$$

for an approximate baudrate delay independent of the crystal frequency. For low frequency/baudrate combinations, some padding with nop instructions to increase the quantum of delay will be necessary.

```
Baud_delay macro
           local  BAUD_LOOP                     ; Local label
           movlw (XTAL*d'3125')/(BAUD/d'10'); The delay const
BAUD_LOOP  addlw -1                             ; Decrement
           btfss STATUS,Z                       ; until zero
            bra  BAUD_LOOP
           endm
```

The constants XTAL and BAUD are defined appropriately at the head of the program using this macro; for instance, Program 12.15. To avoid values that cannot be handled with 16-bit assembler arithmetic,[15] the constant 31,250 has been reduced by a factor of ten, with BAUD also diminished by the same factor. The example shown here with a baud rate of 9,600 and a clock frequency of 20 MHz resolves to a constant K of d'65' and gives the requisite delay of 52 μs.

With our delay macro *in situ*, the basic I/O subroutines of Program 12.15 are analogous to our bit-banging SPI subroutines. The PUTCHAR subroutine simply brings the TX pin Low for two Baud_delay periods and then assigns the pin eight

---

[15] Newer versions of the assembler use 32-bit arithmetic for label arithmetic.

Program 12.15 Asynchronous formatted input and output subroutines

```
                #define XTAL d'20'    ; E.g. 20 MHz 9600 set-up
                #define BAUD d'9600'

; ***********************************************************
; * FUNCTION: Transmits one 8-bit byte in asynchronous format *
; * ENTRY   : 8-bit datum in DATA_OUT, XTAL & BAUD predefined *
; * EXIT    : WREG, STATUS altered. Byte TXed                *
; * RESOURCE: Macro Baud_delay giving a 0.5 bit delay; COUNT *
; ***********************************************************
PUTCHAR         movlw  8              ; Eight data bits
                movwf  COUNT

                bcf    PORTA,TX       ; Start bit
                Baud_delay            ; 2x0.5 bit delay
                Baud_delay
; Now shift out data, LSB first ------------------------------
PUTCHAR_LOOP    rrncf  DATA_OUT,f     ; Rotate right
                btfss  DATA_OUT,7     ; Test what was LSB pre-shift
                bra    ITS_A_0        ; IF 0 THEN output a 0
                bsf    PORTA,TX       ; ELSE output a 1
                bra    PUTCHAR_NEXT   ; and continue

ITS_A_0         bcf    PORTA,TX       ; Output a 0

PUTCHAR_NEXT    Baud_delay            ; One-bit duration
                Baud_delay
                decfsz COUNT,f        ; Repeat eight times
                bra    PUTCHAR_LOOP

                bsf    PORTA,TX       ; Stop bit
                Baud_delay
                Baud_delay
                return

; ***********************************************************
; * FUNCTION: Receives one 8-bit byte in asynchronous format  *
; * ENTRY   : XTAL & BAUD predefined                        *
; * EXIT    : DATA_IN holds the received byte.              *
; * EXIT    : Err is 00 if no Framing error ELSE -1         *
; * RESOURCE: Macro Baud_delay giving a 0.5 bit delay; COUNT *
; ***********************************************************
GETCHAR         movlw  8              ; Eight data bits
                movwf  COUNT
                clrf   ERR            ; Zero Error byte

GETCHAR_START   btfsc  PORTA,RX       ; Poll for 0
                bra    GETCHAR_START

                Baud_delay            ; Hang around for 0.5 bit time
                btfsc  PORTA,RX       ; Check; is it still Low?
                bra    GETCHAR_START
                Baud_delay            ; IF yes THEN hang around
                Baud_delay

GETCHAR_LOOP    bcf    STATUS,C       ; Clear Carry
                rrcf   DATA_IN,f      ; Shift 0 into datum
                btfsc  PORTA,RX       ; Check; is input High?
                bsf    DATA_IN,7      ; IF yes THEN set bit in datum
                Baud_delay
                Baud_delay
                decfsz COUNT,f        ; Do eight times
                bra    GETCHAR_LOOP

                btfss  PORTA,RX       ; Look for a Stop bit (High)
                decf   ERR,f          ; IF Low THEN signal an error
                return
```

times, corresponding to the contents of DATA_OUT, least-significant bit first—the opposite order to SPI/I²C. Finally TX is held High for the same period to give the Stop/Idle condition.

The input GETCHAR counterpart is more complex. After an Idle state, a Low-going voltage at pin RX will be treated as a Start bit. However, if the data stream is subsequentially sampled at intervals of one-bit periods (two evocations of Baud_delay) then as this is just at the transition point of the transmitter, any drift in the two clock rates may cause errors. To avoid this, a half-bit period is evoked and then the state of RX is checked to ensure that the Start bit is still present. If it is, then subsequent samples are taken at two Baud_delay periods, which is approximately at the bit center point. Better noise rejection could be obtained by sampling at a higher rate (over sampling) and then taking a majority decision regarding the logic state of the incoming voltage.

After the eight data bits have been shifted into DATA_IN, the Stop bit is checked for 1. If Stop is 0 then a **Framing error** has occurred. This is signaled by returning a value of −1 in ERR. Other more elaborate schemes may return a variety of error types. For instance, where parity is used then a Parity error can be returned.

As an example, if we wish to transmit the 3-character string "PIC", then the following code fragment would implement our task. For convenience the assembler allows the programmer to represent characters in delimited single quotes to represent their ASCII equivalent, as described on p. 266.

```
    movlw   'P'         ; Same as movlw h'50' (ASCII for P)
    movwf   DATA_OUT ; Put in store
    call    PUTCHAR ; Send it out
    movlw   'I'         ; Same as movlw h'49'(ASCII for I)
    movwf   DATA_OUT ; Put in store
    call    PUTCHAR ; Send it out
    movlw   'C'         ; Same as movlw h'43' (ASCII for C)
    movwf   DATA_OUT ; Put in store
    call    PUTCHAR ; Send it out
```

Handling serial communications this way is only really satisfactory for very simple situations. For example, if the RX pin is not continually monitored, a transmission can be missed or synchronization lost. Also it is difficult to implement a full-duplex link. In addition the procedure is software intensive, with most of the processing power being wasted in delay loops. The situation can be improved somewhat by using an internal timer to generate the baud delay and by using interrupt-driven techniques. For this reason, all PIC18 MCUs have at least one integral communications port to automatically deal with asynchronous transmission.

One of the first applications of the, then new, LSI fabrication techniques in the late 1960s was the implementation of a dedicated hardware asynchronous serial port known as the **universal asynchronous receiver transmitter**. The **UART**[16]

---

[16]Sometimes known as the asynchronous communication interface adapter, or ACIA.

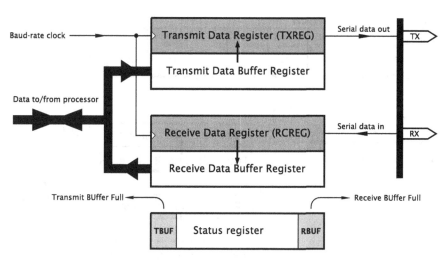

Fig. 12.21 The generic UART

was already in production by the time microprocessors were developed. Virtually all PCs, even in the 1970s, had a serial port implemented by a UART, as do most current desktop systems. As well as dealing with shifting, error checking, and interrupt handling; most UARTs also have an integral baud-rate generator which can be set-up in software to give the correct bit frequency.

The basic structure of a UART is shown in Fig. 12.21. Any given UART circuit will have three core sections. A Transmit shift register will serialize a datum with appended Start and Stop bits to be shuffled out via a TX pin. Associated with TXREG is a Buffer register holding data for onward transmission. A Status register will hold a flag (TBUF in our diagram) indicating when this buffer is free for more data.

The Receive shift register strips the Start and Stop bits off an incoming frame at a RX pin, transferring the parallelized datum when complete into one or more Buffer registers. At the same time, a flag (RBUF in our case) will be set to allow the software to determine that a new datum is ready for collection. This needs to be read before the next frame has been assembled, otherwise an overrun condition will occur and data will be lost. The transmission and reception of a frame is not locked in step; that is, they can overlap, but the baud rates are usually the same.

Real devices are more sophisticated, in that various options, such as the number of bits in a datum, and error condition detection are supported. Thus the associated Control and Status registers are of necessity rather more comprehensive, but the PIC MCU Serial Communication Interface (SCI) module, more usually known as the **USART (Universal Synchronous-Asynchronous Receiver Transmitter** module) shown in Fig. 12.22 is clearly based on the generic UART architecture. This module is actually a dual-purpose SCI in that it supports both the asynchronous protocol described here if the **SYNC** bit in TXSTA[4] is 0 (the default reset value) and a synchronous mode (SYNC = 1) which does not use a Start and Stop delim-ited frame. This latter protocol uses pin RC6/CK as a clock; output if operating as a

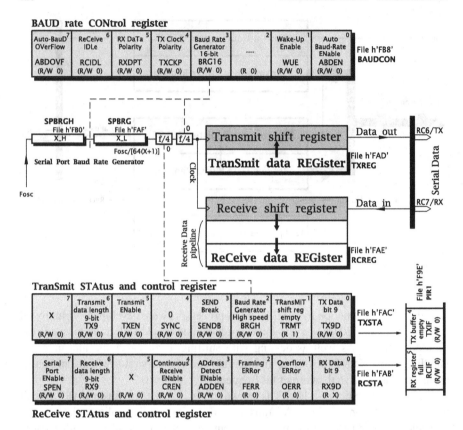

Fig. 12.22 The Enhanced Universal Synchronous/Asynchronous Receiver/Transmitter (USART) module configured for asynchronous serial communication

transmitter and clock input if a receiver. Pin RC7/DT then operates as a I/O data line as appropriate. Synchronous data can be sent either one byte at a time or as a continuous burst. Because of this synchronous function the module is known as a USART rather than a UART. Here we will concentrate on the asynchronous mode, and the three Control/Status registers shown in Fig. 12.22 are illustrated as appropriate to this mode of usage.

There are three version of the USART module in use in the various PIC16/18/24 family devices. The basic module was introduced for use in early PIC16 devices. The Addressable USART (AUSART)[17] is found in later members of that family and early devices in the PIC18 family; such as the PIC18F252. The current version is called the **Enhanced USART** (EUSART), and is a superset of the AUSART with added features of particular use in networks. Where we refer to features common across all versions, we will use the general USART designation.

---

[17]See *The Quintessential PIC® Microcontroller*, Chap. 12, by the author for more details.

The core of the EUSART is the Transmit and Receive shift registers and their associated Buffer and Status registers. To enable the overall module, the **Serial Port ENable (SPEN)** bit in the **ReCeive STAtus register (RCSTA)** (RCSTA[7]) must be set. *Both* RC6 and RC7 pins used for the transmission and reception lines **TX** and **RX**, respectively, have to be set-up as inputs.[18]

Transmission

The transmitter logic is enabled when the **TranSmit ENable (TXEN)** bit in the **TranSmit STAtus register** (TXSTA[5]) is set. To send a character the datum must be moved to the **TranSmit data REGister (TXREG)**, from where it will be transferred to the **Transmit shift register** and shifted out of pin TX. If a 9-bit data format is required the **TX9** bit in TXSTA[6] must be set to 1 and the ninth bit placed in **TXD9** in TXSTA[0] *before* moving the lower eight bits into TXREG. If the Transmit shift register is not empty; that is, it is in the process of shifting out a previous datum, then the new datum will remain in the TXREG buffer register awaiting the completion of transmission before being transferred.

TRMT in TXSTA[1] reflects the state of the Transmit shift register, whilst the **TranSmit Interrupt Flag (TXIF)** in the Peripheral Interrupt Register 1 (PIR1) is automatically set when the TXREG buffer is empty and ready for reloading. If an interrupt on TX buffer is empty is required, the corresponding TXIE mask bit in the Peripheral Enable Register 1 (PIE1) must be set—see Fig. 7.2 on p. 209. TXIF is *automatically* cleared whenever a datum is written into the TXREG, so it doesn't have to be manually cleared in the polling routine or associated ISR.

Reception

If the **CREN** bit in RCSTA[4] is set, the USART's receiver section will be enabled. In this instance, once a Start bit is detected at pin RX then the succeeding eight bits are shifted into the 2-deep **ReCeive data REGister (RCREG)** pipeline, irrespective of what is going on at the transmitter section. When a datum has been received, it is automatically stored in the top RCREG buffer from where it moves to the lower buffer; provided that no datum is still waiting to be read. The **ReCeiver Interrupt Flag (RCIF)** is automatically set whenever a datum is waiting for collection, and this can be used to generate an interrupt if the RCIE mask bit is set; as well as the appropriate global interrupt flags in INTCON[7:6]. RCIF is *automatically* cleared whenever a datum is read. If a datum is waiting in the top buffer, then it then moves down and RCIF is immediately set again, showing that there is another datum ready for collection.

If a third character has been received and the 2-deep Receive pipeline is full, then the **Overflow ERRor (OERR)** bit at RCSTA[1] will be set and this newly received datum will be lost. The RCREG can still be read twice to retrieve the two buffered bytes. However, to clear OERR the receive logic must be reset by clearing CREN and then setting it again. *No further characters will be received until this is done.*

---

[18]Exceptionally, the 18-pin footprint devices, such as the PIC18F1X20, use RB1 and RB4 for TX and RX respectively.

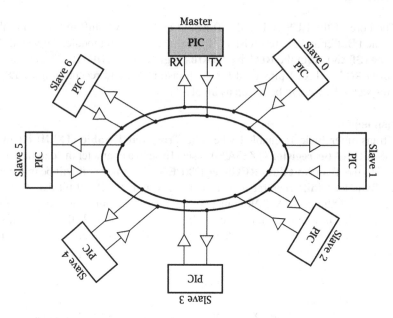

Fig. 12.23 A local area network (LAN) using an asynchronous serial protocol

The **Framing ERRor (FERR)** bit in RCSTA[2] indicates when a Start bit has been detected but at the end of the shift no Stop bit is found. Both FERR (and any ninth received bit) are double-buffered in the same way as the received data, and so should be checked first *before* the datum in RCREG is read, as this will move data down the pipeline and therefore change these auxiliary bits.

All versions of the USART module support the use of an 8- or 9-bit datum in-dividually selectable for both transmission and reception. In the latter case the **RX9** bit in RCSTA[6] must be set to 0. Typically this extra bit, read in **RX9D** in RCSTA[0], will be used to add a parity error-checking capability. However, another use of a 9-bit datum is to implement a network of asynchronous devices. Here the ninth bit can be used to indicate that the other eight bits represent either data or a Slave ad-dress. A basic network using this principle is shown in Fig. 12.23. With an 8-bit address, up to 255 Slaves may be addressed with one address reserved for a general broadcast facility.

To facilitate networking in this manner, newer versions of the USART module can be configured so that the arrival of a set ninth bit sets the RCIF interrupt flag automatically. This is enabled by setting the **ADDEN** bit in RCSTA[3] to 1. With both ADDEN and RX9 set to 1, any frame with its MSB = 0 will be ignored and the datum will not be placed in the Receive Data pipeline. If the MSB is 1, then the datum will be copied out of the Receive Shift Register into the Receive pipeline and RCIF set. The Slave can then read this address from RCREG, with this process clearing RCIF. If a match is found in software, the Slave can clear ADDEN and any subsequent data frames will be captured in the normal way. The Slave can still monitor the ninth bit in RX9D and thus the Master can terminate its conversation by sending a datum with the ninth bit set to 1.

Baud-Rate Generator

This **SPBRG:SPRGH (Serial Port Baud Rate Generator)** register pair is basically paired with a programmable 8/16-bit counter with tappings giving a pair of switchable frequency $\div 4$ rates, which can be set-up to give the appropriate sampling and shifting rates for the desired baud rate. Based on the PIC MCU's clock frequency f_{osc} we have:

$$\text{Baud rate} = \frac{f_{osc}}{64 \times (X+1)} \quad \text{8-bit low-speed: BRG16} = 0 \text{ BRGH} = 0$$

$$\text{Baud rate} = \frac{f_{osc}}{16 \times (X+1)} \quad \text{8-bit high-speed: BRG16} = 0 \text{ BRGH} = 1$$

$$\text{Baud rate} = \frac{f_{osc}}{16 \times (X+1)} \quad \text{16-bit low-speed: BRG16} = 1 \text{ BRGH} = 0$$

$$\text{Baud rate} = \frac{f_{osc}}{4 \times (X+1)} \quad \text{16-bit high-speed: BRG16} = 1 \text{ BRGH} = 1$$

where X is either the 8-bit datum written into SPBRG or the 16-bit word written into SPBRGH:SPBRG.

On reset the BRG operates in its 8-bit mode. This is compatible with earlier versions of the USART module.

For instance, if we require a baud rate of 9600 on a $f_{osc} = 20$ MHz device with an 8-bit low-speed setting (BRG16 : BRGH = 00) then using the formula $X = \frac{f_{osc}}{64 \times \text{BAUD}} - 1$ gives $X \approx 31$. The actual baud rate with SPBRG = h'1F' is 9766, giving an error of $+1.7\%$. If a 16-bit high-speed setting is used (BRG16 : BRGH = 11) we have $X = \frac{f_{osc}}{4 \times \text{BAUD}} - 1$, giving approximately 520, or h'02:08' for SPBRG16:SPBRG. The actual rate here is 9596, giving an error of -0.03. Where there are options for a given baud/clock rate, it may be advantageous to try each valid combination to give the closest fit. Baud rates of well over 1 Mbaud are possible with higher f_{osc} frequencies.

Actually, the BRG produces a frequency of 16 or 64 times the base baud rate, to enable the USART to take three samples around bit midpoints and adopt a majority decision. This increases reliability in a noisy environment.

Apart from the optional 16-bit BRG settings, the EUSART module differs from its AUSART predecessor in that it has extended capabilities for compatibility with the **LIN bus**.[19] The Local Interconnect Network was developed in 1999 by a consortium of European car manufacturers and Motorola/Freescale as a low-cost low-specification adjunct to the more complex and expensive CAN (Control Area Network) serial bus used for vehicle networking. It is designed to operated with any device with a standard UART. One Master and up to 16 Slaves are supported using a single wire of up to 40 meters (130′) length and with a maximum bit rate of 19.2 kbaud.

---

[19]See http://www.lin-subbus.org for more details.

In a quiescent network, all Slaves are normally inactive. The Master device sends a Break character of typically h'000'; which gives 13 0s including the normal asynchronous Start bit, followed by the Stop bit. This longer data field ensures that listening LIN nodes with a UART baud rate not quite at the Master's rate can synchronise their clock. This is done by sending a following Synch data frame, with the datum byte pattern b'01010101', which allows the Slave to measure the time between edges averaged over the frame period. The third frame is the Identifier byte, which carries a 4-bit Slave address and two Parity bits, which can detect two bit errors or correct one bit error. The two remaining bits code the number of data bytes in the message, which can be up to eight bytes. The final byte is a checksum of all the preceding message bytes, which may also optionally include the Identifier byte (version 2.0+).

The EUSART module supports detection or generation of the LIN bus Break frame and auto baud rate synchronization. The BAUDCON and TXSTA register bits relevant to these functions are:

RXDTP, TXCKP

Signals out of/into the TX/RX pins can be individually inverted. This polarity shift can be used to compensate for any inversion caused by line drivers. These pins are respectively Data DT and Clock CK when this module is used in its synchronous mode; hence the name RX/DT Polarity and TX/CK Polarity.

SENB

The SEND Break bit in TXSTA[3] enables the EUSART to transmit a Start bit, followed by 12 0s and then a Stop bit. This special LIN bus character is sent whenever the EUSART is enabled (SPEN = 1) for asynchronous operation (SYNCH = 0). With this setting, the following sequence of events will send a header made up of a Break followed by a Synch byte, as shown in Fig. 12.24.

1. Set TXEN and SENDB bits.
2. Load the TXREG with a dummy character (which is disregarded) when empty (TXIF = 0) to initiate the Break character transmission.
3. Write the byte h'55' to TXREG when empty to load the Synch character into the Transmit Buffer register.
4. After the Break has been sent, the SENDB bit is reset automatically.
5. The Synch character is now transmitted.
6. When the TXREG becomes empty, the next datum byte (usually the Identifier byte) can be sent in the usual way.

WUE, RCIDL

The EUSART module can receive a Break character using the Auto-Wakeup feature. When there is no activity on the LIN bus, all Slave nodes are usually dormant. If they are implemented using PIC MCUs they may well be in a Sleep mode, with the processor clock and therefore Baud Rate Generator inactive. The Auto-Wake-up feature allows the controller to wake up due to activity on the RX pin.

This feature is enabled when the WUE (Wake-Up Enable) bit in BAUDCON[1] is set. To avoid any potential loss of data, the RCIDL bit (ReCeive IDLe) flag in BAUDCON[6] should be checked to ensure that no data is being received. The EUSART is

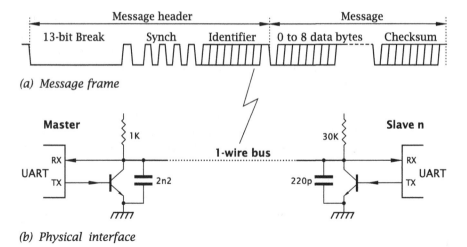

(a) Message frame

(b) Physical interface

Fig. 12.24 LIN bus protocol and physical layer

now in an Idle state. The processor can then optionally be put in its Sleep state; that is the `sleep` instruction executed. Any subsequent ‾‾_ (Start bit) on RX will set RCIF and bring the processor out of its Sleep state. At the end of the break character, the _/‾ of the Stop bit will automatically clear the WUE switch bit. The processor now exits its Idle mode and returns normal running. Doing a dummy read will clear the RCIF interrupt flag in the normal way.

Auto-wake-up only works properly based on a character which is all 0s, to ensure that WUE clears on the Stop bit, and not in the middle of the data. This doesn't have to be twelve 0s, but the longer LIN bus Break gives more time for oscillator start-up.

The alternative method of detecting the Break character is for the Slave device to temporarily alter its baud rate to $\frac{13}{8}$, so that the Stop bit will occur in the correct time slot. However, this presupposes that the Slave baud rate is accurate and relies on all the Slaves being active all the time.

ABDEN, ABDOVF

The EUSART can be configured to automatically set the SBRGH16:SBRG register pair to match the incoming baud rate. If the ABDEN (Automatic BauD rate ENable) bit in BAUDCON[0] is set, then a following Start bit launches the 16-bit SBRG pair as 16-bit counter. After the fifth rising edge, which corresponds to the Stop bit of the Synch character b'01010101' (or ASCII 'U'), the accumulated total is the correct value of X with the chosen setting of the BRG16:BRGH bit setting. ABDEN is now automatically cleared.

The baud rate thus calculated is appropriate for a 16-bit X, that is BRG16 = 1, unless there is an overflow from the counter; in which case ABDOVF (Auto BauD rate OVerFlow) in BAUDCON[7] is set. If the BRG is configured in an 8-bit mode (BRG16 = 0), then the software needs to check that both ABDOVF is 0 and SPBRGH is zero for a valid value of X. ABDOVF should be cleared in software. Doing a dummy read of RCREG will clear the RCIF interrupt flag in the normal way.

Where an auto-baud rate sequence is to follow an auto-wake-up event, as in the LIN bus protocol, then setting ABDEN when WUE is set, will ensure that Auto-Baud Detection (ABD) will occur on the byte *following* the Break character. Only after the ABD process is over, should the EUSART attempt to transmit, as the BRG pair is reversed during this process.

To illustrate how to use the USART, we will repeat our GETCHAR and PUTCHAR subroutines using hardware. Firstly, in the main program we have to set-up the Serial Port Baud Rate Generator and both Transmit and Receive Status/Control registers. Assuming the programmer has defined the constants XTAL and BAUD, then we can let the assembler evaluate the arithmetic to give us the value of X to put in the SPBRG. With this in mind, the initialization code would look something like:

```
          include  "p18f4520.inc"
          config   OSC=INTIO67, WDT=OFF

          #define  BAUD  d'9600'  ; For example 9600 baud rate
          #define  XTAL  d'8'     ; 8MHz crystal
          #define  X     ((XTAL*d'1000000')/(d'64'*BAUD))-1

START     movlw    X              ; Move X to Baud Rate Generator
          movwf    SPBRG
          movlw    b'00100000'    ; 8 data bits, TX enabled
          movwf    TXSTA          ; Low speed SPBRG mode
          movlw    b'10010000'    ; USART enabled, 8 data bits
          movwf    RCSTA          ; Receiver enabled
```

Note that Microchip specifies that both RX and TX pins be configured as inputs for correct operation. As this is the default Reset state, bits TRISC[7:6] have not been explicitly altered in the listing.

With the USART enabled, the subroutines are coded in Program 12.16. PUTCHAR is simply a matter of polling TXIF, which indicates when the TXREG is empty, waiting for it to go to 1. When the go-ahead is given, the datum is copied to the TranSmitter REGister. TXIF is then automatically cleared.

The input GETCHAR is a little more complex if some error checking is to be incorporated. The subroutine polls the state of RCIF, which goes to 1 whenever there is data to be read. Also returned is the variable ERR which is h'00' if there is no problem, −1 if a Framing error occurred, −2 if a Overflow situation is sensed and −3 if both errors occurred. In these latter two situations, OERR is zeroed by resetting the receiver logic. After the error conditions have been checked, the data is read from the ReCeive REGister. Error checking is always done *first* before reading the datum, to avoid altering these Status flags.

Some systems may not wish the processor to hang up waiting for a character which is a long time in coming. In such cases an alternative input subroutine, perhaps called getch, could return an ERR of +1 if the return was empty handed. A better approach would be to generate an interrupt each time an incoming character is sensed rather than using a polling technique.

Program 12.16 The USART-based I/O subroutines

```
; ****************************************************************
; * FUNCTION: Transmits one 8-bit byte in asynchronous format *
; * RESOURCE: PIC MCU USART                                    *
; * ENTRY   : 8-bit datum in DATA_OUT                          *
; * EXIT    : Contents of DATA_OUT unchanged, byte TXed        *
; ****************************************************************
PUTCHAR btfss    PIR1,TXIF     ; Check, is TX buffer full?
        bra      PUTCHAR       ; IF not THEN try again
        movff    DATA_OUT,TXREG; ELSE copy to USART TX register
        return


; ****************************************************************
; * FUNCTION: Receives one 8-bit byte in asynchronous format  *
; * RESOURCE: PIC MCU USART                                    *
; * ENTRY   : None                                            *
; * EXIT    : DATA_IN holds the received byte.                 *
; * EXIT    : ERR is 00 if no error. Framing ERRor only = -1  *
; * EXIT    : ERR = -2 if Overflow ERRor and -3 if both types *
; ****************************************************************
GETCHAR clrf     ERR           ; Zero flag byte
        btfss    PIR1,RCIF     ; Check, is there a char ready?
        bra      GETCHAR       ; IF not THEN try again
; Error return ---------------------------------------------
        btfss    RCSTA,FERR    ; Was there a Framing error?
        bra      CHECK_OERR    ; IF not THEN check for Overflow
        decf     ERR,f         ; ELSE record a Framing error

CHECK_OERR
        btfss    RCSTA,OERR    ; Check for Overflow ERRor
        bra      GET_EXIT      ; IF none THEN complete
        decf     ERR,f         ; Otherwise register error
        decf     ERR,f
        bcf      RCSTA,CREN    ; and reset the logic
        bsf      RCSTA,CREN
GET_EXIT
        movff    RCREG,DATA_IN ; Get datum and put away
        return
```

Asynchronous links can be used in **C** as a standard input/output channel. In the specific case of the CCS **C** compiler, the #use rs232 directive tells the compiler which pins are to be used for RX and TX and the baud rate. The normal **C** I/O functions, such as printf(), use these pins as their link to the standard channel.

As an example, Program 12.17 shows a **C** implementation of an asynchronous 9600-baud duplex link to a terminal; such as shown in Fig. 12.27. A switch is connected to pin RB0, and when the operator sends the character G (for Go) to the PIC MCU it continually monitors this switch. When the switch closes, bringing the pin Low, the terminal is to alert the operator by printing the message "Switch 1 is now closed". The standard **C** I/O functions printf() for output and getch() are used for output and input, respectively.

Program 12.17 Using a duplex asynchronous channel in **C**

```
#include   <18f4520.h>
#use delay (clock = 20000000)/* Tell compiler 20MHz xtal       */
/* Tell compiler baud rate & which pins to use for TX  & RX  */
#fuses HS,NOWDT,NOPBADEN        /* Port B not analog            */
#use rs232(baud=9600, xmit=PIN_A1, rcv=PIN_A2)
#bit SWITCH1 = 0xF81.0          /* Switch connected to RB0      */

main()
{
while(TRUE)
    {
    if(getch() == 'G')
        {
        while (SWITCH1) {;}  /* Do nothing while Switch is hi*/
        printf("Switch 1 is now closed \n");
        }
    }
}
```

As pins RA1 and RA2 are specified as the TX and RX pins, respectively, the compiler will generate a software UART implementation, such as that used in Program 12.15. For this reason the compiler needs to know the crystal frequency so that it can generate the appropriate baud delays. If pins RC6 and RC7 are used with this processor, the compiler will automatically use the USART rather than software for serial interface.

More than one concurrent asynchronous link can be supported, by giving each channel a stream name. For instance:

```
#use rs232(baud=9600, xmit=PIN_C6, rcv=PIN_C7, stream=gps)
#use rs232(baud=1200, xmit=PIN_A0, rcv=PIN_A1, stream=phone)
...............................................................
if(fgetch(gps) == 'T')
    {fprintf(phone,"Start");}
```

sets up a 9600 baud asynchronous channel using the hardware USART named gps and a 1200 baud channel called phone with pins RA0 and RA1 for the transmit/receive lines. Note that the baud rate can be different for each channel, as the latter is entirely software driven.

The body of the code sends the message Start to the phone channel whenever the character T is received from the gps channel. Notice the functions fprintf() and fgetch() for named streams, as opposed to printf() and getch() used for the default stdio channel.

There is more to setting up a communication link than establishing a suitable protocol. PIC microcontrollers have normal logic voltage and current levels which are not intended for connections greater than 30 cm (1′). Although with care,[20] dis-

---

[20]Or sometimes ignorance!

tances considerably in excess of this can be employed; in situations with relatively fast bit rates, different signaling techniques have to be used.

In the era of electromechanical TTYs the 20 mA loop *de facto* standard was in common use. This uses zero and 20 mA current to signal logic 0 and logic 1 respectively. Use of current means that line attenuation is not a problem (as current out must equal current in) and this level of current was sufficient to directly activate the receiver solenoid relay.

Current sources are realized by using high voltages in series with a large resistance. The latter gives long time constants, which, while adequate in the era of 110 baud rates, did not transfer well to the introduction of electronic terminals, UARTs and modems. **RS-232**[21] was introduced in 1969 as the standard interface for connecting an item categorized as a Data Terminal Equipment (DTE), such as a terminal, to approved Data Circuit terminating Equipment (DCE), typically a modem. Thus, not only did it define signaling levels, as shown in Fig. 12.25(a), but also various control and handshake lines, some of which are shown in Figs. 12.25(d) and 12.27. For instance, the modem would signal back to the DTE that a telephone link had been opened with the remote DTE by activating the Clear To Send (CTS) handshake signal. Two data lines plus an optional ground line are needed for a full duplex transmission circuit.

The RS-232 standard has a specified range of 15 m $(50')$ at a maximum rate of 20 kbaud, which it achieves by mapping logic 0 (often called a **space**) to typically $+12$ V and logic 1 (often called a **mark**) to typically -12 V. The receiver can distinguish levels down to ± 3 V. The **RS-423/EIA-423** standard (1978) in Fig. 12.25(b) is similar but can manage 1.2 km $(6000')$ at up to 80 kbaud and 10 Mbaud at 12 m $(40')$ with up to ten receivers.

Both RS-232 and RS-423 are **unbalanced** (or single-ended) standards, where the receiver measures the potential between signal line and local ground reference. Even though the transmitter and receiver grounds are usually connected through the transmission line return, the impedance over a long distance may support a significant difference in the two ground potentials, which will degrade noise immunity and cause significant current flows. Furthermore, any noise induced from outside will affect signal lines differently from ground, due to their dissimilar electrical characteristics; hence the term unbalanced.

The **RS-422/EIA-422** (1978) and **RS-485/EIA-485** (1983) standards are described as **balanced**. Here each signal link comprises *two* conductors, normally twisted around each other, known as a **twisted pair**. The logic level is represented as the *difference* of potential across the conductors, not the difference from ground. Calling the conductors A and B, then logic 0 is represented as $A < B$ and logic 1 by $A > B$. A difference of more than ± 200 mV at the receiver is sufficient to establish the logic level, and the transmitter will typically generate a $\Delta V = \pm 5$ V. As the A and B conductors have the same characteristics and are tightly wound together, they represent similar targets for induced noise. As the same noise voltage appears in *both*

---

[21]Defined in the United States of America as the Electronics Industries Association EIA 232-E standard and in Europe as the V24 interface, by the CCITT.

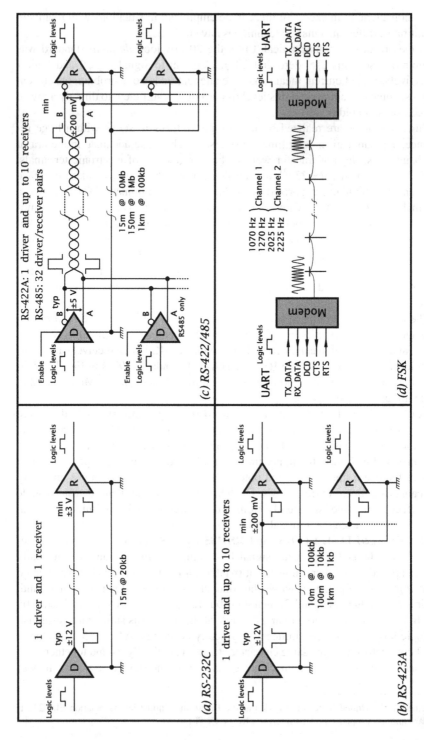

Fig. 12.25 Some signaling configurations

Fig. 12.26 CAT5 4-channel
network cable

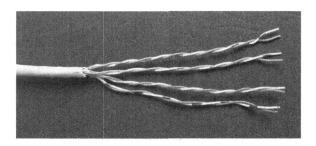

conductors and the receiver only distinguishes *differences*, rejecting common-mode voltages up to ±7 V, then the noise immunity of these balanced links is clearly superior to unbalanced schemes. Commercial twisted-pair cables, used in Local Area Networks (LANs), often carry multiple pairs of conductors, each link having a different twist pitch to reduce cross-talk between links; as shown in Fig. 12.26. PC USB leads also use a balanced signal path.

The main difference between the RS-422 and RS-485 standards is the provision in the latter case for multiple transmitters as well as receivers to implement multi-drop LANs. As only one transmitter can be active at any one time, an RS-485 transmitter buffer must have an enable input, to select the Master device. The single RS-422 transmitter has no need to be disabled.

RS-232 was originally designed for DTE-modem interconnection, although its use is now much more varied; for instance see Fig. 12.27. Figure 12.25(d) shows a simple Frequency Shift Keying (FSK) full duplex system with the mark/space of one channel being represented by the tones 1070/1270 Hz and the other by 2025/2225 Hz; frequencies which fit well inside the normal telephone link bandwidth of 300–3400 Hz. Handshake lines DCD (Data Carrier Detect), CTS (Clear TO Send) and RTS (Ready To Send) are used to control the sequence of modem operations prior to and terminating the communication of data.

Many modem schemes currently use Phase Shift Keying (PSK), where typically at least eight different phases in 45° steps of a single tone are used to encode 3-binary bit code groups (tri-bits) in any one time slot. In this way the baud rate may be increased with the same signaling rate, albeit at the expense of noise immunity; as witnessed by the steady increase in PC-dial-up home telephone Internet data rates from the original 1200 baud systems to 56 kBaud. ADSL broadband systems are based on the same techniques as dial-up, but separate out the voice baseband 300–3.2 kHz channel from the data, which use either 256 or 512 4.3125 kHz parallel channels. This gives a total capacity of up to 2.2 MHz.

As an example, Fig. 12.27 shows the connection between a PIC MCU and the serial port of a PC—or any device with an asynchronous RS-232 port. The Maxim MAX233 dual RS-232 transceiver translates between +12 V and 0 V (logic 0) and between −12 V and +5 V (logic 1). If handshake lines are not being used, as is usual in simple links, the PC can be 'fooled' into treating the interface as ready to accept data by linking, as shown in the diagram. For instance, the serial port UART's RTS

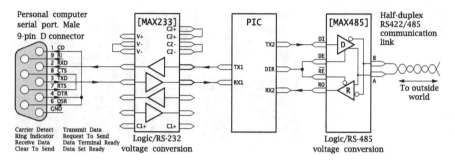

Fig. 12.27 Communicating with a PC via an RS-232 link and the outside world

is looped back to CTS. The MAX233 has two transmit and two receive buffers in all and thus can be used to buffer some additional handshake lines if required.

In Fig. 12.27 the same PIC MCU is shown driving a half-duplex RS-485 link using a Maxim MAX485 voltage converter. Each buffer has a separate Enable of the opposite logic polarity. The PIC MCU can activate the appropriate buffer depending on the communication direction. Alternatively the MAX485 can be used to implement a full duplex channel using two separate links.

The RS-485 link need not use the asynchronous protocol. Any synchronous protocol can be buffered to RS-485, but normally a separate buffered clock channel will be needed.

Examples

Example 12.1 In Example 11.2 we designed a subroutine to compare a *fixed* number TRIP with the byte read in from Port B. In some cases it may be necessary to have the software adapt to changing circumstances, altering the trigger value by reading updates from outside. Rather than using up another eight port lines, it is proposed that the update be fed in from an outside agency in series at pin RA4, with RA3 being used as the clock line. With the assumption that each data bit is stable when the Clock line is High, write a subroutine to read in a new value into memory at TRIP.

Solution One solution is shown in Program 12.18. The Clock line is monitored for a High state, during which time the Data will be stable. By mirroring the state of the Data line into the Carry flag the datum is rotated bit by bit into memory. After each shift, the loop is not completed until the Clock line again goes Low.

This is similar to subroutine SPI_READ in Program 12.3, except that the clock is generated from outside; that is, the PIC MCU is acting as a Slave. This causes a problem in a system where the PIC MCU Slave needs to tell the Master when it wants a new byte. One solution would be to use an additional port line as a Clear To Send handshake.

Program 12.18 Updating Program 11.10's trip value

```
; ************************************************************
; FUNCTION: Shifts in value for TRIP which is subsequently *
; FUNCTION: used as one operand for subroutine COMP        *
; ENTRY   : Data bit changes at RA4 when at RA3 is Low      *
; EXIT    : COUNT is 00, datum is in TRIP                   *
; ************************************************************
SER_TRIP     movlw  8              ; Bit loop count
             movwf  COUNT
SER_TRIP_LOOP1
             btfss  PORTA,3     ; Wait for Clock to go High
             bra    SER_TRIP_LOOP1
             bcf    STATUS,C    ; Carry = 0
             btfsc  PORTA,4     ; Is Data line High?
             bsf    STATUS,C    ; IF yes THEN Carry = 1
             rlcf   TRIP,f      ; Shift bit in from Carry
SER_TRIP_LOOP2
             btfsc  PORTA,3     ; Wait for Clock to go Low
             bra    SER_TRIP_LOOP2
             decfsz COUNT,f
             bra    SER_TRIP_LOOP1
             return
```

Of course the Master could be another PIC MCU and if so, we have an economical way of connecting two PIC MCUs together. If PIC MCUs with integral serial ports are used, then interrupts can be automatically generated and this is a frequently used method of implementing multi-processor networks.

Example 12.2 Design and code the I2C_IN counterpart of the I2C_OUT subroutine of Program 12.9. You may assume that the same variables are available and that the received datum is in DATA_IN on exit.

Solution The I2C_IN subroutine of Program 12.19 shifts the datum in File DATA_IN eight times through the Carry flag, which mirrors the state of pin SDA. At the same time the Clock line SCL is toggled according to the I²C time and protocol specification; as in our I2C_OUT subroutine of Program 12.9. In this protocol the Master signals back to the Slave to stop sending data by letting the SDA line float High in the Acknowledge slot in the ninth clock pulse—see Fig. 12.13. The normal Low state in this slot is called ACK, whilst the deviant High Acknowledge state is called NACK (No ACKnowledge). To cope with both these situations our I2C_IN optionally generates either situation depending on the state of the variable ACKNO, as set by the caller. If File ACKNO is zero on entry, then a normal Low ACK is sent in this slot. Any non-zero value in this variable causes a High NACK to be sent back to the Slave. The Slave then terminates its transmission and listens for the next Stop/Start condition.

Program 12.19 Reading in a byte using the I²C protocol

```
; *************************************************************
; * FUNCTION: Reads in byte from Slave with optional ACK/NACK *
; * ENTRY   : ACKNO = 00 for ACK ELSE NACK                    *
; * RESOURCE: START and STOP subroutines, Delay_600 macro     *
; * EXIT    : DATA_IN holds datum sent from slave             *
; * EXIT    : ACK or NACK sent to Slave, SCL low              *
; *************************************************************
I2C_IN    bcf    TRISA,S_CL   ; Make sure that Clock line is low
          bsf    TRISA,S_DA   ; and DAta pin is input
          movlw  8            ; Loop count = 8
          movwf  COUNT

I2C_IN_LOOP
          bcf    TRISA,S_CL   ; Clock low
          Delay_600           ; For minimum period
          Delay_600
          bsf    TRISA,S_CL   ; Clock high
          bcf    STATUS,C     ; Carry = 0
          btfsc  PORTA,S_DA   ; Check state of incoming bit?
           bsf   STATUS,C     ; IF 1 THEN make Carry = 1
          rlcf   DATA_IN,f    ; and rotate it into the datum
          decfsz COUNT,f      ; Decrement loop count
           bra   I2C_IN_LOOP  ; and repeat eight times

; Now determine if Acknowledge is to sent ---------------------
          bcf    TRISA,S_CL   ; Clock low
          movf   ACKNO,f      ; Test the caller's wish
          btfsc STATUS,Z      ; IF non zero THEN leave as NACK
           bcf   TRISA,S_DA   ; ELSE bring low to signal ACK
          Delay_600           ; Keep Clock low
          Delay_600
          bsf    TRISA,S_CL   ; Now high
          Delay_600
          bcf    TRISA,S_CL   ; Leave with Clock low
          return
```

Example 12.3 Many MCU-based products require storage of data in non-volatile memory for retrieval after the system has been powered down. A typical example is the total distance traveled by a car from new; which should be held independently of the state of the car battery—see Fig. 3.8 on p. 59. Such data is typically held in Electrically-Erasable Programmable Read-Only Memory (EEPROM); as detailed on p. 28. Although PIC18 microcontrollers have an integral EEPROM data module, as described in Chap. 15, capacity is limited to 512 bytes at most. Whilst the Program memory can be used for this purpose, this is also limited and in many cases an external EEPROM memory is required. Most of these devices use an SPI (25XXX family) or I²C interface; specifically the I²C 24AAXXX shown in Fig. 12.28. The 24AAXXX 8-pin serial EEPROMs vary from the 1 kbit 24AA01 to the 512 kbit 24LC512, organized as bytes; i.e., 128 byte to 64 kbyte.

The 24AAXXX serial EEPROMs have the following features.

PIC MCU

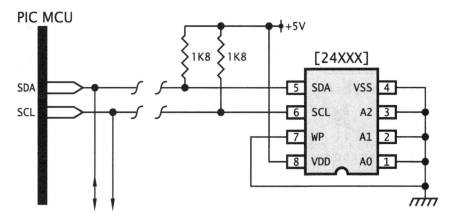

Fig. 12.28 The 24XXX series of I²C serial EEPROMs

- 400 kHz I²C compatible (2.5 ≤ V_{DD} ≤ 5.5 V); 100 kHz 1.8 ≤ V_{DD} ≤ 2.5 V.
- Write protection (ROM mode) using the WP pin.
- 2 ms typical Write cycle time.
- 1,000,000 minimum Write cycle endurance per byte cell.
- Maximum 3 mA Write, 1 mA Read and 1 µA standby current.
- Internal generation of higher programming voltage.

Using a 24AA01 serial EEPROM, show how you could increment a number in the bottom three locations, which represents the total distance. You may assume that the PIC MCU is interrupted on each mile/kilometer and that your software is part of the interrupt handler. You have the resources of the subroutines of Programs 12.9 and 12.19.

Solution Before writing code to implement our specification, we need to look more closely at the protocol used by the 24XXX serial EEPROMs in communicating with the Master PIC MCU. This is encapsulated in the signals shown in Fig. 12.29.

In all cases the Master initiates a data transfer by sending a Start condition followed by a Command byte. The Control byte contains the I²C Slave address 1010; the chip select address A2 A1 A0 and the R/$\overline{\text{W}}$ bit in the order

1	0	1	0	A2	A1	A0	R/$\overline{\text{W}}$

. Although the chip select address is shown as part of the Command byte and the three corresponding pins are shown in Fig. 12.28, newer versions of the smaller EEPROMS do not implement this feature. This is because if EEPROM capacity needs to be expanded then it is more efficient to replace the device by a pin-identical larger version. For example replacing a 24AA01 by a 24AA08 gives an eightfold increase with no hardware alteration. Larger EEPROMS, such as the 24AA256 do implement chip select address pins as the method of expansion, as additional devices will need to be hung on the bus in this situation. Eight 24LC512s will give a capacity of 512 kbyte of non-volatile memory.

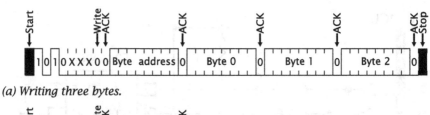

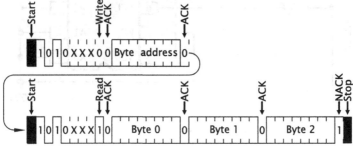

(a) Writing three bytes.

(b) Reading three bytes.

Fig. 12.29 EEPROM Read and Write waveforms

This is normally followed by the address in the EEPROM that data is to be written into or read out of. In the specific case of the 24AA01 the data is arranged as 128 cells, each comprising a byte that can be individually written to or read from. This means that a 7-bit address will fit comfortably in the 8-bit address byte. This scheme will cope with devices up to the 24AA02 but beyond this addresses greater than 8 bit wide are needed. This is done by using the three Chip select bits in the Command byte, giving an address width of 11 bits and a capacity of 2 kbytes (16 kbits). For EEPROMs larger than the 24AA16, two Address bytes are used following the Command byte.

The process of sending the byte address to the EEPROM is implemented as a Write action in Fig. 12.29(a). This is actioned by setting the R/$\overline{\text{W}}$ bit Low in the Command byte. Where a data byte is to be written into the addressed location, this byte comes immediately after the Address byte and then is followed by a Stop condition. If more than one data byte is transmitted before the Stop then this data is stored in a small on-board buffer and the actual programming will not occur until the Stop condition. The 24AA01 can store eight bytes at a time in a single page, with the *lower three* address bits being incremented on each data byte sent. If this address rolls over, earlier addressed data will be overwritten. The size of this page depends on the device; for instance, the 24AA256 uses a 64-byte page. In Fig. 12.29(a), three bytes are shown being written into the 24AA01. As these locations are to be targeted in the bottom three locations, h'00-01-02', then roll-over will not occur.

As soon as the Stop condition is received, the 24AA01 will commence programming the targeted cells with the buffered data. This process takes typically 2–5 ms across the family. If the Master attempts to initiate a process during this time, then the EEPROM will not Acknowledge following the Start-Control byte and this can

Program 12.20 Incrementing the non-volatile odometer count

```
EXTRA_MILE
          call    START       ; Get the three bytes at h'00:01:02'
                              ; Start a transmission packet
; Command byte 1 to initialize address ------------------------
          movlw   b'10100000'; Slave address Master-Write
          movwf   DATA_OUT    ; Copied to pass location
          call    I2C_OUT     ; Send it out
          movf    ERR,f       ; Check for an Acknowledge error
          btfsc   STATUS,Z    ; IF Zero THEN continue
          bra     EXTRA_MILE  ; ELSE try again
; Address 00 -------------------------------------------------
          clrf    DATA_OUT    ; Pass location
          call    I2C_OUT     ; Send it out

; Command byte 2 to change over to Read ----------------------
          call    START
          movlw   b'10100001'; Slave address Master-Read
          movwf   DATA_OUT    ; Copied to pass location
          call    I2C_OUT     ; Send it out

; Now read in three bytes ------------------------------------
          clrf    ACKNO       ; Enable Acknowledge
          call    I2C_IN      ; Read the High byte in 00h
          movf    DATA_IN,w   ; Get byte
          movwf   MSB         ; and put in memory
          call    I2C_IN      ; Read the Middle byte in 01h
          movf    DATA_IN,w   ; Get byte
          movwf   NSB         ; and put in memory
          incf    ACKNO,f     ; Signal a NACK
          call    I2C_IN      ; Read the Low byte in 02h
          movf    DATA_IN,w   ; Get byte
          movwf   LSB         ; and put in memory
          call    STOP        ; End of Read process

; Now increment 3-byte array ---------------------------------
          incfsz  LSB,f       ; Add one
          bra     PUT_BACK    ; IF not THEN continue
          incfsz  NSB,f       ; Increment middle byte
          bra     PUT_BACK    ; IF not zero THEN continue
          incf    MSB,f

PUT_BACK  call    START       ; Start the Write process
          movlw   b'10100000'; Write state
          movwf   DATA_OUT
          call    I2C_OUT
          clrf    DATA_OUT    ; Address 00h
          call    I2C_OUT

          movf    MSB,w       ; Get the new High byte
          movwf   DATA_OUT
          call    I2C_OUT
          movf    NSB,w       ; Get the new Middle byte
          movwf   DATA_OUT
          call    I2C_OUT
          movf    LSB,w       ; Get the new Low byte
          movwf   DATA_OUT
          call    I2C_OUT

          call    STOP
```

be used as a busy indicator. This polling is shown when the first Control byte is sent out in Program 12.20.

The opposite process of reading bytes from the EEPROM, shown in Fig. 12.29(b), is slightly more involved. As in the previous case, an opening address has to be written into the device. After this occurs a repeat Start condition is sent, with the following Control byte having its R/W̄ bit High to indicate Master-Read. The Slave EEPROM then transmits the byte at the specified location to the Master, which Acknowledges receipt and the process continues indefinitely with the address incrementing until the Master does not send an Acknowledge. The Slave then releases the bus and the Master is free to issue a Stop condition. If the initial writing of the first address is omitted, then one beyond the last used address is the first location read from.

The software listed in Program 12.20 follows the process outlined in Fig. 12.29 exactly. Once the initial address h'00' has been sent, the Master PIC MCU goes into a listen mode and three sequential bytes are read from memory; terminated by the Master returning a NACK condition followed by Stop. With the triple-byte distance count in locations MSB:NSB:LSB, the array is incremented in the usual way. Finally address h'00' is again written out to the EEPROM followed by the three updated bytes and the process terminated by the Master transmitting Stop.

Example 12.4 It is possible to combine some of the attributes of synchronous I²C and asynchronous signaling to send data asynchronously in both directions half-duplex along a single link. One example of this is the **1-Wire**™ [22] interface outlined in Fig. 12.30.

In Fig. 12.30(a) a Maxim Integrated Products DS18S20 digital thermometer is shown driven from a single port line, with the MCU acting as a 1-Wire Master.

The DS18S20 has the following features.

- Measures temperature from $-55°C$ to $+125°C$ in $0.5°C$ steps as a signed 16-bit datum.
- $\pm 0.5\%$ accuracy in the range of $-10°C$ to $+85°C$.
- Converts temperature in 750 ms maximum.
- Zero standby current.
- May be powered from the data line; supply range $+3$ V to $+5.5$ V.
- Multidrop capability.

The various DS18S20 functions, such as Convert (h'44') and Read temperature (h'BE'), are initiated by the Master sending the appropriate control data as 8-bit codes, each bit comprising a Start condition (‾_) with eight slots in a frame; as shown in Fig. 12.30(b). As in the I²C case, the data line DQ is pulled into the High state with a pull-up resistor and the Master simulates the logic 1 state by changing

---

[22] 1-Wire® is a registered trademark of Maxim Integrated Products Inc.

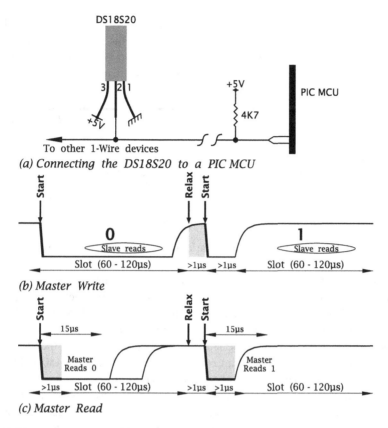

Fig. 12.30   Interfacing the DS18S20 1-Wire digital thermometer

its port line from the Low state to input (see Fig. 12.14(b)). In this state the Master can listen to data sent by the Slave, as shown in Fig. 12.30(c).[23]

For our example we are required to design two subroutines that will respectively write a byte to a 1-Wire Slave and read a byte from the Slave.

Solution From Fig. 12.30(b) we see that writing a bit to a Slave involves the follow-ing tasks:

1. The Master starts the process by forcing the data line Low for at least 1 μs.
2. The Master either keeps the line Low (Write 0) or releases the line (Write 1) for 60–120 μs.
3. The Slave reads the line state between 15–45 μs later.
4. The Master releases the line (if Write 0) for at least 1 μs to relax the system.

___

[23]For more details, see Microchip's application note AN1199 *1-wire® Communication with PIC® Microcontroller*.

The subroutines of Program 12.21 assume that the port line driving DQ has been set-up as described on p. 403 for the I²C bus to give the two states as hard Low and open circuit, pulled up to the High state. Also we assume that we have the delay macro Delay_us *in situ* which gives a K µs delay, where K is the parameter passed to the macro.

```
Delay_us macro K                        ; K is the number of us delay
         local DELAY_US_LOOP
         movlw ((K*XTAL)/d'16')+1; 4~ (4/XTAL us) per loop: 1~
DELAY_US_LOOP
         addlw -1                       ; Decrement count : N~
         btfss STATUS,Z                 ; to zero          : N + 1~
          bra  DELAY_US_LOOP            ;                  : 2(N-1)~
         endm
```

The additional plus one in the formula for loop count ensures that values for K round up.

Both subroutines begin by driving DQ Low for a minimum of 1 µS, defining the Start condition. Writing a single bit to DQ occurs in a slot which has a duration of 60–120 µs, and commences with DQ either Low or released to be pulled High, defining a Write-0 or Write-1 condition. The Slave samples the state of the data line sometime after 15 µs into the slot. Although the duration of the slot is not critical, care needs to be taken as a Low-state duration of between 480 and 960 µs is interpreted by the Slave as a Reset command (see SAQ 12.2).

Eight Write slots are used with a minimum 1 µs relax period interval to transmit the byte; each slot's state following the bit rotated into the Carry flag of the datum byte DATA_OUT. After eight shift/output cycles the process terminates.

Reading from a Slave involves the following tasks:

1. The Master starts the process, forcing the data line Low for at least 1 µs.
2. The Master then listens to data placed on the line by the Slave which is valid for up to 15 µs after the Start edge.
3. The Slave releases the line after 15 µs which should be pulled High by the end of the 60 µs slot.
4. The Master waits a minimum of 1 µs before starting the next slot.

The input subroutine READ_1W follows this task list, sampling the data line sometime before 15 µs into the slot, at which time the Slave's data should have settled to the appropriate voltage level. Each bit is used to appropriately set the Carry flag, which is then shifted right into DATA_IN LSB first. After eight sample/shift loops, DATA_IN has the received byte datum.

Unlike the I²C bus, the 1-Wire architecture is designed for a single Master. However, 1-Wire Slaves have device addresses comprising a 64-bit unique code as part of an internal ROM. The first eight bits are a 1-Wire family code—the DS18S20 code is h'10'. The following 48 bits are a unique serial number and the last eight bits are an error-checking byte.

Program 12.21 Reading and writing on a 1-Wire system

```
;  ***********************************************************
;  * FUNCTION: Writes a byte datum to a 1-Wire slave        *
;  * RESOURCE: macro Delay_us giving N microsecond delay    *
;  * ENTRY    : Datum is in DATA_OUT                         *
;  * EXIT     : DATA_OUT is zero, W, STATUS altered          *
;  ***********************************************************
WRITE_1W movlw     8           ; Loop count
         movwf     COUNT
W_LOOP   bcf       TRISA,DAT   ; Low edge signals Start
         Delay_us  1           ; for 1us
         rrcf      DATA_OUT,f  ; LSB first shift into Carry
         btfsc     STATUS,C    ; Was it a 1?
          bsf      TRISA,DAT   ; IF it was THEN output high
         Delay_us  d'60'       ; Hold for 60us
         bsf       TRISA,DAT   ; Release line to go high
         Delay_us  1           ; Relax for 1us
         decfsz    COUNT,f     ; Repeat eight times
          bra      W_LOOP
         return


;  ***********************************************************
;  * FUNCTION: Reads a byte datum from a 1-Wire slave       *
;  * RESOURCE: macro Delay_us giving N microsecond delay    *
;  * ENTRY    : None                                         *
;  * EXIT     : Datum is in DATA_IN, W, STATUS altered       *
;  ***********************************************************
READ_1W  movlw     8           ; Loop count
         movwf     COUNT
R_LOOP   bcf       TRISA,DAT   ; Low edge signals Start
         Delay_us  1           ; for 1us
         bsf       TRISA,DAT   ; Release line
         Delay_us  8           ; Wait 8us for Slave to O/P data
         bcf       STATUS,C    ; Clear Carry
         btfsc     PORTA,DAT   ; Check input state
          bsf      STATUS,C    ; IF high THEN set Carry
         rrcf      DATA_IN,f   ; Shift bit in -> LSB
         Delay_us  d'48'       ; Wait to end of slot
         decfsz    COUNT,f     ; Repeat eight times
          bra      R_LOOP
         return
```

Self-Assessment Questions

12.1 Show how you could connect four MAX518 ADCs (see Fig. 12.16) on the one I^2C circuit and how channel 1 on the third ADC could be written to.

12.2 Communications along a 1-Wire link begins with a Reset operation, where the Master pulls the line Low for 480–960 µs after which the line is released. The Slave then responds by dragging the line Low after no more than 60 µs delay. This Low state persists for a further 60–240 µs after which the Slave releases this line. Design a subroutine that will do this procedure when called. Assume the resources of Program 12.21 are available to you.

12.3 Parity is a technique whereby the number of digits in a word is always either even or odd. This is accomplished by adding an extra bit which is calculated by the transmission software to be 0 or 1 to meet this overall criterion. For instance, for odd parity of an 8-bit word b'01101111' we have b'1 01101111'. The receiver will check that all nine received bits have an odd count. If one bit (or any odd number) has been corrupted by noise, then a **parity error** is said to have occurred.

Based on the PIC MCU USART, write software to set the asynchronous protocol to 9 bit word and calculate the odd one's parity bit of DATA_OUT which should be placed in TX9D of the TXSTA register prior to the loading of the data into TXREG and transmission.

12.4 Rewrite the subroutine GETCHAR of Program 12.15 as an interrupt service routine called GETCH. Compare the two approaches.

12.5 A certain data logger is to sample temperature once every 15 minutes. The power supply current consumption is reduced by using a Low-voltage part at a V_{DD} of 3 V and a crystal of 32.780 kHz. Under these conditions the current consumption with the Timer 1 running is a maximum of 70 μA (45 μA typical). A I²C EEPROM is to be used to store the data as it is read, but is only powered on at sample time—by using a spare port line as the EEPROM's power supply. The logger is to be left submerged at the bottom of a lake for 6 months before being recovered. Can you choose an appropriate 24LCXXX EEPROM and estimate the required capacity of the 3 V battery in mA-hours?

12.6 A PIC18F4520 is to be used over a mobile phone network, to remotely monitor and control an intelligent home heating system. As shown in Fig. 12.31, the processor communicates with a GM862 GSM (*Groupe Spécial Mobile*) module[24] via a duplex asynchronous serial link. A second simplex asynchronous stream echos commands and gives status messages to a PC with a serial port.

The GSM module is turned on (and also off) with a 1 s duration ⏌‾⎍. The controller must then send the two characters 'A', 'T' (ATtention). If successful, the module will echo back with the message 'A', 'T', '\r', '\n', 'O', 'K', '\r', '\n' (where \r and \n are Carriage Return (0x0D) and New Line (0x0A) respectively).

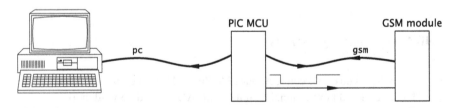

Fig. 12.31 GSM module dual-serial interface

[24]See http://www.roundsolutions.com/techdocs/index.php for more details.

Design software coded in CCS **C** to implement the following task list:

1. Set-up two 2400 baud serial links with the USART being used for the `gsm` stream. The software `pc` stream is to use pin `RA0` as its transmit pin—see p. 436.
2. Use pin `RA1` to pulse the GSM module for 1s.
3. Send the message `'A'`, `'T'` to the GSM module.
4. Wait nominally one minute.
5. As the characters are received in an ISR from the GSM module during the delay, they are stored in a global buffer and echoed to the PC. If too many characters are received, a warning message should be sent.
6. After the delay, if the correct eight characters have been received, then a confirmation message should be sent to the PC, otherwise an error message is transmitted and the process repeated.

Chapter 13
Time Is of the Essence

Of crucial importance in many systems are time-related functions. These may manifest themselves in the measurement of duration, event counting, or control of an external physical event for known periods. An example of the former would be the time between pulses generated by the teeth on a flywheel to measure engine speed for a car dashboard management system—see Fig. 3.8 on p. 59.

Where *time is of the essence* these functions are often best implemented by using hardware counters to time events, rather than software delay routines. In this chapter we will look at the various timer modules which are available to the PIC18 MCU family. After completion you should:

- Know how a Watchdog timer improves the robustness of a MCU-based system and how to use the integral WDT module for both this purpose and to awaken the processor when operating in a power management mode.
- Be able to use the basic 8/16-bit Timer 0 module as both a counter and timer.
- Understand the function of the 16-bit Timer 1/3 modules and their interaction with the Capture/Compare/PWM (CCP) modules.
- Be able to use the 8-bit Timer 2 module together with the CCP modules, to generate pulse-width modulated outputs.

Many MCU-based systems are hosted in an electrically hostile environment, with noise induced outside both through logic lines and the power supply. Our example of an auto dashboard manager is typical of this situation, with induction from the high-voltage ignition sparks and alternator sourced ripple in the battery supply. No matter what precautions in shielding and filtering are taken, it is inevitable that on occasion the MCU will jump out of its proper location in Program memory and 'run amok' with potentially serious consequences on the controlled system.[1] In some cases this is little more serious than requiring a manual reset.[2] However, this is not possible in many situations; for instance, in a pacemaker implanted in the patient's body or a Martian space probe.

---

[1] The same can happen as a result of software bugs.

[2] As in a Window's® PC.

S. Katzen, *The Essential PIC18® Microcontroller,*
Computer Communications and Networks,
DOI 10.1007/978-1-84996-229-2_13, © Springer-Verlag London Limited 2010

One solution to this problem is to use an oscillator/binary counter, which resets the processor when the count overflows.[3] If the software is arranged to clear this counter on a regular basis so that overflow never occurs, then the MCU never resets. If something happens and the MCU jumps out of its normal loop then eventually, without this constant clearing, the counter will overflow and the MCU will be reset to its starting point. This circuit is given the name **Watchdog timer**, as it enhances the system security.

Rather than rely on external Watchdog timers, all PIC MCUs, even the early low-range family, have an integral WDT module. The operation of this function, shown in Fig. 13.1, is essentially:

- A 7-bit counter frequency divides the nominal 31 kHz INTRC internal oscillator (see Fig. 10.11 on p. 321) to give the fundamental Watchdog period of approximately 4.1 ms (3.56 ms minimum, 4.82 ms maximum).
- The Watchdog counter is enabled if the **WDTEN** fuse is 1 (default) or else it can be controlled in software by the **SWDTEN** bit in **WDTCON** (this is the only functional bit in this control register).
- The basic 4.1 ms period is divided by a 16-bit Postscaler counter[4] to give a chain of periods in powers of 2; with a maximum value of 2.18 minutes. Any one of these can be selected with the four WDTPS3:0 fuses in CONFIG2H (see Appendix B) addressing a multiplexer. For instance:

```
config OSC=HS, WDTPS=512, WDT=ON
```

will set up these fuses to b'1001' to give a time-out period of $\times 512 \approx 12$ s. WDT=OFF must be used to disable the hardware control of the WDT. The equivalent for the CCS compiler is:

```
#fuses HS, WDT, WDT512
```

and NOWDT is the counterpart of WDT=OFF.
- The execution of a **clrwdt** (**CLeaR WatchDog Timer**) instruction will reset the WatchDog counter/Prescaler chain to zero. If this happens at a rate faster than the selected period, then the WatchDog Timer will never time out. For the example given, software running in an endless loop with clrwdt being executed more frequently than every 12 seconds, will prevent overflow.
- If the Postscaler does overflow, that is the selected flip flop goes 1, then the $\overline{\text{TO}}$ (**Time Out**) flag in the RCON register will be cleared; its active state. If the processor is running normally (that is, not in a Sleep/Idle state) then this will reset the processor, which begins execution at the Reset vector h'00000'. A WDT reset

---

[3]Other approaches typically are based on a retriggerable monostable.

[4]Early PIC18 devices, such as the PIC18F452, had an 8-bit Postscaler; as did earlier families.

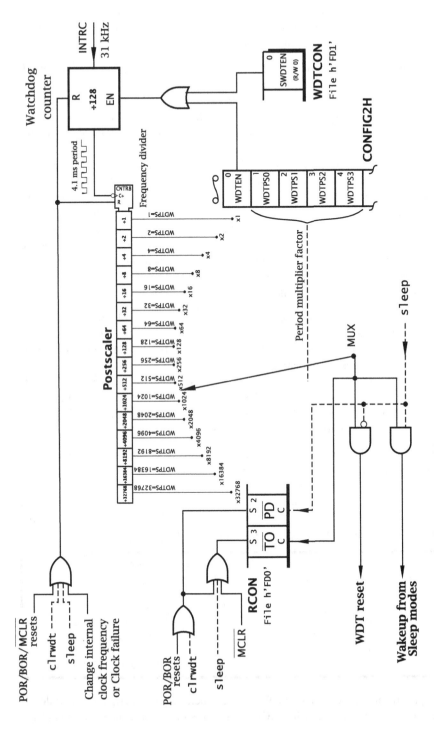

Fig. 13.1   The Enhanced PIC18 MCU Watchdog timer module

does not change the state of $\overline{\text{TO}}$ and so can be checked to distinguish from a 'normal' restart. A clrwdt instruction following this check will set $\overline{\text{TO}}$ and will at the same time clear and restart the WatchDog Timer—see p. 456.

The WDT module can also be used in a secondary role to waken a processor following a sleep instruction. On execution, this instruction will zero and restart the Watchdog counter Prescaler chain. If it should time out, the processor will awaken and execution will restart with the instruction following sleep. The processor does not reset. Normally, sleep is followed by a clrwdt which zeros and restarts the WDT module and deactivates $\overline{\text{TO}}$.

Another way of arousing an idle or sleeping processor is via an external interrupt or internal trigger sourced from a module with a local oscillator; such as Timer 1. Of course, the Primary oscillator will be disabled following a sleep instruction. If interrupts are globally disabled, execution will begin with the instruction following on after sleep. If however, the interrupt is enabled both locally and globally, then the processor will vector to the ISR in the normal way. By examining the $\overline{\text{PD}}$ (**Power Down**) flag, the ISR can determine if it was launched when the process was asleep or idling. $\overline{\text{PD}}$ is set inactive only by executing a clrwdt instruction.

In summary, the WDT module has two functions.

1. Most embedded programs operate in one or more endless loops. By judicious selection of the time-out period, executing clrwdt on each pass will keep the Watchdog counter chain from overflowing. If the software malfunctions, then the processor will reset automatically.
2. The WDT module can be used to awaken the processor from a Sleep or Idle mode, after a nominal delay of between 4.1 ms and 131 seconds. Note that an active WDT timer will add typically 5.5 µA (15 µA maximum) at a V_{DD} of 5 V and 25°C to the power budget—see p. 308.

As an example, consider a system counting cans of beans moving along a conveyer belt in the manner shown in Fig. 13.3, keeping a tally in a File called BEAN_COUNT. On Power-on this tally is to be zeroed. If due to a glitch there is no activity for a period of nominally a minute, the PIC MCU's Watchdog is to reset but keep the tally unchanged. To do this we can use the startup code to check the state of $\overline{\text{TO}}$ and take the appropriate action; for instance:

```
        config   WDT=ON           ; Enable the Watchdog timer
        config   WDTPS=16384      ; Approx 65 second time-out
                                  ;
        org      h'00000'         ; The Reset vector
MAIN    btfss    RCON,TO          ; Was this a Watchdog reset?
        clrf     BEAN_COUNT       ; IF not THEN zero tally
; More initializing code
        clrwdt                    ; Set TO and reset Watchdog
```

Apart from parallel ports and Watchdog timer, the only peripheral module offered by the first line of PIC MCUs was an 8-bit timer called the Real-Time

Counter-Clock. The PIC16 family kept the RTCC, but eventually changed its name to **Timer 0** to line up with the new Timer 1 and Timer 2 modules. Timer 0 in the PIC18 family defaults to an 8-bit counter-timer on reset for compatibility with these earlier devices. However, this version has been expanded with a 16-bit mode of operation.

From Fig. 13.2 we see that the counting chain comprises a cascade of two primary 8-bit counters fronted with an optional 8-bit Prescaler counter. This gives eight selectable clock rates into the primary counter; selected by the **PS2:0** bits in the **Timer 0 CONtrol register T0CON**.[5] If the **PreScaler Assign PSA** bit is 1, its reset value, then the Prescaler flip flop chain is by-passed.

The counting chain can either be clocked by the internal instruction cycle clock $\frac{f_{osc}}{4}$ (that is at the instruction cycle rate) or from an external source via the **Timer 0 Clock Input** $\texttt{TOCKI}$ pin. The **Timer 0 Clock Select T0CS** bit at T0CON[5] is used to select the internal/external mode. When clocked from outside, the active edge is set using the **Timer 0 Set Edge T0SE** bit at T0CON[4].

An event at the $\texttt{TOCKI}$ pin will appear to be random with respect to the internal clock cycle. In order to allow Timer 0 to be read from or written to in the normal way without interaction between the two timing signals, a synchronization stage is necessary. This is done using a 2-stage shift register before the Timer 0 primary counter's clock input. This causes a delay of two instruction cycles; 1 μs with an 8 MHz crystal. Where the primary counter is directly connected to the internal clock, this will cause a 2-cycle delay before anything happens after a datum is written into Timer 0.

How fast can the $\texttt{TOCKI}$ pin be clocked and still reliably be tallied? The answer depends on whether the Prescaler is used to predivide the incoming pulse train. In the former case $PSA = 1$ the external pulse train is directly synchronised by the instruction cycle, and so cannot be more than $\frac{f_{osc}}{4}$; denoted in the data sheet as the period t_{CY}. Actually, a short set-up and hold time is specified, giving a minimum period t_{TOP} of $t_{CY} + 40$ ns; $t_{TOL} = t_{CY} + 20 = t_{TOH}$ for the Low and High durations respectively. For instance, with a 8 MHz f_{osc} clock, the minimum square wave period would be $500 + 40 = 540$ ns (≈ 1.85 MHz).

Where the Prescaler is used, the $\texttt{TOCKI}$ pulse rate can be scaled up to reflect the period multiplication; that is $\frac{t_{CY}+40}{PS}$, where PS is the Prescale ratio. However, there is an absolute minimum High or Low duration of 10 ns. For instance, for a 8 MHz clock and a PS-ratio of 16, the minimum duration is $\frac{540}{16} = 33.75$ ns or ≈ 30 MHz. However, if a 20 MHz clock was used, the overarching period limit of 20 ns or 50 MHz square wave, must be used.

The primary counting chain can be configured as either 8-bit or 16-bit using the **Timer 0 8BIT T0BIT8** bit in T0CON[7]. In the former case, when **TMR0L** overflows, **TMR0IF** in INTCON[2] will be set. In the 16-bit configuration, TMR0L clocks the high byte of Timer 0 and so it is this overflow that is signaled by setting the TiMeR 0 Interrupt Flag. In either case TMR0IF should be cleared in software in any ISR or

---

[5]In earlier families, the Option register $\texttt{OPTION\_REG}$ was used to configure Timer 0, the WDT module and more besides.

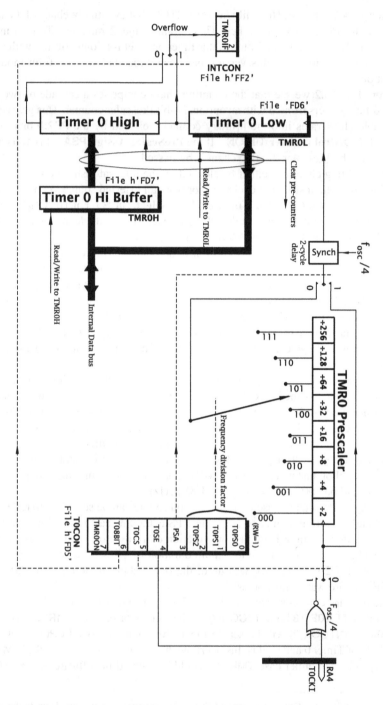

Fig. 13.2 Simplified functional diagram for the Timer 0 module

polling routine in order to catch any subsequent overflow. An interrupt will be set in train if the **TMR0IE** enable bit in INTCON[5] is set, with its appropriate priority set with **TMR0IP** in INTCON2[2].

The low byte of Timer 0 can be read from to written to at any time in the same was as any File. As only eight bits can be accessed by any instruction, there is a potential problem in reading or writing to a 16-bit Timer 0. For instance, consider Timer 0 is h'7F FF' at an instant of time. Reading the low byte gives h'FF'. Perhaps just as the high byte is about to be read, the counter chain increments to h'80 00', then the value obtained will be h'80'. Thus the counter will be erroneously read as h'80 FF'! For this reason, direct access to the high byte of the count is not permitted. Instead its state is accessed via a go-between buffer register **TMR0H**. An instruction reading from TMR0L will automatically copy the high byte of the count into TMR0H, from where it can subsequentially be read. Writing is done the same way, first copying the required high byte into TMR0H and then writing the low byte into TMR0L. Both bytes will then be loaded into the 16-bit Timer 0. For instance, to set-up the timer to h'20 00' we have:

```
movlw  h'20'      ; The high byte
movwf  TMR0H      ; in the High buffer
clrf   TMR0L      ; Low byte = 00 and High byte is h'20'
```

Writing to TMR0L always clears the Prescaler and clock synchronizer, in either 8- or 16-bit modes.

Timer 0 is mainly used either to *count* external events or to measure the *time* between external events. It can also be used to time software toggling port pins for precisely known durations, without tying up the processor in time-wasting delay routines.

We will illustrate the usage of Timer 0 as an *event counter* and *stop clock* with two examples. The first, which also illustrates the Watchdog timer, is to tally cans of baked beans traveling along a conveyer belt, as shown in Fig. 13.3. Each 24 cans passing the sensor should generate a pulse to a packing machine, so that the box can be replaced by a new empty container. This pulse need only be a few microseconds

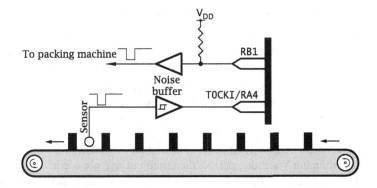

Fig. 13.3 Counting cans of beans on a conveyer belt

in duration. A double-byte count should also to be kept of the number of boxes packed since the last Power-on/Manual Reset. This will be uploaded to the central plant computer at the end of the shift for inventory control.

Our first consideration is the set-up and initialization code. This code, shown below, begins by checking the $\overline{\text{TO}}$ flag at the Reset vector. If zero then the bulk of the initialization code is omitted, as reset was due to a Watchdog time-out. If this was not the case, then Ports A & B pins are configured as digital, with pin TOCKI/RA4 set-up as an input, and RB1 set-up as an output to activate the packing machine.

```
        include   "p18F4520.inc"
        config WDT=ON, WDTPS=128, OSC=XT ; Enable Watchdog

        cblock h'020'
         COUNT:2
        endc

        org      0              ; Reset vector
        btfss    RCON,TO        ; Check if a Watchdog reset
         bra     MAIN_LOOP      ; IF yes THEN no initialization
        bra      MAIN           ; ELSE a fresh start

        org      8              ; Compatible Interrupt vector
        goto     ISR            ; Foreground program

MAIN    setf     ADCON1         ; Make all port pins digital
        bsf      PORTB,1        ; Idle state of the Packing pulse
        bsf      TRISA,4        ; Make sure that TOCK1 is I/P
        bcf      TRISB,1        ; & RB1/Packing machine an O/P
        movlw    b'11101000'    ; Timer on, external rising edge
        movwf    TOCON          ; 8-bit mode, no Prescaler

        movlw    -d'24'         ; Initialize TMR0 to -24 (h'E8')
        movwf    TMR0L
        clrf     COUNT+1        ; Clear the 2-byte score count
        clrf     COUNT
        bsf      INTCON,TMR0IE; Enable Timer0 interrupt
        bsf      INTCON,GIE   ; Enable all interrupts

; The background program which amongst other things - - - - -
MAIN_LOOP
        clrwdt                  ; Regularly resets the wdt
        ...      .....          ; More background code
        goto     MAIN_LOOP      ; DO forever main loop
```

No details were given about the maximum time to execute the Main loop, so for illustration purposes the Watchdog timer has been configured to extend its time-out period by ×128. Timer 0 is enabled in its 8-bit mode to be clocked from TOCKI on a ⌐ with no Prescaler. Finally, Timer 0 itself is set to h'E8' (i.e., −24 decimal) so that 24 can pulses will cause it to overflow and generate an interrupt. Both INTCON flags TMR0IE and GIE are then set to enable the interrupt.

The main background program commences with a clrwdt instruction. Provided that the background endless loop is no longer than nominally 3.56 ms × 128 ≈ 0.455 s, the minimum Watchdog period, then time-out will not occur.

With the initialization code *in situ*, all that remains is to implement the interrupt service routine (ISR) that will be automatically entered after each batch of 24 cans;

Program 13.1 The bean counter interrupt service routine

```
; *****************************************************************
; * FUNCTION: ISR to issue a Packing-machine pulse and re-       *
; * FUNCTION: initialize Timer0 to -24.  Keeps a grand score     *
; * FUNCTION: total in COUNT:2 for background analysis           *
; * RESOURCE: COUNT:2                                            *
; *****************************************************************
ISR     btfss    INTCON,TMR0IF ; Was it a can?
        bra      ISR_EXIT      ; IF no THEN false alarm

; Core code --------------------------------------------------------
        bcf      PORTB,1       ; Pulse packing machine
        movlw    -d'24'        ; Re-initialize Timer0
        movwf    TMR0L
        infsnz   COUNT,f       ; Add one to score count
        incf     COUNT+1,f
        bcf      INTCON,TMR0IF ; Reset interrupt flag
        bsf      PORTB,1       ; End packing machine pulse and
ISR_EXIT
        retfie   FAST          ; return from interrupt with context
```

that is, when Timer0 counts up 24 input pulses and overflows back to zero. When this occurs, Timer0 will set TMR0IF and the PIC MCU will jump to the Interrupt vector at h'00008'. In our initialization code we have placed a `goto ISR` at this point, and so named the routine in Program 13.1. If there are other sources of interrupt then the switch would be to another part of the ISR, as shown in the listing of Program 13.3.

The ISR simply checks the TMR0IF interrupt flag and exits if not set. The core implements the following task list in no particular order:

- Pulse RB1 ⌐_/¯ to signal the packing machine.
- Reset Timer0 to −24.
- Increment the double-byte score Count.
- Reset the Timer0 interrupt flag TMR0IF.

For an alternative approach using Hardware interrupts see Program 7.3 on p. 229. Program 13.2 shows the same task coded in CCS **C**—see also Program 9.3 on p. 293. This follows the structure of the assembly program, using the following CCS-specific functions:

setup_timer_0(mode)

This function takes a list of OR delineated (|) mode commands to set-up Timer0. Each of the timers have an equivalent function to initialize their control register; T0CON in our example. The list of mode commands are defined in the appropriate processor header file; e.g. 18f4520.h. See also SAQ 13.13.6.

set_timer0(value)

Updates the value of the timer. All timers have functions of this form and also **get_timerx()** to read Timer x.

restart_wdt()

Does a `clrwdt` instruction.

Program 13.2 Coding the bean counter in **C**

```
#include <18f4520.h>
#fuses    WDT,WDT128,XT
#bit      PACK_MACHINE = 0xF81.1    /* Activate packing machine */

long int COUNT;                     /* Global 16-bit variable   */

main()
{
/* Do the following initialization if reset was a normal POR  */
if(restart_cause() == NORMAL_POWER_UP)
    {
    COUNT = 0;                      /* Start with a zero count  */
/* Config Timer0 as 8-bit, +ve edge ext clock with no Prescale*/
/* Have to specify 8-bit Timer 0 as 16-bit is default         */
    setup_timer_0(RTCC_EXT_L_TO_H|RTCC_DIV_1|RTCC_8_BIT);
    set_timer0(-24);               /* Timer 0 = -24            */
    set_tris_b(0xFD);              /* RB1 set as output        */
    setup_adc_ports(NO_ANALOGS);   /* All pins digital         */
    enable_interrupts(INT_TIMER0);/* Enable Timer 0 interrupts*/
    enable_interrupts(GLOBAL);     /* Enable interrupt logic   */
    }
while(1)                           /* DO forever               */
    {
    restart_wdt();                 /* Clear WDT                */
    /* DO this; DO that; DO the other; Dummy code              */
    }
}

#int_timer0
isr()
{
PACK_MACHINE = 0;                  /* Pulse the packing machine */
set_timer0(-24);                   /* Timer 0 = -24             */
COUNT++;                           /* One more score            */
PACK_MACHINE = 1;
}
```

restart_cause()
This function interrogates the $\overline{\text{TO}}$ and $\overline{\text{PD}}$ flags in the RCON register. In Program 13.2 the key return value is WDT_TIMEOUT. Other values based on RCON flags are listed in 18f4520.h; for instance, wdt_FROM_SLEEP and BROWNOUT_RESTART.

Notice the use of #int_timer0 to designate the function isr() as a Timer 0 interrupt handler—see p. 292.

Our second example illustrates the use of Timer 0 as a *clock* to measure time between events. The events in question are R-points peaks in the ECG waveform illustrated in Fig. 7.1 on p. 206. In this example, a peak detector interrupts the MCU, which keeps a 2-byte count from a 10 kHz external oscillator. In this manner the period between events can be determined on each event in increments of 100 μs, which we call here **jiffies**. For our example we will modify the specification to eliminate this oscillator and use Timer 0 to keep a nominal 125 μs 2-byte Jiffy tally. The $\frac{1}{8}$ ms tick rate used here is based on a 4.096 MHz main system clock, which is divided down by a Prescaler ratio of 128; i.e. $\frac{4.096}{4 \times 128}$. We will see further on in the chapter, that the more sophisticated timers can be used to give a wider range and precision of clock rates.

As well as setting the INT0IE flag to enable Hardware interrupt 0, which is used to signal an ECG peak event at INT0, the Timer 0 interrupt is also enabled by setting the TMR0IE flag. As a safety feature, if the timer overflows; that is after $65.536 \times 8 \approx 8.2$ s, then we would like to sound the alarm. We will assume a sounder is connected to pin RA0. Whenever an ECG peak is detected, the 16-bit Jiffy count in Timer 0 is copied into two Files JIFFY:2 and sets the File NEW to a non-zero value. Neither this double-byte Jiffy count nor Timer 0 need be cleared as the first reading of the series will always be erroneous—because the patient's heartbeat is not synchronized to the PIC MCU reset! However, File NEW, which is set to non-zero each time an ECG peak is detected, is cleared.

```
          config   WDT=OFF, OSC=XT

          cblock h'020'
           JIFFY:2,NEW:1
          endc

          org    0              ; Reset vector
          bra    MAIN           ; Background program

          org    8              ; Compatible Interrupt vector
          goto   ISR            ; Foreground program

MAIN      movlw b'10010110'     ; INT on -ve edge, internal clock
          movwf T0CON           ; Prescale /128 assigned to Timer0
          setf  ADCON1          ; Make parallel port pins digital
          bcf   PORTA,0         ; Make RA0 start as 0
          bcf   TRISA,0         ; as an Output to alarm
          clrf  NEW             ; Zero the New flag
          bsf   INTCON,TMR0IE   ; Enable the Timer 0 interrupt
          bsf   INTCON,INT0IE   ; Enable the Hardware interrupt
          bsf   INTCON,GIE      ; Enable Interrupt system
          clrf  TMR0H           ; Zero the timer
          clrf  TMR0L

MAIN_LOOP
          ....  .....           ; Background code
          ....  .....           ; More background code
          goto  MAIN_LOOP       ; DO forever main loop
```

The core of the ISR shown in Program 13.3 implements the following task list when an interrupt is received:

Program 13.3 Measuring the ECG waveform period to a resolution of 125 μs

```
; ********************************************************************
; * FUNCTION: IF INT0 interrupt, set NEW, zero TMR0                 *
; * FUNCTION: Also update the two JIFFY bytes                       *
; * FUNCTION: Sound alarm at RA0 IF TMR0 interrupt                  *
; * RESOURCE: JIFFY:2, NEW:1                                        *
; ********************************************************************
ISR          btfss     INTCON,TMR0IF; Was it a heartbeat?
             bra       HEART_BEAT   ; IF yes THEN go to it

; ELSE must have been TMR0 overflowed --------------------------
             bsf       PORTA,0      ; In which case sound the alarm
             bcf       INTCON,TMR0IF; Clear Timer 0's interrupt flag
             bra       ISR_EXIT     ; and return

; This code handles the case when ECG peak detected -------------
HEART_BEAT movff       TMR0L,JIFFY  ; Copy 16-bit count to JIFFY:2
             movff     TMR0H,JIFFY+1
             setf      NEW          ; Tell the world there is new data
             clrf      TMR0H        ; Zero Timer 0
             clrf      TMR0L
             btfsc     INTCON,INT0IF; Reset Hardware0 interrupt flag

ISR_EXIT   retfie FAST             ; and return from interrupt
```

1. IF a Hardware interrupt from peak picker.

   - Copy Jiffy count into memory.
   - Zero Timer 0.
   - Set New indicator.
   - Reset Hardware interrupt flag INT0IF.
   - Return from interrupt.

2. ELSE Timer 0 interrupt.

   - Set alarm.
   - Reset TMR0IF interrupt flag.
   - Return from interrupt.

Both bytes in Timer 0 are copied into the Files called JIFFY+1:JIFFY when a Hardware interrupt is received. Notice how the lower byte of Timer 0 TMR0L is read first, to simultaneously update the high byte buffer TMR0H; which can subsequently be accessed. Conversely, writing to Timer 0, in this case clearing the timer, is done in reverse; that is, TMR0H is updated first. When TMR0L is subsequently cleared, the value in TMR0H is uploaded into the buried high byte of Timer 0.

In setting NEW the ISR is signaling that a fresh period value is ready. When the background program polls File NEW and finds a non-zero datum, then it knows that a fresh count is ready for collection. It then, for instance, could send it to a serial EEPROM as in Example 12.3 on p. 442 or down a serial link to a PC for subsequent storage, processing and display.

The equivalent coding in CCS **C** is shown in Program 13.4. The approach is similar to that in the assembly coding of Program 13.3, but note that a separate function is used for each of the two types of interrupt source. The compiler will generate code

Program 13.4 ECG peak-to-peak timer in CCS **C**

```
#include <18f4520.h>
#fuses    NOWDT,XT
#bit      ALARM = 0xF80.0          /* Overflow alarm            */
long int JIFFY;                    /* Global 16-bit variable    */
int      NEW;                      /* New peak detected         */
main()
{
/* Timer0 internal clocking, prescaled 128. Defaults to 16-bit */
setup_timer_0(RTCC_INTERNAL|RTCC_DIV_128);
set_tris_a(0xFE);                  /* RA0 set as output         */
setup_adc_ports(NO_ANALOGS);       /* All pins digital          */
enable_interrupts(INT_TIMER0);     /* Enable Timer 0 interrupts */
enable_interrupts(INT_EXT);        /* Enable INT0 interrupts    */
enable_interrupts(GLOBAL);         /* Enable interrupt logic    */

while(1)                           /* DO forever                */
    {
    if(NEW == 1)                   /* Dummy code                */
        {
        NEW = 0;
        /* DO this; DO that; DO the other; Dummy code           */
        }
    }
}

#int_timer0
overflow()
{
ALARM = 1;                         /* Sound alarm on overflow   */
}

#int_ext
ecg()
{
JIFFY = get_timer0();
NEW = 1;
set_timer0(0);
}
```

to test each interrupt flag in turn and call up the ISR function as appropriate. Notice the anarchic use of the original name for Timer 0; that is the RTCC (Real-Time Counter Clock). The function `get_timer0();` returns a 16-bit integer, irrespective of the two possible counter sizes. Note how `set_timer0(0);` is used to clear Timer 0.

Extended-range PIC MCUs have three[6] additional timer/counters and associated circuitry with the following properties.

Timer 1

Timer 1 is a 16-bit counter with its own optional dedicated oscillator and programmable Prescaler. Its state can be sampled by an external event and it can control the state of a pin when it reaches a predefined value.

---

[6]The PIC24 family have up to nine additional timers.

Timer 2
This 8-bit counter has both programmable Pre- and Postscaler functions. Its count length can be set by the programmer and it may be used to generate a pulse-width modulated output with no on-going software overhead.

Timer 3
This is a clone of Timer 1 with virtually identical properties. It does not have its own optional oscillator, but can use that provided for Timer 1. Optionally it can provided the timebase for the Capture/Compare/Pulse Width Modulation logic.

Capture/Compare/PWM
Timers can be used in conjunction with additional logic called Capture/Compare/Pulse Width Modulation (CCP) to implement the Timer 1/3 sample instant (Capture), the Timer 1/3 roll-over value (Compare), and the automatic PWM generation from Timer 2.

Timer 1 comprises a primary 16-bit counter implemented as a pair of Files with the lower byte named **TMR1L**. Like Timer 0, the high byte is buried. Normally access to this byte is via the **TMR1H** buffer, which is updated whenever TMR1L is read from or written from in the same manner as the 16-bit mode Timer 0. Thus the state of the 16-bit Timer 1 can be accessed or changed at a single point in time. However, unlike Timer 0 this buffered access can be disabled, effectively coupling TMR1H to the high byte of Timer 1. This unbuffered mode is compatible with earlier versions of this module available in the PIC16 family, but needs to be used with care. For instance, Microchip recommend that the timer be turned off during an unbuffered read of the two bytes to avoid the problems discussed on p. 459. When this counter overflows, then the Interrupt Flag **TMR1IF** in the Peripheral Interrupt Register 1 PIR1[0] is set.

The source of the counting pulses may be external to the device; either events at the **T13CKI** (**Timer 1/3 ClocK Input**) or from a dedicated Timer 1 oscillator. Alternatively, the internal system oscillator $\frac{fosc}{4}$ may be selected as the counting source. In all cases, the counting pulse train can be divided down with a Prescaler counter. External counting pulses default to being synchronized to the system oscillator.

The **Timer 1 CONtrol register T1CON**, shown in exploded form in Fig. 13.4, is used to select the various features of the Timer. All bits in this Control register are 0 on Power-on/Brown-out Resets, which initially disables an unbuffered Timer 1 and its external oscillator, with a Prescaler value of 1:1 and system clock used as the source. All bits can be read and except for T1RUN, be written to.

TMR1ON
Setting T1CON[0] to 1 enables the Timer. In this case the Timer 1 related pins are then automatically set as input, overriding any TRIS settings.

TMR1CS, T1OSCEN
Timer 1 can be configured to measure time from the internal system clock if the **TiMeR 1 Clock Select** switch bit in T1CON[1] is 1 or else use an external source of counting pulses.

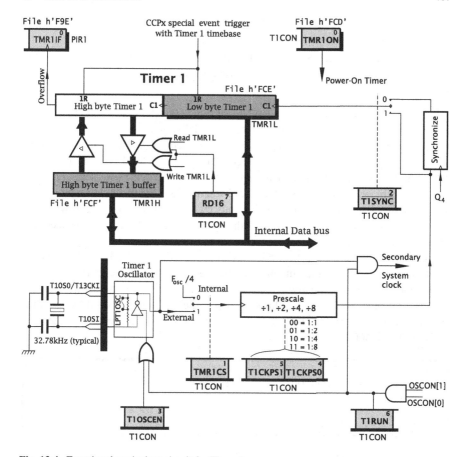

Fig. 13.4 Functional equivalent circuit for Timer 1

External events can be a ⌐/ rising edge (following the first falling edge) on the T13CKI pin, or else if the **Timer 1 OSCillator Enable** switch bit in T1CON[3] is 1, a 'private' oscillator separate from the main PIC MCU oscillator. This avoids having to pick the main crystal to suit the timer, as we did in our Timer 0 ECG peak picker of p. 463. This external oscillator is timed with a crystal across the T1OS0/T13CKI and T1OSC1 pins, with a maximum value of 200 kHz. Typically a 32.768 kHz (2^{15} Hz) watch crystal is used.

T1CKPS1:0

Whatever the source of counting pulses, the primary 16-bit counter may be incremented either directly or on every second, fourth, or eighth event. This is controlled by the setting of T1CON[5:4], as listed in the diagram.

Timer 1 will overflow and set the TMR1IF interrupt flag after $2^{16} = 65,536$ counter events from zero. This in turn can be used to interrupt the processor if the paired TMR1IE mask bit in the Peripheral Interrupt Enable 1 register is set (see

Fig. 7.2 on p. 209), or else polled. In either case the program should manually clear the TMR1IF flag once overflow has been detected.

For instance, if a 32.768 kHz watch crystal is used, then Timer 1 will overflow in two seconds if T1CKPS1 : 0 = 00 and every 16 seconds if T1CKPS1 : 0 = 11.

T1SYNC

Output from the Prescaler is by default synchronized to the system clock, giving a 2-instruction cycle delay. However, unlike Timer 0 this synchronization shift register can be by-passed with T1CON[2] set to 1. The asynchronous mode allows Timer 1 to be used with an external count source when the PIC MCU is asleep. As the synchronizer shift register is clocked from the system clock f_{OSC} (actually Q_4 in Fig. 4.5 on p. 76), which is switched off when in the Sleep mode, a by-pass is necessary in this situation. The asynchronous mode also needs to be used when the external count pulse rate is faster than the system clock, as the synchronizer will then miss some events.

Apart from these cases, $\overline{\text{T1SYNC}}$ should be 0, as the lack of synchronization can lead to an unpredictable outcome if software attempts to write to Timer 1 at the same time as the random external event tries to increment the timer. If the Timer 1 state is to be updated in the asynchronous mode, then it should be disabled and thus stopped by clearing TMR1ON during this process. For instance, to change the state of Timer 1 to h'8000':

```
        movlw   h'80'           ; New high byte
        bcf     T1CON,TMR1ON    ; Stop the timer
        movwf   TMR1H           ; Set Timer1 to 8000
        clrf    TMR1L           ; Update the two bytes
        bsf     T1CON,TMR1ON    ; Restart the timer
```

Altering the state of TMR1L will always clear the Prescale counter and synchronizer.

When the internal system clock is selected (TMR1CS = 0), synchronization is not necessary. In this case the state of $\overline{\text{T1SYNC}}$ is ignored.

RD16

When **ReaD/write 16-bits** is 1, the high-byte buffers are enabled. In this situation both bytes of Timer 1 can be read or written to at the same time as activity at TMR1L.

T1RUN

When the SCS1:0 bits in OSCON[1:0] are set to 01 then the Timer 1's oscillator is used as the system clock—see p. 322. In this situation the **Timer 1 is RUNning as the system clock** bit is set and the state of the T1OSCEN enable bit is overridden; so the running of this oscillator is not controlled by Timer 1. However, Timer 1 can still elect to use the clock if required.

For our example, assume that we require a low-power temperature logger that will read the sensor and transmit its value back to base once every 15 minutes. It is proposed that Timer 1 be used to action this process and that the Timer 1 oscillator with a 32.768 kHz watch crystal is to give the timebase.

Program 13.5 Generating a 15-minute data logger timebase

```
         include "p18f4520.inc"
         config WDT=OFF, OSC=XT, LPT1OSC=ON

         cblock  h'020'
         JIFFY:1
         endc

MAIN     movlw   b'10011111' ; Timer on, external clock, asynch
         movwf   T1CON       ; Extern osc enabled, PS ratio 1:2
         bsf     INTCON,PEIE ; Enable the Peripheral interrupts
         bsf     PIE1,TMR1IE ; Enable the Timer1 interrupt

         clrf    JIFFY       ; Zero Jiffy count
         clrf    TMR1H       ; and zero the counter
         clrf    TMR1L

DOZE     sleep               ; Slumber & wait for Timer1 interrupt
         bcf     PIR1,TMR1IF ; Zero the interrupt flag
         incf    JIFFY,f     ; Record one more Jiffy
         movlw   d'225'      ; Check, 225 Jiffies = 15 minutes?
         cpfseq  JIFFY       ; IF yes THEN skip to take a sample
         bra     DOZE        ; ELSE go back to sleep

; Take a sample -------------------------------------------------
         clrf    JIFFY       ; ELSE reset Jiffy count
         call    SAMPLE      ; Sample temperature and transmit
         bra     DOZE        ; and go back to sleep
```

As the maximum possible overflow time is only 16 s we need to keep a count of overflows to record 900 s in total. Setting the overflow period to 4 s gives us a Jiffy count requirement of $\frac{900}{4} = 225$ to record our 15 minute total. Thus our set-up and main skeleton software would be something like that shown in Program 13.5. Here Timer 1 is set up to use the its external oscillator with a Prescaler ratio of 1:2, giving our 4 s Jiffy.

In order to reduce power consumption the PIC MCU is to be in its Sleep mode, and will be woken up every four seconds. To facilitate this, the TMR1IE mask bit in PIE1[0] is set to 1. As GIE remains in its reset clear state, when awoken, the processor continues onto the instruction following sleep.

After TMR1IF is cleared, one is added onto the Jiffy count. This is tested for 225 and if equal, then it is zeroed and the subroutine to transmit temperature to base is called.

It should be noted that an enabled, Timer 1 adds something of the order of 2 μA (10 μA maximum at 5 V) current drain, which is a consideration that is especially important if it is intended to use Timer 1 to waken the processor from a low-current Sleep state which typically only uses 0.4 μA (2 mA maximum); all figures for the PIC18F4520 at 5 V and 25°C. The **LPT1OSC (Low-Power Timer 1 OSCillator)** fuse can be activated to run the oscillator in a low power mode. Doing this increases the feedback resistor across this oscillator's amplifier. Whilst this reduces power (but no value is given in the data sheet) is also makes the TMR1 oscillator more prone to interference from outside signals.

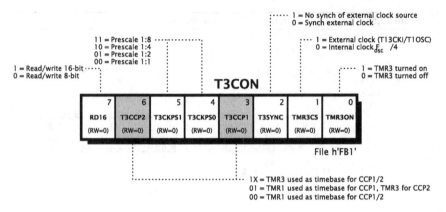

Fig. 13.5 The Timer 3 CONtrol register

Timer 3 is a clone of Timer 1 with T3 replacing T1 in the appropriate SFR and interrupt bits. However, there are some differences, as shown shaded in Fig. 13.5.

- When Timer 3 is configured to use an external counting source, this is shared with the Timer 1's input pin T13CKI (on rising edges following the first falling edge).
- If the Timer 1 oscillator is enabled; that is bit T1OSCEN in T1CON[3] is 1, then this is the source of timing pulses if Timer 3 is configured for external counting. This is true even if Timer 1 is not set-up for an external clock source.

The sharing of resources between the two timers releases two control bits in comparing T1CON with T3CON. These bits, labelled **T2CCP2:1** in Fig. 13.5, are used to assign Timer 1 and Timer 3 as the timebase for the CCP modules; as depicted in Fig. 13.8.

Timer 2 is an 8-bit counter with both a programmable Prescaler and Postscaler; as shown in Fig. 13.6. Input to this counter is always a derivative of the system clock. Unlike the two previous timers, output is not taken from the counter chain but from the **Timer 2 Comparator**. This compares the state of Timer 2 with that in the **Period Register PR2**. On equality an output pulse is generated which resets Timer 2 at the *next* count pulse. This may optionally be used to determine the MSSP port's SPI clock rate, as listed in Fig. 12.9 on p. 392. As determined by the Postscaler, any integer number from 1 to 16 of these reset events will set the Timer 2 Interrupt Flag **TMR2IF** in PIR1[1]. If the Timer 2 mask bit **TMR2IE** is also set, an interrupt is potentially generated.

The value of the Pre- and Postscaler ratio and actuation of Timer 2 is set-up using the **T2CON** Control register as listed below. All bits are cleared on a reset, turning Timer 2 off with 1:1 Pre- and Postscaler ratios. At the same time **TMR2** is cleared as are the Pre- and Postscaler counters. The Period Register is set to all 1s.

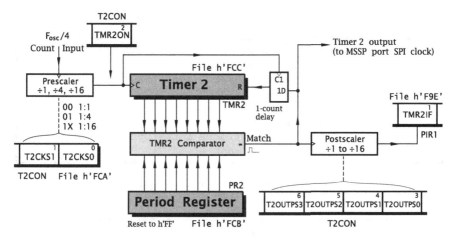

Fig. 13.6 A simplified equivalent circuit for Timer 2

TMR2ON

Setting T2CON[2] to 1 enables Timer 2.

T2CKPS1:0

Timer 2 can be incremented either directly at the instruction cycle rate $\frac{f_{osc}}{4}$ or frequency divided by four or 16. The three settings of T2CON[1:0] are listed in the diagram.

T2OUTPS3:0

The number of Timer 2 overflows activating the TMR2IF interrupt flag can be set to between one and 16 with T2CON[6:3]. This 4-bit code n maps to $n + 1$ periods; from b'0000' = 1:1 to b'1111' = 1:16.

The advantage of this architecture is that time-out can be fine tuned by setting the Period Register to an appropriate value. The delay until TMR2IF is set is given as:

$$\frac{4}{f_{osc}} \times \text{Prescale} \times (\text{PR2} + 1) \times \text{Postscale}.$$

For our example, consider the need for an interrupt 100 times per second as part of a digital real-time clock. Assuming a 4 MHz crystal, choosing a Prescaler ratio of 1:4 gives a clocking period for Timer 2 of 4 μs. If the Period Register is set to 249 then the Timer 2 comparator output period is $250 \times 4 = 1$ ms. Thus setting the Postscaler to 1:10 (1001) will give the required 10 ms (100 Hz) interrupt rate. By varying the Postscaler from 1 to 16 we can have a corresponding interrupt rate from 1 to 16 ms. For fine adjustments a unit change in PR2 alters the rate in $4 \times$ Postscale μs steps.

Set-up code for this example is:

```
movlw  b'01001101' ; Postscale 1:10 (1001). Timer2 on (1)
movwf  T2CON       ; Prescale 1:4 (01)
movlw  d'249'      ; Set up period register to 249
movwf  PR2

bsf    PIE1,TMR2IE ; Enable Timer2 interrupts
bsf    INTCON,PEIE ; Enable Timer2 interrupts
bsf    INTCON,GIE  ; Global enable
```

The setup_timer_2(mode,period,postscale); function is the CCS C equivalent to initialize Timer 2.

```
setup_timer_2(T2_DIV_BY_4,249,10);
enable_interrupts(INT_TIMER2);
enable_interrupts(GLOBAL);
```

Timer 2 can be read from using the get_timer2() function and written to with the set_timer2() function.

Time, the 4th dimension, is an important property of most systems interacting with the physical world. In particular, measuring the span between events and generating precision pulse durations. The various PIC MCU families use **CCP** (**Capture/Compare/Pulse-Width Modulation**) modules in conjunction with the various timers acting as a timebase to implement these functions. Most PIC18 devices offer two CCP modules (exceptionally the PIC18F1X20 with one). Modules **CCP1** and **CCP2** are virtually identical, sometimes even sharing the same timebase, but with separate input/output pins **CP1** and **CP2**. Generic references to the CCP module registers and bit names generally replace the numeral by X; for instance, **CCP1CON** and **CCP2CON** are indicated as **CCPXCON**. Thus, CCPXCON might equally refer to CCP CONtrol register 1 or 2. Any differences will be noted as appropriate.

A CCP module has three main functions.

- When configured in a **Capture** mode, an outside event on the associated CCP pin causes the state of Timer 1 or 3 to be copied into the CCP register. This can be used to derive the time or duration of this event, to a resolution down to 20 ns.
- Configured in a **Compare** mode; when the state Timer 1/3 equals that in the CCP register, the state of the associated CCP pin is changed or Timer 1/3 is reset. This can be used to generate a precisely timed event in hardware with a 100 ns resolution.
- When configured in a **PWM** mode; a CCP module, in conjunction with Timer 2, can generate by hardware a pulse-width modulated output with a variable period and duty cycle of up to 10-bit resolution (0.1%).

In all cases involving Timer 1 or 3, a synchronized clock must be used to guarantee correct operation; that is, $\overline{\text{T1SYNCH}} = 0$ or $\overline{\text{T3SYNCH}} = 0$ as appropriate.

Each CCP module has an associated Control register used to set the mode. In all cases, the appropriate CCP pin needs to be explicitly set-up as an input or output as appropriate. In devices with 28+ pins, the CCP1 and CCP2 pins share with RC2 and RC1 respectively. Pinning on most PIC18 devices corresponds with earlier analogous PIC16 devices. For instance, the PIC18F4520 is a drop-in replacement for the 40-pin PIC16F877. Unfortunately, CCP2 is adjacent to T13CKI/RC0, which is used for Timer 1's local oscillator. Interaction between signals at CCP2 and T13CKI can cause unpredictable operation of this timer. This is particularly the case when this oscillator is configured to operate in its low-power mode; as managed with the LPT1OSC fuse—see p. 469. For this reason pin CCP2 can be moved to an alternative location, sharing with pin RB3. This option is directed with the CCP2MX fuse—see Appendix B.

The Capture mode is illustrated in Fig. 13.7. The various submodes are:

CCPXM3:0 = 0000
On a Power-on/Brown-out Reset all bits are zeroed. This turns off the associated CCP module and clears the Prescaler. The recommended way of avoiding spurious interrupts when changing mode is to turn off the module before making the change.

CCPXM3:0 = 0100
On a ⌐\ edge at the CCPX pin, Timer 1 or Timer 3 is copied into the **CCPRXH:L** pair of Files. At the same time the CCPX Interrupt Flag **CCP1IF** or **CCP2IF** is set, and if corresponding **CCP1IE** or **CCP2IE** mask bit is set, an interrupt will be generated.

CCPXM3:0 = 0101
The time capture described above is triggered on a _/⌐ edge at the relevant CCP pin.

CCPXM3:0 = 0110
Capture is actioned after four rising edges at the CCP pin.

CCPXM3:0 = 0111
Capture occurs after 16 rising edges at the CCP pin.

Once a defined event has taken place, the processor can read this frozen value— that is, the time—either in an ISR or when the appropriate CCPIF flag is polled as a 1. If the timebase timer is reset after each capture, then the sampled datum is the time since the last event. Alternatively, as the timebase timer continues to increment, its captured value can be subtracted from the previous reading to give the difference. As the mode may be altered on the fly, the time between rising and falling edge on CCPX can be measured by toggling CCPXM0 between captures. This may cause the CCPXIF interrupt flag to be set. To prevent false interrupts, CCPXIE should be

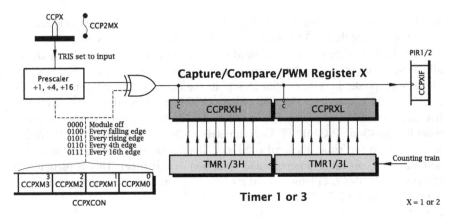

Fig. 13.7 Capturing the time of an event

cleared before the change-over and CCPXIF after the change-over. Alternatively, the
CCP1 module can be used to capture the rising edge and CCP2 the falling edge—see
Example 13.3.

Although it seems perverse; if the CCP pin is set as an output, then under program
control a capture can take place by altering the state of this pin from *inside*. Thus
the time of an *internal* event can be captured.

As our example, consider that we wish to measure the period of our ECG signal
with the peak detector of Fig. 7.1 on p. 206 connected to pin CCP1. If we assume
that we are using Timer 1 synchronously clocked by its own 32.768 kHz watch
crystal as our timebase; our set-up code is something like this:

```
movlw   b'10001011'  ; Timer1 on, external clock, synched
movwf   T1CON        ; Oscillator enabled, PS ratio 1:1
bcf     T3CON,T3CCP2 ; Choose Timer1 as CCP1 timebase

movlw   b'00000100'  ; Capture mode, event = falling edge
movwf   CCP1CON

clrf    NEW          ; Zero NEW flag

bsf     PIE1,CCP1IE  ; Enable the CCP1 interrupt
bcf     PIR1,CCP1IF  ; Ensure that interrupt flag is zero
bsf     INTCON,PEIE  ; Enable Timer/CCP interrupts
bsf     INTCON,GIE   ; Global interrupts enabled
```

The ISR simply reads the contents of the CCPR1H:L register and stores it away
in two temporary locations, setting File NEW to indicate to the background program
that a new time datum exists. Timer 1 is then cleared ready for the next event.

With a crystal of 32.768 kHz, the time resolution of the captured datum is 30.5 μs
with our 1:1 Prescale setting. Timer 1 will overflow in 2 s, which is sufficient to
record a heart rate down to 30 beats per minute. A more robust software system

Program 13.6 Capturing the instant of time an ECG R-point occurs

```
; **********************************************************************
; * FUNCTION : CCP1 ISR to copy CCPR1H:L datum to TEMP+1:TEMP        *
; * ENTRY    : CCP1 interrupt enabled                                *
; * EXIT     : CCPR1H:CCPR1L <- TEMP+1:TEMP. TMR1 zero               *
; * EXIT     : TMR1 <- 0000. NEW <- 1                                *
; **********************************************************************
ISR_ECG    btfss    PIR1,CCP1IF    ; Was it a CCP1 interrupt?
           bra      ISR_EXIT       ; IF no THEN false alarm

           incf     NEW,f          ; Signal a new capture
           bcf      PIR1,CCP1IF    ; Reset interrupt flag
           movff    CCPR1L,TEMP    ; Get captured low byte
           movff    CCPR1H,TEMP+1  ; Get captured high byte
           clrf     TMR1H          ; Zero Timer1's High buffer
           clrf     TMR1L          ; and complete double byte

ISR_EXIT   retfie   FAST           ; and return from interrupt
```

would also enable the Timer 1 overflow interrupt. If this occurs it indicates that the subsequent captured data will be invalid—although time-outs can be counted and thus extend the validity of the captured time. However, in our system it is more likely to be used to set off an alarm!

Modes 0010 and 1000–1011 listed in Fig. 13.8 give five **Compare** modes. Here a 16-bit digital equality comparator detects when the 16-bit Timer 1 or Timer 3 datum equals the setting in the 2-byte CCPRXH:CCPRXL register. When an equality match occurs, the CCPXIF interrupt flag will be set and this can cause an interrupt if the corresponding CCPXIE mask bit, together with the appropriate global enable bits, are set.

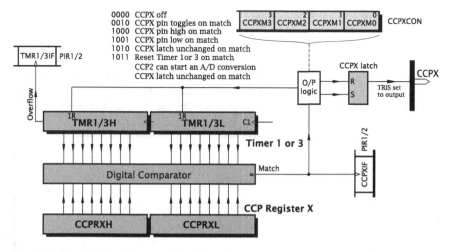

Fig. 13.8 The CCPX module set to Compare mode

Depending on the setting of the CCPXM3:0 mode bits, one of five actions are possible on a match:

CCPXM3:0 = 0010: Toggle Output Pin on Match
Pin CCPX changes state. When this mode is first entered from a CCPX reset (Mode 0000), the initial state of the CCPX latch is 0.

CCPXM3:0 = 1000: Set Output Pin on Match
Pin CCPX is forced High. The CCPX latch can only be cleared by switching the CCPX module to Mode 0000; that is, by turning it off.

CCPXM3:0 = 1001: Clear Output Pin on Match
Pin CCPX is forced into its Low state. The initial state of the associated latch is 1 (the opposite to the match state) when this mode is initially selected from a CCPX reset.

CCPXM3:0 = 1010: Generate Software Interrupt on Match
The CCPX latch remains unchanged and the associated pin can be used as a normal I/O pin. However, the CCPXIF flag is set, effectively generating an internal interrupt (sometimes called a software interrupt) if enabled.

CCPXM3:0 = 1011: Trigger Special Event on Match
The timebase timer is cleared and a potential interrupt is actioned. With CCP2 (this is the only functional difference between CCP1 and CCP2) an ADC module conversion can be optionally initiated—see Fig. 14.12 on p. 510. The CCPX latch remains unchanged and the CCPX pin can be used as a normal I/O pin.

In Modes 0010, 1000 and 1001, the parallel port bits shared with CCP1 and CCP2 should be set-up as outputs. Whilst the CCPX module is off (as it is after any sort of reset) these pins will reflect the state of those port bits. Because clearing CCPXCON to Mode 0000 is the only way of relaxing the CCPX latch to its pre-match value for Modes 1000 and 1001, it is advisable to set each associated port bit to this value to avoid spurious pulses. For instance, if Mode 1001 is being used, then the port bit should be set to 1 in the initialization code to ensure a High-state reset value.

Suppose that we wish to set up Timer 1 as configured in the last example to overflow every 10 seconds. To do this we need to set the timer to roll over after 16 s (Prescaler ratio 1:8) and then shorten the cycle. This is implemented by loading the CCPR1H:L register with the fraction $\frac{10}{16}(2^{16} \times \frac{10}{16})$; which translates to h'A000'. Whenever Timer 1 reaches this value it will automatically be reset and an interrupt will occur if the CCP1IE mask bit (and global PEIE and GIE masks) are set. Initialization code for this is:

```
movlw   h'A0'         ; Set up CCPR1 to h'A000'
movwf   CCPR1H
clrf    CCPR1L
movlw   b'00001011'   ; CCP Compare Mode 1011. Special event
movwf   CCP1CON
movlw   b'00111011'   ; Timer1 on (1), external clock (1)
movwf   T1CON         ; Synched (0), oscillator (1) 1:8 (111)
bsf     PIE1,CCP1IE   ; Enable CCP1 interrupts
bsf     INTCON,PEIE   ; Enable Timer/CCP interrupts
bsf     INTCON,GIE    ; Enable all interrupts
```

The PIC MCU will then be interrupted every 10 seconds.

Because the `CCP1` pin is not changed by Compare Mode 1011, this pin can be used as a normal parallel port input/output independently of the CCP1 module.

One of the more common applications of MCU-based systems is the control of power circuits, such as heating, lighting and electric motor speed control. One approach to this problem would be to use a digital-to-analog converter, such as that discussed in Fig. 12.16 on p. 404, driving a power amplifier. Such linear control is expensive and inefficient due to the large current:voltage products that must be handled by the power amplifier. A rather more efficient and cost effective approach rapidly switches the load on and off at a reasonably fast rate. A power switch, such as a thyristor or power FET, dissipates relatively little power, as when the switch is off no current flows and when the switch is on the voltage across the switch is small—ideally zero.

An example of such waveforms is shown in Fig. 13.9. The average amplitude is simply $A \times N$, where N is the duty cycle fraction of the repeat period. If we vary N from 0 to 100% then the average power will vary in a like fashion—all without the benefit of analog circuitry. This digital-to-analog conversion technique is known as **Pulse-Width Modulation (PWM)**.

The thermal or mechanical inertia of most high-power loads is such that even with a relatively low repetition rate (typically no lower than 100 Hz) the 'bumps' will be smoothed. Low switching rates are more efficient, as each switching action dissipates energy. If PWM is used for more conventional digital-to-analog conversion, such as for audio applications, then a low-pass filter may be utilized to reduce the high-frequency harmonics. In such cases a sampling rate of typically ten times the maximum analog signal should be used to space out the harmonics (see Fig. 14.3 on p. 493) and reduce the necessary filtering burden.

Generating a PWM waveform is conveniently implemented using a counter and digital equality comparator. The output pin is driven from a latch, which is always set as the counter rolls over. The latch is reset when the counter state equals a number

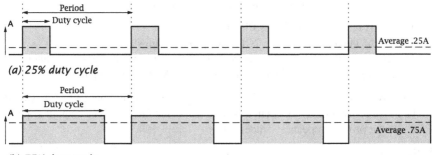

(a) 25% duty cycle

(b) 75% duty cycle

Fig. 13.9 Pulse width modulation

representing the duty cycle. The larger is the Duty number the longer the pin will remain in its High state.

As a simple example, consider a 3-bit count with a Duty number of b'011':

Set High			Reset Low				
000 →	001 →	010 →	011 →	100 →	101 →	110 →	111
			Match				

In this instance the pin will remain in its High state for three counts, giving a duty cycle of $\frac{3}{8}$, or 37.5%. By changing this number, the average power can be altered with a resolution of $\frac{1}{8}$ from zero up to 87.5%.

PIC microcontrollers implement this scheme using the CCP modules. In the PWM mode, Timer 2 is used to implement the timebase counter and the Duty number is fed in via a double-buffered register to a 10-bit PWM comparator. Either or both CCP modules can be used to generate a PWM waveform via their individual Duty number and CCPX pin, but in the latter case they share the same Timer 2 and thus have the same period. Any CCP pin used as a PWM output must be set-up as an output, with associated TRIS bit logic 0.

Period

The timebase is set using Timer 2 in the manner outlined in Fig. 13.6. The roll-over period is a function of the main instruction cycle time $4 \times t_{OSC}$, the Prescaler ratio, and the contents of the Period register PR2. Recalling that Timer 2 resets on the clock pulse *after* equality with PR2 is reached, the total repetitive period is given as:

$$(4 \times t_{OSC}) \times \text{Prescaler ratio} \times (PR2 + 1).$$

For instance, for a 16 MHz crystal, Prescaler ratio 1:16 and PR2 contents of h'63' = d'99' we have:

$$\text{Period} = \left(4 \times \frac{1}{16}\right) \times 16 \times (99 + 1) = 400 \, \mu s.$$

Each time Timer 2 overshoots the Period number, three things happen.

1. Timer 2 is reset to zero (unless PR2 is zero).
2. The PWM latch is set and pin CCPX goes to its High state.
3. The 10-bit content of the Master register is copied into the Slave register and presents the next Duty number to the 10-bit PWM digital comparator.

On its way back up again, when Timer 2 reaches the Duty number (which is stored in the Slave register) the PWM latch is cleared. When the count once again reaches PR2 + 1 the process repeats ... indefinitely.

Duty Cycle

The Duty number is presented to the 10-bit PWM equality comparator in a 2-deep 10-bit wide pipeline. The outer word is located in the 8-bit CCPRXL together with the two lowest bits held in CCPXCON[5:4], which together are labeled in the diagram as the Master register. The contents of the Master register can be altered at any time by the software as two separate movwf instructions. This word is only moved down the pipeline to be presented as the Duty number to the comparator at the end of each period. This reduces the possibility of a mid-period glitch, due to the unsyncronized nature of any changes in the contents of the Master register in relation to the timebase. The Slave register comprises CCPRXH companded with a 2-bit internal latch. While in this mode the CCPRXH register is read-only. This prevents direct access by software to the Duty number.

The core Timer 2 register is only eight bits wide. In order to extend Timer 2 to 10 bits, to match the Duty number, two lower bits are added. These extra two bits either originate from the Prescaler counter which is dividing down the system clock to Timer 2 or else if a Prescaler ratio of 1:1 is chosen, the 2-bit count defining the quadrature clocks of Fig. 4.5 on p. 76. In either case, the result is to give a maximum 10-bit (1:1024) resolution in the Duty cycle, with a counting rate of $\times 4$ of that of the 8-bit Timer 2 core.

When this 10-bit count equals the Duty number, the PWM latch is reset, and the CCPX pin drops to its Low state. It stays low until the next period begins, when Timer 2 rolls over and the cycle repeats *ad infinitum*. In all cases the datum in CCPRXL must be smaller than that in PR2, otherwise the PWM latch will never reset! If PR2 is h'FF' then the resolution of the system is a full 10 bits. Smaller values of Timer 2 period data will reduce this resolution. For instance, if PR2 = h'3F' then the resolution is reduced to 8 bits; six in Timer 2 proper and two extension bits.

For our example, let us assume the situation described previously where our timebase period is 400 μs (2.5 kHz) for a 16 MHz crystal with Prescaler ratio of 1:16 and a PR2 value of h'63'. If we wish to generate a 25% duty cycle, as in Fig. 13.9(a), the set-up code for the CCP1 module would be something like:

```
movlw h'63'        ; Set up Timer2 Period register to d'99'
movwf PR2
bcf    TRISC,2     ; Make CCP1 an output
movlw h'19'        ; Set-up Master to 1/4 full scale (h'63/4')
movwf CCPR1L       ; That is, b'0001 1001'
movlw b'00001100'; CCP1 module PCM Mode (1100)
movwf CCP1CON      ; with CCP1CON[5:4] (00)
movlw b'00000110'; Timer2 Prescale 1:16 (10)
movwf T2CON        ; Timer2 on (1). Start waveform
```

The Timer 2 Postscaler does not affect the PWM generation but still sets the TMR2IF in the normal way. The CCPXIF flag is not affected in this mode.

Many high-power applications, especially dc motor control, require two (half-bridge) or four (full-bridge) switching waveforms to control the load. Most PIC18 devices have an extension of their CCP1 module to add this and related functionality. Of our exemplar devices, this **ECCP1** module is implemented in the PIC18F1X20 and PIC18F4X20, but not the 28-pin PIC18F2X20 range.

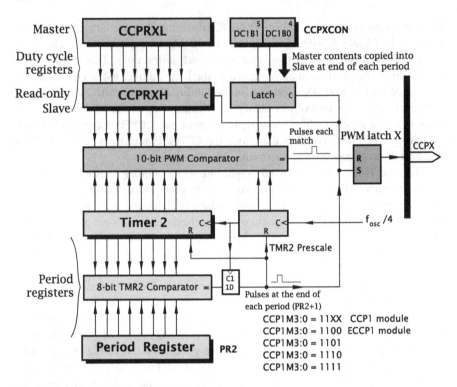

Fig. 13.10 Timer 2 with the CCP1 in its PWM mode

A full treatment of the **Enhanced CCP** module is beyond the scope of this chapter, but as an example, Fig. 13.11 shows a dc motor with both speed and direction controlled from this module. ECCP1 is an upwardly compatible development of the standard CCP1 module, with the latter's PWM mode CCP1 3:0 = 11XX (see Fig. 13.10) being expanded to four submodes using the two unused control bits CCP1CON[7:6].

P1M1:0 = 00
CCP1 compatible, with only the **P1A**/CCP1 pin being modulated. The three other P1 pins can be used as normal port I/O.

P1M1:0 = 01
Full-bridge forward output with pin **P1D** being modulated, P1A active and both **P1B** and **P1C** both inactive. This is the situation shown in Fig. 13.11 where current flows through transistors Q_A, the motor field coils and Q_D.

P1M1:0 = 10
Half bridge output with pin P1A being modulated and P1B modulated in anti-phase.

A variable dead-band control is available in which both pins are inactive for a short variable period at the end of each cycle; controlled using the PWM1CON register. This eliminates 'shoot-through' due to transistors turning off more slowly than turning on. This can give transient short circuits across the power supply.

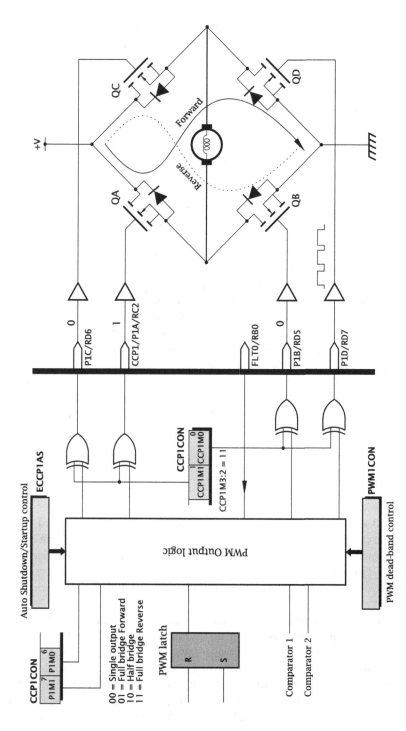

Fig. 13.11 Controlling a dc motor in a full-bridge connection

P1M1:0 = 11

Full-bridge reverse output is the mirror of forward output, using pin P1B as the modulated control and P1C active. The current flow, shown dotted in Fig. 13.11 would result from this mode, and flows through the motor field coils in the opposite direction.

Waveforms can be switched between forward and reverse by toggling the P1M1 bit. When this happens, the modulated outputs are inactive for a short period at the end of the switching cycle to reduce the possibility of shoot-through.

Each pair of PWM pins can be individually inverted; as controlled by the lower two control bits **CCP1M1:0** when the module is in a PWM mode (that is CCP1M3:2 = 11).

When controlling significant power, any malfunction can lead to considerable collateral damage. For instance, a mechanical stall can burn out the motor field windings. To guard against this scenario, the ECCP module provides an **Auto-Shutdown** feature, directed by the **ECCP1AS** register, which immediately places the PWM pins into a defined shutdown state when such an event occurs.

A shutdown event can be signaled by external sensing circuitry via either of the analog Comparator modules (see Fig. 14.7 on p. 499) or else a logic 0 on the **FauLT FLT0** pin. As an example, if the motor is monitored with a Hall-effect sensor, giving a voltage output proportional to current, and this exceeds a threshold, then the analog comparator switches and if enabled (the **ECCPASE** bit in ECCP1AS[7]) an Auto-shutdown process is initiated. Writing to ECCPASE is disabled as long as the cause of the shutdown persists. After the fault clears ECCPASE can be cleared to restart the process. Alternatively, ECCPASE can be automatically cleared in this situation if the **PWM ReStart ENable PRSEN** bit in PWM1CON[7] is 1. Effectively this gives an automatic restart facility.

Examples

Example 13.1 Show how you could use Timer 0, configured for an internal 8-bit count with a maximum prescale value, to generate a PWM waveform via pin RA0. The digital byte in File DATUM is to hold the mark duration in terms of the fraction of the cycle. For instance, if DATUM = d'64', or $\frac{1}{4}$ of full scale, the mark:space ratio should be 1:3. Assuming a 4 MHz crystal, calculate the resulting PWM duration.

Solution Timer 0 will give a time-out related to the number loaded into the timer register at the beginning of the period. If we load in the 2's complement of the byte (the negative value; i.e. $256 - \text{DATUM}$) then the duration will be proportional to this value—the larger it is, the longer the timer has to count before overflowing. Conversely loading in the value of DATUM will give a time-out duration inversely proportional to the value. By alternately loading the 2's complement of DATUM and making the pin High followed by DATUM itself making the pin Low will give us a

Program 13.7 Pulse-Width Modulation using Timer 0

```
MAIN    setf    ADCON1          ; Make ports all digital
        movlw   b'11000111'     ; TMR0 on, Int clock, 1:256 prescale
        movwf   T0CON
        bcf     TRISA,0         ; Make RA0 the PWM output
        bsf     INTCON,TMR0IE   ; Enable Timer 0 interrupt
        bsf     INTCON,GIE      ; Enable all interrupts

; <<<< More background code >>>>

; ******************************************************************
; * FUNCTION   : ISR to generate a PWM waveform at RA0           *
; * FUNCTION   : Set-up TMR0 to either DATUM or 256-DATUM        *
; * FUNCTION   : Depending on the toggling PORTA[0] state        *
; * ENVIRONMENT: DATUM:1                                         *
; * RESOURCE   : Timer0, PORTA[0]                                *
; ******************************************************************
ISR     btfss   INTCON,TMR0IF   ; Has Timer0 overflowed?
        bra     ISR_EXIT        ; IF no THEN false alarm; ELSE

        bcf     INTCON,TMR0IF   ; Reset interrupt flag
        movf    DATUM,w         ; Get datum
        btfsc   PORTA,0         ; Is current output low?
        negf    WREG            ; IF not 2's complement copy of DATUM
        btg     PORTA,0         ; Toggle PWM pin output

        movwf   TMR0L           ; Initialize Timer

ISR_EXIT
        retfie  FAST            ; Return from interrupt with context
```

total period approximately the same as a total Timer 0 time-out as if sequentially counting through all 256 states.

The coding of Program 13.7 sets up Timer 0 in its 8-bit mode to increment at a $\frac{1}{256} \approx 0.00391$ MHz or 3.91 kHz rate with a 1:256 Prescaler setting. When Timer 0 overflows it generates an interrupt. The ISR toggles pin RA0 each time it is executed. The state of this output is a convenient way of determining whether to reload Timer 0

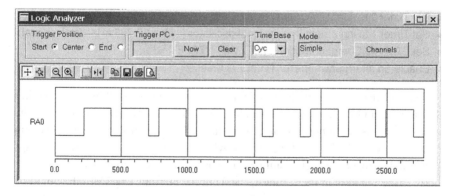

Fig. 13.12 PWM waveform with a DATUM value of h'64', giving a nominal mark:space ratio of 1:3. For clarity, this is shown with a Prescale value of 1:1

with the byte value in DATUM or else its 2's complement value which will also toggle as required.

With the time to count 256 states of $1 \times 256 \times 256 = 65.536$ ms, the cycle rate will be ≈15.26 Hz. The actual measured time was 65.564 ms and mark:space ratio was 1:2.998. This small disparity is due to the time taken to enter the ISR and execute down to the change in Timer 0 state. Increasing the clock frequency will increase the PWM frequency; to a maximum of 152.6 Hz in this case at 40 MHz. Reducing the Prescale value will also increase the PWM cycle frequency, but this instruction offset will increasing reduce the accuracy. Eliminating the Prescaler gives a measured mark:space ratio of 1:2.64 with PWM rate of 35 kHz derived from a 40 MHz system clock; as shown in Fig. 13.12.

Example 13.2 A certain tachometer is to register engine speed in the range 0–12,000 rpm (revolutions per minute). The engine generates one pulse per revolution and it is intended that a PIC18F2420 be used to count the number of pulses each second and calculate the equivalent rpm. Using two of the four available timers, can you design a suitable hardware–software configuration?

Solution One approach to this problem is to create a 1 s gate to count pulses during that time. A speed of 12,000 rpm translates to a maximum pulse count of 200 rps (revolutions per second) and so we can use Timer 0 in its 8-bit mode as the pulse counter driven from pin TOCKI with no Prescaler. The 1 s gate can conveniently be implemented with Timer 1 using its local oscillator with a 32.768 kHz watch crystal. With a Prescale ratio of 1:1 the natural time-out period will be two second. Using this timer as the time base for the CCP1 module in a Compare mode, will enable the count to be reduced to a span of h'0000–7FFF' to give the required gate period.

One possible solution is shown in Program 13.8. Here the initialization code implements the following task list:

- Set Timer 0 to count $\overline{}\underline{}$ events at TOCKI.
- Set Timer 1 to increment from its local oscillator (with a 32.768 kHz watch crystal).
- Set CCP1 to Compare Mode Timer 1 with h'7FFF'.
- Enable an interrupt whenever Timer 1 is reset.

The ISR itself simply multiplies the rps reading from Timer 0 by 60 and copies this into the double File RPM:2. Timer 0 is then zeroed ready for the next count. The variable NEW is set to non zero to flag to the background program that a new rpm reading is available. This will be cleared whenever the new value is used; perhaps sent to a display or maybe transmitted to a computer over a serial link.

To make the system more robust, the Timer 0 interrupt flag TMR0IF should be checked as part of in the ISR to signal overflow and thus to activate an overspeed warning indicator.

Program 13.8 Tachometer software

```
MAIN       movlw    h'FF'           ; Setting CCP1 Register to h'7FFF'
           movwf    CCPR1L
           movlw    h'7F'           ; To reduce timebase from 2 to 1s
           movwf    CCPR1H

           movlw    b'00001011'     ; CCP1 Compare mode 1011
           movwf    CCP1CON

           movlw    b'11111000'     ; Timer0 on, 8-bit, external
           movwf    T0CON           ; No prescale
           setf     ADCON1          ; All port pins digital

           movlw    b'00001011'     ; Timer1 PS1:1, local synched
           movwf    T1CON           ; and enabled

           clrf     NEW             ; Clear the New flag
           bsf      INTCON,PEIE     ; Enable Timer interrupts
           bsf      INTCON,GIE      ; Global enable mask bit on
           bsf      PIE1,CCP1IE     ; Enable CCP1's interrupt
           clrf     TMR0L           ; Tach count zeroed
           clrf     TMR1H
           clrf     TMR1L

; <<<< More background code >>>>

; **********************************************************************
; * FUNCTION    : ISR to measure the number of pulses at T0CKI      *
; * FUNCTION    : in a one-second period generated using TMR1/CCP1*
; * FUNCTION    : as a Timebase                                     *
; * ENVIRONMENT: NEW:1                                              *
; * RESOURCE    : Timer0, Timer1, CCP1                             *
; **********************************************************************
ISR        btfss    PIR1,CCP1IF     ; Did Timer1 reset?
           bra      ISR_EXIT        ; IF no THEN false alarm

           incf     NEW,f           ; Tells world a new reading taken
           movf     TMR0L,w         ; Get totalized pulse count
           clrf     TMR0L           ; Zero pulse count
           mullw    d'60'           ; Multiply by 60 to give rpm
           movff    PRODL,RPM       ; Copy into RPM:2
           movff    PRODH,RPM+1

ISR_EXIT bcf        PIR1,CCP1IF     ; Reset interrupt flag and
           retfie   FAST            ; return from interrupt with context
```

Example 13.3 A PIC18F4520 is to be used to measure the duration of an event. This duration is the time a signal is in its High state, as shown in Fig. 13.13. You can assume that the main crystal is 8 MHz and the duration of the event is guaranteed to be no more than 100 ms.

Solution One way of tackling this problem is to feed the signal shown in the diagram into both pins CCP1 and CCP2 in parallel. Using one CCP module to capture the state of Timer 3 on the rising edge and the other to capture the falling edge gives the duration as the difference in the two captured values. In Program 13.9, Timer 3 is zeroed on a rising edge and thus the second captured Timer 3 state is our duration. If we use a Prescaler ratio of 1:4 and the system machine cycle clock, then we have as

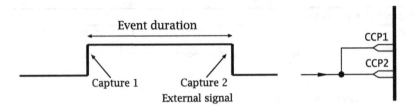

Fig. 13.13 An event manifesting itself as a pulse duration

Program 13.9 Measuring the duration of a pulse

```
MAIN       movlw    b'00000101'    ; CCP1 module captures +ve edge
           movwf    CCP1CON
           movlw    b'00000100'    ; CCP2 module captures -ve edge
           movwf    CCP2CON

           bsf      PIE1,CCP1IE    ; Enable interrupts from CCP1
           bsf      PIE2,CCP2IE    ; Enable interrupts from CCP2

           movlw    b'11100001'    ; Timer3 enabled, 16-bit write,
           movwf    T3CON          ; internal osc, prescale 1:4

           clrf     NEW            ; Clear the New flag

           bsf      INTCON,PEIE    ; Enable Timer/CCP interrupts
           bsf      INTCON,GIE     ; Global enable mask bit on

; <<<< More background code >>>>

; *******************************************************************
; * FUNCTION   : ISR to measure the duration of pulses at CCP1/2*
; * FUNCTION   : in 2us ticks                                    *
; * ENVIRONMENT: NEW:1, TIME:2                                   *
; * RESOURCE   : Timer3, CCP1, CCP2, PORTC[2:1]/CCP1/2           *
; *******************************************************************
ISR        btfsc    PIR1,CCP1IF    ; A CCP1 rising edge capture?
           bra      CAPTURE1       ; IF yes THEN go to it!
           btfss    PIR2,CCP2IF    ; A CCP2 falling edge capture?
           bra      ISR_EXIT       ; IF not THEN false alarm!
CAPTURE2   movff    CCPR2L,TIME    ; Get low byte of captured time
           movff    CCPR2H,TIME+1  ; Get high byte of captured time
           bcf      PIR2,CCP2IF    ; Clear flag
           incf     NEW,f          ; Tell the world: A new time datum
           bra      ISR_EXIT

CAPTURE1   clrf     TMR3H          ; Zero count
           movlw    d'3'           ; Compensates for execution time
           movwf    TMR3L
           bcf      PIR1,CCP1IF    ; Reset interrupt flag and

ISR_EXIT   retfie   FAST           ; Return from interrupt with context
```

---

our counting rate 500 kHz; i.e., the system resolution is 2 μs. The overall maximum duration that can be measured in this way is $2^{16} \times 2 = 131,077$ μs; which is large enough not to overflow for a maximum duration of 100 ms.

The ISR in Program 13.9 simply tests each CCP interrupt flag in turn and goes to the appropriate routine. If CCP1 has signaled a ⟋ event then the timer is zeroed to restart the count. Timer 3 has been configured to increment at a 500 kHz rate and when the next ⌐⌐ occurs the CCP2 module captures the state of this timebase and places it in the 16-bit CCPR2 register. The ISR then copies it into the two Files TIME+1:TIME and this is the period in 2 μs ticks.

Actually resetting Timer 3 on the first event introduces some inaccuracy into the process, as the clearing event takes some 11 instruction cycles to reach this point. The listing shows a starting point of h'0003' for the timebase to mostly trim out this offset error. A slightly more accurate method would be to leave Timer 3 running continually and the two captured 16-bit data subtracted to give the required difference at relative leisure. If the timebase overflows between the first and second reading, then this subtraction will give a 2's complement outcome, and this will need correcting to give the modulus difference.

Self-Assessment Questions

13.1 Using Timer 1 and CCP1, design a system to generate a continuous square wave with a total period of 20 ms from pin CCP1. You may assume that the main crystal is 8 MHz. Hint: Consider using the Compare-Toggle CCP mode described on p. 476.

13.2 The echo sounding hardware shown in Fig. 7.13 on p. 237 uses an external 1.72 kHz oscillator to interrupt the PIC MCU once per 5.813 ms; that is, once every time sound travels 1 cm through air. Assuming that a 20 MHz PIC MCU is used, show how Timer 2 could be used to generate this interrupt rate to an accuracy better than 0.1%.

13.3 The enhanced-range PIC MCU family has three hardware input pins; namely INT0, INT1 and INT2. Suggest some way to use Timer 0 to simulate another hardware interrupt with pin T0CKI.

13.4 As part of a software implementation of an asynchronous serial channel running at 300 baud, a delay of 3.33 ms is to be generated. Assuming that a 8 MHz PIC MCU is the host processor, show how you could use a timer to generate an interrupt each baud period. Extend your routine to enable baud rates up to 19,200 in doubling geometric progression.

13.5 Show how you would use Timer 1 with its separate integral oscillator with a 32.768 kHz watch crystal, to keep the central heating real time clock array HOURS:MINUTES:SECONDS of Example 7.3 on p. 233 up to date.

13.6 The CCS C compiler has integral functions dealing with the Timer and CCP modules. For instance, Timer 1 can be written to using set_timer1 (datum); and read from using get_timer1();. The function setup_timer_1(mode); is used to initialize the timer. Similarly setup_ccp1 (mode); initializes the CCP1CON register. Mode scripts for Timer 1 and the CCP Compare configuration are:

```
T1_DISABLED              T1_INTERNAL   T1_EXTERNAL
T1_EXTERNAL_SYNCH   T1_CLK_OUT    T1_DIV_BY_1
T1_DIV_BY_2              T1_DIV_BY_4   T1_DIV_BY_8
CCP_COMPARE_RESET_TIMER
```

Where separate modes can be separated by the Inclusive-OR | operator.

Show how you would code your solution to SAQ 13.5 in **C**. In CCS **C** a function can be turned into a CCP1 interrupt service routine by preceding it by the directive #INT_CPP1; see Program 9.3 on p. 293 for details. You an also assume that the reserved variable CCP_1 represents the 16-bit CCPR1H:L register.

13.7 Pulse-width modulation can be used to control the speed of a DC motor by altering the average winding current. However, starting up such a motor is problematic, as the current flow is much greater than normal until the motor reaches normal running speed and the consequent back emf reduces winding current. It is proposed that to avoid damage to the current driving transistors, a PWM technique be used to gradually increase the duty cycle from zero to full over a period of several seconds. Show how you could do this assuming a PIC MCU with a 4 MHz crystal and CCP module.

13.8 Light-controlled pedestrian crossings (Pelican crossings) in the UK follow the listed sequence of operations once one of the cross-request switches are closed.

1. Green light only (standby).
2. Amber light only for 3 s.
3. Red light plus buzzer for 15 s.
4. Flashing amber light—five flashes each comprising 3 s on and 3 s off.
5. Return to standby.

Using a suitable PIC microcontroller with a Timer 1 module, design the software to control the lights and buzzer. Although the lights are on both sides of the road, you may assume that they are connected in parallel and are activated by a High state on the relevant parallel port pin. The two input switches, CROSS_REQUEST0 and CROSS_REQUEST1 give 0 when closed. The buzzer is activated by a Low state on the connected parallel port pin.

Chapter 14
Take the Rough with the Smooth

Given that digital microcontrollers are in the business of monitoring and controlling the real environment—which is commonly analog in nature—we need to consider the interaction between the analog and the digital worlds. In some cases all that is required is a **comparison** of two analog voltage levels. However, for more sophisticated situations, analog input signals need conversion to a digital equivalent; that is, **analog-to-digital conversion** (**ADC**). Thereafter the digital patterns can be processed in the normal way. Conversely, if the outcome is to be in the form of an analog signal, then a **digital-to-analog conversion** (**DAC**) stage will be necessary.

Of these various processes, illustrated in Fig. 14.1, A/D conversion is by far the more complex. Most PIC microcontrollers feature integral multi-channel A/D facilities. However, analog outputs frequently require external circuitry to implement the D/A process.

In this chapter we will look at the properties of analog and digital signals and the conversion between them, as relevant to the PIC MCU. After completion you will:

- Understand the quantization relationship between analog and digital signals.
- Be aware of the need to sample an analog signal at least twice the highest frequency component.
- Appreciate how the successive approximation technique can convert an analog voltage to a binary equivalent.
- Understand the operation and be able to configure the Analog Comparator, Voltage Reference and ADC modules.
- Know how to configure I/O pins as either analog or digital.
- Be able to write assembly-level programs to acquire analog data using polling, interrupt-driven, and Sleep techniques, and to interrogate the state of the analog comparators.
- Be able to code high-level **C** programs to set-up and interact with the various analog modules.
- Know how to interface to a proprietary DAC.

The information content of an **analog signal** lies in the continuously changeable worth of some constituent parameters, such as amplitude, frequency, or phase. Although this definition implies that an analog variable is a continuum in the range

S. Katzen, *The Essential PIC18® Microcontroller,*
Computer Communications and Networks,
DOI 10.1007/978-1-84996-229-2_14, © Springer-Verlag London Limited 2010

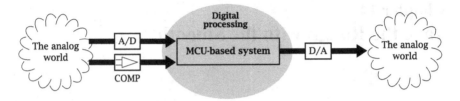

Fig. 14.1 Analog world—digital processing

$\pm\infty$, in practice its span is restrained to between an upper and lower limit. Thus a mercury thermometer may have a continuous range between, say, $-10°C$ and $+180°C$. Below the bottom number mercury disappears into the bulb. Above the highest number and the top of the tube is blown off!

Theoretically the quantum nature of matter sets a lower bound to the smooth continuous nature of things. However, in practice noise levels and the limited accuracy of the device generating the signal sets an upper limit to the resolution that processing needs to take account of.

Digital signals represent their information content in the form of arrangements of discrete characters. Depending on the number and type of symbols making up the patterns, only a finite totality of value portrayals are possible. Thus in a binary system, an n-digit pattern can at the most represent 2^n levels. Although this *rough* grainy view of the world seems inferior to the infinity of levels that can be *smoothly* represented by an analog equivalent, the quantizing grid can be tailored to fit the accuracy of the task to be undertaken. For instance, a telephone speech circuit will tolerate a resolution of around 1%. This can use an 8-bit depiction, which gives up to 256 discrete values with a corresponding $\approx 0.5\%$ resolution. A music compact disk uses a 16-bit scheme, giving a one part in 65,636 grid—around 0.0015% resolution, and a DVD has a 20-bit encoding.

From this discussion it can be seen that any process involving interconversion between the analog and digital domains will involve transition through the **quantization** state. Therefore we need to look at how this affects the information content of the associated signals.

As an example, consider the situation shown in Fig. 14.2, where an input range is represented as a 3-bit code. In essence the process of quantizing a signal is the comparison of the analog value with a fixed number of levels—eight in this case. The nearest level is then taken as expressing the original as its digital equivalent. Thus in Fig. 14.2 an input voltage of 0.4285 of full scale is 0.0536 above quantum level 3. Its quantized value will then be taken as level 3 and coded as b'011' in our 3-bit scheme of things.

The residual error of -0.0536 will remain as quantizing noise, and can never be eradicated—see also Fig. 14.3(d). The distribution of quantization error is given at the bottom of Fig. 14.2, and is affected only by the number of levels. This can

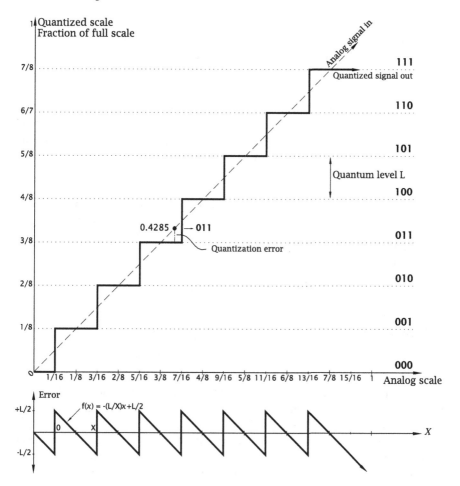

Fig. 14.2 The quantizing process

simply be characterized by evaluating the average of the error function squared. The square root of this is then the root mean square (rms) of the noise.

$$\mathcal{F}(x) = -\frac{L}{X}x + \frac{L}{2}.$$

The mean square is:

$$\frac{1}{X}\int_0^X \mathcal{F}(x)^2\,dx = \frac{1}{X}\int_0^X \left[\frac{L^2}{X^2}x^2 - \frac{L^2}{X}x + \frac{L^2}{4}\right]dx$$

$$= \frac{1}{X}\left|\frac{L^2}{3X^2}x^3 - \frac{L^2}{2X}x^2 + \frac{L^2}{4}x\right|_0^X = \frac{L^2}{12}.$$

Thus the rms noise value is $\frac{L}{\sqrt{12}} = \frac{L}{2\sqrt{3}}$, where L is the quantum level.

Table 14.1 Quantization parameters

Binary bits n	Quantum levels (2^n)	% resolution	Resolution dynamic range	S/N ratio (dB)
4	16	16.25	24.1 dB	26.9 dB
8	256	0.391	48.2 dB	49.9 dB
10	1024	0.097	60.2 dB	61.9 dB
12	4096	0.024	72.2 dB	74.0 dB
16	65,536	0.0015	96.3 dB	98.1 dB
20	1,048,576	0.00009	120.4 dB	122.2 dB

A fundamental measure of a system's merit is the signal-to-noise ratio. Taking the signal to be a sinusoidal wave of peak to peak amplitude $2^n L$, we have an rms signal of $\frac{(\frac{2^n L}{2})}{\sqrt{2}}$; that is, $\frac{\text{peak}}{\sqrt{2}}$. Thus for a binary system with n binary bits, we have a signal-to-noise ratio of:

$$\frac{(\frac{2^n L}{2\sqrt{2}})}{(\frac{L}{\sqrt{12}})} = \frac{2^n \sqrt{12}}{2\sqrt{2}} = 1.22 \times 2^n.$$

In decibels we have:

$$\text{S/N} = 20 \log 1.22 \times 2^n = (6.02n + 1.77) \text{ dB}.$$

The dynamic range of a quantized system is given by the ratio of its full scale ($2^n L$) to its resolution, L. This is just 2^n, or in dB, $20 \log 2^n = 20n \log 2 = 6.02n$. The percentage resolution given in Table 14.1 is of course just another way of expressing the same thing.

The exponential nature of these quality parameters with respect to the number of binary-word bits is clearly seen in Table 14.1. However, the implementation complexity and thus price also follows this relationship. For example, a 20-bit conversion of 1 V full scale would have to deal with quantum levels less than 1 μV apart. Pulse-code modulated telephonic links use eight bits, but the quantum levels are unequally spaced, being closer at the lower amplitude levels. This reduces quantization hiss where conversations are held in hushed tones! Linear 8-bit conversions are suitable for most general purposes, having a resolution of better than $\pm\frac{1}{4}\%$. Actually video looks quite acceptable at a 4-bit resolution, and music can just about be heard using a single bit—i.e., positive or negative!

S/N ratios presented in Table 14.1 are theoretical upper limits, as errors in the electronic circuitry converting between representations and aliasing (discussed below) will add distortion to the transformation.

The analog world treats time as a continuum, whereas digital systems sample signals at discrete intervals. Shannon's sampling theorem[1] states that provided this interval does not exceed half that of the highest signal frequency, then no informa-

---

[1]Shannon, C.E.: *Communication in the Presence of Noise*, Proc. IRE, vol. 37, January 1949, pp. 10–21.

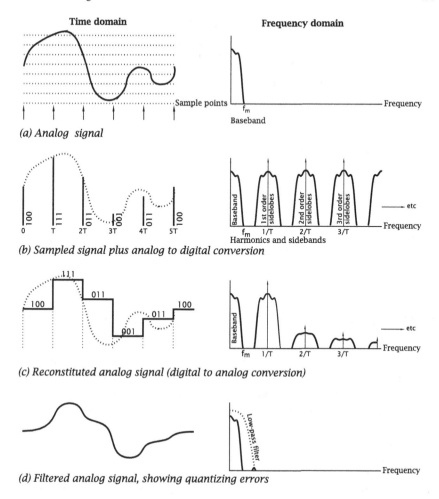

Fig. 14.3 The analog–digital process

tion is lost. The reason for this theoretical twice highest frequency sampling limit, called the Nyquist rate, can be seen by examining the spectrum of a train of amplitude modulated pulses. Ideal impulses (pulses with zero width and unit area) are characterized in the frequency domain as a series of equal-amplitude harmonics at the repetition rate, extending to infinity. Real pulses have a similar spectrum but the harmonic amplitudes fall with increasing frequency.

If we modulate this pulse train by a baseband signal $A \sin \omega_f t$, then in the frequency domain this is equivalent to multiplying the harmonic spectrum (the pulse) $A \sin \omega_h t$ by $B \sin \omega_f t$; giving sum and different components thus:

$$A \sin \omega_h t \times B \sin \omega_f t = \frac{AB}{2}(\sin(\omega_h + \omega_f)t + \sin(\omega_h - \omega_f)t)$$

for each of the harmonic frequencies ω_h.

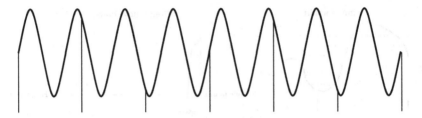

(a) Sampling below the Nyquist rate

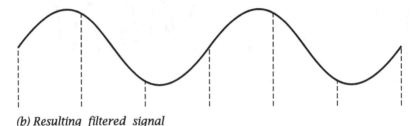

(b) Resulting filtered signal

Fig. 14.4   Illustrating aliasing

More complex baseband signals can be considered to be a band-limited $(0 \rightarrow f_m)$ collection of individual sinusoids, and on the basis of this analysis, each of these pulse harmonics will sport an upper (sum) and lower (difference) sideband. We can see from the geometry of Fig. 14.3(b) that the harmonics (multiples of the sampling rate) must be spaced at least $2 \times f_m$ apart, if the sidebands are not to overlap.

A low-pass filter can be used, as shown in Fig. 14.3(d), to recover the baseband from the pulse train. Realizable filters will pass some of the harmonic bands, albeit in an attenuated form. A close examination of the frequency domain of Fig. 14.3(d) shows a vestige of the first lower sideband appearing in the pass band. However, most of the distortion in the reconstituted analog signal is due to the quantizing error resulting from the crude 3-bit digitization. Such a system will have a S/N ratio of around 20 dB.

In order to reduce the demands of the recovery filter, a sampling frequency some-what above the Nyquist limit is normally used. This introduces a guard band be-tween sidebands. For instance, the pulse code telephone network has an analog in-put band limited to 3.4 kHz, but is sampled at 8 kHz. Similarly the audio compact disk uses a sampling rate of 44.1 kHz, for an upper music frequency of 20 kHz.

A more graphic illustration of the effects of sampling at below the Nyquist rate is shown in Fig. 14.4. Here the sampling rate is only 0.75 of the baseband fre-quency. When the samples are reconstituted by filtering, the resulting pulse train, the outcome—shown in Fig. 14.4(b)—bears no simple relationship to the original. This spurious signal is known as an **alias**. In particular, this will occur where an input analog signal has frequency components *above* half the sampling rate, maybe due to noise. These noise frequencies will alias and will appear as distortion in the reconstituted signal. For this reason analog signals are usually low-pass filtered with hardware at the input of an A/D converter; for instance, see Fig. 14.19. This process is known as **anti-aliasing** filtering.

In dealing with analog inputs, many situations simply need to make a true:false decision on whether a voltage is above or below a reference value V_{ref}. For instance, the signal shown in Fig. 14.5 (see also Fig. 14.21) represents the current during the discharge of an EKG (ECG) diphasic defibrillator, as generated using a Hall effect current to voltage sensor. When nothing is happening the baseline voltage is 2.6 V. When the defibrillator begins its discharge, this voltage rapidly rises to a peak of 3.6 V over a few tens of microseconds. If the MCU is to sample the voltage over the next several tens of milliseconds, say, to calculate the total shock energy, then to begin this process it needs to know when this voltage rises above a threshold. In the diagram this is shown as 3.4 V. It could of course rapidly sample the analog signal using its integral Analog-to-Digital module, as described later on p. 510, but this continuous sample-and-check routine would use most of the processing capability of the processor. It would be much more software efficient to be able to automatically generate an interrupt in hardware when the input voltage V_{defb} rises above this threshold. The resulting ISR could then begin sampling the signal and performing the real-time analysis.

In Fig. 14.5 the analog signal V_{defb} is used as the input to the non-inverting (+) terminal of an **analog comparator**. The inverting terminal (−) is connected to a fixed reference V_{ref} of 3.4 V. Whenever V_{defib} rises above V_{ref}, the comparator's output voltage changes from logic 0 to logic 1, and conversely when $V_{defb} < V_{ref}$ the output goes back to logic 0.

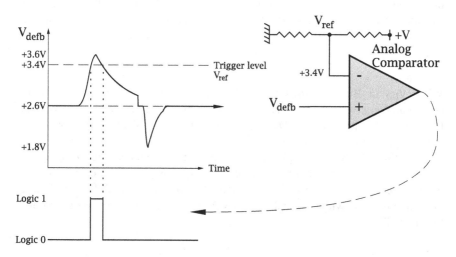

Fig. 14.5 Using an analog comparator to determine the start of the EKG defibrillator discharge

An analog comparator is basically a high-gain analog differential amplifier with no negative feedback. With a very large open-loop gain the amplifier will saturate at either near its positive or negative power supply if the difference between inputs is more than an exceedingly small value. An ordinary operational amplifier can act as an analog comparator, but circuits specifically designed for this purpose give standard logic levels at their output and have a snap action whenever slowly changing noisy inputs cross the differential threshold.

As a simple exemplar of the use of analog comparators, all PIC18[2] devices have an integral **High/Low Voltage Detect** module. The **HLVD** module shown in Fig. 14.6 is mainly used to generate an event whenever the supply voltage V_{DD} drops below or rises above one of up to 15 fixed fractions of the supply range. The actual trip voltages vary with device.

The operation of this module is controlled with the **HLVDCON** register.

HLVDEN
To enable this module, the **High/Low Voltage Detect ENable** bit in HLVDCON[4] must be set to 1. This will power-up the internal 1.2 V Fixed Voltage Reference (FVR) circuit and switch in the resistor chain.

HLVDL3:0
The **High/Low Voltage Detect Limit** bits in HLVDCON[3:0] switch one of sixteen voltages through to the inverting input of the comparator. Fifteen of these are stepped down fractions of V_{DD} derived through a resistor chain. The voltages annotated on the diagram show for the PIC18FXX20 line of devices the value of V_{DD} needed to cause the comparator to switch state. That is, would equal the value of the internal voltage reference fed to the non-inverting comparator inputs.

The 16th analog input, selected with HLVDL3:0 = 1111 effectively connects the non-inverting input of the comparator to the outside world via the HLVDIN pin. If this facility is used, then HLVDIN, which is normally shared with the RA5 pin (but exceptionally RA0 in the PIC18F1X20 devices) needs to be set-up as an input with the appropriate TRIS bit. It will automatically be configured as an analog channel. The module defaults to tapping 5 (HLVDL3:0 = 0101) on a Power-on Reset.

VDIRMAG
When V_{DD} or HLVDIN rises above the specified threshold, the output of the comparator goes to logic 0. This ‾__ edge can be optionally inverted with the Exclusive-OR gate when LHVDCON[7] = 1 to give a potential interrupt whenever the voltage drops below the threshold.

Our PIC18FXX20 exemplar and earlier family members; for instance the PIC18F452, lacked this polarity control and always generated an event whenever the voltage dropped too low. In this case the function was called the Low Voltage Detection module, with LVD replacing HLVD in the associated registers and control bits; e.g. LDVIF.

---

[2]Also the PIC24/30/32 families and some PIC16 devices.

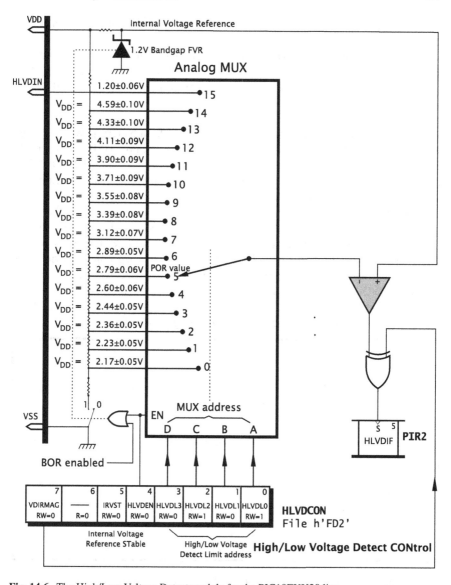

Fig. 14.6 The High/Low Voltage Detect module for the PIC18FXX20 line

IRVST

Both Brown-Out Reset (BOR) and HLVD modules make use of the 1.2 V Fixed Voltage Reference circuit. When enabling either or both of these modules, this reference circuit requires T_{IVRST} (20 μs typical, 50 μs maximum) to stabilize. After this time the **Internal Reference Voltage STable** bit will automatically set to 1 to indicate proper operation of the facility.[3] Interrupts will be disabled until this happens.

Generally if the BOR and HLVD modules are not being used, they should both be disabled to reduce the current consumption. For instance, for the PIC18FXX20 the HLVD module typically uses 30 μA, with a maximum of 70 μA over the operating temperature range and with a V_{DD} of 5 V.

As an example of the usage of this module, consider a battery-powered system with $V_{DD} \approx 3$ V. The battery can be charged by plugging into a USB computer port which gives a 5 V supply. Determine the housekeeping code in a HLVD ISR to detect whenever the USB cable is removed and the supply falls again to below 4 V, and set-up a rising edge detection of nominally 4.5 V on exit.

From Fig. 14.6 we see that tapping 14 will sense the passage about the 4.5 V level and tapping 11 likewise the 4 V threshold. We assume that the background software has enabled the module to set the HLVDIF when the voltage rises beyond the level 14 threshold, and so we have first to reset the trip point and direction before executing the core code. After this kernel has been executed, an endless loop monitors HLVDIF looking for a drop to the lower voltage and then exits with the trip reset to threshold 14. To allow for transients when cables are disconnected, a delay is implemented before finally returning to the background software.

```
ISR_HLVD movlw   b'00011011'   ; Reset to detect falling below
         movwf   HLVDCON       ; 4.5V
         bcf     PIR2,HLVDIF   ; Clear the HLVD flag

< < < < Core   code  > > > >

ISR_HLVD_LOOP
         btfss   PIR2,HLVDIF   ; Don't exit until voltage falls
         bra     ISR_HLVD_LOOP

         movlw   b'10011110'   ; Now set to detect rising above
         movwf   HLVDCON       ; 4V

         call    DELAY_1S      ; Wait for supply to settle
         bcf     PIR2,HLVDIF   ; Clear the HLVD flag again
         retfie  FAST          ; and return to background
```

Most devices (not our exemplar PIC18F1X20) offer the more flexible **Comparator module** shown in Fig. 14.7. Two analog comparators can be configured as directed by the three **CM2:0 Comparator Mode** bits in the **CoMparator CONtrol CMCON** register. The analog inputs can be connected to pins RA0 through RA5, as well as an on-chip reference voltage source. Outputs can be read from the **C2OUT** and **C1OUT** bits in CMCON[7:6] respectively and though digital output pins RA5 : 4 (C2OUT:C1OUT).

---

[3]In the PIC18FXXK20 series, this is replaced by the FVRST (Fixed Voltage Reference STable bit in CVRCON2[6]. CVRCON2[7] holds the FVREN bit to enable this facility.

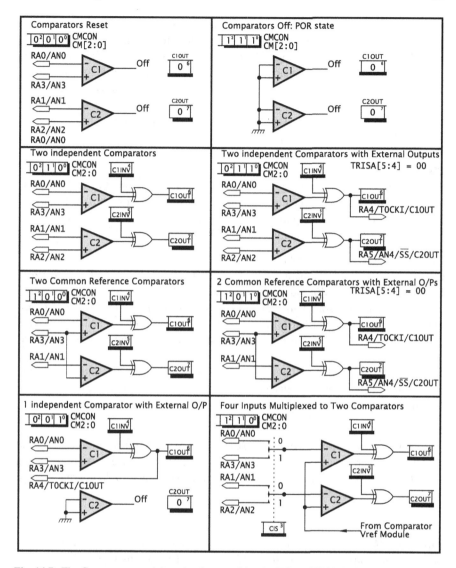

Fig. 14.7 The Comparator module as implemented for the PIC18FXX20 device line

On a Power-on Reset Mode 111, both comparators are off and the Port A pins can be used for other purposes. The seven other modes make use of several Port A pins, which should be set as inputs as appropriate. Any such pins used in a mode will automatically be configured to accept an analog signal ranging between V_{SS} (0 V) and V_{DD}. As a general rule all input pins with an analog function will come out of a POR *configured for analog signals*. This Power-on Reset requirement is to prevent physical damage to the input digital buffers (see Fig. 11.9 on p. 349) if an analog input voltage, say 2.6 V, were present at a pin on powering up. If that pin was set to be a digital input, expecting a voltage around 0 V or V_{DD}, then an intermediate voltage could cause several transistors to conduct at the same time; possibly causing thermal damage. As analog voltages are not well defined, even where a pin is configured as analog, an external resistor is often used to limit current flow if the analog voltage exceeds V_{DD} or goes negative, as shown in Fig. 14.21. Such protection resistors should be kept to less than 10 kΩ to reduce offset voltages due to input leakage currents (typically ±500 nA) and time constants arising from the pin input capacitance of typically 5 pF.

CM2:0
Six active modes, as selected by these three **Comparator Mode** bits CMCON[2:0], basically allow either completely independent operation for one or two Comparators, or both non-inverting inputs can be combined to be used as a common reference input. Mode 000 is a legacy configuration of little utility.

C1OUT, C2OUT, C1INV, C2INV
Outputs of any active Comparator may be read at any time from the **Comparator 1 OUTput** bit in CMCON[6] and likewise in CMCON[7] for Comparator 2. Each output has an associated programmable invertor control **Comparator 1 INVert** and **Comparator 2 INVert** in CMCON[4] and CMCON[5] respectively. When $V_{in+} > V_{in-}$ and the associated INVersion bit is 0 then the Comparator output will read as 1, otherwise as 0.[4]

In Modes 011 and 101 the state of C1OUT and C2OUT as controlled by the programmable invertors are reflected at pins C1OUT and C2OUT respectively. These pins should be set-up as outputs in the normal way.

CIS
The **Comparator Input Switch** in CMCON[3] in Mode 110 is used to connect the comparator inverting inputs V_{in-} to either pins RA0/RA1 or else RA3/RA2. In this state the non-inverting inputs V_{in+} are connected to a configurable internal voltage reference—see Fig. 14.8.

When there is a *change* in an active Comparator output, the **CoMparator Interrupt Flag CMIF** will be set and will generate an interrupt if the associated **CoMparator Interrupt Enable mask CMIE** and global mask bits GIE and PEIE have been set to 1. As each Comparator does not have its own interrupt flag, the software needs to maintain information regarding the previous status of the output bits C1OUT and C2OUT to determine which Comparator actually changed. This status can be updated as part of the ISR. The act of reading CMCON will end the Change mismatch—in the same manner as the Port B Change interrupt described on p. 358. Only then can the CMIF flag can be cleared in software in the normal manner. If the Comparator mode is to be changed 'on the fly' then interrupts should be disabled

---

[4]There is a small uncertainty range in this difference signal of ±10 mV maximum (±5 mV typical) due to Comparator offset voltages.

beforehand. After a delay of not less than 10 µs after the mode change, to allow voltage levels to stabilize, CMCON should be read again to clear any Change mismatch and CMIF cleared afterwards before re-enabling the interrupt system.

As the Comparator module does not depend on the system oscillator, an active Comparator can be used to awaken a sleeping PIC MCU when an external voltage crosses a V_{ref} threshold and sets CMIF. After wakening, the PIC MCU should cancel the Change mismatch and clear CMIF following the `sleep` instruction, or in the ISR if the Comparator interrupt is enabled.

It should be noted that an active Comparator uses considerably more current than the base Sleep value. For instance, the PIC18F45K20 has a typical quiescent current at 3 V and 85°C of 0.5 µA.[5] The Comparator module uses a current of typically 40 µA in the same environment. Thus if Comparators are not being used during the Sleep duration, they should then be disabled.

All family members with a Comparator module have a separate but related **Comparator Voltage Reference CVR** module. As can be seen from Fig. 14.8, this is essentially a resistor chain with an analog multiplexer gating through one of 16 different voltages. The functionality of the module is directed by the **Comparator Voltage Reference CONtrol** register.

CVREN

The **CVR ENable** bit in CVRCON[7] powers on the module, which defaults to off. The CVR module requires a maximum of 10 µs to settle after being enabled. Unless the Comparator module is being used to waken an inactive processor, the CVR module should be disabled during Sleep to reduce quiescent current consumption.

CVR3:0, CVRR

The **Comparator Voltage Reference** bits in CVRCON[3:0] select which analog voltage tapped from node n of the typically 2 kΩ chain of resistors is multiplexed to the V_{ref} output.

Two separate voltage ranges are available, as set with the **Comparator Voltage Reference Range CVRR** bit in CVRCON[5], which switches in or out an extra 8R resistor at the bottom of the chain. The result of this additional resistor is to reduce the span of the selection but to give a finer resolution; i.e. step size. In either case the accuracy is given as better than $\frac{1}{2}$ LSB. When altering this tapping, 10 µs should be allowed for the new output to settle.

The two ranges are:

CVRR	Value V_{ref}	Minimum	In 16 steps of	Maximum
0 (reset)	$\Delta V_{REF} \times (0.25 + n/32)$	$0.25 \times \Delta V_{REF}$	$\Delta V_{REF}/32$	$0.71875 \times \Delta V_{REF}$
1	$\Delta V_{REF} \times n/24$	0 V	$\Delta V_{REF}/24$	$0.625 \times \Delta V_{REF}$

where n ranges from 0 through 15 and ΔV_{REF} is $V_{REF+} - V_{REF-}$.

---

[5]However, note that the Comparator module for this device is somewhat more flexible, with a several control registers to give the required options. For instance, each comparator has its own interrupt flag.

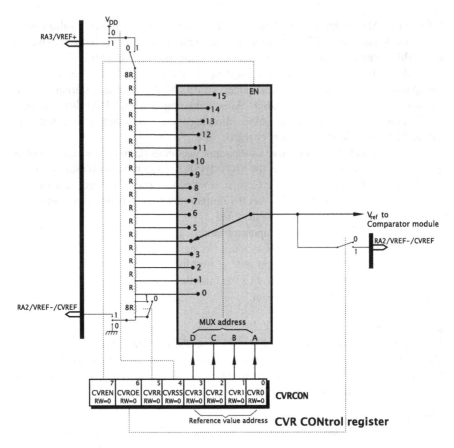

Fig. 14.8 The Comparator Voltage Reference module

CVRSS

By default the Power-on Reset voltage across the resistor chain is $V_{DD} - V_{SS}$. As V_{SS} is normally 0 V, then in this situation replacing ΔV_{REF} in the table above by V_{DD} gives the reset situation.

Using the power supply as the root of a reference voltage is not a good idea where precision is required. Noise from digital circuit switching, power surges and variations due to battery condition will be coupled into the analog circuitry. Early versions of the CVR module gave the designer no choice in the matter. However, subsequent implementations allow the option of a precision external voltage source. Where the **Comparator Voltage Reference Source Select** bit in CVRCON[4] is 1, the voltage at the VREF+ pin is connected to the top of the chain. Normally the bottom of the chain at VREF- is connected to ground, but this is not necessarily the case. Usually the reference voltages should lie within the power supply range; actual values will be specified in the device data sheet. If ΔV_{REF} is a variable analog signal, then if used in conjunction with the ADC module of Fig. 14.10, a limited range analog multiplication can be implemented.

CVROE

When the **CVR Output Enable bit CVROE** in CVRCON[6] is set to 1, the selected analog voltage V_{ref} is gated through to the appropriate pin. As this pin, usually RA2, is shared with the V_{REF+} input, this connection should only be set whenever an external voltage reference is not used. When chosen, the setting of TRISA[2] is overridden. However, current consumption will be less if this is set-up as an output.

Due to the relatively low value of resistance, which also depends on the selected tap, Microchip recommend that this external reference voltage be buffered—typically with an operational amplifier. If necessary, the amplifier gain can be altered to give finer control over the V_{REF} value and filtering can be added to reduce high-frequency noise. Used in this way, the Voltage Reference module can be employed as a simple 4-bit digital-to-analog converter.

As our example, assume that V_{DD} is 5 V and we are going to generate our threshold voltage of 3.4 V for Fig. 14.5. We will need to use the high range; that is, CVRR $= 0$, and calculate a value for CVR[3:0]:

$$5 \times (0.25 + n/32) = 3.4,$$

$$0.25 + n/32 = 3.4/5,$$

$$n = (3.4/5 - 0.25) \times 32 = 13.76,$$

giving $n = 14$ as our closest approximation. Making CVR[3:0] = b'1110' gives an actual V_{ref} of 3.4375 ± 0.078128 V.

The code to set-up the Comparator and CVR modules for our defibrillator example using Comparator 1 with RA3 as the analog input is then:

```
include   "p18f4520.inc"
movlw  b'00001110'  ; Comparator mode 110
movwf  CMCON        ; Switch to RA3 (CIS = 1)

movlw  b'10001110'  ; CVR module on (1), not external (0)
movwf  CVRCON       ; Hi internal range (00), CVR3:0 = 1110

bsf    PIE2,CMIE    ; Enable Comparator interrupts

call   DELAY_10US   ; Allow 10us for voltages to settle
movf   CMCON,f      ; Read CMCON to clear any Change state

bcf    PIR2,CMIF    ; Zero the Comparator interrupt flag
bsf    INTCON,PEIE  ; Enable Peripheral interrupt group
bsf    INTCON,GIE   ; & Globally enable interrupt system
```

Notice especially that before enabling the interrupt system a delay of 10 μs is executed to allow internal analog voltages to attain equilibrium. Reading the CMCON register then clears any Change situation, after which the Comparator Interrupt Flag CMIF is cleared. The general interrupt system can then be enabled by setting the PEIE and GIE mask bits in the usual way in the INTCON register.

In many situations more information on the analog signal is needed than a bang-bang comparison with a reference voltage. For instance, in the waveform shown in Fig. 14.5 the deviation of the voltage squared from the baseline, integrated with time, would be required to measure power. In such a situation the incoming signal would have to be sampled and converted from an analog amplitude to a digitized equivalent.

The mapping function from an analog input quantity to its digital equivalent can be expressed as:

$$V_{in} \mapsto V_{ref} \sum_{i=1}^{n} k_i \times 2^{-i}$$

where k_i is the ith binary coefficient having a Boolean value of 0 or 1 and $V_{in} \leq V_{ref}$, where V_{ref} is a fixed analog reference voltage. Thus V_{in} is expressed as a binary fraction of V_{ref}, and the Boolean coefficients k_i are the required binary digits of the series $\frac{1}{2}, \frac{1}{4}, \frac{1}{8}, \ldots$.

To see how we might implement this in practice, consider the following mechanical successive approximation analogy. Suppose we have an unknown weight W (analogous to V_{in}), a balance scale (equivalent to an analog comparator) and a set of precision known weights 1, 2, 4, and 8 g (analogous to a V_{ref} of 15 g). A systematic technique based on the task list might be:

1. Place the 8 g weight on the pan. IF too heavy THEN remove ($k_1 = 0$) ELSE leave ($k_1 = 1$).
2. Place the 4 g weight on the pan. IF too heavy THEN remove ($k_2 = 0$) ELSE leave ($k_2 = 1$).
3. Place the 2 g weight on the pan. IF too heavy THEN remove ($k_3 = 0$) ELSE leave ($k_3 = 1$).
4. Place the 1 g weight on the pan. IF too heavy THEN remove ($k_4 = 0$) ELSE leave ($k_4 = 1$).

This technique will yield the nearest lower value as the sum of the weights left on the pan. For instance, if W were 6.2 g then we would have a weight assemblage of $4 + 2$ g, or b'0110' for a 4-bit system.

The electronic equivalent to this **successive approximation** technique uses a network of precision resistors or capacitors configured to allow consecutive halving of a fixed voltage V_{ref} to be switched in to an analog comparator; which acts as the balance scale.

Most MCUs use a network of capacitors valued in powers of two to subdivide the analog reference voltage, such as shown in Fig. 14.9. Small capacitance values are easily fabricated on a silicon integrated circuit and although the exact value will vary somewhat between different batches of ICs, within the one device all capacitor values will closely match and track with changes in temperature and supply voltage. Multiples of this base value can be fabricated by paralleling unit devices—typically FET gate-source capacitance.

Before the conversion process gets underway, the network has to be primed with the unknown analog input voltage V_{in}, as shown in Fig. 14.9(a). The dynamics of this

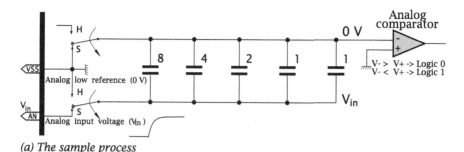

(a) The sample process

(b) The hold process

Fig. 14.9 Initializing the 8-4-2-1 capacitor network for a 4-bit convertor

sampling acquisition process involves charging up this capacitance network through both internal and external resistance; allowing for the settling time of the internal analog switches. If we take the 10-bit ADC module of Fig. 14.2 as an example, then the parallel capacitor network appears at the AN pin as a 25 pF (25×10^{-12}F) capacitor. Internal switch resistance is of the order of 2 kΩ with a 5 V V_{DD}, but is rather temperature and supply voltage dependent (e.g. 4 kΩ at 3 V). Contact resistance is given as 1 kΩ. Externally the maximum recommended value is 2.5 kΩ in order to keep any ohmic voltage offset due to the pin leakage current of ± 100 nA and time constants small. A smaller external source resistance will of course reduce the time constant.

The time constant τ (CR) with the values given here is $25 \times 10^{-12} \times 5.5 \times 10^3 \approx$ 140 ns for a total resistance of $2.5 + 3.0 = 5.5$ kΩ. In order to get within 0.05% of the final voltage; that is, $\frac{1}{2}$ of a 10-bit quantum level, takes approximately $8 \times \tau \approx 1.2$ µs. The data sheet gives the maximum amplifier/switch settling time of 0.2 µs. Taking a worse-case scenario at 25°C, a sampling time of 1.4 µs should ensure stability before the conversion. For temperatures above this, an additional 0.02 µs per °C should be added to this figure. For instance, at 85°C an extra 1 µs should be added; giving a total of 2.4 µs minimum acquisition time. Of course, to evaluate the maximum rate of samples that can be taken, the actual conversion time must be added to this acquisition time.

During the sample (S) period the top capacitor electrodes are held to 0 V and bottom electrodes are charged to V_{in}. The change-over to the hold (H) position, shown in Fig. 14.9(b), grounds the bottom electrodes and allows the top electrodes to

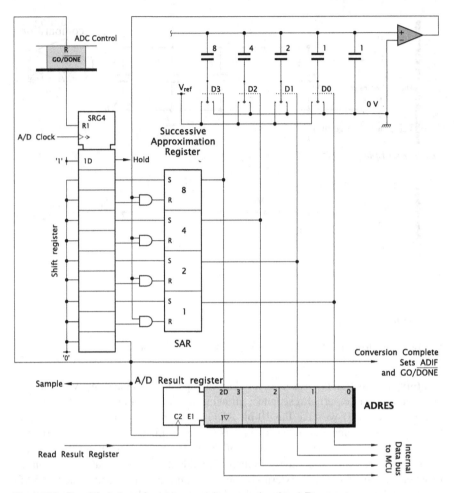

Fig. 14.10 Simplified view of a 4-bit successive approximation A/D converter

float. The voltage across a capacitor can only change if charge is transferred across electrodes, $\Delta Q = C\Delta V$. Thus the change in voltage $\Delta V = -V_{in}$ at the bottom electrodes is matched at the top floating electrodes; which now become $0 - V_{in}$, as charge cannot flow in or out of the floating top electrodes. Thus at the start of the conversion process the inverting input of the analog comparator is $-V_{in}$.

A 4-bit version of the successive approximation network at the heart of the ADC module is shown in a simplified form in Fig. 14.10. The step-by-step process is sequenced by a shift register (SRG4—see Fig. 2.22 on p. 36) when the programmer sets the GO/DONE bit in the ADC Control register. As the Control shift register is clocked, a single 1 moves down to activate each step in the sequence:

| Hold | bit 3 | bit 2 | bit 1 | bit 0 | Complete/Sample |

The capacitor network is switched to Hold and each capacitor, beginning with the largest value, is switched to V_{ref} in turn. The outcome of the comparator then determines the state of the corresponding bit in the Successive Approximation Register (SAR). The process is detailed in Fig. 14.11. After four set–try–reset actions, the outcome in the SAR is transferred to the **Analog-to-Digital RESult** register. The GO/$\overline{\text{DONE}}$ switch/flag bit is now cleared to indicate end of conversion and the **Analog/Digital Interrupt Flag ADIF** set. Finally, the analog input is again switched back into the capacitor network (Sample) which then charges up ready for the next conversion after a suitable settling period.

The total conversion time is approximately six times the clock period T_{AD} of the sequencer shift register—one period for each bit plus one each for the Hold and Complete/Sample slots. In the case of a 10-bit module, this will be approximately 12 times the clock period. For the PIC18FXX20 devices, the minimum clocking period is 0.7 µs ($\approx$1.4 MHz) for $V_{DD} \geq 3$ V. The PIC18LFXX20 with $V_{DD} = 2$ V has a minimum figure of 1.4 µs and older devices, such as the PIC18F/LF1X20, have values 1.6/3.0 µs. For the PIC18FXX20 devices there is a 25 µs lower clocking period, as charge slowly leaks away from the network capacitors; so a clocking frequency of less than 40 kHz should be avoided. Other devices will have other values, and very high temperatures increase leakage currents. For instance, the PIC18FXXK20 at 125°C has a lower boundary of 4 µs (250 KHz). From Fig. 14.12 we see that the ADC clock can be derived from one of seven sources. Six these are fractions of the system clock rate and the seventh is a stand-alone CR oscillator with a nominal T_{AD} of 1.2 µs for the PIC18FXX20 (6/9 µs for the PIC18F/LF1X20).

The conversion process, where each successive half-fraction of V_{ref} is added to and conditionally taken away from the initial value is illustrated in Fig. 14.11. As we have seen in Fig. 14.9, at the end of the acquisition period the top plates of the capacitor array are at $-V_{in}$. As an example, let us assume that V_{in} is 0.4285V_{ref}.

1. The process begins by switching in V_{ref} into the lower plate of the largest capacitor, as controlled by the SAR_8 latch in Fig. 14.10. This causes an injection of charge $\Delta Q = C_{total}V_{ref}$, which is identical across both the 8-unit capacitor C_1 and the rest of the capacitors. These latter also have a parallel value of 8 units in Fig. 14.11. Thus the voltage at node N rises by $V_{ref}/2$ to $-0.485 + 0.5 = +0.07125V_{ref}$. In general $\Delta V_N = V_{ref}C_k/C_{total}$. The comparator output is now logic 0 and thus the SAQ_8 latch is consequently cleared, reversing the $V_{ref}/2$ step.
2. SAQ_4 switches V_{ref} into the next highest capacitor, giving a $V_{ref}/4$ step at N ($\frac{4}{16}$). The resulting voltage of $-0.485 + 0.25 = -0.178V_{ref}$ gives a comparator output of logic 1 and SAR_4 remains set with the node voltage staying at $-0.1785V_{ref}$.
3. SAQ_2 switches V_{ref} into the second lowest capacitor, giving a $V_{ref}/8$ step at N ($\frac{2}{16}$). The resulting voltage of $-0.1785 + 0.125 = -0.0535V_{ref}$ gives a comparator output of logic 1 and SAR_2 remains set with the node voltage staying at $-0.0535V_{ref}$.
4. SAQ_1 switches V_{ref} into the lowest capacitor, giving a $V_{ref}/16$ step at N ($\frac{1}{16}$). The resulting voltage of $-0.0535 + 0.0625 = +0.009V_{ref}$ gives a comparator output of logic 0 and SAR_1 is cleared reversing the $V_{ref}/16$ step.

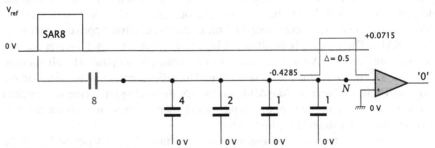

(a) The most significant bit

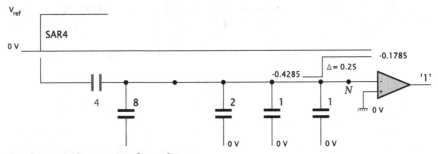

(b) The second most significant bit

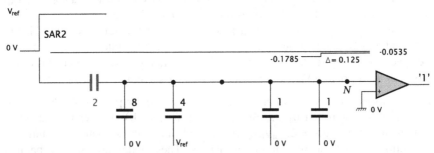

(c) The third most significant bit

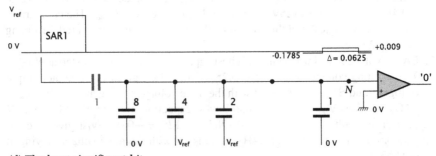

(d) The least significant bit

Fig. 14.11   The successive approximation process

The state of the SAR of b'0110' or $0.375V_{ref}$, represents the best 4-bit fit to $V_{in} = 0.4285V_{ref}$. The residue $0.0535V_{ref}$ is the quantizing error.

Most MCUs use an 8- or 10-bit capacitor array. In principle the technique can readily be extended to higher resolutions, but in practice the difficulty in matching ever greater capacitors and internal logic noise means the majority of processors are limited to 12-bit resolution.[6] External high-speed successive-approximation devices with 12+ bit resolution, usually using a resistor ladder network, are readily available, but can be expensive. A low cost example is the Microchip MCP3301, which features a 13-bit resolution with a single differential voltage input and a SPI interface.

Matching of the array capacitors, offsets, and resistance of internal switches, leakage currents, and analog comparator non-linearities all contribute to errors in the conversion process. The four types of error specified in the data sheets relate to the analog-digital transfer function illustrated in Fig. 14.2. Ideally this mapping should be a staircase with equal steps in a straight line from the first transition at 0.5 of a step to 1.5 of the step before the maximum analog voltage.

Integral non-linear
Maximum deviation of the actual transfer characteristic over the whole range.

Differential non-linear
Maximum deviation of any step from the ideal.

Gain
Deviation of the slope from the ideal.

Offset
Shift of transfer characteristic left or right from the ideal voltage giving the first step; that is first non-zero code.

Our exemplar devices give a value for the first three of these errors, which are non cumulative, as $< \pm 1$ LSB. Offset error is quoted as a maximum of ± 2 LSB. However, it is guaranteed that the transfer is monotonic; that is, the binary code will never move in the reverse direction for any change ΔV_{in} of input voltage. These error figure are for a given $\Delta V_{REF} \geq 3$ V.

Virtually all the PIC18 family members feature an integral 10-bit ADC module, based on a capacitor network with characteristics previously described. From the user's perspective the details of the conversion process are less important than the system aspects integrating this module into software. The architecture of the **ADC** module is shown in Fig. 14.12. There are only relatively small differences across the range.

Although there is a single analog-to-digital converter, this is fronted with an analog multiplexer. This allows the software to select up to 13 separate analog voltages

---

[6]For instance the PIC24H family.

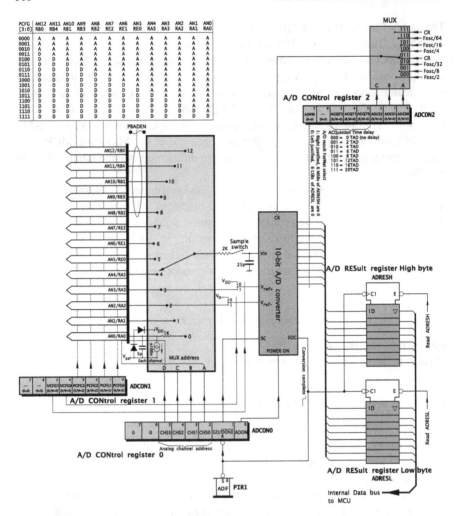

Fig. 14.12 The PIC18FXX20 10-bit multi-channel analog-to-digital convertor module

one at a time.[7] Three Control registers allow the program to select any one channel for sampling and to determine the source of the sequence clock. In addition the appropriate pins may be set-up as either analog or digital, with control over the internal/external nature of the reference voltages. The conversion is initiated via the **GO/DONE** bit, which also indicates when the process is complete and the 10-bit outcome can then be read from the two Result registers.[8]

Our description of the ADC module can be split into the initial set-up and the conversion process.

---

[7] 18- and 24-pin devices, lacking a Port E have up to 10 possible analog input channels.

[8] Early PIC16 device ADC modules use one AD Result register to hold one 8-bit outcome.

Initialization

In setting up your module you need to consider the following points:

1. How do I enable the module?
2. Which channels am I going to use?
3. How am I going to clock the module?
4. What delay am I going to use to allow for stabilization after starting a conversion?
5. Do I only need an 8-bit outcome?

All these choices, together with operational matters, are set-up using the various AD Control registers.

ADCON0

ADCONtrol register 0 enables the module, initiates and indicates the completion of the process, and picks the input channel for conversion.

ADON

On a Power-on Reset the ADC module is disabled. Setting **ADON** in ADCON[0] to 1 turns the module on. An enabled module typically uses 1μA even when idling, so it should be disabled when not in use where power consumption is a consideration. During a conversion the average current is 180 μA. The GO/$\overline{\text{DONE}}$ switch bit should not be set to 1 in the same instruction as the module is enabled to avoid beginning a conversion at the same time as the module is being started up.

GO/$\overline{\text{DONE}}$

Setting ADCON0[1] initiates a conversion. GO/$\overline{\text{DONE}}$ remains 1 until the conversion is complete, which is signaled by a 0 state. At the same time the **ADIF** flag will be set, which can trigger an interrupt.

CHS3:0

These three **CHannel Select** bits in ADCON0[5:2] pick which pin is gated through to the A/D converter logic.

Early versions of these modules use five Port A and in 40+ footprint devices, three Port E pins. Newer ADC modules have added an optional five Port B pins to this complement. The **PBADEN** fuse must be active (PBADEN=ON) to enable these potential additional analog inputs.

ADCON1

ADCONtrol register 1 configures blocks of pins shared with the digital parallel ports to accept analog range voltages and selects between internal and external reference voltages.

PCFG3:0

From Fig. 14.12 we see that analog channels coexist with normal digital port lines. Figure 14.13 shows one of these shared port lines in a little more detail. Here the connection to the analog ports is pictured in parallel to the ordinary

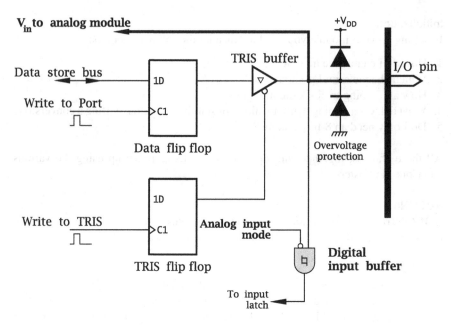

Fig. 14.13 Simplified logic of a combo analog/digital port line

digital connections. The major problem with this arrangement is that the digital input buffer can be damaged by prolonged analog voltages midway between the logic levels causing excessive current flows.

On a Power-on Reset all potentially shared A/D port pins are configured as analog-friendly; that is with disabled digital input buffers. In the case of the five Port B A/D pins, this is only the case when the PBADEN fuse is on. If the number of analog channels required for a particular application is less than the maximum available, some unused channels can be reclaimed for use as ordinary digital I/O. This is accomplished using the appropriate settings in the **Port ConFiGuration** bits. The actual choices, number, and location of these bits are device specific,[9] but those patterns applicable to the PIC18FXX20 group are shown in Fig. 14.12. For instance, if you only require a single analog channel for your project, the pattern b'1110' will leave pin RA0/AN0 as analog and the rest for other digital purposes.

In previous chapters, in examples where no analog activity was required (such as Program 11.1 on p. 338) the ADCON1 register still needed to be set to b'11111111' (setf ADCON1) to configure all pins as digital. Failure to do this is one of the more common errors, as all devices have analog module(s) and

---

[9]For instance, the PIC18F1X20 use ADCON1 to individually enable each of the seven available analog inputs, and as a consequence the two VCFG bits move to ADCON0. The PIC18FXXK20 group use the two ANSELH:ANSEL registers to configure each of the 13 channels individually.

all relevant pins will therefore always default to analog on a Power-on Reset or avoid possible damage.

From the diagram we can make the following deductions.

- A port pin configured as analog will read as logic 0 due to the disabled digital input buffer.
- The TRIS buffer is not affected and thus the appropriate TRIS bits should be 1; that is, the direction of the port pins configured as analog should be set to input to prevent contention between the analog V_{in} and the digital state of the Data flip flop.
- The ADC can read an analog voltage at the pin even if that pin has not been configured as analog. However, the still active digital input buffer may consume an excessive current outside of the device's specification.

VCFG1:0

Analog channels 2 and 3 are dual-purpose,[10] in that they may optionally be used to supply external reference voltages, as a less noisy and more accurate alternative to the power-supply lines V_{DD} and V_{SS}. Each **Voltage ConFiGure** bit gates one of these channels through as V_{ref+} and V_{ref-} to the ADC logic. For instance, for a single analog channel and an external V_{REFH} ADCON1 should be b'0001 1011'—actually this will give three analog input channels.

If external references are used, then V_{REFH} should normally be between $\frac{V_{DD}}{2}$ and $V_{DD} + 0.6$ V and V_{REFL} between $V_{DD} - 3$ V and $V_{SS} - 0.3$ V (PIC18FXXK20 figures). In any case ΔV_{REF} should be not less than 3 V for $V_{DD} \geq 3$ V and 1.8 V for supply voltages below this.

ADCON2

The **A/D CONtrol register 2** selects the source of the ADC sequence clock; delay in beginning a conversion after a Start Conversion command and how the 10-bit outcome is aligned within the 16-bit result field.

ADCS2:0

The ADC module needs a clock signal in order to time the set and test sequence of Fig. 14.11. If the clock rate is too fast, changes in switching voltages will not have time to settle. Our exemplar devices quotes a minimum clock period T_{AD} as 0.7 μs; corresponding to a clocking rate of 1.43 MHz.

Typically the A/D clock is derived as a fraction of the system clock. The **A/D Clock Select** bits in ADCON2[2:0] can be set to give processor frequency division ratios from ÷2 to ÷64. Normally the smallest ratio is chosen to give a F_{AD} less than the quoted value. For instance, a PIC18FXX20 with an 8 MHz crystal needs a ÷8 ratio (ADCS[2:0] = 001), giving a F_{AD} of 1 MHz. Choosing ÷4 would give 2 MHz; which is beyond the 1.43 MHz limit. Table 14.2 shows the maximum possible system clock rate for the various division values.

---

[10]In the PIC18FXXK20 channel 15 is also special, connected to the FVR—see Fig. 14.6.

Table 14.2 Maximum device frequency F_{AD} against the six division ratios

To generate ADC clock		Maximum device frequency to give	
Division ratio	ADSC2:0	$T_{AD} \geq 0.7\ \mu s$	$T_{AD} \geq 1.4\ \mu s$
$f_{osc}/2$	000	2.86 MHz	1.43 MHz
$f_{osc}/4$	100	5.71 MHz	2.86 MHz
$f_{osc}/8$	001	11.43 MHz	5.72 MHz
$f_{osc}/16$	101	22.86 MHz	11.43 MHz
$f_{osc}/32$	010	40.00 MHz	22.86 MHz
$f_{osc}/64$	110	40.00 MHz	22.86 MHz
CR[1]	X11	1.2 μs typ	2.5 μs typ

Note 1: Normally used in the Sleep mode and in any case not recommended for frequencies above 1 MHz

To allow operation in a low-speed system clock environment; for instance, when a 32.768 kHz watch crystal is used, a separate internal Capacitor-Resistor (CR) oscillator is provided. As this stand-alone oscillator is separate from the system clock, a conversion can be completed while the PIC MCU is in its Sleep state. In this situation, the End Of Conversion interrupt can be used to waken the processor. Doing a conversion with the system clock turned off makes sense, as this gives a quiet environment with little digital noise. If the separate CR clock is used with a system clock of greater than 1 MHz, then Microchip recommends using a Sleep conversion; as the lack of synchronization between the two clock sources increases noise induced into the analog circuitry.

Our example was based on a minimum A/D clock period of 0.7 μs. The actual value will vary with device, and will be quoted in the appropriate data sheet. Table 14.2 also shows the maximum possible device frequency for the PIC18LFXX20, which has a minimum T_{AD} of 1.4 μs.

ACQT2:0

We have already seen on p. 504 that whenever the capacitor network is connected to an analog voltage, finite time is required to charge up to its equilibrium value. Typically this will occur whenever the input channel changes or after a new conversion is started. Thus for full accuracy, the software should wait an appropriate time after a conversion is complete or change in channel before starting again. In our exemplar modules the **ACQuisiTion** bits in AD-CON[5:3] can be set to automatically insert a fixed delay after the GO/$\overline{\text{DONE}}$ bit is set.

Figure 14.14 shows a timeline for a conversion process.

1. Once the go-ahead is given, a fixed number of T_{AD} cycles (between 2 and 20) is inserted before the capacitor network is disconnected from the analog input pin. Typically a four cycle delay (ACQT2:0 = 010) covers most situations.

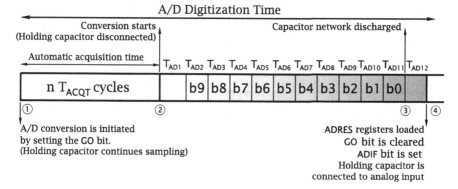

Fig. 14.14 Timeline for the conversion process

2. With the analog voltage held, the conversion begins. After a further T_{AD} period, each bit is built up from the MSB downwards.

3. After the LSB has been computed, the capacitor is discharged for one period, giving a total of $T_{AD} \times 12$ A/D clocks to complete the process.

4. The 10-bit digitized word is copied to the two ADRES registers, GO/$\overline{\text{DONE}}$ is cleared in parallel with setting the ADIF interrupt flag to indicate the end of the conversion. Concurrently, the analog voltage is again gated through from the selected pin.

The Power-on Reset of these bits is not to insert any fixed delay. This is the legacy condition, as older ADC modules do not provide this facility. In this situation the conversion process is delayed by one instruction cycle. This facilitates a conversion process when the MCU is asleep, to allow time for the `sleep` instruction to be executed.

ADFM

Our example ADC module needs two Files to hold the 10-bit outcome. As the total capacity of ADRESH:ADRESL is 16 bits, there are two ways of aligning these ten bits.

Many applications only require 8-bit resolution and processing. Where this is the case, the bottom two bits of the outcome word can be thrown away. From Fig. 14.15(a) we see that this is facilitated by left alignment. The content of ADRESL is simply ignored.

Where a full 10-bit word is necessary, setting the **A/D ForMat** b it in AD-CON2[7] to 1 will right-align the datum. As can be seen from Fig. 14.15(b), the outcome is a 10-bit datum extended to a 16-bit format by padding with leading zeros. Normal 16-bit arithmetic and other processing algorithms can then be used.

Conversion Process

After the module has been configured, from the user's perspective digitizing a selected analog channel is relatively straightforward. Assuming first that interrupts are

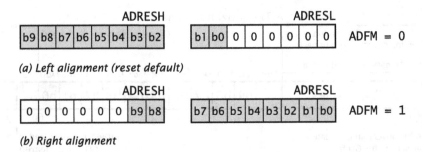

(a) Left alignment (reset default)

(b) Right alignment

Fig. 14.15 Aligning the 10-bit digital outcome in a 16-bit field

not being used, the following steps can be identified (including, for completeness, the initialization process) and is visualized with the timeline of Fig. 14.14.

1. Configure ADC module.

   - Set up port pins as analog/voltage reference (ADCON1).
   - Select ADC conversion clock source (ADCON2).
   - Select fixed A/D acquisition delay; if any (ADCON2).
   - Select initial ADC input channel (ADCON0).
   - Turn on ADC module (ADCON0).

2. Start conversion by setting the GO/$\overline{\text{DONE}}$ bit to 1.
3. Wait for ADC conversion to complete by polling the GO/$\overline{\text{DONE}}$ bit for logic 0.
4. Read the ADRES registers.
5. For next conversion go to step 1 or step 2 as required.

As an example, consider that we wish to continually read each of the eight analog pins AN0 through AN7 of a PIC18FXX20 in turn, while outputting the most significant eight digitized bits to Port B and the channel number to the lower three bits of Port D. The main crystal is 20 MHz and the power supply is to be used for the reference voltages.

The listing of Program 14.1 assumes that the ADC module has been initialized at reset with start-up code of the form:

```
include "p18f4520.inc"

config WDT = OFF, PBADEN=OFF

movlw    b'00000111' ; Make pins AN7:0 analog
movwf    ADCON1      ; I.e. shared Ports A & E all analog
clrf     TRISB       ; All Port B Output
movlw    b'11111000' ; Low 3 bits of Port D are Outputs
movwf    TRISD

movlw    b'00000001' ; CH0 (0000) for 1st conversion
movwf    ADCON0      ; No conversion (0), ADC turned on (1)

movlw    b'00010101' ; Right justified (0), 4TAD delay (010)
movwf    ADCON2      ; FAD = FOSC/16 (101)
```

Program 14.1 Scanning an 8-channel data acquisition system

```
MAIN        clrf    CHANNEL      ; Use a GPF to hold the channel count
MAIN_LOOP   movf    CHANNEL,w    ; Get the Channel number
            andlw   b'00000111'; Zero the top five bits
            movwf   CHANNEL
            movwf   PORTD        ; Copy to Port D

            call    GET_ANALOG ; Digitize it; returned in W
            movwf   PORTB        ; and copy to Port B

            incf    CHANNEL,f  ; Advance to next channel
            bra     MAIN_LOOP  ; and DO forever

; ***********************************************************
; * FUNCTION    : Analog/digital conversion at channel n    *
; * ENTRY       : Channel number in W                        *
; * EXIT        : Digitized 8-bit analog value in W          *
; ***********************************************************

GET_ANALOG  rlncf   WREG,w       ; Shift channel number left >>2
            rlncf   WREG,w       ; with outcome in W
            bcf     ADCON0,CHS0; Zero channel bits
            bcf     ADCON0,CHS1
            bcf     ADCON0,CHS2
            bcf     ADCON0,CHS3
            addwf   ADCON0,f   ; Moves channel number to ADCON0[5:2]
            bsf     ADCON0,GO  ; Start conversion after fixed delay
GET_ANALOG_LOOP
            btfsc   ADCON0,GO  ; Check for End Of Conversion
            bra     GET_ANALOG_LOOP
            movf    ADRESH,w   ; Fetch datum when GO/NOT_DONE is zero
            return
```

which configures the module to enable all eight analog channels with internal reference voltages with ADCON1 set to $\boxed{00\,|\,00\,|\,0111}$. ADCON2 is initialized to $\boxed{0\,|\,0\,|\,101\,|\,101}$ to select a left-aligned outcome with a fixed acquisition delay of $4 \times T_{AD}$ and a F_{AD} rate of $\frac{20}{16} = 1.25$ MHz ($T_{AD} = 0.8$ μs). Channel zero is set-up in ADCON0[5:2] (the initial value is irrelevant) and the module turned on. With an initial zero value of GO/$\overline{DONE}$ no conversion is actioned.

With the module initialized, the main software of Program 14.1 spends all its time in a loop reading the digitized equivalent of each channel in turn from ADRESH (the top eight bits) and copying it in turn to Port B. Before the digitization, the Channel counter is sent to Port D as a modulo-3 number.

The acquisition itself is implemented using the GET_ANALOG subroutine, to which is passed the desired channel number in the rightmost three bits of the Working register. This is logic shifted two places to the left to align the channel number with the CHSn bits in ADCON0[5:2]. After clearing the CHS3:0 bits, the shifted Channel number can then be added into ADCON0 to set CHS3:0 to the appropriate channel.

After the channel number has been set-up, the GO/$\overline{DONE}$ bit in ADCON0[1] is set to initiate a conversion.[11] A fixed delay of $4 \times T_{AD}$ has been actioned before the

---

[11] A conversion may be aborted at any time by clearing GO/$\overline{DONE}$.

conversion proper begins, to allow for switch delay and stabilization—see p. 505. The completion of the process can be monitored by polling GO/DONE until this goes to 0. At this point the content of ADRESH is the 8-bit outcome of the conversion.

Each actual conversion takes around $4 + 12 \times 0.8 \approx 13$ µs. Taking into account the housekeeping software, an 8-channel scan and display takes around 132 µs to complete. That is, around 7500 scans per second.

Rather than polling for completion, the end of conversion can be used to generate an interrupt. In particular if a conversion is to be done in the Sleep mode then this interrupt can be used to waken the device. The ADC module can operate when the PIC MCU is in its Sleep state as it has the option of its own private oscillator to sequence the conversion; even if the system oscillator is disabled. The main advantage of a conversion while asleep is the electrically quiet environment when the system oscillator is off. Against this is the considerably longer conversion time, as when the PIC MCU is wakened, there will be the normal 1024-cycle delay to restart the system oscillator—see p. 319.

This personal oscillator may be used even where the PIC MCU is not put to sleep. However, as there is no synchronization between the system and local oscillators, clock feedthrough noise becomes a problem; especially with system clock rates above 1 MHz.

The following task list outlines the Sleep state conversion process.

- The ADC clock source must be set to CR, ADCS2 : 0 = X11.
- The ADIF flag must be cleared to prevent an immediate interrupt.
- The ADIE and PEIE mask bits must be set to enable the ADC interrupt to awaken the processor.
- The GIE mask bit must be 0 unless the programmer wishes the processor to jump to an ISR when it awakens.

Program 14.2 Scanning an 8-channel data acquisition system

```
; ************************************************************
; * FUNCTION    : Analog/digital conversion at channel n    *
; * ENTRY       : Channel number in W                        *
; * EXIT        : Digitized 8-bit analog value in W          *
; ************************************************************

GET_ANALOG rlncf  WREG,w     ; Shift channel number left >>2
           rlncf  WREG,w     ; with outcome in W
           bcf    ADCON0,CHS0; Zero channel bits
           bcf    ADCON0,CHS1
           bcf    ADCON0,CHS2
           bcf    ADCON0,CHS3
           addwf  ADCON0,f   ; Moves channel number to ADCON0[5:2]
           bcf    INTCON,GIE ; Disable all interrupts
           bcf    PIR1,ADIF  ; Ensure that AD Int flag is 0 before
           bsf    ADCON0,GO  ; Starting the conversion after 2TAD

           sleep             ; Doze in peace while converting

           bsf    INTCON,GIE ; Re-enable interrupts (optional)
           movf   ADRESH,w   ; Fetch datum when GO/NOT_DONE is zero
           return
```

- The GO/$\overline{\text{DONE}}$ bit in the ADCON0 register must be cleared to initialize the conversion; followed immediately by the `sleep` instruction.
- On wakening, the ADRESH:L registers hold the digitized value.

For instance, consider a Sleep state version of the GET_ANALOG subroutine of Program 14.1. This time the initialization code must set up the interrupt system as specified in the task list, to ensure that when the AD Interrupt Flag ADIF is set at the end of the conversion (at the same time as the GO/$\overline{\text{DONE}}$ flag goes to 0) the PIC MCU is woken up.

```
include "p18f4520.inc"

config WDT = OFF, PBADEN=OFF

movlw    b'00000111' ; Make pins AN7:0 analog
movwf    ADCON1      ; I.e. shared Ports A&E all analog
clrf     TRISB       ; All Port B Output
movlw    b'11111000' ; Low 3 bits of Port D are Outputs
movwf    TRISD

movlw    b'00000001' ; CH0 (0000) for 1st conversion
movwf    ADCON0      ; No conversion (0), ADC turned on (1)

movlw    b'00010101' ; Right justified (0), 4TAD delay (010)
movwf    ADCON2      ; FAD = FOSC/16 (101)

bcf      PIR1,ADIF   ; Zero the AD interrupt flag
bsf      PIE1,ADIE   ; Enable AD interrupts
bsf      INTCON,PEIE; & enable the Peripheral interrupt group
```

Apart from the initialization of the interrupt system, the only change is to the setting of ADCON2[3:0], which is made b'111' to select the internal CR ADC oscillator as the clock.

The Sleep version of GET_ANALOG shown in Program 14.2 is virtually identical to the original version, with the following changes.

1. GIE may need to be cleared if other devices can request an interrupt.
2. Before the conversion is started, the ADIF flag is cleared to ensure that the Sleep state is not prematurely terminated.
3. A `sleep` instruction follows the setting of the GO/$\overline{\text{DONE}}$ switch. In our example a $T_{AD} \times 2$ delay is inserted before the conversion proper starts. The internal clock gives a $T_{AD} = 1.2$ μs step rate for this conversion for the PIC18FXX20 (2.5 μs for the PIC18FXXLF20 line). If a zero fixed acquisition time is selected, a delay of one instruction cycle is added before the ADC clock starts, to allow the `sleep` instruction to be executed before starting a conversion.
4. There is no need to poll the GO/$\overline{\text{DONE}}$ status flag, as the PIC MCU will only restart after the conversion has completed and will then execute the following instruction. In our example the GIE mask bit has been cleared, and it should then be set again to 1 if there is to be interrupt activity from other sources. If GIE is permanently left at 1 then the processor will automatically jump to an ISR after it awakens.

For our final example we are going to code a 20 MHz PIC18F4520 in CCS **C** to act as a magnitude comparator in the manner of Example 11.2 on p. 364. Here we want to measure up the parallel-input 8-bit word N at Port B against an analog input at Channel 0. Outputs at pins RC2 : 0 are to represent Analog Lower Than N (b'001'), Equivalent (b'010') and Higher Than N (b'100') respectively. The comparator is to have a hysteresis of $\Delta = \pm 1$ bit; called delta in our program. That is, if a previous comparison showed Analog $< N$ then the new trigger level is $N + 1$. Similarly, on a downward trajectory the trigger level is decreased to $N - 1$.

The function compare() of Program 14.3 assumes that initialization code of the form:

```
#include <18f4520.h>
#byte PORT_B = 0xF81
#byte PORT_C = 0xF82
#device ADC=8               /* Configure for an 8-bit outcome  */
/* Declare function to which is sent delta (+1 or -1)
and which returns updated value +1 or -1                       */
unsigned int compare(unsigned int delta);

int main()
{
unsigned int hysteresis = 0;
set_tris_c(0xF8);
setup_adc(ADC_CLOCK_DIV_16);
setup_adc_ports(AN0);
set_adc_channel(0);
```

has already been executed.

The key internal functions used here are:

setup_adc(ADC_CLOCK_DIV_16)

This function configures bits ADCS2:0 in ADCON2[2:0] to select the module's clock source; here the processor oscillator/16. The internal CR oscillator is selected with the script ADC_CLOCK_INTERNAL.

setup_adc_ports(AN0)

This configures bits PCFG3:0 in ADCON1[3:0] to select which port pins are analog, which are digital, and if external reference voltages are to be used. The script AN0 indicates that port pin RA0 is to be analog with internal reference voltages; with the rest being digital; PCFG3:0 = b'1110'—see Fig. 14.12. The equivalent script using an external V_{REF+} at RA3 is AN0 | VSS_VREF. Scripts appropriate to any particular device are stored in the corresponding header file; in this case 18f4520.h. For instance, for the situation outlined in Program 14.1, the script AN0_TO_AN7 would be applicable. All devices with an ADC module have scripts ALL_ANALOG and NO_ANALOGS; for instance see Program 9.3 on p. 293.

set_adc_channel(n)

This is used to set up the channel number bits CHS3:0 in ADCON0[5:2].

Program 14.3 A digital/analog comparator with hysteresis

```
unsigned int compare(unsigned int delta)
{
unsigned int analog;
analog = read_adc();
if(analog > PORT_B + delta) {PORT_C = 0x04; delta = 0xff;}
else if(analog == PORT_B) {PORT_C = 0x02;}
else {PORT_C = 0x01; delta = 1;}
return delta;
}
```

read_adc()

This activates GO/$\overline{\text{DONE}}$ in ADCON0[1] and returns with the digitized value from ADRESH:L when GO/$\overline{\text{DONE}}$ goes to 0.

#device ADC=8

This directive configures the ADC module to left align the 10-bit outcome (see Fig. 14.15) and is used by the function read_adc() to return an 8-bit int, which it gets from ADRESH. The directive device ADC=10 returns a long int from ADRESH:L.

The function compare() in Program 14.3 expects the value of the hysteresis, called delta, which here is either +1 or −1 (h'FF'). After the ADC module is read, the digitized value analog is compared with the contents of Port B plus delta and the three Port C bits RC2:0 set to their appropriate state.

At the same time as the comparison is resolved, delta will be updated to reflect the outcome (i.e., +1 if analog < (PORT_B + delta), −1 if analog > (PORT_B + delta)). The value delta is returned by the function to allow the caller function to update its variable; called, say, hysteresis. Thus to activate the comparator outputs and also update hysteresis at the same time, the caller might have a statement such as hysteresis = compare(hysteresis);. An alternative would be to define the variable hysteresis before the main function main(), making it global; that is, known to all functions. In this situation its value need not be passed by the caller back and forth to any appropriate function.

Conversion from a digital quantity to an analog equivalent is somewhat simpler than the converse and not so commonly required. Perhaps for these reasons digital-to-analog converters (DACs) are not often found as an integral function in most MCU families. We have already seen that one way of providing this mapping is to vary the mark:space ratio of a pulse train of constant repetitive duration; as shown in Fig. 13.9 on p. 477. Here a small digital number gives a skinny pulse, which when smoothed out by a low-pass filter (which gives the average or d.c. value) translates to a low voltage. Conversely, a large digital number leads to a correspondingly large mark:space ratio; which in turn, after smoothing, yields a higher voltage.

PWM conversion can be very accurate and is simple to implement. However, extensive filtering is required to remove harmonics of the pulse rate and this makes the conversion slow to respond to changes in the digital input. Normally PWM is used to control heavy loads, such as motors or heaters, where the inertia of these devices inherently provides the smoothing action. Furthermore, the pulsed nature of the signal is ideally suited to power control, activating thyristor firing circuits.

Another way is to switch in a tapping on a chain of resistors, each adding one least significant bit increment to the grand total. This is the principle used in the Comparator Voltage Reference module of Fig. 14.8. However, rather a lot of resistors are needed; e.g., 1024 for 10-bit resolution.

Many commercial DAC devices are available which can be controlled externally. Two examples were given in Figs. 12.3 and 12.5 on pp. 384 and 388, where the MCU transferred digital data in series. Here for completeness, we will look at an example where parallel data transfer is used.

The majority of proprietary devices are based on an R-2R ladder network, such as that shown in Fig. 14.16(a). Voltage appearing at any bit switch node emerges at the output node in an attenuated form. As our analysis will show, each move to the left attenuates this voltage b_i by 50%, which is the binary weighting relationship:

$$V = \sum_{i=0}^{N+1} b_i \times 2^i$$

for an N-bit word.

In Fig. 14.16(b), at node A looking to the left we see a resistance of R $(2R//2R)$ and the voltage b_0 is attenuated by two. As we move to the right the process is repeated, with each voltage divided by two. Thus, at node B the voltage $b_0/2$ is further divided by two as is voltage b_1, giving $V_B = b_0/4 + b_1/2$. As the network is symmetrical the resistance looking right at any mode is also 2R. This means that as seen from *any* digital switch, the total resistance is $2R + 2R//2R = 3R$. This is important, as the characteristics of a transistor switch, such as resistance, are dependent on current, and keeping this the same reduces error.

For clarity our analysis has been for three bits. This can be extended by simply moving the leftmost terminating resistor over and inserting the requisite number of sections. This does not affect the resistance as seen left of the mode, and therefore does not change the conditions of the rightmost sections. An inspection of our analysis shows that nowhere does the absolute value of resistance appear. In fact the accuracy of the analysis depends only on the R:2R ratio. While it is relatively easy to fabricate accurate rationed resistors on a silicon die, this is certainly not the case for absolute values. For this reason R:2R networks are the standard technique used for most integrated circuit DACs.

The Maxim MAX506 of Fig. 14.17 is an example of a commercial D/A converter (DAC). This 20-pin footprint device contains four separate DACs sharing a common external V_{REF}. Digital data is presented to the D7 : 0 pins and one of four latch registers selected with the A1 : 0 address inputs. Once this is done, the datum byte is loaded into the selected register n and appears at the corresponding output VOUTn.

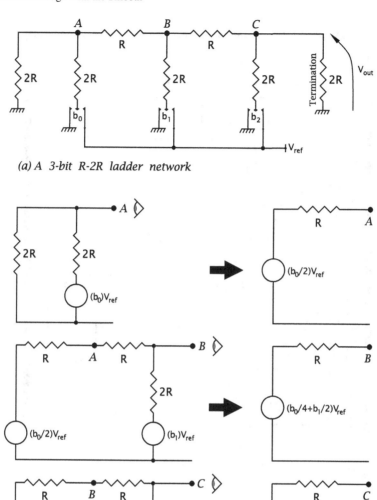

(a) A 3-bit R-2R ladder network

(b) Reducing the circuit

Fig. 14.16 R-2R digital-to-analog conversion

This output analog voltage ranges from zero (Analog GrouND) for a digital input of h'00' through to V_{REF} for a digital input of h'FF'. Where V_{SS} is connected to ground, then V_{REF} can be anything between 0 V and V_{DD} (+5 V). However, V_{SS} can be as low as −5 V and in this situation V_{REF} can be anywhere in the range ±5 V. If

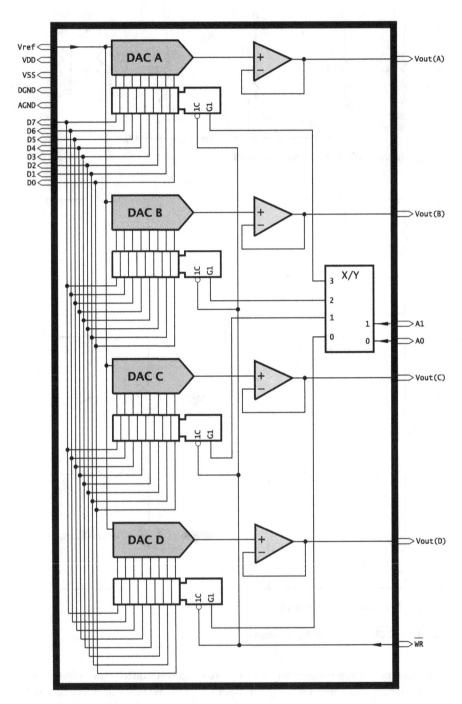

Fig. 14.17 The Maxim MAX506 quad 8-bit D/A converter

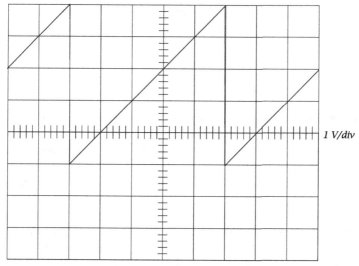

TIME BASE   0.1 ms/div

Fig. 14.18 Generating a continuous sawtooth using a MAX506 DAC

V_{REF} is negative for dual supplies then the output voltage will also be negative. In either case, effectively the output can be treated as the product $D \times V_{REF}$; where D is the digital input byte scaled to the range 0–1 (h'00–FF').

The MAX505 is a 24-pin variant which permits separate reference voltages to be used for each of the four DAC channels. In addition, the MAX505's DAC latches are isolated from the converter ladder circuits by a further layer of latches, all clocked at the *same* time with a $\overline{\text{LDAC}}$ ($\overline{\text{Load DAC}}$) control signal. This double buffering permits the programmer to update all four DACs simultaneously after their individual latches have been set-up.

As an example, consider that a MAX506 quad DAC has its Address selected via RA1:0 and RA2 drives the $\overline{\text{WR}}$ input to latch in the addressed data from Port B. We need to generate the continuous staircase sawtooth waveform shown in Fig. 14.18 from DACD. A suitable software routine would be something like the following listing:

```
        movlw   b'0111'     ; DACD is channel 3 (b'11'), WR = 1
        movwf   PORTA       ; To MAX506 WR, A1:0
LOOP    movwf   PORTB       ; Datum to MAX506's D7:0
        bcf     PORTA,2     ; WR = 0; Latch datum in
        bsf     PORTA,2     ; WR = 1; by pulsing WR
        addlw   1           ; Increment staircase count
        bra     LOOP        ; and repeat forever
```

where we are assuming that PORTB and PORTA[2:0] have been set-up as outputs.

The typical DAC staircase output waveform shown in the oscillogram in Fig. 14.18 is based on a 12 MHz crystal clocked PIC MCU. With a loop cycle count of six cycles gives a sawtooth duration of $(256 \times 6)/3 \approx 0.5$ ms, at 2 μs per step.

Examples

Example 14.1 The analog input channel voltage range for most ADC modules is limited to the supply voltage rails at most. Many situations require a digitized mapping from bipolar analog signals. Design a simple resistive network to translate a bipolar voltage range of ± 10 V to a unipolar range of 0–5 V, assuming V_{REF+} is $+5$ V. Extend the design to give an anti-aliasing filter, assuming a sampling rate of 5000 per second.

Solution One possibility is shown in Fig. 14.19. The value of the three resistors must be such that the input voltage V_{in} range of ± 10 V will be shifted so that the midpoint of 0 V gives half-scale ($V_{REF+}/2 = 2.5$ V) at the input pin AN. The range at this pin must also be attenuated by a factor of 4. A more general way of expressing this is given by the relationship $V_{out} = V_{in}/K + V_{REF+}/2$.

1. When V_{in} is zero, the voltage at the summing node is half-scale, which maps to b'10000000'. To do this, R_1 paralleled with R_2 must have the same resistance as R_3, i.e.,

$$R_3 = R_1 // R_2.$$

2. The attenuation of the network is a function of the potential divider between R_1 and $R_2 // R_3$. This gives us the value of K as:

$$K = (R_1 + (R_2 // R_3))/(R_2 // R_3) = 4.$$

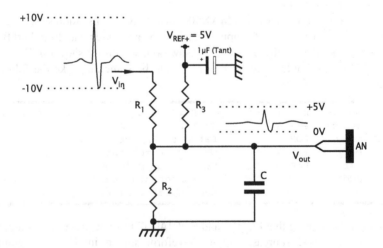

Fig. 14.19 A level-shifting resistor network

After some manipulation we have:

$$R_1 = (K/2 - 1) \times R_2,$$
$$R_2 = K/2 \times R_3.$$

Of course we have three unknowns and only two equations, so we have to start off by choosing a value for one of them. If we pick a value of 5 kΩ for R_3, then we have $R_2 = 2 \times 5 = 10$ kΩ and $R_1 = 10$ kΩ.

The resistance looking out from the pin is all three resistors in parallel; which in our case is 2.4 kΩ. This meets the maximum to keep within a LSB leakage error for a 10-bit conversion. For 8-bit resolution, the resistor values could be increased by a factor of four.

A small capacitor at the summing node can be used to implement a simple first-order low-pass filter to attenuate high frequencies from external sources, such as the MCU's system clock, and act as an anti-aliasing filter; as described in Fig. 14.4. With a sampling rate of 5000 per second, then ideally the filter break frequency should be no more than 2.5 kHz—half the sampling frequency. As this filter has an attenuation of only 6 dB per octave, choosing a break frequency $\frac{1}{2\pi CR}$ of 1 kHz provides a generous margin. We then have:

$$\frac{1}{2\pi CR} = 1000,$$
$$C = \frac{10^{-6}}{4.8 \times \pi},$$
$$C \approx 66 \text{ nF}.$$

To further reduce noise, the filter capacitor should have good high-frequency characteristics; e.g., polyester (capacitors become inductors at high frequencies) and together with the resistors, be physically as close as possible to the pin and not adjacent to any digital lines. It is always good practice to decouple the reference voltage and power supply with a low-value Tantalum electrolytic capacitor or/and a 0.1 μF ceramic capacitor to reduce switching noise from the MCU and other devices taking power from the same source. Using a separate supply and ground connection to the power supply to the PIC MCU should also reduce noise from this source.

Example 14.2 As part of a smart biomedical monitor, the peak analog value of an electrocardiogram (EKG) signal is to be determined anew for each cycle. This R-point (see Fig. 7.1 on p. 206) maximum value is to be output from Port B and RA0 is to be pulsed High whenever this value is being updated. Assuming that a 16 MHz PIC18F1220 is used to implement the intelligence, and the EKG signal (conditioned as shown in Fig. 14.19) is connected to Channel 1 RA1, devise a possible strategy. Timer 0 is being used to interrupt the processor at nominally 2000 times per second—see Program 13.3 on p. 464. Design a suitable ISR to implement your strategy.

Solution As in any biomedical parameter, the EKG signal will vary from cycle to cycle in gain, shape, and period. Even if this were not so, imperfections in the data

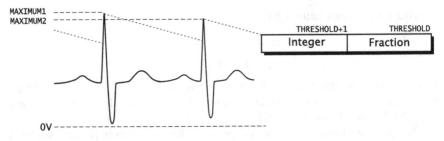

Fig. 14.20 EKG detection strategy

acquisition system, notably the skin electrodes, can cause slow baseline (d.c.) drift. Thus the threshold at which the signal is to be tracked to its peak R-value must be reset at some sensible fraction of its previous peak during the period following the last update.

One possibility is shown in Fig. 14.20. Here the threshold is slowly decremented after the peak to ensure that a following peak of lower amplitude is not missed. On the basis of a lowest EKG rate of 40 beats per minute (period 1.5 s), if we reduce the threshold by $\frac{1}{64}$ of a bit every sample, then the maximum reduction would be a count of $\approx$47 at a sample rate of 2000 per second. To do this the threshold value THRESHOLD in Program 14.4 is stored as a double-byte number of form integer:fraction and $\frac{1}{64}$ of an integer (i.e., fraction = b'00000100') subtracted in each sample where the peak value MAXIMUM is not updated. This droop rate can be altered by changing the subtracted fraction.

The task list implemented by this listing is:

1. DO a conversion to get ANALOG.
2. IF (ANALOG > THRESHOLD)

   - MAXIMUM = ANALOG.
   - THRESHOLD = ANALOG.
   - PORTB = ANALOG.
   - RA5 = 1.

3. ELSE

   - Reduce THRESHOLD by $\frac{1}{64}$.
   - RA5 = 0.

In updating THRESHOLD (where ANALOG > THRESHOLD) the integer byte takes the new value of MAXIMUM whilst the fractional byte is zeroed. Treating this byte pair as a 16-bit word, this effectively equates the threshold as MAXIMUM $\times$ 256 or THRESHOLD = MAXIMUM << 8, where MAXIMUM has been shifted left eight places. We are assuming that THRESHOLD has been zeroed in the background program during the initialization phase and that we are doing an 8-bit conversion.

If the digitized analog sample is less than the threshold trip value then h'04' = b'00000100' is subtracted from the lower byte at THRESHOLD and if this produces a borrow, then the upper byte at THRESHOLD+1 is decremented. This subtract $\frac{1}{64}$

Program 14.4 EKG peak picking

```
; ****************************************************
; * FUNCTION: ISR to update the EKG parameters        *
; * ENTRY   : On a Timer0 interrupt                   *
; * EXIT    : Update MAXIMUM and THRESHOLD:THRESHOLD+1 *
; * RESOURCE: GET_ANALOG subroutine gets 8-bit digitized data*
; ****************************************************
EKG_ISR    btfss   INTCON,TMR0IF ; Was this a Timer0 interrupt?
           bra     EKG_EXIT     ; IF not THEN exit

           bcf     INTCON,TMR0IF ; ELSE clear flag
           movlw   1            ; Initiate a conversion of
           call    GET_ANALOG   ; Channel 1

           cpfslt  THRESHOLD    ; Skip IF Threshold < Analog
           bra     BELOW        ; ELSE don't update MAXIMUM

           movwf   MAXIMUM      ; Digitized byte is new MAXIMUM
           movwf   PORTB        ; made visible to outside
           bsf     PORTA,0      ; which is signaled
           movwf   THRESHOLD+1  ; Now update double-byte
           clrf    THRESHOLD    ; threshold
           bra     EKG_EXIT     ; and finish

; Land here if the input is below the threshold ---------------
BELOW      bcf     PORTA,0      ; Signal no update

; Now reduce the threshold by 1/64 unless it is zero ----------
           movf    THRESHOLD+1,f ; Is integer threshold zero?
           bz      EKG_EXIT     ; IF it is THEN leave alone

           movlw   h'04'        ; 1/64 = b'000001000'
           subwf   THRESHOLD,f  ; Take away from fraction byte
           bc      EKG_EXIT     ; Skip if no borrow
           decf    THRESHOLD+1,f ; ELSE decrement integer threes

EKG_EXIT   retfie  FAST         ; Return from interrupt
```

routine is skipped if the integer threshold has reached zero; thus preventing underflow.

Program 14.4 uses the subroutine GET_ANALOG of Program 14.1. As the PIC18F1220 has no CHS3 in ADCON0[5], this line of code should be removed.

Program 14.5 gives the **C**-coded version implementing our task list. The #int_timer0 directive tells the compiler to treat the following function as a Timer 0 ISR. In function ecg_isr(), the variables threshold and maximum are declared static. This means that their value will be retained after the function has exited and will thus be available next time on entry. The default way of treating **C** function variables is to hold their value only for the duration of the function. An alternative way of dealing with this problem is to declare such variables outside any function; in which case they will be global and retain their value during the run.

Program 14.5 An implementation of the EKG peak picker in **C**

```
#bit  RA0    = 0xF80.0       /* Pin RA0 is bit 0 of Port A       */
#byte PORT_B = 0xF81         /* Port B is File h'F81'            */

#int_timer0
void ecg_isr(void)
{
unsigned int analog;
static unsigned long int threshold = 0;
static unsigned int maximum;
analog = read_adc();

if(analog > threshold>>8)
  {
  maximum = analog;          /* New maximum value               */
  PORT_B = analog;           /* Show the outside world          */
  threshold = maximum << 8;  /* New 2-byte threshold            */
  RA0 = 1;                   /* Tell outside world              */
  }
else if(threshold >= 0x0004)/* IF threshold not less than h'0004'*/
  {
    threshold = threshold - 0x0004; /* THEN reduce by 1/64      */
    RA0 = 0;                 /* Signal no update                */
  }
}
```

The threshold variable is defined as a long int and the CCS compiler will then treat this datum as a 16-bit variable as required. The definition in equating threshold to zero will only initialize it once when the program begins its run, as the variable is static. Again this is not the normal behavior of a default auto variable.

In equating threshold to the new maximum value, the latter is multiplied by 256 by shifting left eight times. A good compiler will automatically change a N*256 to N<<8; or even better just take the upper byte of the pair as the outcome. This double-byte form allows for the reduction by $\frac{1}{64}$ of a bit h'0004' to give the specified falling trip level.

Example 14.3 A microcontroller is to be used to calculate a measure of power discharged by the diphasic defibrillator of Fig. 14.5. When the MCU detects the beginning of the discharge, 256 samples are to be taken at a nominal rate of 20,000 per second; with the sum of the squares of the deviation from the baseline voltage being an analog measure of the power—assuming that the resistance of the patient's chest/electrodes remains constant whilst all this is going on!

A 4.096 V voltage reference device is to act as an external reference voltage for a ADC module, giving a 16 mV resolution for an 8-bit conversion. After the process begins, pin RA4 is to be pulsed as a trigger for a storage oscilloscope, which allows the waveform to be captured for archiving purposes. When the process has been completed, the top byte of the power summation is to be output via Port B for display.

Show how you might use a 20 MHz PIC18F4520 device to implement the logic of the measurement system. You can assume that the voltage reference device can be biased as for a Zener diode. In practice an optional potentiometer can be used to trim the voltage slightly for more accurate results.

Solution Figure 14.21 shows a suitable hardware configuration, from which we can estimate the peripheral budget. The signal itself ranging between +1.8 and +3.6 V (see Fig. 14.5) is connected to Analog channel 0 at pin RA0/AN0. The 10 kΩ resistor protects the analog input against overvoltage as well as implementing an anti-aliasing filter, with the 3.3 nF capacitor giving a 450 kHz nominal breakpoint. As the actual defibrillator uses very large voltages (of the order of 25 kV) the two 1N4004 diodes act as additional protection against high-voltage spikes, supplementing the internal diodes shown in Fig. 14.13.

The external 4.096 V source is directly connected to pin RA3; which will require VCFG0 in ADCON1[4] to be set to use this as the positive reference voltage. Both the V_{DD} and V_{REF+} voltages are decoupled with 1 μF tantalum capacitors to reduce noise at this point.

An internal analog comparator is used to detect the initial rise of the discharge voltage as described in Fig. 14.5. If Comparator module mode b'110' is used, then the CVR module can be used to generate an internal reference voltage as described

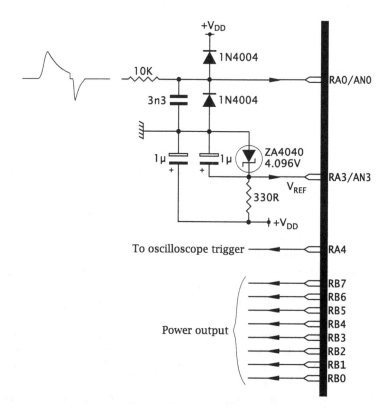

Fig. 14.21 Measuring the discharge power for an EKG defibrillator

on p. 503. With the CIS bit cleared (see Fig. 14.7) Comparator 1 can share Analog channel 0 with the ADC module.

Finally, both RA4 and all of Port B must be configured as digital outputs. The former is going to be used to generate a synchronization pulse, and the latter to output the end result of the analysis.

```
            org     0            ; Reset vector
            bra     SET_UP       ; Go to the background routine
            org     8            ; Interrupt vector
            goto    ECG_ISR      ; Service a Comparator interrupt

SET_UP movlw   b'00000110'  ; Comparator mode 110 CIS = 0
            movwf   CMCON
            call    DELAY_10US   ; Allow 10us for voltages to settle
            movf    CMCON,f      ; Read CMCON clears any Change state
            bsf     PIE2,CMIE    ; Enable Comparator interrupts

            movlw   b'10001110'  ; CVR on (1), not ext (0), Hi range
            movwf   CVRCON       ; Supply ref (0), CVR3:0 = 1110

            movlw   b'101111'    ; Make RA4 an output
            movwf   TRISA
            clrf    TRISB        ; PortB is all output

            movlw   b'00000001'  ; Turn on ADC module, Channel 0
            movwf   ADCON0
            movlw   b'00011110'  ; RA0 analog input
            movwf   ADCON1       ; RA3 is VREF+ input
            movlw   b'00010101'  ; Clock/16, 4TAD sampling delay
            movwf   ADCON2

            bcf     PIR2,CMIF    ; Zero the Comparator Interrupt flag
            bsf     INTCON,PEIE  ; Enable Peripheral Interrupt group
            bsf     INTCON,GIE   ; & Globally enable interrupt system
```

Based on our analysis, the initialization code is shown above. The modules are set-up as follows:

1. The Analog comparator module is turned on in Mode b'110' with CIS = 0. A 10 μs delay is also employed to allow the module to settle. After this, the CMCON register is read to clear any Change condition. This allows the CMIF flag to be subsequently zeroed and interrupts enabled.
2. The CVR module is enabled and set to tapping b'1110' in the high range to give a 3.4375 V reference.
3. The ADC module is enabled and configured to use pin RA0 as an analog channel and to use RA3 for the external V_{REF+}. Alignment is set-up to facilitate an 8-bit outcome. The ADC clock is sourced as the system 20 MHz frequency divided by 16 and a 4 × T_{AD} sampling delay.
4. PORTA[4] is set-up as an output, as required for AN0. Other Port A pins are left as inputs. All of Port B is configured as output.

Program 14.6 Gauging the defibrillator discharge power

```
MAIN        sleep                    ; Idle
            movff   POWER,PORTB ; Output top byte of power
            bra     MAIN


; **********************************************************
; * FUNCTION: ISR to begin the defibrillator analysis      *
; * ENTRY   : On a Comparator module interrupt             *
; * EXIT    : Update POWER:3                                *
; * RESOURCE: GET_ANALOG subroutine gets 8-bit digitized data *
; * RESOURCE: SQUARE subroutine does 8x8 multiplication     *
; **********************************************************
ECG_ISR     btfss   PIR2,CMIF   ; Was this a Comparator interrupt?
            bra     ECG_EXIT    ; IF not THEN exit

            clrf    POWER+2     ; Zero the 3-byte grand total
            clrf    POWER+1
            clrf    POWER       ; LSB
            clrf    COUNT       ; Prepare to do loop 256 times

            bcf     PORTA,4     ; Pulse RA4
            bsf     PORTA,4     ; to generate a synch pulse
            bcf     PORTA,4

ACQUIRE     clrf    WREG        ; Channel 0 (W is h'00')
            call    GET_ANALOG ; Do a conversion
            addlw   -BASELINE   ; Difference from baseline voltage
            call    SQR         ; Square it

            movf    SQUARE,w    ; Get LSB of squared voltage
            addwf   POWER,f     ; Add it to the low byte of Power
            bnc     NEXT_BYTE   ; IF no Carry THEN next byte
            incf    POWER+1,f   ; Increment the high byte of Power
            bnc     NEXT_BYTE   ; IF no Carry THEN next byte
            incf    POWER+2,f   ; IF yes THEN increment upper byte

NEXT_BYTE   movf    SQUARE+1,w ; Get MSB of squared voltage
            addwf   POWER+1,f   ; Add it to the high byte of Power
            bnc     CONTINUE    ; IF no Carry THEN finished
            incf    POWER+2,f   ; THEN increment the Upper byte

CONTINUE    call    DELAY_460US; Wait around for the next sample
            incfsz COUNT,f      ; Increment the loop count and do
            bra     ACQUIRE     ; another acquisition if not zero

ECG_EXIT    movf    CMCON,f     ; Change situation
            bcf     PIR2,CMIF   ; and clear the interrupt flag
            retfie FAST         ; and return from interrupt
```

The actual software itself is shown in Program 14.6. The Main routine simply sleeps until the Comparator module changes state and generates an interrupt. When control returns to the background routine, the top byte of the triple-byte Power accumulator is copied to Port B and the process repeated for the next run.

The foreground routine first confirms the source of the interrupt and then clears a loop counter and the three bytes used to store the grand sum of the 256 squares of the sampled voltage. Pin RA4 is then pulsed to tell the outside world that the discharge is beginning.

The GET_ANALOG subroutine listed in Program 14.1 is used to acquire an 8-bit digitized sample. The difference between the baseline voltage of 2.6 V (see Fig. 14.5) is then determined. This signed voltage is then squared using the SQR routine of Program 8.3 on p. 259. The two global return bytes SQUARE+1:SQUARE are then added to the triple-byte total POWER+2:POWER+1:POWER array.

This is repeated 256 times, with an extra loop delay of 460 µs to give an approximate 500 µs total delay necessary to give the specified 2 kHz sampling rate. If a greater sampling rate accuracy is required, then a Timer can be used to generate a High-priority interrupt at this point. When this has been completed, taking a total time of around 128 ms, the Comparator module is read to clear the difference condition. This is done at the end of the process, rather than at the beginning, as the input voltage will fall back through the 3.4375 V Comparator threshold part way through the process and trigger another change! The CMIF flag is then cleared and the ISR quit.

Of course this is rather rudimentary. For instance, the baseline voltage may vary with time, so a learning run prior to an analysis may be necessary. If fairly stable, this value can be burnt into non-volatile memory as described in the next chapter. The use of a fixed number of samples can be restrictive, and additional loops can be implemented until the voltage difference drops below a certain threshold.

Example 14.4 Using C coding, show how a 10-bit digitized reading from Channel 3 of a PIC18F4520 can be acquired with the processor in its Sleep state.

Solution The CCS compiler uses the sleep() function to put the MCU to sleep; this simply translates to the sleep instruction. A Sleep conversion cannot be implemented using the read_adc() function of Program 14.3 as no processing is done in the Sleep state. Instead we need to set and clear individual interrupt related bits before going to sleep in the manner outlined in the assembly-level Program 14.2. On wakening the state of ADRESH:L registers can then be read 'manually' and combined to give the 10-bit outcome.

Coding for this specification is shown in Program 14.7. Here the PEIE, ADIF and GO/DONE bits are defined using the #bit directive. This time the script ADC_CLOCK_INTERNAL is used with the setup_adc() internal function to select the internal CR clock for the DAC module; as necessary for the Sleep conversion.

The internal function disable_interrupts(GLOBAL) clears *both* GIE and PEIE mask bits. The complementary enable_interrupts(GLOBAL) sets both bits, but we need to enable the PEIE only and leave GIE cleared. This is implemented by the 'bit-twiddling' statement PEIE=1;. Similarly, clearing the ADIF

Program 14.7 Sleep conversion in **C**

```
#include <18f4520.h>
#device ADC=10            /* Configure for a 10-bit outcome          */
#use delay(clock=8MHZ)    /* Tell compiler its an 8MHz clock        */

#bit ADIF = 0xF9E.6       /* The A/D Interrupt Flag in PIR1[6]      */
#bit PEIE = 0xFF2.6       /* The group interrupt flag in INTCON[6]  */
#bit GO   = 0xFC2.1       /* The Go/NOT_DONE bit in ADCON0[1]       */

#byte ADRESH = 0xFC4      /* The Result registers                   */
#byte ADRESL = 0xFC3

main()
{
unsigned long int result; /* 16-bit digitized outcome              */
set_tris_a(0x0E);
setup_adc(ADC_CLOCK_INTERNAL|ADC_TAD_MUL_2);
setup_adc_ports(AN0_TO_AN3);
set_adc_channel(3);
disable_interrupts(GLOBAL);/* Disable all ints (GIE & PEIE=1)      */
ADIF = 0;
enable_interrupts(INT_AD);
PEIE = 1;                  /* Enable the auxiliary group interrupts */
/*    Code                                                          */
GO = 1;
sleep();

result = ((long)ADRESH<<8)+ADRESL; /* When awake read each byte */
return;
}
```

flag is directly actioned by ADIF=0;. Before calling sleep(), the statement GO=1; manually starts the conversion. After sleep() the ADRESH register is read and cast to a long int to ensure that the compiler treats it as a 16-bit object. Multiplying by 256 tells an intelligent compiler to treat it as the top byte of a 16-bit object. Adding ADRESL puts this in the low byte of the 2-byte outcome.

Self-Assessment Questions

14.1 In Example 14.2 the decay of the threshold level was linear. Although this is fairly effective for situations where the nominal period is known *a priori* and does not vary greatly, an exponential decay would be better suited where this is not the case. To generate this type of relationship a fixed *percentage* of the value at each sample point should be subtracted to give the new outcome rather than a constant. Show how you could modify Programs 14.4 and 14.5 to decrement at a rate of $\frac{1}{4096}$ ($\approx 0.025\%$) on each sample and determine the time constant in terms of the number of samples.

14.2 Real-world analog signals are noisy. In practice this means that some form of filtering or smoothing is frequently required. In any circumstance, noise coming in from outside should not have any appreciable frequency components above half the sampling rate, since such noise will be frequency shifted back into the baseband; as shown in Fig. 14.4. Such low-pass filtering must be applied to the signal *before* the A/D conversion, as shown in Fig. 14.21.

Although this external anti-alias filter must by definition be implemented using hardware circuitry (such as a CR network), noise within the passband can be smoothed out using software filtering routines. One simple approach to digital filtering is to take multiple readings and average them to give a composite outcome. For example, 16 readings summed and shifted right four times ($\div16$) would reduce random noise by a factor of $\sqrt{16} = 4$—see also SAQ 6.9 on p. 203.

Another approach well known to statisticians, is to take a moving average; for example, of a stock price over a month interval. A comparatively efficient algorithm of this type is a 3-point average:

$$\text{Array[i]} = \frac{S_n}{4} + \frac{S_{n-1}}{2} + \frac{S_{n-2}}{4},$$

where S_n is the nth sample from the analog module.

Show how you could modify the GET_ANALOG subroutine to remember the last samples from the two previous calls and return the smoothed value.

14.3 It has been proposed that as part of the EKG monitor of Example 14.2 that the MAX506 DAC of Fig. 14.17 be used to introduce an automatic gain control (AVC) function preceding the PIC MCU's analog input. The aim of the AVC is to keep the peak of the analog input between $\frac{3}{4}$ and $\frac{7}{8}$ full scale. How might you go about implementing this subsystem? *Hint*: Recall that the output of each channel of a MAX506 is the product of its digital input and V_{REF}, and that the latter can vary between 0 V and V_{DD}.

14.4 An input analog sinusoid signal, conditioned as shown in Fig. 14.19, is to be full-wave rectified; that is, voltages that were originally negative are to have their sign changed. Design a routine to do this, assuming that the 8-bit digitized input voltage is available at ADRESH and the processed output is to be presented via Port B to a DAC.

14.5 Figure 14.22 is based on Fig. 10 of Microchip's application note AN546 *Using the Analog-to-Digital (A/D) Converter*, as a means of providing an external voltage reference source for power-sensitive applications. How do you think the circuit works and what factors govern the choice of current limiting resistor?

Fig. 14.22 A controllable
external voltage circuit

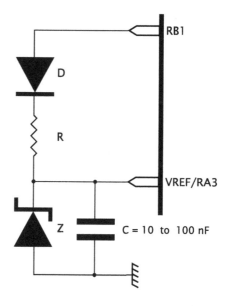

Chapter 15
To Have and to Hold

Many applications involving digital circuitry involve the storage of semi-permanent information. Such **non-volatile** data is distinguished from that held in RAM, in that it will remain unaltered during periods when the system is powered down. Typically the nature of such data will be look-up tables (e.g. see Program 6.13 on p. 196), alphanumeric strings and data gathered from the outside world. A good example of this is the odometer tally in a car; which needs to be retained in the absence of a power supply—see Example 12.3 on p. 442. Although this facility can be implemented using an external EEPROM memory, such as the 24XXX of Fig. 12.28 on p. 443; where only a modest amount of non-volatile data needs to be kept, integral EEPROM storage increases reliability and reduces cost, size, and power requirements.

Our objective here is to examine these integral non-volatile storage facilities. After reading this chapter you will:

- Be familiar with the characteristics of the EEPROM Data module.
- Know how to both read and write data to the EEPROM Data module.
- Understand how the main Flash EEPROM Program memory can be used to retrieve, and in some devices to store, non-volatile data.
- Be able to contrast the EEPROM Data module and Flash Program memory as a location for non-volatile data.

The now obsolete PIC16C84, introduced in 1994, had several firsts. The Program store used electrically erasable PROM technology, which meant that UV radiation was not needed to erase data—see Fig. 2.13 on p. 27. Along with this innovation, an EEPROM Data peripheral module was featured; which enabled the programmer to store up to 64 bytes of non-volatile data independently of the normal Data store.

The PIC16C84 and its analogous Flash EEPROM Program store successor, the PIC16F84, remained the only EEPROM family member until the introduction of the PIC16F87X in 1998. This innovative device and many of its successors, including the PIC18 family, blurred the Harvard architecture's separation between Program and Data, by allowing data to be read from the Program store and in some devices, to be written into it.

S. Katzen, *The Essential PIC18® Microcontroller,*
Computer Communications and Networks,
DOI 10.1007/978-1-84996-229-2_15, © Springer-Verlag London Limited 2010

Before examining the details, it is instructive to look at an application requiring the use of non-volatile storage. A good example of this is the smart card of Fig. 12.1 on p. 380. Here we need to store, amongst other things, the card account number, PIN number, start and expiry dates. Some of this data, such as the account number, is essentially fixed. Security data may need to be altered occasionally by the user from a terminal. If the card is used as a cash card, its credit will need to be charged via an ATM and discharged when payments are made. The size and cost sensitivities of a smart card processor is such that *integral* EEPROM data storage is vital.

Figure 15.1 shows the logic organization of the PIC18 EEPROM Data module.[1] The memory matrix is not part of the normal Data and Program stores but is indirectly accessed via four SPRs, which address the target byte, collect/hold data, and control the Read and Write processes.

EEPROM Matrix

The enhanced-range EEPROM Data module architecture supports up to 256 byte-sized cells. Key features are:

- 1,000,000 typical (10^5 minimum) Erase/Write cycle endurance for each cell at 5 V and 25°C.[2]
- Maximum Erase/Write cycle time 4.86 ms; 4.11 ms typical.
- Data retention greater than 40 years.

EEADR

The **EEPROM ADdRess Register** addresses the target cell holding the EEPROM data.

EEDATA

The **EEPROM DATA Register** either holds the 8-bit datum read out of the addressed cell or the byte the programmer wishes to write to the target EEPROM cell.

EECON1

The EEPROM Data module has two modes of operation, with **EEPROM CONtrol Register 1** controlling and monitoring the Read and Write actions. EECON1 is also used to switch between the EEPROM Data module, Flash EEPROM or Configuration data (such as the Fuses).

RD

The **ReaD control** bit initiates a Data EEPROM read action. After one instruction cycle RD will automatically be cleared. This bit is not used when accessing the Program store EEPGD = 1 or CFGS = 1.

WR

Setting the **WRite control** bit initiates a Data EEPROM or Flash memory erase or write operation. WR automatically clears when the process has completed.

---

[1] A few family members have an extended architecture. For instance, the PIC18F2620 supports 1 kbyte.

[2] Compare with 10,000 to 100,000 for Flash Program memory.

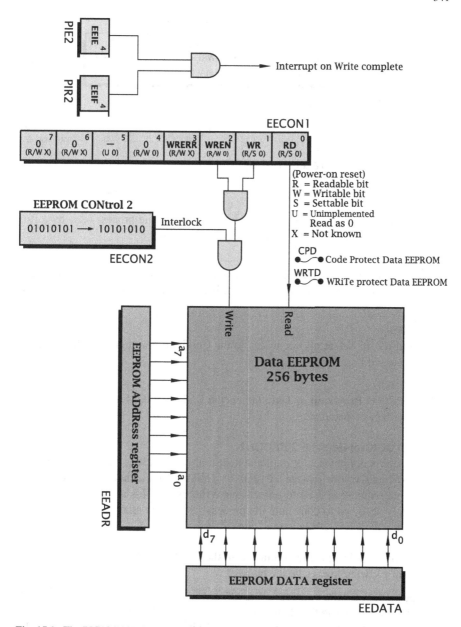

Fig. 15.1 The PIC18 EEPROM Data module

WREN

WRite ENable enables a write process by gating WR. This reduces the probability of an erroneous destruction of data.

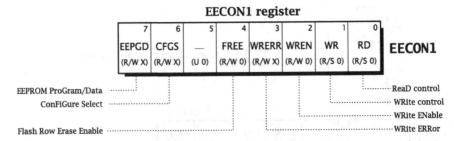

Fig. 15.2 The PIC18FXX20 EECON1 register

WRERR

WRite ERRor signals a prematurely terminated write process (e.g. by a Reset). When 0, a successful write process is signaled.

FREE

Flash Row Erase Enable causes a row of Program memory to be erased whenever a write process is actioned. FREE is cleared automatically on completion of the process.

CFGS

ConFiGure Select allows access to the Configuration region of the Program store—see Fig. 15.4.

EEPGD

EEPROM ProGram or Data targets the Data EEPROM module when 0 otherwise Flash memory.

EEPROM CONtrol Register 2 EECON2

This register is not physically implemented; it always reads as zero. Rather the action of writing the code pattern b'01010101' followed immediately by b'10101010' *with no interruption* is used to unlock the Write cycle. This arcane incantation is deliberately designed to convolute the process, as security against accidental alterations in the data. EECON2 gives all zero if read.

In order to read a specified datum from the EEPROM Data module we have to implement software to execute the task list:

1. Copy the target cell's address to EEADR.
2. Set RD to 1 to initiate the Read cycle.
3. RD is automatically cleared and the target 8-bit datum can be read from EEDATA any time from the next instruction cycle, as convenient.

Subroutine READ_EEPROM in Program 15.1 directly implements this process and illustrates the return of the datum from the EEPROM cell via the Working register. The datum will remain in EEDATA until the register is reused.

Writing data to the EEPROM Data module is deliberately made more Byzantine to reduce the chance of a spurious Write corrupting the data due to a software bug or

Program 15.1 Retrieving a byte from the EEPROM Data module

```
; **********************************************************
; * FUNCTION: Gets one byte from the EEPROM Data module    *
; * ENTRY   : Address in EEADR                             *
; * EXIT    : Datum in W and in EEDATA. System in Bank0    *
; **********************************************************
READ_EEPROM   movlw   b'00000001' ; Set RD for Read cycle
              movwf   EECON1      ; Read datum into EEDATA
              movf    EEDATA,w    ; Copy into W
              return              ; for return
```

processor malfunction because of, say, a power glitch. The task list to write a datum to a specified cell is:

1. Copy the target cell address to EEADR.
2. Set WREN in EECON1[2] to enable the Write process.
3. Disable all interrupts.
4. Send h'55' to EECON2.
5. Send h'AA' to EECON2.
6. Set WR to initiate the Write cycle.
7. Clear WREN.
8. Enable interrupts.
9. Wait until WR returns to zero, signaling the completion of the Write cycle, and exit.

The Write cycle will not initiate if the interlock sequence in items 4–6 is not exactly followed without interference. For instance, an interruption during the interlock sequence will abort the Write cycle. Thus, in this situation interrupts should be disabled by clearing GIE until the Write cycle has been initiated.

If desired, the completion of the Write cycle can be used to interrupt the processor. This is enabled by setting the EEIE mask bit. When the interrupt flag EEIF, is set on completion of the Write action, then the interrupt is generated in the normal way. EEIF should be cleared in the ISR.

It is possible that the processor is reset; for instance, by a Watchdog overflow, before the Write cycle is complete. In this situation, the EEPROM datum may be corrupt. The WRERR flag in EECON1[3] will be set if the Write operation has been prematurely terminated with a Reset action. If this is not the case, when the cycle is complete the datum may be read back and verified to give extra security. The WREN bit may be cleared at this point to help prevent an accidental Write. Doing this before the Write is complete will not affect the operation.

Program 15.2 implements this task list. Both EEDATA and EEADR are set up by the caller program with the byte datum and address. The subroutine is not exited until the Write cycle has completed; ≈4 ms. This ensures that these SPRs will not be altered during the cycle, which may possibly give an erroneous outcome.

In order to illustrate these concepts, we will repeat Example 12.3 on p. 442, which incremented a non-volatile triple-byte odometer total in external serial EEPROM, but this time using the internal Data EEPROM module. We will assume that the odometer count is located at EEPROM cells h'10–12'.

Program 15.2 Putting a byte into the EEPROM Data module

```
; ************************************************************
; * FUNCTION: Writes one byte into the EEPROM Data module   *
; * ENTRY   : Datum byte in EEDATA, module address in EEADR *
; * EXIT    : Interrupts disabled for 7 instructions        *
; ************************************************************
WRITE_EEPROM
        clrf    EECON1
        bsf     EECON1,WREN ; Enable for Write cycle
        bcf     INTCON,GIE  ; Disable all interrupts

        movlw   h'55'       ; Now do the interlock
        movwf   EECON2
        movlw   h'AA'
        movwf   EECON2

        bsf     EECON1,WR   ; Initiate the Write cycle
        bcf     EECON1,WREN ; Optionally disable any other Writes
        bsf     INTCON,GIE  ; Re-enable interrupts

EE_FINI btfsc   EECON1,WR   ; Check, has the Write completed?
        bra     EE_FINI     ; IF not THEN retry
        return              ; and return when cycle has finished
```

The coding shown in Program 15.3 makes use of the two subroutines READ_ EEPROM and WRITE_EEPROM to read and subsequently write the three odometer bytes from/to the EEPROM module. The address of the first (highest) byte is copied into EEADR at the beginning of the subroutine and is subsequently incremented and decremented *in situ* to point to the appropriate datum.

After the 3-byte odometer state has been fetched and copied into memory it is incremented in exactly the same manner as in Program 12.20 on p. 445. The augmented array is then written back into EEPROM in the opposite sense as it was read, with EEADR being decremented. The WRITE_EEPROM subroutine checks that the Write cycle has been completed before returning and thus timing need not be checked by the calling program.

Program 15.4 gives an equivalent coding in CCS **C**. The approach here is to build up a 3-byte object, called mile, from the three individual EEPROM bytes. After incrementing the mileage, it is taken apart into its constituent bytes, which are put back into the Data EEPROM. CCS **C** has special library functions which facilitate reading from and writing to Data EEPROM, as well as putting together and taking apart bytes which are more efficient than shifting and ANDing with standard **C**. Such functions used in this program are:

read_EEPROM(addr)
This function returns an 8-bit integer, being the contents of the Data EEPROM cell who's address has been passed to the function.

write_EEPROM(addr,value)
A single byte datum is written into Data EEPROM at the specified address.

Program 15.3 Incrementing the non-volatile odometer count in Data EEPROM

```
; ************************************************************
; FUNCTION: Adds one onto the triple-precision odometer total *
; RESOURCE: Subroutines WRITE_EEPROM and READ_EEPROM          *
; ENTRY   : Current total in EEPROM module at h'10:11:12'      *
; EXIT    : Incremented total back in EEPROM module            *
; EXIT    : also available in RAM at MSB:NSB:LSB               *
; ************************************************************
EXTRA_MILE ; Get the three bytes from the Data EEPROM ---------
           movlw  h'10'         ; Address of high-byte odometer total
           movwf  EEADR         ; Copy into EEPROM address register
           call   READ_EEPROM   ; Read byte from EEPROM module
           movwf  MSB           ; and put into File register MSB

           incf   EEADR,f       ; Address of middle byte odometer
           call   READ_EEPROM   ; Read byte from EEPROM module
           movwf  NSB           ; and put into File register NSB

           incf   EEADR,f       ; Address of low byte odometer
           call   READ_EEPROM   ; Read byte from EEPROM module
           movwf  LSB           ; and put into file register LSB

; Now increment 3-byte array -----------------------------------
           incfsz LSB,f         ; Add one & skip IF zero
            bra   PUT_BACK      ; IF not THEN continue
           incfsz NSB,f         ; Increment middle byte
            bra   PUT_BACK      ; IF not zero THEN continue
           incf   MSB,f

; Put the augmented odometer count back in Data EEPROM -------
PUT_BACK movff  LSB,EEDATA    ; Copy new odo low byte to EEDATA
           call   WRITE_EEPROM ; Write to EEPROM cell h'12'

           decf   EEADR,f       ; Address of middle byte
           movff  NSB,EEDATA    ; Get new odo mid byte to EEDATA
           call   WRITE_EEPROM ; Write to EEPROM cell h'11'

           decf   EEADR,f       ; Address of high byte
           movff  MSB,EEDATA    ; Get new odo low byte to EEDATA
           call   WRITE_EEPROM ; Write to EEPROM cell h'10'

           return
```

make32(val1, val2, val3, val4)

This function builds up a 32-bit number out of up to four 8- and 16-bit numbers. Up to four parameters are passed, with the rightmost being the lower byte or word of the returned value.

In standard **C** this function is equivalent to a process of shifting and addition. For instance, the make32() function in our program is equivalent to:

```
mile = (read_eeprom(0x10)<<16)
       +read_eeprom(0x11)<<8 read_eeprom(0x12);
```

There is also an analogous make16() function to build up a 16-bit word from two bytes.

make8(value,offset)

This function is the reverse counterpart of make32() int that it returns a sin-

---

Program 15.4 Incrementing the odometer in **C**

```
void extra_mileage(void)
{
int32  mile;                           /* 32-bit variable        */
/* Build up a 3-byte word from three individual EEPROM bytes    */
mile=make32(read_eeprom(0x10),read_eeprom(0x11),read_eeprom(0x12));
mile++;                                /* One more mile (or km)  */
write_eeprom(0x12,make8(mile,0)); /* Mile byte0 to EEPROM @ 0x12 */
write_eeprom(0x11,make8(mile,1)); /* Mile byte1 to EEPROM @ 0x11 */
write_eeprom(0x10,make8(mile,2)); /* Mile byte2 to EEPROM @ 0x10 */
}
```

---

gle byte from a 16- or 32-bit number. For the latter, the `offset` parameter is 0, 1, 2, 3 and specifies the byte number to be extracted. For instance, in our program `make8(mile,2)` is equivalent to the standard **C** expression `(mile>>16)&0xFF`.[3]

As well as altering data under program control it is possible to initialize the state of the EEPROM Data module at the same time as the executable program is being externally blasted into the Program memory. To facilitate this, the EEPROM Data module can be treated as if it overlays the Program store, with cell h'00' mapping into h'F00000'. For instance, to store the value of sine every 10° between 0° and 90° as part of the program source code we have:

```
        org h'F00000'       ; The EEPROM Data module
SINE de  0,      h'2C', h'57', h'7F', h'A4'
     de  h'C4', h'DD', h'F0', h'FB', h'FF'
```

where the assembler directive **de** (Data EEPROM) specifies the comma-delimited list of data. This data is normalized as $\sin\theta \times 256$. After the PIC MCU has been programmed, the contents of the EEPROM Data module will look like Fig. 15.3.

Any data programmed in this way can be subsequently read by the program. For instance, to read sin(50) the contents of EEPROM Data module location h'05' ($\frac{50}{10}$) is read; giving from our diagram h'C4' or decimal 196 ($\frac{196}{256} = 0.76525 = \sin(50)$).

Writing to the EEPROM causes some wear and tear to the thin layer of insulation between gate and substrate; as illustrated in Fig. 2.13 on p. 27. The actual mecha-

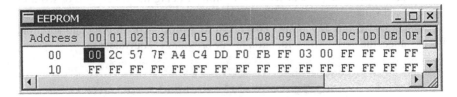

Fig. 15.3 The first 32 bytes of EEPROM holding the sine look-up table

---

[3]For another approach to this example without using these `makex()` functions, see Example 15.1 in my *Quintessential PIC® Microcontroller*.

nism is disputed, but perhaps may be charges lodged in the insulation during the quantum tunneling process. Damage does not occur during reading, which can be done indefinitely, but the Data sheet gives two **endurance** figures for writing.

Byte Endurance

Parameter D120 quotes a typical figure of 10^6 erase/writes before a cell becomes unusable. This is over the normal ambient temperature range of $-40°C \leq T_A \leq +85°C$ and a V_{DD} of 5 V. The minimum endurance is 100,000 writes.

Total Endurance Before Refresh

Parameter D124 is more difficult to understand. During Write actions, a small perturbation occurs to other cells in the module. This does not cause damage, as such, but may eventually corrupt the logic state. The figure given is typically 10^7, with a minimum of 10^6.

As no destruction is done to the physical operation of the cells, writing data back again; that is **refreshing** it, on a regular basis will restore it. In practice, problems are only likely to occur when the EEPROM holds a mixture of constant data never written to (such as our table of sines) and others that are frequently updated—such as our odometer. If there are many of the latter, it could be that the cumulative number of Write processes will exceed this parameter.

Generally endurance figures rise significantly with lower V_{DD} values and ambient temperatures. More details are given in Microchip's application note AN1019 *EEPROM Endurance Tutorial*. Apart from a regular refresh, it may be possible to segregate the constant and infrequently accessed data into the Flash EEPROM Program store.

It is possible to prevent access to the Data EEPROM module during the initial programming blasting action. The default setting of the **CPD** fuse allows external access for both reading and writing. The CPU can always read data from this module, but internal writes can be inhibited by clearing the **WRTD** fuse.

Many projects make use of a fraction of the Program store memory. For instance, if a PIC18F4520 uses 10 kbytes worth of executable software, that leaves 22 kbyte of wasted resource. It is possible to blast data into the Program store at the same time as the executable code; in a similar manner to that shown in Fig. 15.3. However, because of the Harvard structure, which completely separates the Data and program memories, such data cannot be read using normal software—see Fig. 3.2 on p. 43.

Early PIC MCU devices partly circumvented this restriction by using lists of `retlw` instructions to fabricate subroutines to return one byte constant from effectively a look-up table; for instance, Program 6.6 on p. 175. This stratagem is wasteful of storage, as each byte is stored as a 16-bit instruction in the PIC18 family. Furthermore, there is no mechanism to alter such data after the initial programming process.

Some later PIC16 devices introduced additional logic to modify the Harvard structure and gain access to the Program store; which was treated as an extension of

Table 15.1 Table Read and Table Write instructions

Operation	Mnemonic	Description
Read from Program store as pointed to by TBLPTR[21:0] into TABLAT		
TaBLe ReaD		
TABLPTR:3 unchanged	tblrd*	[TABLAT] <-<TBLPTR>
TABLPTR:3 post incremented	tblrd*+	[TABLAT] <- <TBLPTR++>
TABLPTR:3 post decremented	tblrd*-	[TABLAT] <- <TBLPTR->
TABLPTR:3 pre incremented	tblrd+*	[TABLAT] <- <++TBLPTR>
Write to Program store as pointed to by TBLPTR[21:0] from TABLAT		
TaBLe WriTe		
TABLPTR:3 unchanged	tblwt*	<TBLPTR ><- [TABLAT]
TABLPTR:3 post incremented	tblwt*+	<TBLPTR++> <- [TABLAT]
TABLPTR:3 post decremented	tblwt*-	<TBLPTR -> <- [TABLAT]
TABLPTR:3 pre incremented	tblwt+*	<++TBLPTR> <- [TABLAT]

Note that no Status flags are altered by these instructions

TBLPTR:3	3-byte pointer	TABLAT	1-byte data latch
[ ]	Contents of	< >	Pointed to by
++	Incremented	- -	Decremented

the data EEPROM module.[4] The PIC18 family uses a slightly different approach, making use of two new instructions tblrd and tblwt to streamline Program store manipulation; as detailed in Table 15.1.

The technology used for Flash EEPROM is different to that for the Data EEPROM, with a much smaller cell size. The notably smaller cell size is necessary to economically cope with the vastly greater capacity of the former, but it does impact with endurance ratings. Key parameters are:

- 100,000 typical (10^4 minimum) Erase/Write cycle endurance for each cell at 5 V and 25°C.
- Typical Write cycle time 2 ms.
- Data retention minimum of 40 years; typically 100 years.

Access to the Program store involves two sets of control registers in addition to the EECON1 (Fig. 15.2) and EECON2 registers used for the Data EEPROM module.

TABLAT

TABle LATch is an 8-bit SFR used to hold the byte data read from the Program store. It performs a similar role during a Write process.

---

[4]For details see Fig. 15.4 in my *The Quintessential PIC® Microcontroller*, 2nd edn. Springer, 2005.

TBLPTRL, TBLPTRH, TBLPTRU

The **TaBLe PoinTeR** comprises three SFRs, holding the low, high and upper portions of the 22-bit address of the Program store. TBLPTR[20:0] allows access to potentially 2 Mbyte of program memory; although our exemplars vary from 8 through 32 kbyte (4 to 16 kword). Effectively, TBLPTR[21] selects the address range holding Device/User ID and the Configuration fuses—see p. 316.

Figure 15.4 shows the PIC18F4520 Flash Program store as seen from the internal perspective of the `tblrd` instruction. Essentially this memory space is functionally sectioned into two zones. The Special zone holds the eight Configuration fuse bytes at h'30000–7' (see Appendix B) and two read-only device ID bytes set by Microchip to identify the device and silicon revision code. Also there are eight user accessible bytes at 20000–7' which can be used by the end user to store identity codes; such as software revision code. To gain access to this zone, the CFGS bit (see p. 542) in EECON1[6] must be set to 1. If CFGS = 0, its default reset value, then EEPGD in EECON1[7] can be used to select between the reset default Data EEPROM module or the general code area of the Program store.

The task list for reading from the Program store is:

1. Ensure that the EEPGD and CFGS are set-up to target either the Special or Normal zones as appropriate.
2. Copy the target cell's address to the three TBLPTR registers.
3. Execute a `tblrd` instruction.
4. The datum byte can be read from TABLAT by the following instruction onwards.

To illustrate the process, the subroutine READ_PROG__EEPROM_PLUS of Program 15.5 reads one byte from the cell address which is passed in the TBLPTR:3 registers and returns with the requested byte in TABLAT. EEPGD:CFGS are set to 10 on entry, to target the normal zone of the Program store, and cleared on exit to the default reset state targeting the Data EEPROM module.

We see from Table 15.1 that `tblrd` has four variants. The normal `tblrd*` leaves the pointer unchanged, whilst it is possible to automatically do a 3-byte increment or decrement *after* use, or an increment *before* use. This facilitates access to character strings or tables which are resident in sequential cells. Our subroutine code shows the post-increment version of the access.

Program 15.5 Reading a byte from the Normal Flash EEPROM zone

```
; ********************************************************************
; * FUNCTION: Gets one byte from the Flash Program store          *
; * ENTRY    : Address in TBLPTRU:TBLPTRH:TBLPTRL (TBLPTR:3)      *
; * EXIT     : Datum in TABLAT, TBLPTR incremented               *
; ********************************************************************
READ_PROG_EEPROM_PLUS
        bsf       EECON1,EEPGD ; Enable access to Program memory
        bcf       EECON1,CFGS  ; Ensure not Configuration memory
        tblrd*+                ; Get pointed-to datum & inc pointer
        bcf       EECON1,EEPGD ; Back to default condition
        return                 ; and return with the datum
```

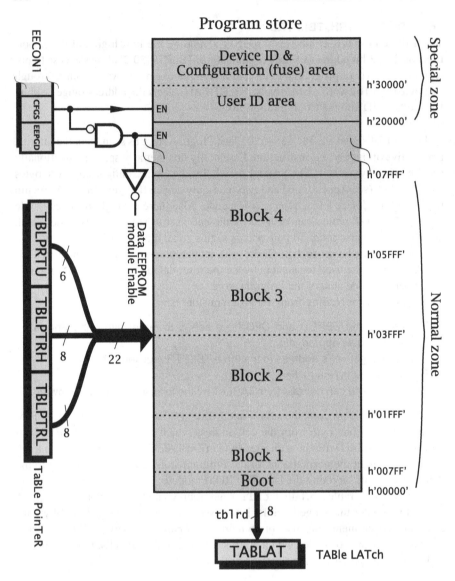

Fig. 15.4 Reading a byte from the PIC18F4520 Flash EEPROM storage system

As an example of the use of READ_PROG_EEPROM_PLUS we will design a subroutine that will return the cosine of an integer degree between 0 and 90, normalized to the range h'0000–FFFF' (treated as a fraction, this corresponds to 0–0.999).

Essentially we need to load in a table of 91 2-byte constants at the same time as the executable code is blasted into the Program store. In Program 15.6 the directive **dw** (**Data Word**) is used with a word datum in a comma separated list. For convenience the **radix** directive is used to specify constants which by default are treated

Program 15.6 Generating the cosine of an angle

```
          radix   decimal    ; All following number are decimal
;  *************************************************************
;  * FUNCTION: Generates cosine of an integer degree 0 -- 90  *
;  * RESOURCE: Subroutine READ_PROG_EEPROM_PLUS               *
;  * ENTRY   : Integer in WREG range 0 -- 100                 *
;  * EXIT    : 16-bit cosine in COSH:COSL.                    *
;  *************************************************************
TABLE      ; Table of constants expressed in decimal
 dw 65535,65525,65495,65445,65375,65286,65176,65047,64897,64728
 dw 64539,64331,64103,63855,63588,63302,62996,62671,62327,61965
 dw 61583,61182,60763,60325,59869,59395,58902,58392,57864,57318
 dw 56755,56174,55577,54962,54331,53683,53019,52339,51642,50930
 dw 50203,49460,48702,47929,47142,46340,45524,44695,43851,42995
 dw 42125,41243,40347,39440,38521,37589,36647,35693,34728,33753
 dw 32768,31772,30767,29752,28729,27696,26655,25607,24550,23486
 dw 22414,21336,20251,19161,18064,16962,15854,14742,13625,12505
 dw 11380,10252,9121,7987,6850,5712,4571,3430,2287,1144,0


; Build up address of start of table in TBLPTR:3 --------------
COSINE clrf    TBLPTRH
       clrf    TBLPTRU
       bcf     STATUS,C    ; 1st multiply by two to give word offset
       rlcf    WREG,w
       movwf   TBLPTRL     ; Low byte
       btfsc   STATUS,C    ; Skip IF no Carry from x2
        incf   TBLPTRH,f   ; ELSE make high byte 01

       movlw   low TABLE   ; Get low byte of table address
       addwf   TBLPTRL,f   ; Add it to offset low byte
       movlw   high TABLE  ; Now get high byte of table address
       addwfc  TBLPTRH,f   ; and add it plus any carry to high byte
       movlw   upper TABLE ; Now get upper byte of table address
       addwfc  TBLPTRU,f   ; and update upper byte of pointer

; Now read pointed-to word and return in COSH:COSL ------------
       call    READ_PROG_EEPROM_PLUS
       movff   TABLAT,COSL ; The low byte
       call    READ_PROG_EEPROM_PLUS
       movff   TABLAT,COSH ; The high byte

       return
```

as decimal. The de directive (see p. 546) can be used for individual bytes, but note that an odd length list is padded out with h'00' to always give an even block.[5] This ensures that any subsequent instructions will start on an even boundary!

Directly following the table is the executable code. The subroutine itself effectively copies WREG × 2 into TBLPTR:3 and then adds the 3-byte address TABLE to it to point to the relevant table entry. Note the use of the three assembler operators upper, high and low to extract the corresponding bytes out of an assembler label—see p. 196. Each byte is then read, with TBLPTR:3 being automatically incremented before exiting.

In Fig. 15.4 the Normal Program store zone is shown divided into five blocks. The areas relate to **code protection**.

---

[5]The db directive pads out *each* byte with a zero byte, and thus is rather wasteful.

Any one of these blocks can be made private, in as much as they cannot be read using a `tblrd` instruction from outside that block. For instance, if the lowest 2 kbyte of the Flash memory space; that is the Boot area,[6] wishes to be confidential, then the **EBTRB (External Block Table Read Boot)** fuse should be cleared—EBTRB=ON. As listed in Fig. 15.5, there is a like fuse for each block; e.g. **EBTR1** for Block 1.

The great majority of PIC18F devices[7] support self writing into the Flash EEPROM space. This is a somewhat convoluted operation, not least because execution of program code cannot continue during the several millisecond hiatus where data is being changed.

As shown in Fig. 2.13 on p. 27, programming an EEPROM bit involves tunneling charge onto the floating gate of a transistor. The erased state of this transistor is no charge on this gate. In terms of the Microchip implementation, this reads as logic 1 and thus any bit can only be changed from 1 to 0. In practical terms, an EEPROM byte must first be erased to all 1s and then any 0s blasted in. In the case of the Data EEPROM module, single bytes can be written to as this erase-before-write action is done automatically. Thus the necessity to erase is hidden from the software designer. However, this is not true for the much larger PIC18 Program stores, which are organized typically in rows 64 bytes. Thus an erase carried out before any byte is written to will return 64 cells to h'FF'. An erase takes around 4 ms to perform, during which time code execution ceases.

Single byte writing is not supported, and programming is carried out as a block of typically 8 bytes (e.g. PIC18F1X20) through to 64 bytes (e.g. PIC18FX6K20). In Fig. 15.5 the 32-cell block write architecture of the PIC18FXX20 devices is shown. Thus the `tblwt` instruction's interaction with the Program store is more complex than its `tblrd` counterpart.

The `tblwt` instruction implements a **Short Write** into the **Write buffer**, which in our diagram is a bank of 32 registers. TABLAT holds the single byte to be copied into the Write buffer and TBLPTR[4:0] holds the address of the target buffer register. A Short Write operation takes one instruction cycle.

The actual **Long Write** into the Program store blasts all 32 bytes from the Write buffer based on a block starting at TBLPTR[21:5]00000. A Long Write is initiated using the same EECON1 and EECON2 interlock process utilized for the Data EEPROM module, but with the EEPGM or CFGS set as appropriate—see Program 15.2. A Long Write process takes around 2 ms; during which time code execution ceases.

As an Erase process obliterates a block of data in Flash EEPROM, a Write sequence of events must first copy the 64-byte block contents into RAM before erasing. This image data can then be copied 32 bytes at a time (or whatever is appropriate for the processor's architecture) into the Write buffer; edited as required, and then

---

[6]So called because this restart area often contains code to initialize and start up the application code; that is boot up.

[7]There are exemptions; such as the PIC16F4510.

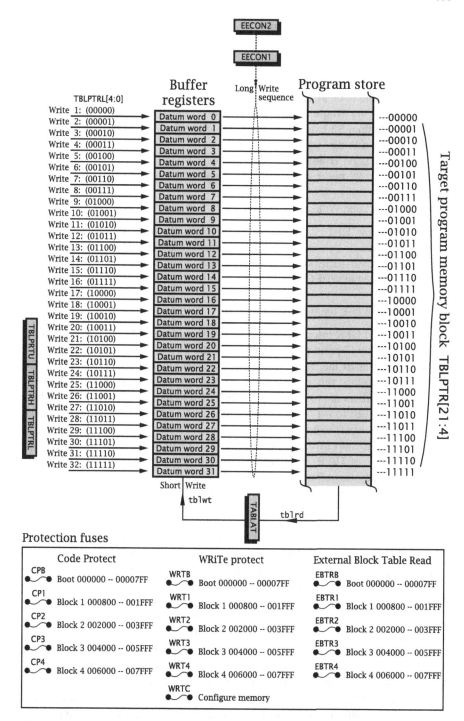

Fig. 15.5 Writing to Flash memory with the PIC18FXX20 group

Long written back into the Program store. This gives the task list for altering up to 64 bytes in the Program store as:

1. Copy target 64-byte Flash memory block into RAM using the `tblrd*` + instruction.
2. Edit data in the RAM image as appropriate.
3. Point back at the Program store and execute an Erase procedure.
4. Copy the bottom 32 bytes from the RAM image into the Write buffer using 32 Short Writes with `tblwt`.
5. Implement a Long Write sequence.
6. Copy the top 32 bytes from the RAM image into the Write buffer using Short Writes.
7. Implement a Long Write sequence.

The subroutine `WRITE_PROG_PLUS` listed in Program 15.7 is an example which writes a single byte into the Program store. It is designed for processors (such as the PIC18FXX20 series) with a 64-byte Erase and 32-byte Long Write structure. The byte is passed in WREG and the location is preloaded into the triplet TBLPTR:3 SFR.

Following the task list we have:

1. A block of 64 GPR Files called IMAGE is reserved using the `cblock` directive. FSR0 is set-up to point to this block using the `low` and `high` assembler directives to pull apart the 2-byte address IMAGE and initialize FSR0L and FSR0H respectively. Once this is done, the `tblrd*+` instruction is used inside a 64-pass loop with an auto incremented FSR0 used to indirectly copy each the byte in the Flash EEPROM block via TABLAT into the 64-byte shadow RAM image from IMAGE to IMAGE+31.

   In order to start at the bottom of the Flash EEPROM block, the lower six bits ($2^6 = 64$) of TBLPTRL is zeroed. However, the entry value of this register is saved for later use before being modified.

2. The offset from the bottom of the block in which the new datum byte is to be written to is the lower six bits of the entry value of TBLPTRL. Adding this to the bottom address of the RAM image; i.e. FSR0 = IMAGE + TBLPTRL[5 : 0], gives the target address. Into this pointed-to location is copied the new byte; which has been temporarily stored in the File DATUM from WREG on entry.

3. Before writing back the image RAM into the Program store, a 64-byte Erase needs to be executed. Again, the bottom of this block is computed by zeroing the lower six bits (strictly the Erase process will automatically ignore these lower bits, but this base value is used for the Write processes following).

   Erase uses the subroutine BLAST_FLASH. This is the same as the interlock Data EEPROM WRITE_EEPROM subroutine in Program 15.2, but with the EEPGD and FREE bits in EECON1 set to 1. All cells in the target 64-byte Program block are erased to h'FF' after approximately 4 ms.

4. The bottom 32 bytes of shadow RAM are copied into TABLAT and transferred in succession into the Write Buffer registers of Fig. 15.5. An auto incrementing FSR0 initialized to IMAGE points into the shadow RAM to extract the datum.

Program 15.7 Writing a byte into the Flash Program store

```
; **********************************************************
; * FUNCTION    : Writes a single byte into Flash Program store*
; * FUNCTION    : with a 64-byte Erase and 32-byte Long Write *
; * ENTRY       : Address in TBLPTR:3; byte in WREG          *
; * EXIT        : Byte updated in Flash memory, TBLPTR++     *
; * RESOURCE    : SFR FSR0, subroutine BLAST__FLASH_EEPROM   *
; * ENVIRONMENT : IMAGE:64, COUNTER:1, DATUM:1, POINTERL     *
; **********************************************************
            cblock h'020'
            IMAGE:64
            COUNTER:1, DATUM:1, POINTERL:1
            endc

WRITE_PROG_PLUS
            bsf    EECON1,EEPGD      ; Point to Flash memory
            bcf    EECON1,CFGS

            movff  TBLPTRL,POINTERL  ; Copy Low address byte
            movwf  DATUM             ; and data

; Now read all 64 bytes from Flash memory into RAM image -------
            movlw  d'64'             ; Counter for 64 byte block
            movwf  COUNTER
            movlw  low IMAGE         ; Set up FSR0 to point to
            movwf  FSR0L             ; the RAM image block
            movlw  high IMAGE
            movwf  FSR0H
            movlw  b'11000000'       ; Zero the bottom 6 bits
            andwf  TBLPTRL,f         ; of Flash address

READ_LOOP   tblrd*+                  ; Get byte from Flash
            movff  TABLAT,POSTINC0   ; Copy byte to RAM image
            decfsz COUNTER,f         ; Record one more byte
             bra   READ_LOOP

; Now modify the one byte in the RAM image @ Base + offset -----
            movlw  low IMAGE         ; Set up FSR0 to point to
            movwf  FSR0L             ; the RAM image block
            movlw  high IMAGE
            movwf  FSR0H
            movf   POINTERL,w        ; Now add the 6-bit offset
            andlw  b'00111111'       ; POINTERL[5:0]
            addwf  FSR0L,f           ; to the base RAM address
            btfsc  STATUS,C          ; to target the byte within
             incf  FSR0H             ; the RAM image
            movff  DATUM,INDF0       ; and update it

; Now erase the Flash block of 64 bytes ------------------------
            movf   POINTERL,w        ; Get back entry address
            andlw  b'11000000'       ; Remove bottom 6 bits
            movwf  TBLPTRL           ; to point to Flash block
            bsf    EECON1,FREE       ; Set up for an erase
            call   BLAST_FLASH       ; and go to it
            bcf    EECON1,FREE       ; and now set for a Write
```

(*continued on the next page*)

The tblwt* + instruction gives the incrementing counterpart doing a Short Write into the buffer; starting at the entry value of TBLPTRL with the bottom six bits zeroed. This takes around 2 ms.

Program 15.7 (*Continued*)

```
; Now write back the first 32 bytes into Flash EEPROM memory  -
; TBLPTRL is still at bottom of 64-byte block at this point ---
                movlw   d'32'               ; Set up 32 loop count
                movwf   COUNTER
                movlw   low IMAGE           ; Set up FSR0 to point to
                movwf   FSR0L               ; the RAM image block
                movlw   high IMAGE
                movwf   FSR0H
WLOOP1          movff   POSTINC0,TABLAT     ; Get data byte from buffer
                tblwt*+                     ; Do a short write
                decfsz  COUNTER,f           ; Record one more
                bra     WLOOP1
                movf    POINTERL,w          ; Get back entry address
                andlw   b'11000000'         ; Bottom of block
                movwf   TBLPTRL
                call    BLAST_FLASH         ; Do a Long Write

; Now write back the last 32 bytes into Flash EEPROM -----------
; TBLPTRL is at bottom of 64-byte block at this point ----------
                movlw   d'32'               ; Set up 32 loop count
                movwf   COUNTER
WLOOP2          movff   POSTINC0,TABLAT     ; Get datum from buffer
                tblwt*+                     ; Do a short write
                decfsz  COUNTER,f           ; Record one more
                bra     WLOOP2

                bsf     TBLPTRL,5           ; Point to top 32-byte block
                call    BLAST_FLASH         ; Do a Long Write

                movff   POINTERL,TBLPTRL    ; Now increment entry pointer
                tblrd*+                     ; Dummy read; TBLPTR:3++
                return

; ************************************************************
; * FUNCTION: Blasts code into Flash memory from Write buffer  *
; * ENTRY  : Data in Write buffer; base address in TBLPTR:3    *
; * ENTRY  : EEPGD or CFGS set as appropriate                  *
; * EXIT   : Interrupts disabled for 7 instructions            *
; ************************************************************
BLAST_FLASH
                bsf     EECON1,WREN ; Enable for Write cycle
                bcf     INTCON,GIE  ; Disable all interrupts

                movlw   h'55'       ; Now do the interlock
                movwf   EECON2
                movlw   h'AA'
                movwf   EECON2

                bsf     EECON1,WR   ; Initiate the Write cycle
                bcf     EECON1,WREN ; Optionally disable any other Writes
                bsf     INTCON,GIE  ; Re-enable interrupts

                return              ; and return when cycle has finished
```

5. The second block of 32 bytes is written into the Program store in a similar fashion. FSR0 is already pointing to the start of the top half of the image RAM, and TBLPTRL[5:0] is zeroed to form the start pointer of the Write Buffer register block.

After 32 copy/Short Writes, TBLPTR:3 is now pointing half way up the 64-cell Program store block. Initiating a Long Write now transfers the 32-byte Write Buffer register contents into the Program store. This also takes around 2 ms.

6. Finally, the entry value of TBLPTRL is restored and one added to the triplet register using a dummy `tblrd*+`.

Any of the blocks illustrated in Fig. 15.4 can be protected against alteration using a Long Write. For instance writing to Block 1 is regulated by the **WRT1 (WRiTe 1)** fuse, which defaults to off. The **WRTC** (WRiTe Configuration) fuse protects the Configuration fuses from internal change. If this fuse is set to 0 during the external programming process, it cannot be changed to a 1 internally; that is, active Configuration code protection cannot be internally removed.

Finally, the various blocks can be individually protected against either interrogation or alteration by the external Device programmer. For instance, to code protect the Boot block, the **CPB (Code Protect Boot)** fuse should be cleared; i.e. `config cpb=on`. However, the Device programmer can always do a memory erase to set all zones to h'FF'.

Examples

Example 15.1 The CCS compiler has the following built-in functions dealing with the Flash Program store:

`read_program_eeprom(address)`
Reads a 2-byte word from the specified Program store code zone address.

`write_program_memory(address, dataptr,count)`
Writes `count` bytes pointed-to by `dataptr` beginning at `address`. An Erase followed by a Long Write is done whenever the function is about to write into a multiple of a Write block.

Based on these routines, write a **C** function to keep a 3-byte count of pages printed by a laser printer.

Solution This coding is broadly similar to the odometer of Program 15.4. The function `read_program_memory()` returns a double-byte, and so is cast to an `unsigned int` (byte) when building up the 24-bit `page` using the `make32()` function. Once this is done, and the count incremented, it is split into an array of three bytes. The name of this array `p_count` is the address of the first element of the array, and this is the parameter `dataptr` passed to the `write_program_memory()` function. As address 0x2000 is the bottom of a Flash EEPROM Write-to block, an erase and Long Write is performed. Note that unlike the assembly-level subroutine of Program 15.7, non of the other data that may lie in the 64-byte Erase block is preserved!

Program 15.8 C-based coding for the laser printer

```
#include     <18f4520.h>
#FUSES NOWDT, NOEBTR, NOWRT
#ROM int8 0x2000 = {0,0,0}

void page_count(void);  /* Global variable holding page count */

void main(void)
{
/* Main routine code

}

void page_count(void)
{
unsigned int32  page;       /* 32-bit variable              */
unsigned int p_count[3];   /* Individual bytes             */
/* Build up a 3-byte word from 3 individual bytes from EEPROM */
page=make32((unsigned int)read_program_eeprom(0x2002),
(unsigned int)read_program_eeprom(0x2001),
(unsigned int)read_program_eeprom(0x2000)));
page++;                      /* One more page               */
p_count[0] = make8(page,0);
p_count[1] = make8(page,1);
p_count[2] = make8(page,2);
write_program_memory(0x2000,p_count,3); /* Write 3 bytes back */
}
```

Actually this is not a good use of the Flash program store, as the endurance of any cell may be as low as 10,000 and is typically only 100,000. This data is likely to wear out faster than a good printer! Writing data to the Program store should only be used for data that rarely changes.

Example 15.2 In Example 14.3 the discharge energy of a defibrillator was calculated by calculating the sum of the squared voltage differences of the sampled inputs from a baseline value. In this case we used a baseline value of 2.6 V, from observation of the waveform. This average value may vary from instrument to instrument and over time with usage. It is proposed to enhance the software by introducing a learning feature, which could be called up when a switch connected to, say, pin RA4 is closed. This subroutine will sample the quiescent voltage 256 times to give a double-byte total. Taking the upper byte is tantamount to dividing by 256 and thus gives an average value. This datum is to be burnt into location h'00' of the Data EEPROM, and this can subsequentially be used as a learnt baseline value, which if necessary can be updated at regular intervals. Assuming that the GET_ANALOG subroutine of Program 14.1 on p. 517 is available, show how a suitable subroutine could be coded.

Solution From Fig. 14.21 on p. 531 we see that the voltage from the defibrillator's Hall effect current sensor is connected to the RA0/AN0 pin. With the assumption

Program 15.9 Learning the baseline

```
; ******************************************************************
; * FUNCTION: Sums 256 analog samples to find an average byte *
; * FUNCTION: value for the Baseline voltage which is blasted *
; * FUNCTION: into the EEPROM Data module                     *
; * RESOURCE: GET_ANALOG, WRITE_EEPROM subroutines            *
; * ENTRY   : None                                            *
; * EXIT    : Average of Channel 0 in EEPROM location h'00'   *
; ******************************************************************
LEARN clrf    ACCUMULATOR+1   ; Zero double-byte sum MSB
      clrf    ACCUMULATOR     ; Zero LSB
      clrf    COUNT           ; Loop count zero

LEARN_LOOP
      movlw   0               ; Start an Analog channel 0
      call    GET_ANALOG      ; Digitization
      addwf   ACCUMULATOR,f   ; Add to LSB of total
      btfsc   STATUS,C        ; Was there a Carry
       incf   ACCUMULATOR+1,f ; IF yes THEN increment MSByte

      decfsz COUNT,f          ; Count down one
       bra    LEARN_LOOP

; Burn in datum into EEPROM Data module ----------------------
      movff   ACCUMULATOR+1,EEDATA  ; Get the average value
      clrf    EEADR           ; into Data EEPROM @ location h'00'
      call    WRITE_EEPROM    ; Blast it in

      return                  ; All done
```

that the ADC module has been enabled, as described in Program 14.6 on p. 531, our task is to read the digitized Channel 0 byte 256 times inside a loop, accumulating to give a 16-bit total sum. Taking the top byte of this pair effectively gives an average value for this analog input (that is, divides by 256). If the defibrillator is quiescent during this learning run, this average gives the baseline voltage at this time.

When we have a baseline value, this byte can be burnt into the EEPROM Data module in the normal way. This can be subsequently read and treated in the same way as the constant BASELINE in Program 14.6.

Program 15.9 uses the File COUNT to count 256 loop passes. Each pass adds the digitized byte to the double-byte Accumulator File pair. On exit from the loop, subroutine WRITE_EEPROM burns this top Accumulator byte into location h'00' in the EEPROM Data module.

In a real situation a better outcome could be obtained by sampling 65,536 times and accumulating a triple-byte sum. The top byte of this triplet would again represent an average.

Example 15.3 As an alternative to the approach to Programs 6.6 and 6.13 on pp. 175 and 196, construct a 7-segment active-low decoder based on a look-up table located in Flash EEPROM.

In addition to the 16 hexadecimal characters illustrated in Fig. 6.8 on p. 173, the input code b'10000' is to blank out all segments and b'10001' is to illuminate a decimal point only, connected to bit 7.

Solution Essentially Program 15.10 is a byte-sized version of Program 15.6, with de replacing the 16-bit dw directive. TBLPTR:3 is set-up to point to the initial address of the table, and as only a single byte is pointed to, the mapping of the 5-bit input code in WREG on entry into the subroutine is a simple offset to TBLPTR:3 and does not need multiplied by two. In addition, the READ_PROG_EEPROM_PLUS subroutine is only called once.

Note that each invocation of de (and also the equivalent db directive) adds a h'00' Null to the line. Thus, for instance, using a separate de for each of the last two bytes of data, in order to make the documentation look more attractive, would have a detrimental effect!

In comparing the outcome with the legacy table of retlw instructions; the table itself takes 18 bytes as compared to 36 (each retlw takes two bytes. However, the overhead code needs nine instructions compared to 15 in Program 6.13 on p. 196. Here we are assuming that the subroutine READ_PROG_EEPROM_PLUS comes free, in that it is necessary for other purposes. From this it is clear that the overwhelming advantage of our approach here lies with larger tables, such as that of Program 15.6.

Program 15.10 Generating an extended active-low 7-segment code based on Flash EEPROM

```
; *****************************************************************
; * FUNCTION: Decodes to active-low hexadecimal 7-seg code      *
; * FUNCTION: plus Blank and decimal Point                      *
; * RESOURCE: Subroutine READ_PROG_EEPROM_PLUS                  *
; * ENTRY    : Integer in WREG range b'00000 - 01111' (0 -- F)  *
; * ENTRY    : b'10000' (Blank) and b'10001' for dP             *
; * EXIT     : pgfedcba 7-segment code in WREG                  *
; *****************************************************************
TABLE7                        ; Table of constants
 de h'C0',h'F3',h'A4',h'B0',h'E7',h'92',h'82',h'F8' ; 0 -- 7
 de h'80',h'98',h'88',h'83',h'C6',h'A1',h'86',h'8E' ; 8 -- F
 de h'FF',h'7F'                              ; Blnk & dP

; Build up address of start of table in TBLPTR:3 --------------
SVN_SEG clrf    TBLPTRH
        clrf    TBLPTRU
        movwf   TBLPTRL       ; Low byte

        movlw   low TABLE7    ; Get low byte of table address
        addwf   TBLPTRL,f     ; Add it to offset low byte
        movlw   high TABLE7   ; Now get high byte of table address
        addwfc  TBLPTRH,f     ; and add it plus any carry to high byte
        movlw   upper TABLE7  ; Now get upper byte of table address
        addwfc  TBLPTRU,f     ; and update upper byte of pointer

; Now read pointed-to word and return in WREG ----------------
        call    READ_PROG_EEPROM_PLUS
        movf    TABLAT,w

        return
```

Self-Assessment Questions

15.1 Good program practice dictates that the datum written into Data EEPROM should be verified as the value that was intended to be written. Show how you could modify the WRITE_EEPROM subroutine of Program 15.2 to return a value of −1 in a File called ERROR if the action is not successful, otherwise zero.

15.2 Repeat Example 15.2 but using **C** coding.

15.3 Microchip-compatible assemblers have the directive da (DAta) which can be used to store strings of character codes in Program memory. For example:

```
MESSAGE   da "Hello world\n",0
```

which places the characters inside quotes, coded in 7-bit ASCII code in each byte, followed by an all zeros byte. The \n escape character means New Line; ASCII code h'0A'. As for the de and db directives, an evocation of da will be padded out to an even length.

Assuming that this is done in a PIC18 device; write a subroutine called PDATA (Print DATA) to fetch each character from Program memory and transmit to a terminal using the subroutine PUTCHAR of Program 12.15 on p. 425.

15.4 A certain hotel security system is to use a PIC-based reprogrammable smart card for electronic guest room locks. On registration the card is to be charged up with the following details:

1. A 4-digit room number, e.g., 1311.
2. Start data, e.g., 15112009.
3. End date, e.g., 16112009.

Assume that the PIC MCU has an integral EEPROM Data module and communicates with the receptionist's terminal via a serial input subroutine, such as described in Program 12.15 on p. 425. Data is coded in ASCII in the order outlined, preceded with the character STX, terminated by ETX and delimited by SP; see Table 1.1 on p. 5. Design a routine to extract the two dates and store them in Data EEPROM.

Chapter 16
A Case Study

Up to this point our microcontroller material has been presented piecemeal. To complete our study we are going to put much of what we have learned to good use and design both the hardware and software of an actual widget (gadget). This is not an easy task to do in a single short chapter. However, very little new material needs to be presented at this point, rather a process of coalescence.

We begin with our specification. Students invariably talk too long during their oral presentations. It is proposed that a dedicated embedded microcontroller-based system be designed to act as a time monitor. This monitor should default to a time-out of 10 minutes, but should have the provision to vary the allotted time from 1 to 99 minutes.

Once triggered, the monitor should perform the following sequence of operations:

1. When the RESET switch is closed, a green lamp will illuminate and a dual 7-segment display will show a count-down from the time-out value to `03` at 1-min intervals.
2. After a further minute, an amber lamp only will illuminate; the count of `02` will be displayed and a buzzer will sound for nominally one second.
3. After a further minute, a red lamp only will illuminate, together with a display of `01`. The buzzer will sound for 2 seconds.
4. Finally, after another minute the display will show `00`; the red lamp will continue to be illuminated and the buzzer will sound continuously until the STOP switch is pressed. This will halt the timer and turn off all displays, lamps, and the buzzer. Indeed, closing the STOP switch at any time during the sequence above will cause the system to permanently halt. The system may be restarted from the time-out value by resetting the processor.
5. At any time the sequence can be frozen by toggling the PAUSE switch. When toggled again, the sequence will continue on from where it left off.
6. In order to alter the time-out from the default value of `10`, the SETT switch must be closed when the system is reset. The display will then show `99` and will count down slowly. The value showing when the SETT switch is released will be the new time-out and will be retained indefinitely until another Set Time process.

S. Katzen, *The Essential PIC18® Microcontroller,*
Computer Communications and Networks,
DOI 10.1007/978-1-84996-229-2_16, © Springer-Verlag London Limited 2010

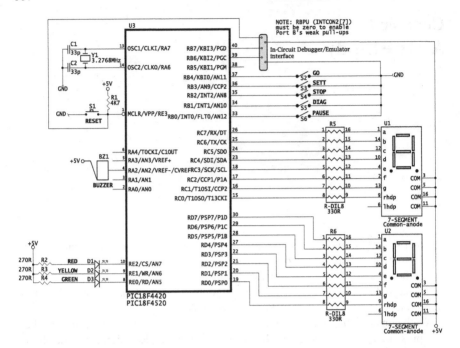

Fig. 16.1 The annunciator hardware

The first decision to be made is the choice of microcontroller (MCU). In this case we are constrained by the need to use one of our book's model device, i.e., PIC18F1X20 or PIC18FXX20. Choosing the 40-pin PIC18F4420 or 4520 device will give the most economical outcome; having sufficient port pins to directly drive the peripheral switches, displays and buzzer. The book's website shows an alternative solution using the 18-pin PIC1X20 device with serial data transmission to the display devices.

Based on this decision the final target hardware is shown in Fig. 16.1. The port pin budget is allocated as follows:

Switches

The five switches S2 . . . S6 requesting the functions Go, Set-Time, Stop, Diag, Pause are read from Port B at pins RB4:0. By using this port's internal pull-up resistors (see Fig 11.12 on p. 352) no external resistors are required.

S1 with R1 provides a Manual reset in order to restart the count. In the PIC18FXX20, pin 1 can optionally be configured as an additional Port E line RE3. On a Power-on Reset this defaults to $\overline{\text{MCLR}}$, but in Program 16.1 the MCLRE fuse is explicitly set-up to enable this function.

All six switches can be conveniently implemented as momentary contact keyboard switches.

Lamps

Three suitably colored 10 mm (0.4″) high-brightness LEDs D3 . . . D1 driven from pins RE2 : 0 provide the light signals. Port E is chosen, rather than the more obvious Port B, to leave pins RB7 : 5 free for use in programming and in-circuit debugging. The 330 Ω series resistors limit the current to nominally 10 mA.

Buzzer

The buzzer should be a miniature solid-state device. A typical piezo-electric implementation will operate over a wide dc voltage range of typically 3 to 16 V and require little more than 1 mA at 5 V.[1] The buzzer is driven via pin RA2.

Numerical Display

Two 7-segment displays give the required 2-digit read-out, facilitating the maximum specified period of 99 minutes. These are connected directly to Port C and Port D, giving the least-significant and most-significant digits respectively. Both ports need to be set for digital output.

The common-anode 7-segment display pinning shown in the diagram is that of the 16-pin Dual In Line (DIL) footprint with both left and right decimal points— lhdp and rhdp. Only the latter is used here, to indicate that the system has paused. Alternative 16- and 14-pinouts are commonly available and even dual-digit packages. However, even the 16-pin footprint pinout is not standardized.

Smaller-sized displays, typically below 20 mm/0.8″, use a single LED for each bar, with a conducting voltage drop of around 2 V.[2] The DIL 330 Ω series resistors R5 and R6 limit the current to around 10 mA. The common anodes are connected directly back to the normal +5 V power supply to avoid current surges affecting the logic circuits, and should be decoupled by small tantalum capacitors. Although the displays are normally rated for 20 mA, restricting the current to this value gives sufficient illumination.[3]

Crystal

A 3.2768 MHz crystal provides the timing for the MCU's clock oscillator, giving an instruction cycle rate of 819.21 kHz. A typical crystal of this value has a frequency tolerance of ±30 ppm and temperature coefficient of ±50 ppm over the operating range.

This odd frequency choice is $2^{16} \times 50$; so if we use the 8-bit Timer 0 with a Prescaler value of 1:64 then we can create an interrupt 50 times per second—see p. 569. An alternative low-power configuration would be to use a 32.768 kHz crystal and generate an interrupt every 2 seconds using the 16-bit Timer 1. However, compared to the current consumption of the LED display components, the MCU's power dissipation is minor. A cheaper approach would be to use the internal oscillator in conjunction with Timer 1. In the context of this project, the slightly lower precision is of little significance. The book website looks at these options in more detail.

---

[1] If you want to preclude any possibility of the speaker continuing, a piezo-electric sound bomb producing 110 dB at 1 m distance needs a 12 V d.c. supply at 200 mA.

[2] Larger displays, e.g., 2.24″/56 mm, have typically two or four LEDs in series. In the latter case a separate 12 V supply and current buffering would be needed.

[3] Alternatively low-current 7-segment displays are available.

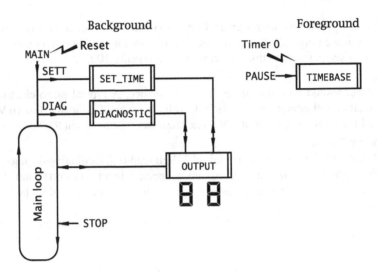

Fig. 16.2 The modular software structure

With the hardware environment designed, we can now concentrate on the software.

Figure 16.2 shows the basic modular structure for our system. Here the distinctive double right/left edged box denotes a subroutine or interrupt service routine (ISR). Three distinct processes can be identified together with two major supporting tasks.

Timebase Process

All processes are time related. Timekeeping is implemented in hardware by generating an interrupt 50 times each second. By keeping a Jiffy count, seconds and minutes are updated and are used to sequence the appropriate process.

By monitoring the PAUSE switch this decrementing time chain can be bypassed, hence freezing the countdown for as long as necessary.

Output Display Process

All processes need to output the state of the count or status information to the two 7-segment displays. This typically involves code conversion and copying this data to the appropriate port. If a smaller footprint processor is used, this may involve parallel-to-serial conversion and shifting. This the task is better gathered into one module or driver to hide the complexity of the actual hardware used in any particular implementation.

Main Process

The Main process is a loop displaying the 2-digit Minute count until it reaches zero, with a premature break if the STOP switch is closed.

Set-Time Process

If the SETT switch is closed when the PIC MCU comes out of a Manual reset, then the SET_TIME subroutine quickly decrements the display count until the switch is released.

This displayed value is then written into Data EEPROM and is used by all subsequent runs as the starting value for the Minute count. The default value when the MCU is programmed for the first time is ten.

Diagnostic Process

If the DIAG switch is closed on coming out of a Manual reset, the system enters a diagnostic subroutine. This essentially exercises each peripheral device in a manner calculated to ease hardware fault finding.

Before looking at the these tasks in more detail, we will briefly consider the software environment of the system when the Program store is loaded, and the initialization code which is performed at run-time after a reset. This is codified in Program 16.1.

Load-Time

The config directive sets up the processor to use a medium frequency external crystal, disable the Watchdog timer, disable Port B analog functionality and enables the PoWer-on Reset Timer. We have already observed on p. 315 that the normal supply voltage can be used to program a device. Here we have turned this facility off, so the PGM/RB5 pin can be used as a normal Port B I/O. If this were not the case, PGM should be tied low for normal software execution and high when code is blasted in. Also enabled is an external $\overline{MCLR}$ facility, to allow a Manual reset to act as a system start or Go function.

When the code is blasted into the main flash EEPROM Program, coincidentally store location 0 of the Data EEPROM is set to ten. This means that a freshly programmed PIC MCU will default to a 10-min count down. This value can subsequently be altered using the Set-Time process described in Program 16.5.

Run-Time

The code executed each time a reset is actioned is used to initialize the run-time environment.

Vectors

To initialize the Reset vector at h'00000' to point to MAIN and Interrupt vector at h'00008' to point to the ISR. We are not using the Priority interrupt facility.

Parallel Port Configuration

Ports C, D and E as well as pin RA2 are set-up as outputs, to drive the 7-segment displays, status LEDs and buzzer. Any remaining pins are left as inputs. In addition Port A and E's Power-on Reset analog capabilities are disabled.

Timer 0 Setting

To set-up the Prescaler ratio to 1:64 and Timer 0 clock source to internal with an 8-bit count. The Timer 0 interrupt is also enabled.

Program 16.1 The initialization code

```
BUZ     equ 2           ; Buzzer activated at PORTA[2]
GREEN   equ 0           ; Green LED activated at PORTE[0]
YELLOW  equ 1           ; Yellow LED activated at PORTE[1]
RED     equ 2           ; Red LED activated at PORTE[2]
PAUSE   equ 0           ; Pause switch read at PORTB[0]
DIAG    equ 1           ; Diagnostic switch read at PORTB[1]
STOP    equ 2           ; Stop switch read at PORTB[2]
SETT    equ 3           ; Set switch read at PORTB[3]
GOO     equ 4           ; Go switch read at PORTB[4]
LSD     equ PORTC       ; PORTC is connected to Least Sig Digit
MSD     equ PORTD       ; PORTD is connected to Most Sig Digit

        cblock  h'020'
        MINUTE:1, SECOND:1, JIFFY:1, NEW_SEC:1
        COUNT:1, UNITS:1, TENS:1, DATA_L:1, DATA_H:1
        TIME_OUT:1, TEMP:1, Pause:1
        endc

        config MCLRE=ON, PBADEN=OFF, OSC=XT, WDT=OFF, LVP=OFF
        config PWRT=ON
        org    h'F00000'   ; The EEPROM Data module
        de     d'10'       ; Default value is 10 minutes

RST     org    0           ; Reset vector
        bra    MAIN
        org    8           ; Interrupt vector
        goto   ISR_TMR0

MAIN    bcf    TRISA,2     ; RA2 Output to Buzzer
        setf   TRISB       ; PortB connected to switches
        clrf   TRISC       ; Port C Output to LSD display
        clrf   TRISD       ; Port D Output to MSD display
        clrf   TRISE       ; PORTE[2:0] drives LEDs
        bcf    INTCON2,RBPU ; Enable Port B's internal pull-ups
        movlw  b'11000101' ; Enable (1) 8-bit TMR0 (1) internal (1)
        movwf  T0CON       ; Enable PS set to 1:64.

; PORTA & E analog inputs disabled ----------------------------
        setf   ADCON1      ; by putting 11111111 into the ADCON1
        clrf   Pause       ; The PAUSE switch toggle
        clrf   NEW_SEC     ; Reset NEW_SEC second flag

        clrf   TMR0L
        bcf    INTCON,T0IF
        bsf    INTCON,T0IE ; Enable Timer0 interrupts
        bsf    INTCON,GIE  ; Enable all interrupts

        btfss  PORTB,SETT  ; Check the Set Time switch
        call   SET_TIME    ; IF closed THEN set total time
        btfss  PORTB,DIAG  ; Check the Diagnostic switch
        call   DIAGNOSTIC  ; IF closed THEN DO diagnostic routines
```

Process Select

To check the state of the DIAG and SETT switches to optionally choose either the Diagnostic or Set-Time processes. If neither switch is closed the normal Main process is entered.

Looking at each of the tasks outlined in Fig. 16.2 in some detail.

ISR_TMR0

All processes are dependent on the foreground Timebase task to update the real-time clock information. As shown in Program 16.2, this is interrupt driven and is based on the Timer 0:Prescaler dividing down the 3.2763 MHz crystal-driven oscillator to give overflow every $\frac{1}{50}$ s. As can be seen in Program 16.1, the Timer 0 interrupt is enabled and thus the PIC MCU will enter interrupt handler ISR_TMR0 whenever the timer overflows—every 256 outputs from the Prescaler. Remembering that the instruction cycle is $\frac{1}{4}$ of the crystal frequency, a Prescaler ratio of 1:64 will give a timebase rate of 50 per second; that is, $\frac{3.2763 \times 10^6}{4 \times 64 \times 256} = 50$.

The task list for this function is then:

1. IF PAUSE switch open THEN
    (a) Decrement the time chain by one Jiffy.
    (b) IF new second THEN flag it.
2. ELSE
    (a) Toggle the Pause flag.
    (b) IF set THEN tell the world that the system is paused.
    (c) ELSE display time to indicate normal running.
    (d) Wait until SETT switch is released.
3. Return from interrupt.

Time Chain Decrementation

From Program 16.2 we see that time is kept as a 3-byte count chain located in Files MINUTE, SECOND and JIFFY to hold the total. Assuming that the state of bit 0 of File Pause is 0, then one is added to the Jiffy count. Normally the ISR then exits, but when Jiffy reaches 50 it is reset to zero and the Seconds count decremented. The File NEW_SEC is also made non-zero to indicate to background software that a second has elapsed. In the situation where the Second count reaches zero then it is reset to 59 and the Minute count is decremented. The procedure is similar to the incrementing count of Example 7.3 on p. 233.

Freeze handling

The Timebase task also handles the Pause function. The simplest approach would be to skip over the time decrement code if the PAUSE switch is closed. However, the necessity to keep the switch closed would be irksome if the period is more than a few minutes.

Program 16.2 The timebase software

```
; *****************************************************************
; * The ISR to decrement the real-time clock                    *
; * Adding a 20ms Jiffy on each entry                           *
; * Sets NEW_SEC to a non-zero value each Second update         *
; *****************************************************************
ISR_TMR0 btfss    INTCON,T0IF    ; Was it a Timer0 time-out?
         bra      ISR_TMR0_EXIT  ; IF no THEN false alarm

         btfsc    Pause,0        ; Check the Pause flag
         bra      ISR_TMR0_EXIT  ; IF closed THEN don't increment

         incf     JIFFY,f        ; Record one more 1/50 second
         movlw    d'50'          ; Has Jiffy count reached 50?
         cpfseq   JIFFY          ; Skip IF not
         bra      ISR_TMR0_EXIT  ; ELSE THEN finished
         clrf     JIFFY          ; ELSE zero Jiffy count

         movf     SECOND,f       ; Test for Seconds count = 00?
         bz       NEW_MIN        ; IF it is THEN a new minute
         decf     SECOND,f       ; ELSE decrement Seconds count &
         incf     NEW_SEC,f      ; tell background prog new second
         bra      ISR_TMR0_EXIT  ; and exit

NEW_MIN  movlw    d'59'          ; Reset Seconds to 59 seconds
         movwf    SECOND
         movf     MINUTE,f       ; Test for Minutes count = 00?
         bz       ISR_TMR0_EXIT  ; IF it is THEN no more decrement
         decf     MINUTE,f       ; ELSE decrement Minutes

; *****************************************************************
ISR_TMR0_EXIT
         btfss    PORTB,DIAG     ; IF in Diagnostic mode
         bra      ISR_TMR0_FINI  ; ignore Pause facility

         btfss    PORTB,PAUSE    ; ELSE check the PAUSE switch
         rcall    FREEZE         ; IF closed THEN update Pause flag

ISR_TMR0_FINI
         bcf      INTCON,T0IF    ; Clear interrupt flag and return
         retfie   FAST           ; from interrupt with context
```

(continued on the next page)

Implementing a push-on push-off scenario is ergonomically superior and can be more economically implemented in software rather than using a toggling switch. In Program 16.2 the Pause handling code is located in the separate subroutine FREEZE. It is permissible to call a subroutine from an ISR in the same manner as calling one subroutine from another; that is, nesting. The hardware stack allows nesting up to 31 deep. In our situation only two stack locations are used.

Program 16.2 (*Continued*)

```
;  *********************************************************
;  * FUNCTION: Increments the Pause flag.                 *
;  * FUNCTION: IF = 1 THEN displays the decimal points     *
;  * FUNCTION: IF = 0 THEN displays the normal count       *
;  * RESOURCE: Subroutine DISPLAY. Vars Pause, TENS, UNITS *
;  * ENTRY    : PAUSE switch closed                        *
;  * EXIT     : IF Pause[0] is 0 display dP ELSE Minute count *
;  *********************************************************
FREEZE    btg     Pause,0         ; Toggle Pause flag, bit 0
          btfss   Pause,0         ; Check status of Pause flag
          bra     UNFREEZE        ; Change 1 -> 0, unfreeze

; Display freeze ---------------------------------------------
          movlw   b'00010001'     ; Code for decimal point
          movwf   TENS            ; In situ for display
          movwf   UNITS
          call    DISPLAY         ; Display
          bra     FREEZE_EXIT

; Land here if Pause 0 -> 1 ---------------------------------
UNFREEZE  movf    MINUTE,w        ; Display the normal Minute count
          call    OUTPUT

FREEZE_EXIT
          btfss   PORTB,PAUSE     ; Wait until switch is opened again
          bra     FREEZE_EXIT
          clrf    TMR0L           ; Reset TMR0 to give debounce delay

          return
```

Subroutine FREEZE is only entered if the PAUSE switch is closed. On each entry the value of bit 0 of the File Pause is toggled. Pause[0] is tested and if it is 1 then the code to illuminate only the two decimal points is sent to the DISPLAY subroutine. This is an arbitrary indicator display; another possibility would be PR. If Pause[0] is 0 then the state of the Minute count is sent to the OUTPUT subroutine and indicates to the user that the Pause function has ended.

Finally, the subroutine does not exit until the user releases the PAUSE switch. This is important, as on exit the ISR will be re-entered again at the next Timer 0 overflow, and this would cause Pause to be repeatedly retoggled. Some measure of switch debounce is obtained by zeroing Timer 0 and the Prescaler when the switch is released and only then clearing T0IF. This means that the switch will not be retested for a whole $\frac{1}{50}$ second.

OUTPUT

The Output task acts as a device driver, interfacing to the two 7-segment displays. Tasks implemented by Program 16.3 are:

1. Conversion to 7-segment code.
2. Presentation to the two display devices.
3. Convert a binary byte limited to decimal 99 to 2-digit BCD.

Program 16.3 The display driver

```
;  ***********************************************************
;  * FUNCTION: Displays datum as a 2-digit decimal output, with *
;  * FUNCTION: direct access to look-up table for non-BCD glyphs*
;  * FUNCTION: and to the 7-segment display devices           *
;  * RESOURCE: Subroutines BIN_2_BCD, SVN_SEG                 *
;  * RESOURCE: Vars NEW_SEC, TENS, UNITS                      *
;  * ENTRY   : Natural binary byte 0 - 99 in WREG             *
;  * EXIT    : TENS & UNITS data displayed, NEW_SEC zeroed    *
;  ***********************************************************
OUTPUT_BCD call    BIN_2_BCD     ; Convert to BCD

; Direct entry to look-up table for non-BCD displays -----------
OUTPUT      movf    TENS,w        ; Get MSD for display
            call    SVN_SEG       ; Convert to 7-segment code
            movwf   DATA_H        ; Ready to display Tens digit
            movf    UNITS,w       ; Get LSD for display
            call    SVN_SEG       ; Convert to 7-segment code
            movwf   DATA_L        ; To  Significant Digit

; Direct entry to the 8 segments of both displays --------------
SEND        movff   DATA_L,LSD    ; Out to hardware
            movff   DATA_H,MSD
            clrf    NEW_SEC       ; Reset NEW_SEC flag

            return
```

(continued on the next page)

Binary to 7-Segment Decoder

The kernel of this module entered at OUTPUT converts the 5-bit code in both Files TENS and UNITS to 7-segment code; subsequently located in Files DATA_H and DATA_L respectively. The code shown in Program 16.3 is similar to the subroutine SVN_SEG in Program 15.10 on p. 560. However, the look-up table is different; reflecting the mirror image connections of the segments in Fig. 16.1. That is the segments a through g and dP are connected to port lines 7 though 0 respectively.

Display

The Display routine, which sends the data to the actual display devices, is particularly simple with our choice of hardware. The byte data in DATA_H and DATA_L are copied to the two appropriate parallel ports; both of which have been set-up as output. This is the only part of the driver software that needs alteration if the hardware is changes; e.g. to a serial architecture.

If desired; it is possible to directly connect to each display, to by-pass the 7-segment decoder by entering at SEND. For instance, to display the message PA, we have:

```
movlw  b'00111111'    ; Code for P (no dP)
movwf  DATA_H
movlw  b'00011000'    ; Code for A.
movwf  DATA_L
call   SEND
```

Program 16.3 (*Continued*)

```
; *****************************************************************
; * FUNCTION: Decodes to active-low hexadecimal 7-seg code      *
; * FUNCTION: plus Blank and decimal Point                      *
; * RESOURCE: Subroutine READ_PROG_EEPROM_PLUS                  *
; * ENTRY   : Integer in WREG range b'00000 - 01111' (0 -- F)   *
; * ENTRY   : b'10000' (Blank) and b'10001' for DP              *
; * EXIT    : abcdefgP 7-segment code in WREG                   *
; *****************************************************************
TABLE7                        ; Table of constants
 de h'03',h'CF',h'25',h'0D',h'E7',h'49',h'41',h'1F' ; 0 -- 7
 de h'01',h'19',h'11',h'31',h'63',h'85',h'61',h'71' ; 8 -- F
 de h'FF',h'FE'                              ; Blnk & dP

; Build up address of start of table in TBLPTR:3 --------------
SVN_SEG clrf    TBLPTRH
        clrf    TBLPTRU
        movwf   TBLPTRL        ; Low byte

        movlw   low TABLE7     ; Get low byte of table address
        addwf   TBLPTRL,f      ; Add it to offset low byte
        movlw   high TABLE7    ; Now get high byte of table address
        addwfc  TBLPTRH,f      ; and add it plus any carry to high byte
        movlw   upper TABLE7   ; Now get upper byte of table address
        addwfc  TBLPTRU,f      ; and update upper byte of pointer

; Now read pointed-to word and return in WREG ------------------
        call    READ_PROG_EEPROM_PLUS
        movf    TABLAT,w

        return

; *****************************************************************
; * FUNCTION: Converts a binary byte to a two BCD digits        *
; * RESOURCE: vars UNITS, TENS                                  *
; * ENTRY   : Binary byte in W range 00 - 63h (0 - 99d)         *
; * EXIT    : Ten's digit in TENS, Unit's digit in UNITS        *
; *****************************************************************
; Divide by ten
BIN_2_BCD clrf  TENS           ; Zero the Ten's count

LOOP10    incf  TENS,f         ; Record one ten subtracted
          addlw -d'10'         ; Subtract decimal ten
          bc    LOOP10         ; IF no borrow (C==1) THEN DO again

          decf  TENS,f         ; Compensate for one inc too many
          addlw d'10'          ; Add ten to residue to give units
          movwf UNITS          ; The residue is the Units

          return               ; and return to caller
```

Binary to BCD Conversion

The display normally reflects the current state of the Minute countdown. As this is kept in a naturally coded 8-bit binary datum in MINUTE; as defined in Program 16.2, we need to convert to two BCD digits. These in turn are used as the input to the OUTPUT routine. The BIN_2_BCD subroutine listed here is based on the routine described in Program 5.11 on p. 149. This subroutine

is restricted to the range 0–99, which is not a problem with our 2-digit display
hardware.

Diagnostic

If the DIAG switch is closed when the PIC MCU comes out of reset, then the code
transfers to the subroutine DIAGNOSTIC in Program 16.4. The Diagnostic process
aims to exercise the various peripheral devices interfaced to the process in order to
verify in a reproducible manner the status of the interconnection and the devices
themselves.

Program 16.4 The Diagnostic process

```
; ********************************************************************
; * FUNCTION: Checks each switch and activates a corresponding *
; * FUNCTION: LED or buzzer. Continually activates a unary     *
; * FUNCTION: pattern to both 7-segment displays               *
; * RESOURCE: Subroutine OUTPUT:SEND                           *
; * RESOURCE: Vars TEMP, DATA_H, DATA_L                        *
; * ENTRY   : DIAG switch closed                               *
; * EXIT    : DIAG switch open                                 *
; ********************************************************************
DIAGNOSTIC
          movlw  b'11111110'  ; The initial 7-segment pattern
          movwf  TEMP         ; in memory
D_LOOP    movlw  b'11111111'  ; Turn off all LEDs and buzzer
          movwf  PORTE
          bsf    PORTA,BUZ
; Now scan switches ------------------------------------------------
          btfss  PORTB,PAUSE  ; IF Pause switch closed
           bcf   PORTE,GREEN  ; THEN Green LED
          btfss  PORTB,STOP   ; IF Stop switch closed
           bcf   PORTE,YELLOW ; THEN Yellow LED
          btfss  PORTB,SETT   ; IF Set switch closed
           bcf   PORTE,RED    ; THEN Red LED
          btfss  PORTB,GOO    ; IF Go switch closed
           bcf   PORTA,BUZ    ; THEN Buzzer
; Now turn on each segment in turn of both displays ------------
          movff  TEMP,DATA_L  ; Get pattern
          movff  TEMP,DATA_H
          call   SEND

          btfsc  PORTB,DIAG   ; IF Diagnostic switch open
           return             ; THEN exit the diag subroutine
          clrf   NEW_SEC      ; Reset the New Second flag
; Now move the display pattern on one and wait for a second ----
          rlncf  TEMP,f       ; Rotate it <<

D_LOOP2   tstfsz NEW_SEC      ; Wait for the new second
           bra   D_LOOP       ; IF yes THEN repeat routine
           bra   D_LOOP2      ; ELSE try again
```

Switches

Each of the five switches input via Port B are checked in turn. If a switch is closed, either one of the LEDs or the buzzer is activated. In this manner both the switches and the listed output devices are tested. The DIAG switch is of course verified by moving the system into this Diagnostic process and the RESET switch is tested by initiating the startup process.

If there were more switches than output devices then either combinations of the latter could be activated or else one or more segments in the numerical display could be pressed into service.

LEDs and Buzzer

The static output devices are tested in conjunction with the switch test listed above. Of course the failure of a LED to light or buzzer to sound may be due to either the input or output device circuit. Determining which has failed is easily accomplished by using a voltmeter or logic probe. Also remember that all LEDs should be illuminated during the Set-Time process.

Display

Each of the display devices is tried out by lighting one segment in turn at a 1-s rate, in an endless loop. This is implemented by generating a walking unary pattern b'11111110 → 11111101 → ··· 01111111' sent out to the output sub-routine SEND once each time NEW_SEC is non-zero. NEW_SEC is incremented

---

Program 16.5 The Set-Time process

```
; *****************************************************************
; * FUNCTION: Slowly counts down from 99-00. When SETT switch   *
; * FUNCTION: released EEPROM is Written with displayed count    *
; * RESOURCE: Subroutines DISPLAY, WRITE_EEPROM, ISR_TMR0        *
; * RESOURCE: var TIME_OUT, SECOND                               *
; * ENTRY   : SETT switch is closed on Reset                     *
; * EXIT    : EEPROM Data address 00 is updated                  *
; *****************************************************************
SET_TIME  movlw  d'99'             ; Start count at 99 seconds
          movwf  SECOND
          clrf   JIFFY             ; and no Jiffies
          movlw  b'000'            ; All LEDs on
          movwf  PORTE
SET_LOOP  movf   SECOND,w          ; Get Second count
          call   OUTPUT_BCD        ; Display it and reset NEW_SEC

          btfsc  PORTB,SETT        ; Check; does user want to stop?
          bra    UPDATE            ; IF yes THEN update EEPROM & exit
          movff  SECOND,TIME_OUT   ; Make a temporary copy
S_LOOP    tstfsz NEW_SEC           ; NEW_SEC is non-zero for a new sec
          bra    SET_LOOP          ; A new second means DO display
          bra    S_LOOP            ; ELSE check again for a new sec

UPDATE    movf   TIME_OUT,w        ; Get the value
          movwf  EEDATA            ; Set up EEPROM
          clrf   EEADR
          call   WRITE_EEPROM      ; Program EEPROM
          return                   ; and return to main program
```

in the Timer 0 interrupt-handling routine each time the Seconds count is incremented and cleared in the Diagnostic procedure code. This acts as a ratchet,
giving only one new display each second.

Set Time Process

This subroutine is entered when the SETT switch is closed whenever the processor
comes out of a Manual reset. Its function is to allow the operator to change the
contents of the EEPROM Data module location h'00' to any value up to 99. This
location holds the initial count-down value used by the Main process to determine
the length of the procedure.

The strategy behind the coding shown in Program 16.5 is to initialize the Second count to 99 and then let it decrement at a 1-s rate, as determined by the fore-

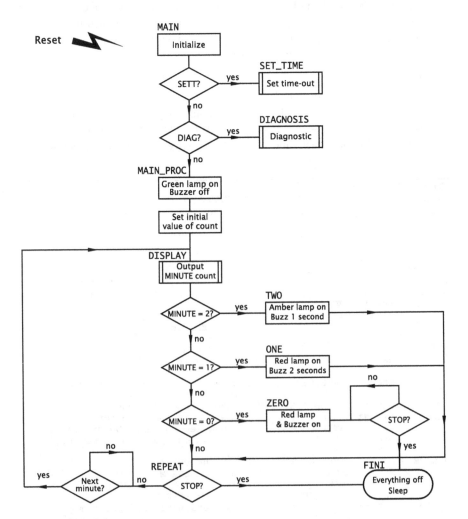

Fig. 16.3 The Main process

ground ISR. The value of SECOND is displayed each time the ISR sets the flag File NEW_SEC to a non-zero value; that is, once per second. Subroutine OUTPUT clears NEW_SEC, so the net effect is to update the display only once each second. Each second the SETT switch is checked, and when re-opened, the state of the Seconds count is transferred to the EEPROM Data module at UPDATE using the WRITE_EEPROM subroutine of Program 15.2 on p. 544.

Main Process

The complete background system flow chart is shown in Fig. 16.3. This shows in

---

Program 16.6 The Main process

```
MAIN_PROC movlw   b'11111110'  ; Green LED on
          movwf   PORTE        ; Red and Yellow off
          bsf     PORTA,BUZ    ; Buzzer off

; Get start value from EEPROM ----------------------------------
          clrf    EEADR        ; EEPROM address zero
          call    READ_EEPROM  ; Get the start value
          movwf   MINUTE
          movlw   d'59'        ; Initial value for seconds
          movwf   SECOND       ; is 59
          clrf    JIFFY        ; and zero Jiffies

DISPLAY   movf    MINUTE,w     ; Get Minute count
          call    OUTPUT_BCD   ; Output to display

; The 2-minutes-to-go phase ----------------------------------
; At a count of two sound the buzzer for one second and turn on
; the Yellow lamp ----------------------------------
TWO       movlw   2            ; Minute count = 2?
          cpfseq  MINUTE
          bra     ONE          ; IF not THEN try for one minute
          movlw   b'101'       ; ELSE Yellow LED on
          movwf   PORTE
          bcf     PORTA,BUZ    ; Buzzer on
TWO_LOOP  movf    NEW_SEC,f    ; Check NEW_SEC status
          bz      TWO_LOOP     ; IF still zero THEN try again
          bsf     PORTA,BUZ    ; ELSE Turn off buzzer after one sec
          bra     REPEAT       ; repeat display

; The 1-minute-to-go phase ----------------------------------
; At a count of one sound the buzzer for two second and turn on
; the red lamp ----------------------------------
ONE       movlw   1            ; Minute count = 1
          cpfseq  MINUTE
          bra     ZERO         ; IF not THEN try for zero minutes
          movlw   b'011'       ; Red LED on
          movwf   PORTE
          bcf     PORTA,BUZ    ; Buzzer on
ONE_LOOP  movf    NEW_SEC,f    ; Check NEW_SEC status
          bz      ONE_LOOP     ; IF still zero THEN try again
          clrf    NEW_SEC      ; Again clear NEW_SEC flag
UN_LOOP   movf    NEW_SEC,f    ; Again check NEW_SEC status
          bz      UN_LOOP      ; IF still zero THEN try again
          bsf     PORTA,BUZ    ; Turn off buzzer after two seconds
          bra     REPEAT       ; Repeat display
```

(continued on the next page)

Program 16.6 (*Continued*)

```
; The Timed-Out phase -----------------------------------------
; When the Minute count reaches zero, sound the buzzer ---------
; until the Stop switch is closed -----------------------------
ZERO       movf    MINUTE,f    ; Minute count = 0?
           bnz     REPEAT      ; IF not THEN repeat after minute
           bcf     PORTA,BUZ   ; Buzzer on
ZERO_LOOP
           btfsc   PORTB,STOP  ; Check the Stop switch
           bra     ZERO_LOOP   ; and continue until closed
FINI       movlw   b'111'      ; Turn lamps off
           movwf   PORTE
           bsf     PORTA,BUZ   ; and buzzer
           movlw   b'00010000'; Code for blank
           movwf   TENS        ; Blank both displays
           movwf   UNITS
           call    OUTPUT

           sleep               ; and await another reset

REPEAT     btfss   PORTB,STOP  ; Check the Stop switch
           bra     FINI        ; IF closed THEN freeze
           movf    SECOND,f    ; Wait 'til Second count is again zero
           bnz     REPEAT      ; IF not THEN wait again
           clrf    NEW_SEC     ; ELSE wait one more second
R_LOOP     tstfsz  NEW_SEC     ; Check NEW_SEC status
           bra     DISPLAY     ; IF non zero THEN repeat display
           bra     R_LOOP      ; ELSE repeat display
```

outline the decision stream taken after a reset, and in more detail, the Main process. Although this looks rather complex, it may be broken down into five phases, with the corresponding coding shown in Program 16.6.

Preamble

On reset if neither SETT or DIAG switches are closed, the Main procedure code is entered at MAIN_PROC. This reads the initial value of the countdown period from Data EEPROM location h'00' and initializes the count chain. The green lamp is illuminated and other lamps and buzzer are turned off.

Countdown

The Countdown phase continually displays the Minute count—updated behind the scenes by the ISR. The green lamp remains illuminated as long as this display does not drop below 𝟬𝟯. This phase is complete whenever the count drops below 3 minutes or else the STOP switch is closed. In the latter case all displays are blanked and the PIC MCU is put into its Sleep state.

In all situations, except where the STOP command is issued, the Minute count is displayed at minute intervals. The routine at REPEAT checks the Second count and if zero the loop is repeated; that is, once per minute. Repeating the loop each minute eases the task of sounding the buzzer once only when the Minute count drops to both two and one.

Two Minutes to Go

When the display is 𝟬𝟮 the amber lamp is illuminated and the buzzer sounded for one second. The latter is timed using the NEW_SEC variable. Again the loop can be prematurely exited if the STOP switch has been closed.

One Minute to Go

When the display is ⬛⬛ the loop diverts to illuminate the red lamp. The buzzer is sounded for 2 s; implemented in code as two 1-second buzzes.

Timed Out

When the Minute count reaches zero, not only is ⬛⬛ displayed but also the buzzer sounds continually. This cacophony can only be silenced by pressing the STOP switch, or by resetting and starting again. As in previous situations when the STOP switch is closed, all displays are blanked out and the PIC MCU is placed in its Sleep state.

After the source code has been assembled and where possible simulated (see Fig. 8.8 on p. 265) it can then be burned into the PIC MCU's Program store. In the first instance only the diagnostic software and associated tasks need to be programmed in order to check the target hardware. The precise details will depend somewhat on the PIC MCU Device programmer being used and its associated software.

The screen shot shown in Fig. 16.4 shows the situation where the Microchip PICSTART® Plus development programmer (see Fig. 16.5) is used in conjunction with the MPLAB® IDE. Communication with the host computer is via a RS-232 serial port. Once the programmer is specified via the Programmer | Select Programmer menu and contact is made from the drop-down menu shown at the top left of Fig. 16.4. The Output window shows the progress and status of the burn-in process and the Configuration Bits window, activated via the Configure menu, shows the state of the option fuses. Optionally these fuses can be manually set-up before burning, but it is best to embed these options in the source program. The Program Memory window, enabled from the View menu, shows the contents of the actual device as long as the code protection remains off. When the processor is configured to turn on Code Protect, neither Program, Read nor Verify tasks can be carried out. The complete process takes less than a minute to burn-in the 476 bytes that this case study generates.

Once a project is ready for production, using a dedicated programmer to replicate the software is a legitimate approach. To that end, gang programmers are available that can clone several devices at a time from a master device. However, as a development tool while the project is evolving, this technique is not ideal. After software simulation, prototype software is burnt-in and MCU physically transferred to the target board. In the event that there are functional problems, there are few tools that can be easily used to trace the bug. Whilst software simulation is a powerful tool, there is no interaction with the real hardware.

An extension to this simulate-program-insert process is to have the IDE control the target hardware, and the simulation software interact with the real world. The traditional approach to this problem, as depicted in Fig. 8.6 on p. 263, is to use an

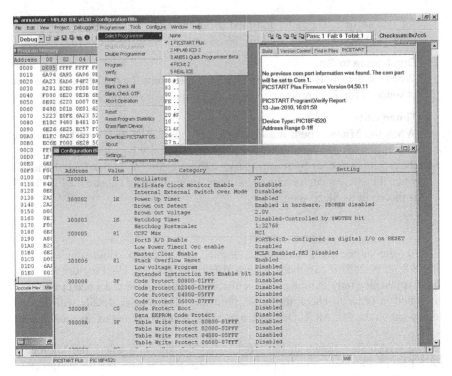

Fig. 16.4 Programming the PIC MCU using the PICSTART® Plus from the MPLAB® Version 8 IDE

In-Circuit Emulator. Here, an ICE hardware pod replaces the target MCU with similar circuitry and communicates with the computer running the IDE. In this manner, the application software runs in real silicon but still under control of the IDE, so that the designer can monitor activity at the various ports and registers as execution progresses.

The traditional ICE is an expensive development tool, costing several thousands of dollars. As an alternative approach, in the late 1990s Microchip introduced a more cost effective approach, where each MCU carries additional logic which allows it to execute in a Debug mode. Enabled by the **DEBUG** fuse, this **In-Circuit Debugger** facility replaces much of the expensive ICE circuitry. An **ICD** module connects to the PC, usually via a USB port; which powers the pod as well as supporting communication to MPLAB® or a compatible third-party IDE. This pod in turn interfaces to the application board via a 6-wire cable connecting the target processor's **PGD**/RB7 (**ProGram Data**) and **PGC**/RB6 (**ProGram Clock**) programming pins together with $\overline{\text{MCLR}}$/V$_{PP}$, V$_{DD}$ and V$_{SS}$. The target board will usually have a RJ11 telephone socket wired to these on-chip hardware resources, or else a ICD header carrying the processor can be plugged into the processor socket on the board.

Fig. 16.5 The Microchip
PICSTART® Plus
programmer

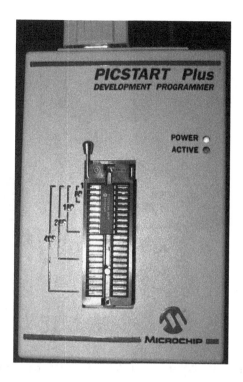

In-Circuit Debugging primarily differs from traditional In-Circuit Emulation in
the intrusion these techniques make on the target resources. For instance, the ICD
technique required as a minimum the exclusive use of:

- Pins RB7 and RB6.
- Two levels of hardware stack.
- 512 bytes of Program memory.
- 14 bytes of Data memory.

In addition, the target board needs to be capable of running without the pod; that is,
with a functioning clock and power-supply set-up. The target processor should have
its configuration fuses set correctly with no code protection and a disabled Watchdog
timer. In addition Low-Voltage Programming (Single-Supply Programming) should
be disabled—see p. 315.

When the Debugger | Program menu option is selected, the application soft-
ware is burnt into Program memory. In addition, a small Debug operating system,
called a **Debug Executive** is loaded into the top of the Program store. This reduces
the allowable size of the application software. This Executive then monitors and executes the application software and communicates with MPLAB® via the PGD and
PGC pins when $\overline{\text{MCLR}}$ is released.

The facilities available during an ICD run depend on the sophistication and cost
of the pod. For instance, for the ICD3 pod these are:

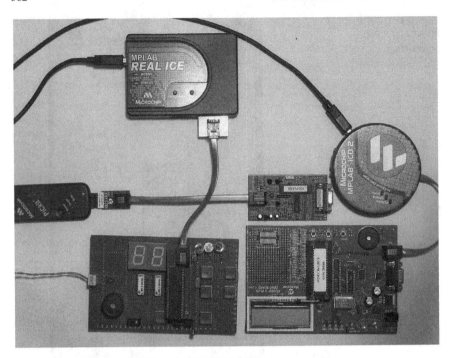

Fig. 16.6 The PICkit™ 2 (left), MPLAB® ICD 2 (right) and Real ICE™ (top) pods

- Real-time execution until Breakpoint or Halt.
- Several Complex Breakpoints.
- Single step execution with Watch window.

The Microchip Real ICE™ (top of Fig. 16.6) is connected and driven in a similar manner to an ICD, but provides a much richer and faster interaction with less disturbance with the target processor. However, transparency comes at a price, which in this case is *circa* $500+ compared to $200 for the MPLAB™ ICD 3 and $70 for the PICkit™ 3. The *quid quo pro* is a much faster operation, with many of the functions being taken over by the ICE pod's hardware. Additional functions made possible with this arrangement are software traces, real-time Watch windows and interaction with actual signals on the board with logic analyser probes.

As these ICD/ICE pods are capable of burning-in code to the target processor, they can be used as a simple programmer—see also Fig. 10.8 on p. 315. From Fig. 16.5 the `Programmer|Settings` menu enumerates modules, such as the ICD2, in its list of possible programmers. Programming in this manner differs from that via the `Debugger` menu, in that the Executive is not downloaded; only the application software. Using an ICD simply as a straight programmer can be a cost effective alternative to a dedicated programmer, such as a PICSTART® Plus, where small numbers of devices are to be programmed. For instance, a PICkit™ 3 costs around $70 compared to the latter's $200.

The hardware and software circuits have been presented here as a simple illustrative case study to integrate many of the techniques described in the body of the text. If you decide to build your own version, files, **C** coding, PCB, comparison with a Motorola 68,000 MPU version and other ideas for experimentation, which you are welcome to contribute, are given on the associated website detailed in the Preface. Good luck!

Appendix A
Acronyms and Abbreviations

ABDEN	Automatic BauD rate ENable; BAUDCON[0]
ABDOVF	Automatic BauD rate OVerFlow; BAUDCON[7]
ACK	ACKnowledge state in the I^2C protocol
ACQT2:0	ACQuition Time delay bits; ADCON2[5:3]
ADC (A/D)	Analog-to-Digital Converter/Conversion module
ADCON*n*	A/D CONtrol 0 register 2, 1, & 0
ADCS2:0	A/D Clock Select, bits; ADCON2[2:0]
ADDEN	ADDress ENable bit; RCSTA[3]
ADFM	A/D module outcome ForMat bit; ADCON2[7]
ADIE	A/D Interrupt Enable mask bit; PIE1[6]
ADIF	A/D Interrupt Flag bit; PIR1[6]
ADIP	A/D Interrupt Priority bit; IPR1[6]
ADON	A/D module ON bit; ADCON0[0]
ADRESH	A/D RESult High byte register
ADRESL	A/D RESult Low byte register
ALU	Arithmetic Logic Unit
AN*n*	ANalog input pin *n*
ANSI	American National Standards Institution
ASCII	American Standard Code for Information Interchange
BAUDCON	BAUD CONtrol register
BRG16	Baud Rate Generator 16-bit mode bit; BAUDCON[3]
BRGH	Baud Rate Generator High mode bit; TXSTA[2]
BSR*n*	Bank Select Register bits; BSR[3:0]
BCD	Binary Coded Decimal
BF	Buffer Full bit; SSPSTAT[0]
C	Carry flag bit; STATUS[0]
C1OUT	Comparator 1 OUTput bit; CMCON[6]
C2OUT	Comparator 2 OUTput bit; CMCON[7]
C1INV	Comparator 1 INVertor bit; CMCON[4]
C2INV	Comparator 2 INVertor bit; CMCON[5]
C1OUT	Comparator 1 OUTput pin
C2OUT	Comparator 2 OUTput pin
CAN bus	Control Area Network bus
CCPX	Capture/Compare/PWM module X

S. Katzen, *The Essential PIC18® Microcontroller*,
Computer Communications and Networks,
DOI 10.1007/978-1-84996-229-2, © Springer-Verlag London Limited 2010

CCP1	CCP1 input/output pin
CCPR1H	CCP Register 1 High byte
CCPR1L	CCP Register 1 Low byte
CCP1CON	CCP1 CONtrol register
CCP1IE	CCP1 Interrupt Enable mask bit; PIE1[2]
CCP1IF	CCP1 Interrupt Flag bit; PIR1[2]
CCP1IP	CCP1 Interrupt Priority bit; IPR1[2]
CCP1M3:0	CCP Mode control bits; CCP1CON[3:0]
CCP2	CCP2 input/output pin
CCPR2H	CCP Register 2 High byte
CCPR2L	CCP Register 2 Low byte
CCP2CON	CCP2 CONtrol register
CCP2IE	CCP2 Interrupt Enable mask bit; PIE2[0]
CCP2IF	CCP2 Interrupt Flag bit; PIR2[0]
CCP2IP	CCP2 Interrupt Priority bit; IPR2[0]
CFGS	ConFiGure Select bit; EECON1[6]
CHS3:0	ADC CHannel Select bits; ADCON0[5:2]
CIS	Comparator Input Switch bit; CMCON[3]
CISC	Complex Instruction Set Computer
CK	USART synchronous ClocK I/O pin
CKE	ClocK Edge bit; SSPSTAT[6]
CKP	ClocK Polarity bit; SSPCON1[4]
CMIE	CoMparator change Interrupt Enable mask bit; PIE2[6]
CMIF	CoMparator change Interrupt Flag bit; PIR2[6]
CMIP	CoMparator change Interrupt Priority bit; IPR2[6]
CM2:0	Comparator (analog) Mode bits; CMCON[2:0]
CMOS	Complimentary Metal-Oxide Semiconductor
CMCON	CoMparator (analog) CONtrol register
CPU	Central Computing Unit
CREN	Continuous Receive ENable bit; RCSTA[4]
$\overline{\text{CS}}$	Chip Select pin (active low)
CTS	Clear To Send; RS-232 handshake signal
CVR3:0	Comparator Voltage Reference mode bits; CVRCON[3:0]
CVREN	Comparator Voltage Reference ENable bit; CVRCON[7]
CVRCON	Comparator Voltage Reference CONtrol register
CVROE	Comparator Voltage Reference Output Enable bit; VRCON[6]
CVRR	Comparator Voltage Reference Range select bit; CVRCON[5]
CVRSS	Comparator Voltage Reference V_{REF} Source Select bit; CVRCON[4]
D/$\overline{\text{A}}$	Data/Address bit; SSPSTAT[5]
DAC (D/A)	Digital-to-Analog Converter/Conversion module
DC	Digit Carry flag bit; STATUS[1]
DC1B1:0	Duty Cycle 1 Bits; CCP1CON[5:4]
DCE	Data Circuit terminating Equipment
DSP	Digital Signal Processing
DT	USART synchronous DaTa pin
DTE	Data Terminal Equipment
ea	Effective Address
ECCP	Enhanced CCP module
ECCP1AS	ECCP1 module Auto Startup/Shutdown register

EEADR	EEPROM ADdress Register
EECON*n*	EEPROM CONtrol registers 1 & 2
EEDATA	EEPROM DATA register
EEIE	EEPROM Interrupt Enable mask bit ; PIE2[4]
EEIF	EEPROM Interrupt Flag bit; PIR2[4]
EEIP	EEPROM Interrupt Priority bit; IPR2[4]
EEPGD	EEPROM ProGram/$\overline{\text{Data}}$ bit; EECON1[7]
EEPROM	Electrical Erasable PROM
EPROM	Erasable PROM
EUSART	Enhanced USART module
FERR	Framing ERRor bit; RCSTA[2]
FLT0	FauLT input pin 0 for ECCP1 module
FREE	Flash Row Erase Enable bit; EECON1[4]
FSR*n*	File Select Register 2, 1 & 0
FVR	Internal 1.2 V Fixed Voltage Reference
GIEH/GIE	Global Interrupt Enable mask (High-priority) bit; INTCON[7]
GIEL/$\overline{\text{PEIE}}$	Global Interrupt Enable mask (Low-priority) bit; INTCON[6]
GO/$\overline{\text{DONE}}$	ADC Start Convert (GO)/End Of Conversion (DONE) bit; ADCON0[1]
GPR	General-Purpose File Register
GSEN	General SENd receive enable bit; SSPCON2[7]
HLVDCON	High/Low Voltage Detect CONtrol register
HLVDEN	High/Low Voltage Detect ENable bit; HLVDCON[4]
HLVDIE	High/Low Voltage Detect Interrupt Enable mask bit; PIE2[5]
HLVDIF	High/Low Voltage Detect Interrupt Flag bit; PIR2[5]
HLVDIP	High/Low Voltage Detect Interrupt Priority bit; IPR2[5]
HLVDIN	High/Low Voltage Detect INput pin
HLVDL3:0	High/Low Voltage Detect Limit bits; HLDVCON[3:0]
IC	Integrated Circuit
ICSP™	In-Circuit Serial Programming
I²C	Inter-Integrated Circuit serial protocol
IDE	Integrated Development Environment
IEC	International Electrotechnical Commission
ICD	In-Circuit Debugger
ICE	In-Circuit Emulator
INDF*n*	INDirect File registers 2, 1, 0
INT2:0	External INTerrupt input pins
INTCON	INTerrupt CONtrol Register
INTCON*n*	INTerrupt CONtrol Registers 2 & 3
INTEDG2:0	External INTerrupt EDGe polarity selection bits; INTCON2[4:6]
INT2:0IE	INTerrupt Enable mask bits; INTCON[4], INTCON3[3:4]
INT2:0IF	INTerrupt Flag bitss; INTCON[1]; INTCON3[0:1]
INT2:1IP	INTerrupt Priority bits; INTCON3[7:6]
INTSRC	INTernal SouRCe bit; OSTUNE[7]
I/O	Input/Output
IPEN	Interrupt PRiority ENable bit; RCON[7]
IPR*n*	Interrupt Priority Interrupt Registers 2 & 1
IRVST	Internal Reference Voltage STable bit; HLVDCON[5]
ISR	Interrupt Service Routine
ksps	Kilo samples per second

LAT*X*	LATch*X* (Parallel I/O port LATch register *X*); e.g., LATA
LED	Light-Emitting Diode
LIN bus	Local Interconnect Network bus
LSB	Least-Significant Bit or Byte
LSD	Least-Significant Digit
LSI	Large-Scale Integration
LVP	Low-Voltage Programming (c.f. High-Voltage Programming)
$\overline{\text{MCLR}}$	Master CLear Reset pin (active low)
MCU	MicroController Unit
MPU	MicroProcessor Unit
μs	Microsecond (10^{-6} s)
ms	Millisecond (10^{-3} s)
MSB	Most Significant Bit or Byte
MSD	Most Significant Digit
MSI	Medium-Scale Integration
MSSP	Master Synchronous Serial Port
N	Negative flag bit; STATUS[4]
NACK	Not ACKnowledge state in the I^{2}C protocol
ns	Nanosecond (10^{-9} s)
$\overline{\text{OE}}$	Output Enable pin (active low)
OERR	Overflow ERRor bit; RCSTA[1]
OS	Operating System
OTP	One-Time Programmable (EPROM)
OSCON	OScillator CONtrol register
OSCTUNE	OSCillator TUNE register
OV	OVerflow flag bit; STATUS[3]
P	StoP condition bit; SSPSTAT[4]
P1*X*	PWM ECCP1 pins P1A,B,C,D
PC	Program Counter or Personal Computer
PCFG3:0	ADC Port ConFiGuration bits; ADCON1[3:0]
PCL	Program Counter Low byte register
PCLATH	Program Counter LATch High byte register
PCLATU	Program Counter LATch Upper byte register
$\overline{\text{PD}}$	Power Down sleep mode bit; RCON[2]
PEIE/GIEL	PEripheral Interrupt Enable mask bit; INTCON[6]
PGC	ProGram Clock pin; shared with RB6
PGD	ProGram Data pin; shared with RB7
PGM	ProGram Mode pin; shared with RB5
PIC	Peripheral Interface Controller
PIPO	Parallel-In Parallel-Out register
PIE*n*	Peripheral Interrupt Enable register 2 & 1
PIR*n*	Peripheral Interrupt Register 2 & 1
PISO	Parallel-In Serial-Out shift register
POR	Power-on Reset
PORT*X*	Port*X* (Parallel I/O port register *X*); e.g., PORTA
PR2	Period Register for Timer 2
PRNG	Pseudo Random Number Generator
PRODH	PRODuct High byte register
PRODL	PRODuct Low byte register

PROM	Programmable ROM
PRSEN	PWM ReSet ENABLE bit; PWM1CON[7]
PS2:0	Prescale rate Select bits for TMR0; T0CON[2:0]
PSA	Prescale Scaler Assign bit; T0CON[3]
PSP	Pseudo Stack Pointer
PWM	Pulse Width Modulation
PWM1CON	PWM ECCP1 module CONtrol register
PWRT	PoWer-on Reset Timer
R*Xn*	Register (Parallel I/O port) *X* pin *n*; e.g., RA0, RE3
RAM	Random Access Memory
RBIE	Register port B change Interrupt Enable bit; INTCON[3]
RBIF	Register port B change Interrupt Flag bit; INTCON[0]
RBIP	Register port B change Interrupt Priority bit; INTCON2[0]
R̄B̄P̄Ū	Register port B Pull-Up bit; INTCON2[7]
RCIDL	ReCeive IDLe bit; BAUDCON[6]
RCIE	ReCeive register Interrupt Enable mask bit; PIE1[5]
RCIF	ReCeive register Interrupt Flag bit; PIR1[5]
RCIP	ReCeive register Interrupt Priority bit; IPR1[5]
RCON	Reset CONtrol register
RCREG	ReCeive data REGister
RCSTA	ReCeive STAtus register
RD	ReaD bit; EECON1[0]
RD16	ReaD/write 16-bits; T1CON[7] & T3CON[7]
R/W̄	Read/Write packet in MSSP module bit; SSPSTAT[2]
RISC	Reduced Instruction Set Computer (see CISC)
ROM	Read-Only Memory
rtl	Register Transfer Language
RTCC	Real-Time Counter Clock; anarchic name for Timer 0
RTS	Ready To Send: RS-232 handshake signal
RX	ReCeive pin for USART
RX9	ReCeive 9-bit data control bit; RCSTA[6]
RX9D	Ninth bit received by the USART bit; RCSTA[[0]
RXDTP	Receive/DaTa Polarity bit; BAUDCON[5]
S	Start condition bit; SSPSTAT[3]
SAR	Successive Approximation Register
SCK	Serial ClocK in SPI protocol
SCL	Serial CLock in I^{2}C protocol
SDA	Serial DAta bidirectional I^{2}C pin
SDI	Serial Data Input pin in SPI protocol
SDO	Serial Data Output pin in SPI protocol
SEN	Stretch clock ENable bit; SSPCON2[0]
SENDB	SEND Break bit; TXSTA[3]
SIPO	Serial-In Parallel-Out shift register
SISO	Serial-In Serial-Out shift register
SMP	SaMPle incoming data bit; SSPSTAT[7]
SP	Stack Pointer; STKPTR[4:0]
SPBRG	Serial Port Baud-Rate Generator low byte
SPBRGH	Serial Port Baud-Rate Generator High byte
SPEN	Serial Port ENable bit; RCSTA[7]

SPI	Serial Peripheral Interface protocol
SFR	Special-Function File Register
SSP	Synchronous Serial Port
SSPADD	SSP ADDress register
SSPBUF	MSSP BUFfer register
SSPCONn	MSSP CONtrol register 2 & 1
SSPEN	MSSP Enable bit; SSPCON1[5]
SSPIE	MSSP Interrupt Enable mask bit; PIE1[3]
SSPIF	MSSP Interrupt Flag bit; PIR1[3]
SSPIP	MSSP Interrupt Priority bit; IPR1[3]
SSPM3:0	MSSP Mode control bits; SSPCON1[3:0]
SSPOV	MSSP OVerflow bit; SSPCON1[6]
SSPSR	MSSP Shift Register
SSPSTAT	MSSP STATus register
STATUS	Status register
STKPTR	STacK PoinTeR register
STKFUL	STacK FULl bit; STKPTR[7]
STKUNF	STacK UNderFlow bit; STKPTR[6]
STVREN	STacK oVer/underflow Reset ENable configuration bit (also STVR)
SWDTEN	Software WatchDog Timer ENable bit; WDTCON[0]
SYNC	SYNChronous mode in the USART bit; TXSTA[4]
T08BIT	Timer 0 8 BIT/16 bit; T0CON[7]
T0CKI	Timer 0 ClocK Input pin (normally shared with RA4)
T0CS	Timer 0 Clock Select bit; T0CON[5]
T13CKI	Timer 1/3 ClocK Input pin
T1CKPS1:0	Timer 1 ClocK Prescale bits; T1CON[5:4]
T1RUN	Timer 1 oscillator is RUNning the system bit; T1CON[6]
T2CKS2:0	Timer 2 ClocK Source Prescale ratio bits; T2CON[2:0]
T2OUTPS3:0	Timer 2 OUTput Post Scaler bits; T2CON[3:0]
T0CON	Timer 0 CONtrol register
T1CON	Timer 1 CONtrol register
T1SYNC	Timer 1 SYNChronize bit; T1CON[2]
T1OSCEN	Timer 1 OSCillator ENable bit; T1CON[3]
T2CON	Timer 2 CONtrol register
T3CON	Timer 3 CONtrol register
T3CCPn	Timer 3 CCP2 & 1 timebase select; T3CON[6 & 3]
TABLAT	TABle LATch register
TBLPTR	TaBLe PoinTeR (TBLPTRU:TBLPTRH:TBLPTRL registers)
TMRn	TiMeR 0, 1, 2 & 3 registers
TMR0IE	Timer 0 Interrupt Enable mask bit; INTCON[5]
TMR0IF	Timer 0 Interrupt Flag bit; INTCON[2]
TMR0IP	Timer 0 Interrupt Priority bit; INTCON2[2]
TMR0H	TiMeR 0 High byte buffer register
TMR0L	TiMeR 0 Low byte register
TMR1CS	TiMeR 1 Clock Select bit; T1CON[1]
TMR1H	TiMeR 1 High byte buffer register
TMR1L	TiMeR 1 Low byte register
TMR1IE	Timer 1 Interrupt Enable mask bit; PIE1[0]
TMR1IF	Timer 1 Interrupt Flag bit; PIR1[0]

TMR1IP	Timer 1 Interrupt Priority bit; IPR1[0]
TMR1ON	TiMeR 1 ON bit; T1CON[0]
TMR2IE	Timer 2 Interrupt Enable mask bit; PIE1[1]
TMR2IF	Timer 2 Interrupt Flag bit; PIR1[1]
TMR2IP	Timer 2 Interrupt Priority bit; IPR1[1]
TMR2ON	TiMeR 2 ON bit; T2CON[2]
TMR3H	TiMeR 3 High byte buffer register
TMR3L	TiMeR 3 Low byte register
TMR3IE	Timer 3 Interrupt Enable maskbit; PIE2[1]
TMR3IF	Timer 3 Interrupt Flag bit; PIR2[1]
TMR3IP	Timer 3 Interrupt Priority bit; IPR2[1]
$\overline{\text{TO}}$	Watchdog Time Out bit; RCON[3]
TRISX	TRIState$X$ (Data Direction register$X$); e.g., TRISA
TRMT	TRansMiT shift register empty bit; TXSTA[1]
TUN4:0	INTOSC TUning bits; OSCAL[4:0]
T0SE	Timer 0 Set Edge bit; T0CON[4]
TTL	Transistor Transistor Logic family
TTY	TeleTYpewriter
TX	Transmit pin for USART
TX9	TranSmit 9-bit data in USART; TXSTA[6]
TX9D	Ninth bit for transmission in USART; TXSTA[0]
TXCKP	Transmit/ClocK Polarity bit; BAUDCON[4]
TXEN	Transmit register ENable bit; TXSTA[5]
TXIE	Transmit register Interrupt Enable mask bit; PIE1[4]
TXIF	Transmit register Interrupt Flag bit; PIR1[4]
TXIP	Transmit register Interrupt Priority bit; IPR1[4]
TXREG	Transmit data REGister
TXSTA	Transmit STAtus register
UA	Update slave 10-bit Address in MSSP bit; SSPSTAT[1]
UART	Universal Asynchronous Receiver Transmitter
USART	Universal Synchronous-Asynchronous Receiver Transmitter
VCFG1:0	A/D Port reference Voltage ConFiGuration bits; ADCON1[4:3]
V_{DD}	Positive (Drain) supply voltage
V_{EE}	Earth (0 V) supply voltage
V_{PP}	Positive Programming voltage
VLSI	Very Large-Scale Integration
W	Working register (alternative to WREG)
WCOL	Write COLlision bit; SSPCON1[7]
WDT	WatchDog Timer
WDTCON	WatchDog Timer CONtrol register
WR	WRite bit; EECON1[1]
WREG	Working REGister in Data store
WREN	WRite ENable bit; EECON1[2]
WRERR	WRite ERRor bit; EECON1[3]
WUE	Wake Up Enable bit; BAUDCON[1]
Z	Zero flag bit; STATUS[2]

Appendix B
Configuration Registers and Bits for the PIC18FXX20

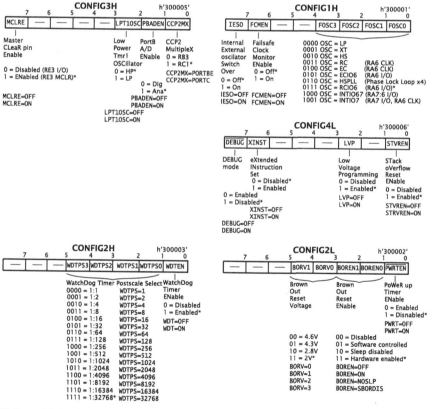

* Indicates default/unprogrammed value

S. Katzen, *The Essential PIC18® Microcontroller,*
Computer Communications and Networks,
DOI 10.1007/978-1-84996-229-2, © Springer-Verlag London Limited 2010

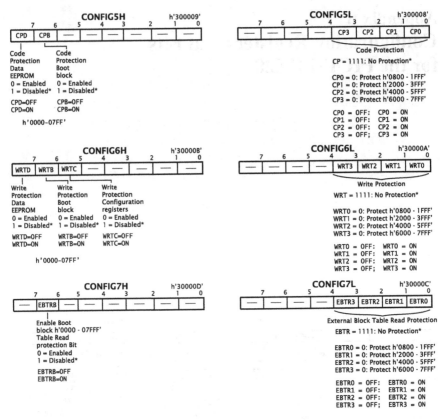

* Indicates default/unprogrammed value

Appendix C
C Instruction Set

C Operators, Their Precedence, and Associativity

Operator	Operation	Example
Top priority		
Direction (associativity) $\Rightarrow$		
( )	Function call	`sqr()`
[ ]	Array element	`x[6]`
.	Structure element	`PIA1.CRA`
->	Structure element using a pointer	
Unary operators		
Direction (associativity) $\Leftarrow$		
!	Logical NOT	`!x`
~	Inversion (1's complement)	`~x`
-	Negative	`y=-x`
+	Unary plus	`y=x- +(y+z)`
++	Increment	`x++ or ++x`
-	Decrement	`x- or -x`
&	Address of	`&x`
*	Contents of address	`*address`
(type)	Cast	`(long)x`
sizeof	Size of object in bytes	`sizeof x`

(*continued on the next page*)

S. Katzen, *The Essential PIC18® Microcontroller,*
Computer Communications and Networks,
DOI 10.1007/978-1-84996-229-2, © Springer-Verlag London Limited 2010

Operator	Operation	Example
Arithmetic		
Direction (associativity) ⇒		
*	Multiplication	`z=x*y`
/	Division	`z=x/y`
%	Remainder	`z=x%y` (Integer types only)
+	Addition	`z=x+y`
−	Subtraction	`z=x-y`
Shift		Integer types only
Direction (associativity) ⇒		
»	Shift left	`z=x»3`
«	Shift right	`z=x«3`
Relational operators		Boolean objects
Direction (associativity) ⇒		
<	Less than	`while (x<3 )`
<=	Less than or equal	`while (x<=3)`
>	Greater than	`while (x>3 )`
>=	Greater than or equal	`while (x>=3)`
==	Equivalent	`while (x==y)`
!=	Not equivalent	`while (x!=0)`
Bitwise logic		Integer types only
Direction (associativity) ⇒		
&	AND	`x&0xFE` (Clear bit 0)
^	Exclusive-OR	`x^0x01` (Toggle bit 0)
|	OR	`x|0x01` (Set bit 0)
Objectwise logic		Boolean objects
Direction (associativity) ⇒		
&&	Logical AND	`x&&y` is True if both x and y are True
||	Logical OR	`x||y` is True if both or either x and y are True
? :	Conditional	`x=(y>z)?5:10` x=5 if y>z True else x=10

(*continued on the next page*)

	(*Continued*)	
Operator	Operation	Example

Assignment

Direction (associativity) $\Leftarrow$

Operator	Operation	Example
=	Simple	x=3
+=	Compound plus	x+=3 e.g. (x=x+3)
-=	Compound minus	x-=3 e.g. (x=x-3)
*=	Compound multiply	x*=3 e.g. (x=x*3)
/=	Compound divide	x/=3 e.g. (x=x/3)
%=	Compound remainder	x%=3 e.g. (x=x%3)
&=	Compound bit AND	x&=3 e.g. (x=x&3)
^=	Compound bit EX-OR	x^=3 e.g. (x=x^3)
|=	Compound bit OR	x|=3 e.g. (x=x|3)
«=	Compound shift left	x«=3 e.g. (x=x«3)
»=	Compound shift right	x»=3 e.g. (x=x»3)

Direction (associativity) $\Rightarrow$

Operator	Operation	Example
,	Concatenate	if(x=0,y=3;x<10,x++)

Lowest priority

Index

S. Katzen, *The Essential PIC18® Microcontroller,*
Computer Communications and Networks,
DOI 10.1007/978-1-84996-229-2, © Springer-Verlag London Limited 2010